The
Astronomy
Revolution

The
Astronomy
Revolution

400 Years of Exploring the Cosmos

Edited by
Donald G. York
Owen Gingerich
Shuang-Nan Zhang

CRC Press
Taylor & Francis Group
Boca Raton   London   New York

CRC Press is an imprint of the
Taylor & Francis Group, an **informa** business

CRC Press
Taylor & Francis Group
6000 Broken Sound Parkway NW, Suite 300
Boca Raton, FL 33487-2742

First issued in paperback 2019

© 2012 by Taylor & Francis Group, LLC
CRC Press is an imprint of Taylor & Francis Group, an Informa business

No claim to original U.S. Government works

ISBN-13: 978-1-4398-3600-2 (hbk)
ISBN-13: 978-0-367-38209-4 (pbk)

Visit the Taylor & Francis Web site at
http://www.taylorandfrancis.com

and the CRC Press Web site at
http://www.crcpress.com

*The contemplation of celestial things will make a man
both speak and think more sublimely and magnificently
when he descends to human affairs.*

—Marcus Tullius Cicero, Roman Philosopher and Statesman (106–43 BCE)

*Upon one tree there are many fruits, and in one kingdom many people.
How unreasonable it would be to suppose that besides the heaven and earth
which we can see there are no other heavens and no other earths!*

—Deng Mu (邓牧), Chinese Philosopher (1247–1306 CE)

Contents

PART I Creativity and Technology in Astronomical Discovery

PART II Impact of Telescopes on Our Knowledge of the Universe

PART III Some Near-Term Challenges in Astronomy

PART IV   Technologies for Future Questions

PART V   Intellectual Impact of the Telescope on Society

PART VI   "Big Questions" Raised by New Knowledge

Preface

This book is a product of the *New Vision 400* (*NV400*) conference held in Beijing in October 2008 in conjunction with the widely celebrated 400th anniversary of the invention of the telescope in 1608 by Hans Lipperhey (see http://nv400.uchicago.edu/). Like the conference, this book emphasizes the effects of technology on society and the origin of our understanding of numerous deep questions that arise out of scientific research, specifically astronomy and our knowledge of the cosmos. Looking beyond science questions to the role of moral responsibility in human civilizations, this volume offers the unique vantage points of contributions from both Eastern and Western cultures, which often differ dramatically in worldview and in knowledge. A Chinese-language edition of this book, to be published by Peking University Press, is also in development.

Part I focuses on the general theme of creativity and technology in scientific—particularly astronomical—discovery and is based on presentations that were primarily aimed at young people at the public event preceding the *NV400* conference. These discussions will be accessible to many readers, regardless of their technical training. The editors structured the specific topics covered in Parts II through V around selected examples of well-recognized areas of astronomical knowledge, modern challenges, new technologies, and historical impact. The book concludes with Part VI, an investigation of "big questions": What is the origin of the laws of physics as we know them? Why do these specific laws exist? Are these laws the same everywhere? How do these scientific laws relate to the moral laws of society? Does what we know depend on cultural ways of asking the questions? Is there life elsewhere? What about the questions that science cannot answer? The Introduction that follows the Acknowledgments provides in-depth background information on the structure and scope of this volume. The Appendix presents more information about the October 2008 conference in Beijing.

We hope we have succeeded in shaping a book that celebrates the historical significance of the telescope, informs the question of how we came to know what we know about the Universe, and inspires young astronomers to deepen our understanding of the cosmos and of ourselves as we continue the quest to unveil the heavens.

Donald G. York, Chief Editor
Department of Astronomy and Astrophysics
The University of Chicago
Chicago, Illinois

Owen Gingerich, Co-Editor
Harvard-Smithsonian Center for Astrophysics
Cambridge, Massachusett

Shuang-Nan Zhang, Co-Editor
Institute of High Energy Physics
Chinese Academy of Sciences
Beijing, China
and
Department of Physics
University of Alabama
Huntsville, Alabama

Acknowledgments

The editors wish to acknowledge the sponsors, committees, and staff who made the *New Vision 400 (NV400)* conference on which this book is based possible. Please see the Appendix for more information about the conference and listings of those to whom we owe a debt of gratitude for their efforts in bringing this historic October 2008 event to fruition in Beijing.

The editors especially wish to acknowledge the following people, who made major contributions to the *NV400* project, and especially to the evolution of this book:

Charles L. Harper, Jr., president of Vision-Five.com Consulting, who served as one of the original project developers in his former role as senior vice president and chief strategist of the John Templeton Foundation (JTF), working with Donald York and Hyung Choi.

Hyung S. Choi, director of Mathematical and Physical Sciences at JTF, who played an integral role in developing the academic program for the symposium in conjunction with Charles Harper and Donald York.

Jian-Sheng Chen of Peking University and the National Astronomical Observatories, Chinese Academy of Sciences (NAOC), who collaborated with Donald York as co-principal investigator in organizing the *NV400* conference, which formed the basis for this book.

Sui-Jian Sue of NAOC, who served on all of the conference committees, interfaced with all of the local Chinese institutions, handled all of the financial arrangements in China, and, in the end, became a highly respected friend of all the organizers of the conference and of the editors of this book.

Xiao-Chun Sun of the Institute for the History of Natural Sciences, Chinese Academy of Sciences, and Xiao-Wei Liu of Peking University (and both contributors to this volume), who provided continuous assistance and advice to the conference organizers and to the editors of this book on matters of Chinese culture, language, and science.

Pamela Bond Contractor, president and director of Ellipsis Enterprises Inc., who served as developmental editor of this volume. Working closely with the volume editors, authors, and publisher, she managed the book development process from the initial proposal to the finished product.

The Ellipsis editorial staff, Robert W. Schluth, senior editor, and Matthew P. Bond, associate editor, who were responsible for issuing instructions to authors and copyediting and preparing the manuscript for submission to the publisher.

Finally, the editors thank John Navas, senior acquisitions editor in physics at Taylor & Francis/CRC Press, London, who supported this book project from inception to completion.

Contributors

Charles A. Beichman, Executive Director, NASA ExoPlanet Science Institute, California Institute of Technology, and Jet Propulsion Laboratory, Pasadena, California.

Elliott D. Bloom, Professor of Particle Astrophysics, Kavli Institute for Particle Astrophysics and Cosmology and SLAC National Accelerator Laboratory, Stanford University, Menlo Park, California.

Alan H. Bridle, Astronomer, National Radio Astronomy Observatory, Charlottesville, Virginia (co-author with K. Y. Lo).

Paul C. W. Davies, Director, The Beyond Center for Fundamental Concepts in Science, Arizona State University, Tempe, Arizona.

George F. R. Ellis, Emeritus Distinguished Professor of Complex Systems, Department of Mathematics and Applied Mathematics, University of Cape Town, South Africa, and G. C. McVittie Visiting Professor of Astronomy, Queen Mary, London University, United Kingdom.

Richard S. Ellis, Steele Professor of Astronomy, Department of Astronomy, California Institute of Technology, Pasadena, California.

Glennys R. Farrar, Professor of Physics, Center for Cosmology and Particle Physics and Department of Physics, New York University, New York City, New York.

Alexei V. Filippenko, Professor of Astronomy, Richard and Rhoda Goldman Distinguished Professor in the Physical Sciences, Department of Astronomy, University of California, Berkeley, California.

Riccardo Giacconi, University Professor, Department of Physics and Astronomy, Johns Hopkins University, Baltimore, Maryland.

Owen Gingerich, Professor of Astronomy and History of Science, Emeritus, Harvard–Smithsonian Center for Astrophysics, Cambridge, Massachusetts.

Peter Harrison, Andreas Idreos Professor of Science and Religion and Director, Ian Ramsey Centre, Harris Manchester College, University of Oxford, Oxford, United Kingdom.

Yi-Long Huang, University Professor, Institute of History, National Tsing Hua University, Hsinchu City, Taiwan, Republic of China.

Renata Kallosh, Professor of Physics, Department of Physics, Stanford University, Stanford, California (co-author with Andrei Linde).

Yung Sik Kim, Professor, Department of Asian History and Program in History and Philosophy of Science, Seoul National University, Seoul, Korea.

Tsung-Dao Lee, University Professor in Theoretical Physics, Department of Physics, Columbia University, New York, and Director, China Center of Advanced Science and Technology (World Laboratory), Beijing, China.

Andrei Linde, Professor of Physics, Department of Physics, Stanford University, Stanford, California (co-author with Renata Kallosh).

Xiao-Wei Liu, Professor of Astronomy, Kavli Institute for Astronomy and Astrophysics, Peking University, Beijing, China.

K. Y. Lo, Director and Distinguished Astronomer, National Radio Astronomy Observatory, Charlottesville, Virginia (co-author with Alan H. Bridle).

Geoffrey W. Marcy, Professor of Astronomy, Department of Astronomy, University of California, Berkeley, California.

Ben Moore, Director, Institute for Theoretical Physics and Professor of Astrophysics, University of Zurich, Zurich, Switzerland.

Sara Seager, Ellen Swallow Richards Associate Professor of Planetary Science and Associate Professor of Physics, Departments of Earth, Atmospheric and Planetary Sciences and of Physics, Massachusetts Institute of Technology, Cambridge, Massachusetts.

Michael Shao, Project Scientist, Space Interferometry Mission and Keck Interferometer, and Director, Interferometry Center of Excellence, Jet Propulsion Laboratory, Pasadena, California.

Mark Sullivan, Royal Society University Research Fellow, Department of Physics (Astrophysics), University of Oxford, Oxford, United Kingdom.

Xiao-Chun Sun, Professor of the History of Science and Associate Director, Institute for the History of Natural Sciences, Chinese Academy of Sciences, Beijing, China.

Naoki Yoshida, Associate Professor, Institute for the Physics and Mathematics of the Universe, University of Tokyo, Chiba, Japan.

Shuang-Nan Zhang, Professor of Physics and Director, Key Laboratory of and Center for Particle Astrophysics, Institute of High Energy Physics, Chinese Academy of Sciences, Beijing, China, and Research Professor of Physics, Department of Physics, University of Alabama, Huntsville, Alabama.

Introduction: The *New Vision 400* Project

CONTENT AND SCOPE OF THIS VOLUME

As noted in the Preface, this book originated with the *New Vision 400 (NV400)* conference held in Beijing in October 2008 in conjunction with the widely celebrated 400th anniversary of the invention of the telescope in 1608 by Hans Lipperhey. This volume has benefitted greatly from the scientific and cultural perspectives related to astronomy from both the Eastern and Western traditions. The conference poster preceding Part I highlights some of these important contributions from both sides of the world. The Appendix and the conference website (http://nv400.uchicago.edu/) provide further information about the Beijing meeting.

Also, as noted in the Preface, the three opening talks from the Beijing conference provide the first three chapters in Part I. Different in purpose and scope from the rest of the science meeting, and therefore from the other chapters in this volume, these more personal reflections will serve to expand the book's readership to a more general, less technical audience. In the opening chapter, which initially appeared as a paper in the Chinese journal *Physics*, Nobel Laureate Tsung-Dao (T. D.) Lee relates modern astronomy to the grand sweep of physics (contrasting the very large and the very small) and emphasizes the international, multicultural nature of the modern enterprise of science. Nobel Laureate Riccardo Giacconi relates the increase in understanding of astronomical objects that has come from the augmentation of data from ground-based facilities with that from multi-wavelength observatories in space. He emphasizes the technological innovations in planning and software that enabled, first, x-ray astronomy, then ultraviolet, then optical telescopes on the grandest scales. Shaw Prize Winner Geoffrey Marcy discusses the discovery of extrasolar planets, the rapid growth of the numbers of known planets around other stars, and the analogies to the planets of our solar system (presaging the detailed later discussions by Sara Seager and Charles Beichman that promise new insights into the origins of life). The technology of grand machines is at the heart of the three opening stories, and the three authors who lived those stories know them and their impact on our knowledge and worldviews better than anyone.

Part II includes four chapters that discuss the impact of telescopes on our knowledge of the Universe. Galaxies were, of course, unheard of in Galileo's time, but today they form the focus of virtually all questions in astrophysics. Ben Moore provides an overview of the latest views on how these objects formed and evolved to their current appearances and where current research is headed as computer simulations strive to catch up with the facts obtained from the use of telescopes. The beginning of it all often forms a key question. Naoki Yoshida presents the modern view of how the large-scale structure of the Universe, attributed to early organization of dark matter, led to the first light of stars, a topic integrally associated with Moore's chapter. Stars, and ultimately the galaxies that contain them, eventually cease to be as luminous objects because they either die with a whimper over millions of years or with a bang in mere seconds. Alex Filippenko discusses our expanding knowledge of supernovae, especially interesting because of their, as yet, poorly understood relation to gamma-ray bursts. He also discusses their use as tracers of the expansion history and the structure of the Universe. (See the "Supernovae" box for more on this topic, which is a central theme of this volume.) Of course, many time-consuming and detailed questions of astrophysics underlie these first three thematic areas. Xiao-Wei Liu discusses one such theme, the anomalous abundance of the elements that seemingly exist in two forms in the same objects—planetary nebulae, one of the forms of death-by-whimper of the lower-mass stars.

SUPERNOVAE*

Supernovae, the death explosion of massive stars, relate to the content of most of the chapters in this book, in addition to "starring" in a chapter about which they are the central topic. They are a common element of Chinese and Western astronomy, but were actually better recorded historically in China (as "guest stars"). Supernovae presumably will help us understand what kinds of stars formed the first stars. They formed the heavy elements that seeded the galaxies with carbon, nitrogen, and oxygen that form the basis for life. Currently, they are believed to be related to gamma-ray bursts, apparently the most powerful explosions in the Universe. Dark energy came to the fore with evidence that the supernovae, treated as standard candles, are more distant than expected unless the Universe is undergoing acceleration. They leave black holes distributed in each galaxy, possibly forming the massive nuclei of quasars. Whatever their source, the black holes may be responsible for the highest-energy cosmic rays.

Despite the fact that hundreds are now known and have been well studied, supernovae continue to yield surprises to challenge astronomers. About 150,000 years ago, a blue star exploded in a distant satellite galaxy of the Milky Way. It took but a few seconds. The light from that explosion reached Earth at the end of February 1987. Conveniently, within a few years, the *Hubble Space Telescope* was launched (just in time!), ready to take the dramatic images of Supernova 1987A (SN 1987A) shown on the back cover. During its lifetime, the star had previously been a red supergiant, which then transitioned into a blue supergiant before exploding. About 20,000 years before exploding, a fast wind from the blue supergiant expanded outward, sweeping up material around it that had been lost from the star in the previous red supergiant stage, thus forming an hourglass-shaped "cocoon," with a bright equatorial ring and two rings at higher latitudes (as imaged with *Hubble*).

The three-ring "sky" image filling the back cover shows a large area of the host galaxy of the supernova, the Large Magellanic Cloud, with the tiny, three-ring image of the supernova effluent near the center. Over time, small structures appeared around the inner ring, and they continue to change appearance each year, to this day. The image in the upper-right corner of the back cover zooms in on the region around the inner ring. Figure 11a in Chapter 6 shows the three rings seven years after the supernova was discovered, and Figure 11b shows the bright dots on the ring 12.5 years later.

When the supernova exploded, the intensity of the explosion caused an extremely fast-moving shock wave to expand outward. Around 1997, the shock wave began to impact finger-like protrusions extending from the inner edge of the equatorial ring. As these are impacted by the shock, they become heated and light up, forming the bright spots on the ring. The regularity in spot structure is not clearly understood. Note: The persistent dot at the lower right and on the ring is a star, unrelated to the explosion. The ring is expected to continue to brighten for several more decades at least, then eventually become less visible as the shock expands beyond the ring. Such behavior may be common throughout the Universe, but the proximity of SN 1987A and the success of *Hubble* have allowed us to observe it for the first time in unprecedented detail.

The availability of *Hubble* images is an example of the unexpected benefits of new technology on our understanding of the Universe and of the individuals who make and operate our wonderful observing machines.

---

* Vikram Dwarkadas, senior research associate in the Department of Astronomy and Astrophysics at The University of Chicago, contributed to this discussion of supernovae.

Back Cover Images:

Background: Large Magellanic Cloud in the Region of Supernova 1987A, February 4, 1999. Glittering stars and wisps of gas create a breathtaking backdrop for the self-destruction of a massive star, called Supernova 1987A, in the Large Magellanic Cloud, a nearby galaxy. Astronomers in the Southern Hemisphere witnessed the brilliant explosion of this star on February 23, 1987. Shown in the *Hubble* telescope image, the supernova remnant, surrounded by inner and outer rings of material, is set in a forest of ethereal, diffuse clouds of gas. From the *Hubble*site NewsCenter Archive. Credit: The *Hubble* Heritage Team (AURA/STScI/NASA). Available at: http://hubblesite.org/newscenter/archive/releases/star/supernova/1999/04/image/a/.

Upper-right Corner: Three Rings of Gas Surround Supernova 1987A. From the *Hubble*site Gallery Picture Album. Credit: Peter Challis (Harvard-Smithsonian Center for Astrophysics). Available at: http://hubblesite.org/gallery/album/pr1998008g/).

Following these examples of our knowledge of the Universe based on extensive use of telescopes, the four chapters in Part III ask: What large questions can be pursued on a grand scale in the future and using what instruments and what strategies? Elliott Bloom gives the first results and talks about the future prospects of the Fermi satellite for investigating various forms of dark matter, which was first suspected in the 1930s, and relates that work to other experiments aimed at its direct detection. The even grander enigma of dark energy is taken up by Mark Sullivan, who discusses the evidence we need to confirm the current most popular views on its origin. Black holes (BHs) consistently form an ever-richer part of the evolving explanation of the Universe and all that is in it, and Shuang-Nan Zhang discusses how close we are to confirming the existence of these most mysterious of small objects in the Universe. Finally, the enigma of cosmic rays, known since 1911 but still unexplained, is taken up by Glennys Farrar in a discussion of the very latest attempts to solve the puzzle of the most energetic single particles in the Universe.

Attacking these challenging questions will certainly require new technologies and new facilities, as discussed in the three chapters in Part IV. Radio astronomy has led the way in finding the boundaries of the Universe through identifying quasi-stellar objects, the compact forms of stars (neutron stars and BHs), and is poised to do the same for explaining many of the above issues. K. Y. (Fred) Lo and Alan Bridle discuss the newest plans for enhancing our facility base in this field. Over the 400 years of the existence of the telescope, the instrument makers strove to improve images in order to obtain sharper views of the Universe, largely without much success. Michael Shao outlines the use of adaptive optics, which promises to allow the construction of ground-based instruments that rival the *Hubble Space Telescope* in image quality. Richard Ellis discusses the very large telescopes that promise, with adaptive optics, to lead us to new revolutions in our understanding, revolutions that will likely dwarf the one started by Galileo and that, by all accounts, will not require another 400 years to change the face of our body of knowledge.

The intellectual impact of the telescope and of the discoveries made with it are well known, as explored in the three chapters in Part V. Yi-Long Huang highlights how the cosmos was perceived in China in the days before telescopes by focusing on the calendars and astrology used in those days. Although less well known in the West, his story certainly has parallels in other societies and serves as a good example of the "big picture" before the telescope arrived. Owen Gingerich isolates the 200-year period in the West after the telescope was first used to discuss the dramatic impact of this invention that we realize today; but he also emphasizes the inertia that Galileo's ideas encountered: the question of whether Earth moves took a while to settle. Xiao-Chun Sun describes the impact of the telescope on Chinese society with its own time-honored astronomical tradition. Better eclipse observations using the telescope led to better calendars. The new discoveries through telescopic observations also intrigued Chinese intellectuals and stimulated speculations on a number of cosmological issues.

In Part VI, the last section, the seven chapters delve into the "big questions" left to humankind with the knowledge gained from the telescope. First, Sara Seager describes the recent and quite unexpected improvement in our ability to detect the atmospheres and biomarkers of the planets discussed in Marcy's chapter at the beginning of the book.

Even the question, Does life exist elsewhere?, can now be addressed broadly with straightforward applications of instruments already on the drawing boards, as discussed by Charles Beichman. The three chapters together (including Marcy's in Part I) highlight the fortuitous existence in the national science facility plans (largely made before the discovery of the first extrasolar planet) of instruments optimized for planet detection and the consequent optimism about rapid advances to come.

In contrast to the high optimism about the possibility of discovering (or not) life elsewhere, less accessible "big questions," such as: What is the true geometry of the Universe? What is the origin of the physical laws of the Universe? are more tenuous. Dealing with the Multiverse, Paul Davies in his chapter and Renata Kallosh and Andrei Linde in their chapter discuss whether the Universe we see is as small and deterministic as it appears in the first seven chapters of this book. These two chapters in tandem show the evolution of ideas in this relatively new field, not only because of the authors' approach, but because the Davies chapter is a modified version of a paper first published in 2004, and the Kallosh–Linde chapter gives very recent results.

The last three chapters of the book tackle the relation among the cosmos inaccessible via our transportation systems, the evidently reachable cosmos of our imaginations, and the human condition. Yung-Sik Kim discusses the relation between the cosmos and humanity in traditional Chinese thought. Some ancient writers chose to emphasize observation, some theory. The balance between the two approaches, which drove science forward so rapidly, was not reached until later times, primarily in the West. Peter Harrison addresses the apparent intelligibility of the cosmos, reflecting on humanistic and scientific traditions in the West that led to the successful "tension between a kind of optimistic rationalism and critical empiricism" that is essentially the missing balance noted by Kim and the cultivation of the idea that science was a useful pursuit. The arguments for the empirical approach that developed are interestingly similar to arguments for a Multiverse. George Ellis asks the ultimate philosophical question: How does one account for the qualities of life such as purpose, ethics, aesthetics, and meaning, issues normally considered outside the realm of science and not explained by its current conclusions?

Among the oral presentations at the meeting were two that we were unable to include as chapters in this volume: (1) "The Cosmic Microwave Background Radiation—A Unique Window on the Early Universe" by Gary Hinshaw, NASA/Goddard Space Flight Center, Greenbelt, Maryland, USA, and (2) "The Development of Large-Scale Structure in the Universe" by Simon White, Max Planck Institute for Astrophysics, Garching, Germany. Unfortunately, neither author was available to prepare a manuscript for this book in time for publication. Material related to the cosmic background radiation is found in the Kallosh–Linde and Sullivan chapters.* Also, Yoshida kindly agreed to expand his manuscript in order to include some of the key ideas in the area of large-scale structure in the Universe in his chapter on the first stars, and Moore touches on the matter as well.

---

* Older but still relevant material, written at the same level as the chapters in this volume, can be found in a recent book in honor of Charles H. Townes: *Visions of Discovery: New Light on Physics, Cosmology, and Consciousness.* Edited by Raymond Y. Chiao, Marvin L. Cohen, Anthony J. Leggett, William D. Phillips, and Charles L. Harper, Jr. Cambridge: Cambridge University Press (2011): http://www.cambridge.org/gb/knowledge/isbn/item2709757/?site_locale=en_us and http://www.cambridge.org/gb/knowledge/isbn/item2709757/?site_locale=en_gb. See the following chapters: "The Microwave Background: A Cosmic Time Machine" by Adrian T. Lee (pp. 233–46); "Dark Matter and Dark Energy" by Marc Kamionkowski (pp. 247–93); and an 'Ultrasonic' Imageof the Embryonic Universe: CMB Polarization Tests of the Inflationary Paradigm" by Brian G. Keating (pp. 382–409).

IN CONCLUSION

The editors trust that the chapters in this book represent a moderately coherent, truly global perspective on the state of astronomical knowledge—its depth and its mysteries—at the benchmark 400th year after the invention of the telescope. We are the recipients of the current state of cosmic knowledge, and continue our pursuit of further knowledge with our probing questions, because of a simple combination of some lenses that took place 400 years ago. History records that virtually all of modern science arose with the telescope, and its use has forced big questions on us regarding our place in the Universe; whether the instrument itself will lead to answers to those questions is yet to be seen.

Front Cover Images:

Upper-left corner: Bust of Zhang Heng 张衡 (78–139). A mathematician, astronomer, geographer, engineer, poet, and artist, Heng became chief astronomer in 112 under Emperor An of the Han dynasty, serving for 24 years. Among his many contributions to astronomy, mathematics, and technology were accurately estimating the value of pi (π), inventing the seismometer and odometer, and correcting the calendar to bring it into alignment with the seasons. Heng also explained lunar eclipses and demonstrated that the Moon was illuminated not by an independent light source, but by the reflected light of the Sun. The chapter by Xiao-Chun Sun discusses more about Heng's contributions. Credit: Image of bust in the courtyard of the Ancient Observatory, Beijing, XIAO Jun, Director. Reproduced with permission.

Upper-right corner: Bust of Galileo Galilei (1564–1642). An astronomer and physicist, Galileo published his epochal *Sidereus nuncius* ("*Starry Messenger*") in 1610, the first scientific treatise based on observations made with the telescope, a new invention that he had transformed from a carnival toy into a discovery machine. His book described the Earth-like mountains and plains on the Moon, the multitude of stars invisible to the unaided eye, and his discovery of the moons of Jupiter. This was the launch of telescopic science and the consequent revolution in astronomy. Galileo's passionate defense of the heliocentric Copernican cosmology brought him into conflict with the Roman Catholic Church, and in 1633 he was sentenced to house arrest. In 1993, Pope John Paul II acknowledged that the Church had done Galileo a grave injustice. Chapter 1 by T. D. Lee discusses this episode. Image copyright Museo Galileo – Institute and Museum of the History of Science, Florence. Reproduced with permission.

Center: Galileo's Two Surviving Telescopes. Only two existing telescopes have a reasonable Galilean pedigree, both housed in the Galileo Museum in Florence. The chapter by T. D. Lee also shows one of them, and the chapter by Owen Gingerich shows the other. Image copyright Museo Galileo – Institute and Museum of the History of Science, Florence. Reproduced with permission.

The *New Vision 400* conference poster is a montage of Eastern and Western scientific and cultural images spanning almost 2,000 years of humankind's endeavor to make sense of the cosmos. *Designed by Steven Lane, The University of Chicago.*[*]

---

* Readers can download the poster and poster description here: http://nv400.uchicago.edu/logo.html. Note that the speaker listings in the poster were accurate as of the date it was originally printed, but that these evolved over time; please see http://nv400.uchicago.edu/ and the Appendix in this volume for more information about the symposium program and speakers, as well as for background information important to the development of this book.

NEW VISION 400 CONFERENCE POSTER DESCRIPTION

The *New Vision 400* conference poster is a montage of Eastern and Western scientific and cultural images spanning almost 2,000 years of humankind's endeavor to make sense of the cosmos.

BACKGROUND

The background image for the poster shows NGC 602, a young, bright, open cluster of stars located in the Small Magellanic Cloud. The image is a composite of many separate exposures made by the Advanced Camera for Surveys (ACS) instrument on the *Hubble Space Telescope* using several different filters.

UPPER LEFT

Bust of Zhang Heng

Zhang Heng (78–139) was a mathematician, astronomer, geographer, engineer, poet, and artist. He became chief astronomer in 112 under Emperor An of the Han dynasty, serving for 24 years. Among Heng's many contributions to astronomy, mathematics, and technology were accurately estimating the value of pi (π), inventing the seismometer and odometer, and correcting the calendar to bring it into alignment with the seasons. He also explained lunar eclipses and demonstrated that the Moon was illuminated not by an independent light source, but by the reflected light of the Sun.

BELOW THE BUST OF ZHANG HENG

Galileo's Moon

This watercolor of the Moon by Galileo Galilei (1564–1642) was published in *Sidereus nuncius* ("*Starry Messenger*") (1610), the first scientific treatise based on observations made with a telescope. Galileo's watercolor shows the surface characteristics of the Moon as "uneven, rough, and strewn with cavities and protuberances, [not unlike the surface] of the Earth." The color of the Moon is likely darker than Galileo originally intended, the result of aging paper.

TOP CENTER

Suzhou Astronomical Chart

The Suzhou Astronomical Chart is the best-known star map from Chinese history. It was engraved in stone in 1247 by Wang Zhi Yuan, according to the design by Huang Shang. The chart is based on observations made in the 11th century.

UPPER RIGHT

Frontispiece from *Principia*

The image of Urania, "the muse of astronomy," revealing the heavens to the mathematician (Isaac Newton) is from the frontispiece of the first English translation of the *Principia*, Volume 1 (published in 1729). Regarded as one of the most important scientific works ever written, Newton's *The Mathematical Principles of Natural Philosophy*, better known as the *Principia* (originally published in 1687), contains the statement of Newton's laws of motion, his law of universal gravitation, and a derivation of Kepler's laws for the motion of the planets.

BELOW THE *PRINCIPIA* FRONTISPIECE

Diagram of a Reflecting Telescope

This diagram from Sir Isaac Newton's *Opticks* (1704) illustrates how a reflective telescope works. Newton communicated the details of his telescope to the Royal Society in 1670, but it did not become widely known until the publication of *Opticks* more than 30 years later.

CENTER, IN THE WORD "VISION"

Sun

The Sun, captured during a large coronal mass ejection by the *Solar and Heliospheric Observatory* (*SOHO*), a project of international cooperation between the European Space Agency (ESA) and the National Aeronautics and Space Administration (NASA).

BELOW THE SUN

The Blue Marble

Earth at night, taken as part of NASA's Earth Observatory program dedicated to the science of global warming and climate change, part of the Earth Observing System (EOS) Project Science Office at NASA's Goddard Space Flight Center.

BOTTOM RIGHT

Goldstone Apple Valley Radio Telescope

The Goldstone Apple Valley Radio Telescope (GAVRT) project, a partnership involving NASA, the Jet Propulsion Laboratory, and the Lewis Center for Educational Research in Apple Valley, California, allows students to use a dedicated 34 m (111 ft) radio astronomy telescope at NASA's Deep Space Network Goldstone Complex. Connected via the Internet, students point the massive dish at targets in space and record their findings.

BOTTOM LEFT

Wilkinson Microwave Anisotropy Probe

The Wilkinson Microwave Anisotropy Probe (WMAP) satellite, a NASA Explorer mission, revealed conditions as they existed in the early Universe by measuring the properties of the cosmic microwave background radiation (CMBR) over the full sky. Launched in June 2001, *WMAP* continued to collect data until the mission concluded in August 2010.

SURROUNDING WMAP

Lagrange Points

The Italian–French mathematician Joseph Lagrange (1736–1813) discovered five special points in the vicinity of two orbiting masses where a third, smaller mass can orbit at a fixed distance from the larger masses. These Lagrange Points mark positions where the gravitational pull of the two large masses precisely balances the centrifugal force required to rotate with them.

ALONG BOTTOM EDGE

Fraunhofer Lines

The set of spectral lines named for the German physicist Joseph von Fraunhofer (1787–1826) was first observed as dark features (absorption lines) in the optical spectrum of the Sun.

Part I

Creativity and Technology in Astronomical Discovery

1 From the Language of Heaven to the Rationale of Matter*

Tsung-Dao Lee

CONTENTS

In this chapter, I will introduce a few of the most important historical developments in the study of astronomy, then focus on some concepts of physics that resulted from our desire to understand the function of the Heavens, and finish with a discussion of what we can look forward to as we continue our quest to answer life's most challenging questions.

ASTROPHYSICS IN ANCIENT CHINA

The ancient Chinese made many important scientific breakthroughs, and their impact on astronomy is certainly among their many significant contributions. Some of the most interesting of these are their discovery of novae and supernovae (SNe), their creation of some of the first devices used to track the movements of the Heavens, and their observation of sunspots.

Among the different kinds of celestial objects, the most striking are novae and SNe. A nova is typically several tens-of-thousands times as bright as the Sun, while a supernova (SN) is several tens-of-billions times as bright as the Sun. Both of them were first discovered in China and mark an important development in the evolution of astronomy.

The earliest discovery of a nova dates back to the 13th century BCE. The event was recorded in an oracle bone inscription (see Figure 1.1) containing characters that mean that on the seventh day of the month, when the Moon rose, a great new star appeared in company with Antares: "新 (new) 大 (big) 星 (star) 并 (in company with) 火 (Antares)" (Needham and Wang, 1959, p. 424). Another piece of oracle bone inscription indicated that within a few days, the luminosity of the star had decreased substantially, a characteristic feature commonly associated with novae.

The earliest discovery of a SN was also made in China in the year 185 CE and recorded in the Book of Later Han. Much later, Chinese astronomers also found another particularly famous SN in year 1 of the Greatest Harmony during the reign of Emperor Rencong 仁宗 from the Song 宋 dynasty (1054). The historical record says that on August 27, 1054 CE, a very bright new star as big as an egg suddenly appeared in the sky. In fact, the brightness of the star was recorded for nearly

* Translated by Zu-Hui Fan, Professor of Cosmology and Galaxy Formation, Department of Astronomy, School of Physics, Peking University, Beijing, China, from the article based on Professor Lee's October 2008 *New Vision 400* symposium presentation (see the Preface): "From the Language of Heaven to the Rationale of Matter," originally published in the Chinese journal *Physics,* 37 (2008), 831–35; reprinted with permission.

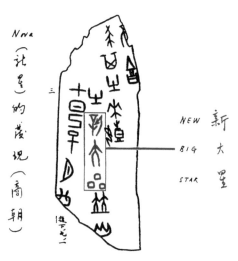

FIGURE 1.1   The oracle bone carved with "新 (new) 大 (big) 星 (star) 并 (in company with) 火 (Antares)." (Courtesy of T. D. Lee from his private collection.)

2 years from its first appearance up until it vanished in July 1056 (Needham, 1959, pp. 426–7). However, these discoveries were not the only ones for which the ancient Chinese were responsible.

The Yan Huang 炎黄 civilization developed in the central part of China about 5,000 years ago and then expanded across the continent. It was quite different from the contemporary, ocean-based Western civilizations. As our ancestors began to observe the Heavens, they discovered that all of the stars revolved slowly around the sky. They started developing tracking systems, using a period of 12 double-hours (*shi chen* 时辰) to log the movements. They theorized that the sky revolved around an axis, which pointed to the existence of a Celestial Pole. This would provide the impetus to create some of the earliest astronomical devices.

In the *Zhou Rituals* (*Zhou li* 周礼), it is said that "the Blue Round Jade (*cang bi* 苍璧) was used to pay homage to Heaven, and the Yellow Rectangular Jade (*huang cong* 黄琮) was used to pay homage to Earth (*yi cang bi li tian, yi huang cong li di* 以苍璧礼天, 以黄琮礼地)." What is *cang bi* and what is *huang cong*? They are jade objects, *cang bi* being round in shape, representing Heaven, and *huang cong* being rectangular in shape, representing Earth. Both have a round-shaped hole in the middle. These objects represented the first steps in the development of yet another Shang 商 dynasty jade object called *xuan ji* 璇玑 (see Figure 1.2).

According to the *Canon of Shun* 舜, in the *Book of Documents* (*Shu jing shun dian* 书经舜典), *xuan* 璇 means "beautiful jade," and *ji* 玑 means "a rotatable instrument." But this jade object was an astronomical instrument (璇美玉也，玑为转远，径八尺，圆周二丈五尺强，玉者正天文之器). While the *xuan ji* was normally quite large, about 2.4 m (8 ft) in diameter and 7.6 m (25 ft) in circumference, the only surviving object of a similar nature is from the Shang dynasty, measuring only about 30 cm (11.8 in) diameter. It is very likely that this *xuan ji* was the symbolic representation of an actual instrument created before the Shang dynasty.

My hypothesis is that the pre-Shang *xuan ji* was an instrument for determining the position of the Celestial Pole (see Figure 1.3). Its rotatable disk had three notches on the rim to register the locations of three individual stars, and it rotated in such a way that the three stars were always aligned with the three notches. The rotational axis of the disk was a bamboo tube about 4.6 m (15 ft)–6 m (20 ft) long that had a small hole through its center by which the location of the Celestial Pole was determined. Assuming that the diameter of the small hole was about 2 mm (0.079 in), the position of the Celestial Pole could be determined with an accuracy of 0.013 degrees. In order to support the bamboo tube, a heavy stone casing was built around it. This long stone casing later evolved into what they referred to as a *cong,* and the large disk was called a *bi* or *xuan ji*.

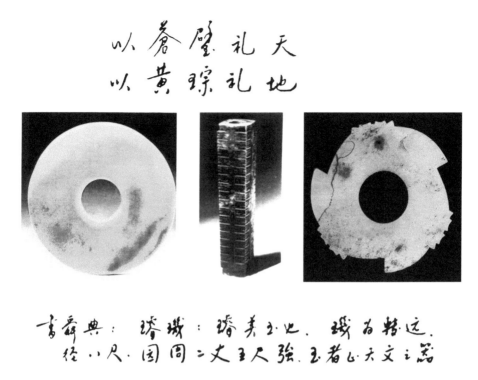

以蒼璧礼天
以黄琮礼地

書舜典：璿璣：璿美玉也．璣為轉運．
徑小尺．圓周二丈王人強．玉者山天文之器

FIGURE 1.2   *Cang bi* 苍璧, left, was used to pay homage to Heaven, and *huang cong* 黄琮, middle, was used to pay homage to Earth; the image on the right is *xuan ji* 璇玑. (Courtesy of T. D. Lee from his private collection.)

We now know that the precession of the axis of Earth's rotation has a period of 25,000 years, which results in slow movement of the Celestial Pole in the sky. At present, the Celestial Pole is near α Ursa Minor, the so-called North Pole star, but during the time that the *xuan ji* was first constructed, the Celestial Pole's position was about a fifth of a cycle from this (see Figure 1.4). Assuming that the three notches on the *xuan ji* identified the stars marking the location of the Celestial Pole, the next step was to discern the period in which the three bright stars, 120 degrees apart in right ascension, were exactly in line with these notches. By consulting the astronomical almanac, we find that the period around 2700 BCE met this requirement. At that time, α Draco, one of the stars in the Chinese constellation *Ziwei* 紫微 (in the Western constellation Little Dipper), was very close to the Celestial Pole, even closer than α Ursa Minor is to the Celestial Pole at present. Surrounding α Draco were three luminous stars, η and λ Draco and η Ursa Major (see Figure 1.5), which were exactly in alignment with the three notches on the *xuan ji*. If its function was as I suggest, then this means that as early as 2700 BCE our ancestors had already used an astronomical instrument to determine the location of the Celestial Pole with an accuracy of 0.013 degrees. At that time, the star α Draco in the constellation *Ziwei* was the Celestial Pole. This is probably the reason why *Ziwei* had long been astrologically associated with the rise and fall of the empire and its rulers. The differences between the sky that the ancient Chinese astronomers observed and what we can see today are a product of 4,700 years of the precession of the Earth's rotational axis. Now, the position of *Ziwei* in the sky no longer has special meaning for us.

China is also among the earliest countries to have observed sunspots. According to Needham and Wang (1959), the Europeans rarely paid any attention to strange astronomical phenomena such as sunspots because they had the preconception that the Earth's rotational axis was perfect. It was Galileo, who in 1610 became the first European to observe sunspots with his telescope (see below), some 1,600 years after the first Chinese observations of them in 28 BCE during the reign of Liu

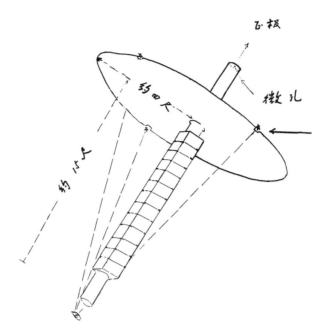

FIGURE 1.3   The hypothetical instrument *xuan ji*, which was used to determine the position of the fixed point in the sky (the Celestial Pole). The length of the bamboo tube was about 4.6 m (15 ft)–6 m (20 ft), and the diameter of the disk was about 2.4 m (8 ft); the three notches, or gaps, used to identify the stars marking the location of the Celestial Pole are shown at the edge of the disk. (Courtesy of T. D. Lee from his private collection.)

Xiang 刘向. Chinese official histories kept about 120 records of sunspots for the period from 28 BCE to 1638 CE. Additional records can also be found in local histories, biographies, and other historical documents. The Chinese often referred to sunspots as "Dark Qi" (*hei qi* 黑气), "Dark Seed" (*hei zi* 黑子), or "Dark Crow" (*wu* 乌), and they described their sizes in such terms as "coins" (*qian bi* 钱币) (Needham, 1959, pp. 434–36).

THE INVENTION OF THE TELESCOPE

The telescope was invented 400 years ago. By the autumn of 1608, news of this invention had spread throughout Europe, and less than 2 years later Galileo Galilei published his epoch-making *Sidereus nuncius* (*The Starry Messenger*). He wrote in his book: "About ten months ago, a report reached my ears that a certain Fleming had constructed a spyglass by means of which visible objects, though very distant from the eye of the observer, were distinctly seen as if nearby" (pp. 28–9). Shortly afterward, Galileo constructed his own telescope (see Figure 1.6) and turned it toward the sky.

At that time, the cosmological theory sanctioned by the Catholic Church was the geocentric one, in which the Earth was considered the center of the Universe and all celestial bodies were thought to orbit it. From his observations of Jupiter with his telescope, however, Galileo derived a completely different cosmological theory. On January 7, 1610, Galileo discovered that beside Jupiter were one star on the right-hand side and two stars on the left-hand side (see Figure 1.7). The next day, the three stars were all on the right-hand side, and none were on the left-hand side. Just two days later, the stars on the right-hand side had all disappeared, and two faint ones showed up on the left-hand side again. On the following day, one of the faint stars on the left-hand side became brighter. A day later, all the faint stars were gone, and only one star was on the right-hand side, while two were on the left-hand side. Then, on January 13, the configuration of the stars changed significantly again, with three on the right and one on the left! From these observations, Galileo reasoned that these stars behaved as is they were "moons" of Jupiter—i.e., four satellites

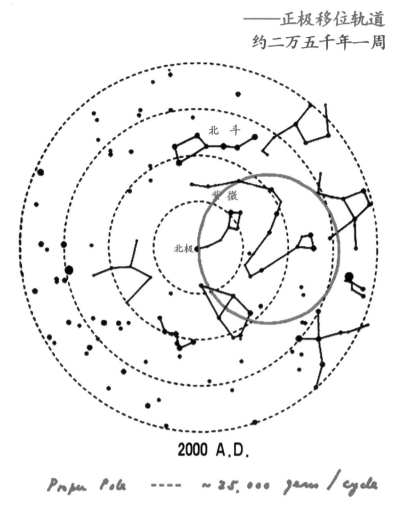

——正极移位轨道
约二万五千年一周

2000 A.D.

Proper Pole ---- ~25,000 *years / cycle*

FIGURE 1.4 The current star chart about the star α Draco of the Chinese constellation *Ziwei* 紫微 (in the Western constellation Little Dipper). (Courtesy of T. D. Lee from his private collection.)

that orbited it, but not Earth! This showed that not all celestial objects were revolving about Earth, which refuted the theory backed by the Catholic Church.

Other important discoveries by Galileo were the equality of inertial mass and gravitational mass discovered in 1591, lunar mountains and craters discovered from 1609 onward, the phases of Venus, and so on. In 1632, he published his famous book *Dialogue Concerning the Two Chief World Systems*.

In 1633, Galileo was sentenced to house arrest by the Roman Catholic Church. By order of the pope, he was prohibited from publishing, giving lectures, teaching, and discussing academic topics with his friends. Four years later, in 1637, Galileo lost his sight. He died on January 8, 1642. His death might have stunted the growth of modern science were it not for Isaac Newton, born in the same year, on December 25. Fortunately, the power of the Pope did not extend to England at that time, and Newton was able to further advance modern science following in Galileo's footsteps.

In China, when the last emperor, Chongzhen, committed suicide in 1644, the Ming dynasty met its demise. Science in China was stagnant at that time, with no significant developments of which to speak.

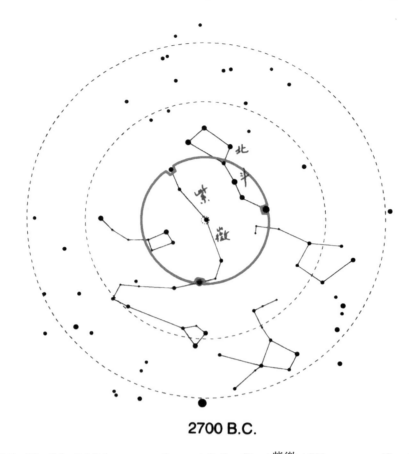

2700 B.C.

FIGURE 1.5   The Celestial Pole was near the constellation *Ziwei* 紫微 4,700 years ago. (Courtesy of T. D. Lee from his private collection.)

In 1991, China released a postcard in order to commemorate the 400th anniversary of the discovery of the "equality of inertial mass and gravitational mass" by Galileo. I was lucky enough to be tasked with its design (see Figure 1.8).

In 1993 in Vatican City, Pope John Paul II acknowledged that the Roman Catholic Church had done Galileo a grave injustice (see Figure 1.9). During that event, I made a speech attempting to

FIGURE 1.6   The telescope constructed and used by Galileo. (Photo by Franca Principe, Galileo Museum, Florence. Reproduced with permission.)

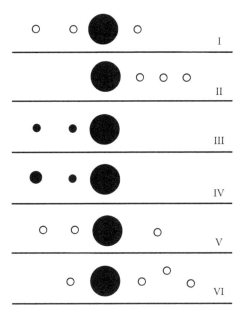

FIGURE 1.7   Galileo's preliminary "decoding" of "the language of Heaven" from his observations of Jupiter using his new telescope in 1610. (Courtesy of T. D. Lee from his private collection.)

伽利略发现"惯性质量和引力质
量等价"400 周年（1591－1991）

400th Anniversary of the Discovery of the "Equality of
inertial mass and gravitational mass" ($m_i \equiv m_g$) by Galileo
Galilei (1591–1991)

$m_i \equiv m_g$

中华人民共和国邮电部发行
Released by the Ministry of Posts and Tele-
communications of the People's Republic of China

FIGURE 1.8   The postcard designed by the author in 1991. (Courtesy of T. D. Lee from his private collection.)

FIGURE 1.9   At the Vatican on May 8, 1993, Pope John Paul II admitted that Galileo was right and apologized to scientists everywhere. On that occasion, the author (right) made a speech on behalf of scientists all over the world. (Courtesy of T. D. Lee from his private collection.)

summarize the views of scientists all over the world, in which I said to the pope that either argument (that the Earth orbits the Sun or that the Sun orbits the Earth) could be correct because their motions were relative to each other. This concept of relativity was not realized in Galileo's time. However, it was wrong of the pope to force Galileo to recant his scientific findings, prohibit him from lecturing, and put him under house arrest, and I was happy to see that the Church had taken the requisite steps to exonerate him and clear his name.

THE DEVELOPMENTS OF PHYSICS IN THE 20TH CENTURY

The most important developments in physics during the 20th century were the theory of special (and later general) relativity, quantum mechanics, and the understanding of nuclear energy. In 1905, Albert Einstein published five seminal papers, one of which contained his proposal of the special theory of relativity. This would later be commemorated in 2005, which the United Nations declared the World Year of Physics, recognizing Einstein's many important scientific contributions. The theory of quantum mechanics was discovered in the 1920s. This and the work of Einstein were epoch-making in the history of modern physics.

On August 2, 1939, Einstein wrote a letter to President F. D. Roosevelt in which he pointed out that "the element uranium may be turned into a new and important source of energy in the immediate future." On December 2, 1942, a team of scientists led by Enrico Fermi achieved the first self-sustaining nuclear chain reaction. For the first time, humankind was able to obtain energy from the same processes that create solar energy. On that day, Professor Arthur Compton of the University of Chicago made a famous phone call to Dr. James Conant, the Science Advisor to the President, to inform him of the success of the nuclear reactor using a coded message. Compton said: "The Italian navigator has landed in the new world." Conant asked: "How were the natives?" "Very friendly," Compton replied. At that time, Italy and the United States were at war, so their conversation had to be very brief. Here, the Italian navigator referred to Fermi, and the story of the discovery of the new world by Columbus alluded to the successful operation of the first nuclear reactor.

The discovery and use of fire signified the beginning of civilization. The energy of fire comes from solar energy, which in turn is nuclear energy. The Sun itself is a huge reactor of hydrogen nuclei. The success of the first direct production of controllable nuclear energy by the group of scientists led by Fermi made it possible to extract energy from sources other than the Sun. This is

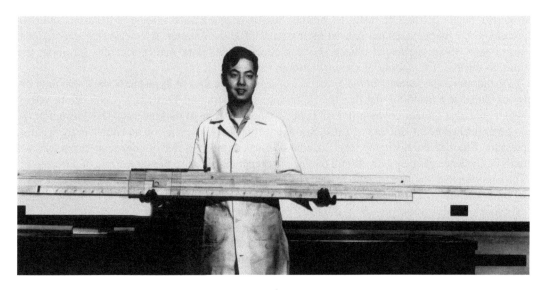

FIGURE 1.10 The special slide rule, handmade by Enrico Fermi and the author (shown) in 1948, used for calculating the inner temperature distribution of the main sequence of stars. The upper scale is 18Log, and the lower one is 6.5Log. (Courtesy of T. D. Lee from his private collection.)

one of the most important advancements in science and technology in the 20th century, as well as in the history of humankind.

I still remember that when I was Fermi's Ph.D. student in the 1940s, each week he spent half a day discussing physics with me personally. One day, he asked me whether I knew the central temperature of the Sun. I answered, "It is about 10 million degrees." He then asked whether I had ever calculated it. I said no. He said to me, "You must check it yourself before accepting other people's conclusions." I replied that it was somewhat complicated. The calculations involve two equations. One has a term of T^{18}, and the other has a term of $T^{6.5}$, where T is the temperature. Thus, the computations are very complex. Fermi said, "I will help you to make a special slide rule to simplify the calculations." Then, we worked together to make a large wooden slide rule (see Figure 1.10). The upper scale is 18Log and the lower one is 6.5Log. With this big "toy," I finished the calculations easily. This is a good example of how a teacher trains his student—leading by example. That is the mark of a great professor, and this resulted in an experience that I could never forget.

THE FUTURE OF THE 21ST CENTURY

As mentioned in the previous section, in 1905 Einstein published five important papers that have deeply influenced the development of global civilization. Einstein's first paper was "A new determination of molecular dimensions." In his second paper, he developed the concept of the light quantum. The third paper was about Brownian motion. The fourth was about the special theory of relativity. And the famous concept of mass-energy equivalence ($E = mc^2$) is laid out in his fifth paper. I believe that the scientific achievements of Einstein will have even more significant consequences in the 21st century than they have already had. At present, dark energy, which was first proposed by Einstein, has extremely important functions in our Universe; another reason that I suspect his influence on the development of science in the 21st century could be even more profound than that on the 20th century. Understanding the nature of dark matter and dark energy is a great challenge for the scientists of our generation, and I believe that we will eventually reach our goals.

In our "Big Bang" Universe, matter contributes only about 5% to the total energy. The rest consists of about 25% dark matter and about 70% dark energy. This sounds very strange. What is dark matter? We do not know. What is dark energy? We do not know that either. What we refer to as known matter consists of electrons, protons, neutrons, and a trace amount of positrons and anti-protons. Dark matter does not consist of any parts of known matter. By measuring gravitational interactions between celestial objects, we can conclude that dark matter accounts for about five times as much as the amount of known matter in the Universe.

Furthermore, the recent *Hubble Space Telescope* observations of Type la SN show that not only is our Universe expanding, but that the expansion is accelerating. This acceleration can be attributed to negative pressure, which in turn is due to the existence of dark energy. The concept of the cosmological constant, another of Einstein's many contributions, relies on an equivalently negative pressure. The dark energy component is about 14 times as much as that of known matter in terms of mass equivalence of energy. In 2004, I published a paper discussing a possible origin of dark energy. The idea *tian wai you tian* 天外有天 literally means "heavens beyond Heaven." So what do I mean by this? The theory I proposed is that the existence of dark energy is possibly an indication of the existence of a Multiverse outside of our "Big Bang" Universe.

In 2005, I wrote another paper exploring the generation and the state of the strongly interacting quark-gluon plasma. The idea *he tian xiang lian* 核天相连 literally means "the nuclei and the heavens are connected." In this theory, I suggest that some new matter can be generated as the result of the negative pressure of dark energy and that this may be related to nuclear energy. Recently, at Brookhaven National Laboratory, scientists tried to generate this new type of matter with the collisions of highly energetic gold ions. They discovered the occurrence of a new type of nuclear matter, the strongly interacting quark-gluon plasma (sQGP). Because of the natural existence of negative pressure for nuclei within the quark model, it is possible that the nuclear energy is associated with the phase transition of dark energy.

I would like to end with a personal recollection. In 1952, C. N. Yang 杨振宁 and I wrote two papers on statistical mechanics. Having read our papers, Einstein asked us (through his assistant, Bruria Kaufman) if we would like to have a discussion with him. We then went into his office and found that our papers were right there on the desk. He said that the papers were interesting and then asked some details about the lattice gas. His questions were mostly about the basic concepts of physics, and he was quite satisfied with my answers. He spoke English slowly with a strong German accent. We had extensive discussions for more than an hour. At the end, he stood up and shook my hand. He said to me, "I wish you future success in physics." I recalled that his palm was big, thick, and warm. For me, this was truly a most unforgettable experience, and I was deeply touched by his good wishes.

In this volume, we commemorate the invention of the telescope 400 years ago. But I would also like to emphasize the contributions of Galileo, Einstein, and the Yan Huang civilization for their dedication to humankind and to science. Earth is not the largest planet in our Solar System. Our Sun is not especially luminescent among the 40 billion stars in the Milky Way. The Milky Way is not the largest galaxy in the Universe. But thanks largely to the achievements of these great scientists, Earth, with its yellow land and blue water, has borne witness to the beauty and evolution of the human spirit and its system of morality and has given rise to an incredible system of thought and investigation that will hopefully result in a thorough understanding of the Universe's greatest mysteries.

REFERENCES

Galileo Galilei (1610). *Sidereus nuncius (The Starry Messenger)*.
Galileo Galilei (1632). *Dialogue Concerning the Two Chief World Systems*.
Needham, J. and Wang, L. (1959). *Science and Civilization in China*, Vol. 3: *Mathematics and the Sciences of the Heavens and the Earth*. Cambridge: Cambridge University Press.

2 The Impact of Modern Telescope Development on Astronomy

Riccardo Giacconi

CONTENTS

INTRODUCTION

The introduction of new technology in astronomical instruments has always resulted in important new advances in our knowledge. In the last 50 years, the pace of technological development has been so great as to lead to a revolution in observational capabilities. The development of space-borne instrumentation, better optics and detectors, and computers has made it possible to create modern telescopes and observatories that have revolutionized our understanding of the Universe.

The main approach has been to apply modern technology to astronomy as advocated by George Ellery Hale as early as 1929. In an article for *Harper's Magazine*, he wrote: "From an engineering standpoint our telescopes are small affairs in comparison with modern battleships and bridges." He ended up by building the 60 and 100 in (152 and 254 cm) telescopes on Mount Wilson and initiating the construction of the 200 in (508 cm) telescope on Palomar, which dominated world of optical astronomy for decades.

The space race in the second half of the 20th century and generous support for astronomy from private and government funds have created the conditions for the construction of great and expensive new telescopes. In the 10-year period between 1990 and 2001, powerful new astronomical facilities have become operational: the *Hubble Space Telescope* (*HST*) was launched in 1990; the Keck I and Keck II telescopes were commissioned in 1992 and 1996, respectively; the European Southern Observatory's (ESO) Very Large Telescope (VLT) was commissioned in 1998; the *Chandra X-ray Observatory* was launched in 1999; and the infrared *Spitzer Space Telescope* was launched in 2001.

These new facilities have required very significant efforts for their development, often carried out over one or two decades, but their advent has also required astronomers to change the way they do astronomy in order to cope with the increased complexity of construction and operations and to effectively utilize the huge flow of data produced by these telescopes.

This chapter briefly discusses some particular aspects of these changes: technological advances open up the most distant reaches of the Universe to exploration in all wavelengths; automation and new methodology allow calibration, analysis, and archiving of terabytes of data; data from different observatories and wavelengths are available worldwide; rapid distribution of these data allows analysis and follow-up of important new discoveries; and data and images have become readily available to the public and to educational institutions.

As an example of how technological advances open up the Universe at all wavelengths, I will summarize the development of x-ray astronomy over the last 50 years and its remarkable discoveries and also follow up its methodological heritage to *HST* and the ESO VLT.

I will discuss how the new computer power allows for sophisticated modeling that permits the selection of efficient designs and operational approaches. I will also show how the development of end-to-end data systems makes it possible to effectively utilize terabytes of data per year.

I will show how data from different observatories and wavelengths are available worldwide and will discuss *HST*, VLT, and *Chandra* as examples. I will show how the rapid distribution of data maximizes scientific returns.

As I develop these themes, I will highlight some of the major findings obtained with these observatories, which are among the most unexpected and baffling results in astronomy. They include the study of intergalactic plasmas, where most of the mass of the Universe resides (in the form of baryons); the study of stellar-mass and supermassive black holes (BHs); the properties of dark matter; and the discovery of dark energy. We now believe that dark matter and dark energy constitute most of the matter-energy content in our Universe. Given that their nature is not yet understood, astronomy is posing some of the most fundamental questions about the physical Universe we live in.

THE SCIENTIFIC AND TECHNOLOGICAL DEVELOPMENT OF X-RAY ASTRONOMY

To illustrate the path followed in the development of the new branches of astronomy, which include millimeter wave, infrared (IR), ultraviolet (UV), x-ray, and gamma ray, I have chosen x-ray astronomy as my example because it is the one I know through firsthand experience. It is also the one that has had the greatest impact in changing methodology and transmitting this experience to all of astronomy.

The Earth's atmosphere absorbs most of the wavelengths of the light reaching us from the stars (Figure 2.1), and to observe them without its absorption or scattering effects we must go into space. The first experiments in x-ray astronomy were undertaken by Herbert Friedman of the Naval Research Laboratory. He used captured German V-2 rockets to start the study of the Sun in 1948 and carried out observations throughout the next decade (Mandelshtam and Efremov, 1958). His efforts to extend these studies to other stars were, however, unsuccessful.

When Bruno Rossi and I started our work in 1959, we were primarily interested in extending these observations to extrasolar sources, first by improving the traditional instrumentation used by Friedman and later by use of focusing x-ray telescopes (XRTs) (Giacconi and Rossi, 1960).

To search the sky for stellar x-ray sources, my group at American Science and Engineering (AS&E) developed a rocket payload that included larger detectors, with anticoincidence background suppression and a much wider field of view. After two unsuccessful attempts, we obtained our first results on June 18, 1962, when our instruments, carried aloft by an Aerobee rocket, reached 100 km altitude (Figure 2.2). The plot shows how the counting rate of the counters varies as the rocket spins on itself, thus scanning different regions of the sky. We observed, during the 300 sec at altitude, an unexpected high flux of x-rays in the direction of the constellation Scorpio, as well as an isotropic background. We named this first source Sco X-1 (Giacconi et al., 1962).

Sco X-1 was an extraordinary object whose emission was 1,000 times greater than that of the Sun at all wavelengths and 1,000 times greater than its own optical emission. This discovery pointed to the existence of a new class of celestial objects and new processes for x-ray emission different from those known in the laboratory. It therefore generated a great deal of interest in this new field and led to a lively competition between many groups both in the United States and abroad (Hirsh, 1979).

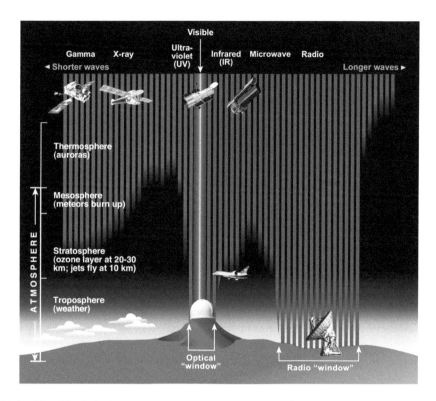

FIGURE 2.1 The altitude in the atmosphere to which different wavelengths can penetrate.

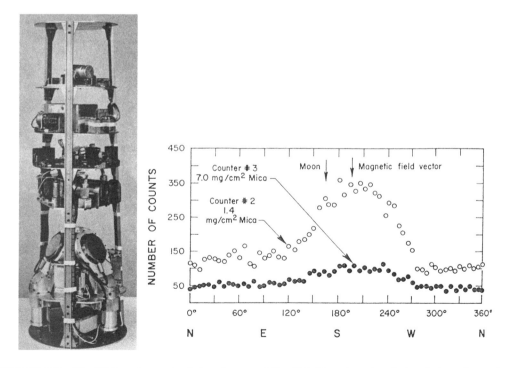

FIGURE 2.2 The 1962 rocket payload that discovered Sco X-1. Note the azimuthal variation of counting rates as the detectors sweep through Sco X-1.

The next major step was the use of detectors of the same type (thin window proportional counters) on an orbiting satellite, such as *Uhuru*. While a rocket flight provided only 300 sec of observation, a satellite provided years. The satellite *Uhuru* was designed and built by our group at AS&E; it was launched in 1970 and provided a sensitivity that was 10,000 times greater than the discovery rocket (Figure 2.3).

Observations from *Uhuru* increased the number of known sources from about 30 (discovered from rockets over the previous eight years) to about 360. The map in Figure 2.3 is in Galactic coordinates, the horizontal axis is the Milky Way, and most sources at high galactic latitudes are extra-Galactic. The size of the dot is proportional to the logarithm of their intensity, and some of the interesting sources are identified. The sources were found to be Sco X-1-like, supernova remnants (SNRs), binary x-ray sources, galaxies, quasars, clusters of galaxies, intergalactic plasmas, and the background (Giacconi et al., 1971).

The study of binary x-ray stars led to the discovery of systems containing a normal star and a star at the end of its evolution: a neutron star or a BH (the source Cyg X-1 contained the first identified BH of solar mass) (Oda et al., 1971; Shreier et al., 1972; Tananbaum et al., 1972). They emit x-ray radiation by converting the energy, acquired in the fall of gas from the normal star into the deep potential well of the companion, in the heating of high-temperature plasmas. The energy generated per nucleon is 100 times greater than that generated by fusion, and this mechanism is also the one responsible for the emission from the nucleus of quasars.

The discovery of intergalactic plasmas in clusters of galaxies was also completely unexpected. Such plasmas could not be observed except through their x-ray emission, their characteristic emission at a temperature of tens of millions of degrees. Although of very low density, they fill the enormous empty space between galaxies and their total mass exceeds that in galaxies by a factor of 10. We know today that most of the normal matter in the Universe is in this form (Gursky et al., 1972).

A new technological advance of great importance was the development of XRTs. Since 1960, Rossi and I had proposed the use of grazing incidence XRTs to increase the sensitivity of our planned observations by orders of magnitudes. However, their development took several years. They

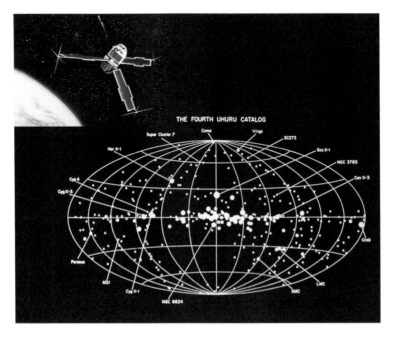

FIGURE 2.3   The *Uhuru* observatory launched in 1970 and the fourth *Uhuru* x-ray sky map. The size of the dots is proportional to the logarithm of their intensity.

were first used in the period from 1964 to 1973 in rocket studies of the solar corona and culminated in 1973 in the *Skylab* mission (Figure 2.4).

Skylab, which contained a solar observatory, was the first manned space station launched by the United States. Several visits by the astronauts permitted the use of retrievable film to obtain x-ray pictures of the Sun over several solar rotations with a resolution of 5 arcsec. The data forced a major rethinking of the theories of heating and containment of plasmas in the solar corona (Vaiana and Rosner, 1978).

The use of XRTs for stellar astronomy had to await the development of higher efficiency optics and the development of high-resolution electronic detectors. This was achieved in 1978 with the launch of the *Einstein Observatory* (Giacconi et al., 1981). It was the first stellar x-ray observatory using an imaging XRT (Figure 2.5). The *Einstein Observatory* could achieve an angular resolution of 4 arcsec and included two imaging detectors and two spectrometers. Its sensitivity was 100 times greater than that of *Uhuru* and a million times greater than that of the discovery rocket. In the inset in Figure 2.5 one can see the binary x-ray sources and the supernovae (SNe) in the center of M31, our neighbor galaxy.

The *Einstein Observatory* opened up x-ray observation in all classes of celestial objects (Elvis, 1990). They include auroras on planets, main-sequence stars, novae and SNe, pulsars, x-ray binaries and SNe in external galaxies, normal galaxies, active galactic nuclei, quasars, groups and clusters of galaxies, and the sources of the x-ray background. This was a turning point for x-ray astronomy, which changed from a subdiscipline of interest mostly to physicists to an important observing tool for all astronomers. The *Chandra X-ray Observatory*, which I will discuss later, was launched in 1999 and provided further large gains in angular resolution and sensitivity, permitting the detection of sources some tens-of-billions of times fainter than Sco X-1 and extending the x-ray observations to objects at cosmological distance.

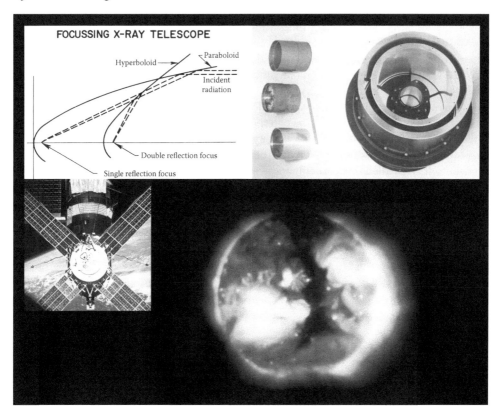

FIGURE 2.4 Schematic diagram of a grazing incidence x-ray telescope. The largest telescope was a beryllium mirror of 30 cm diameter launched in 1973 on *Skylab*. A front view of the solar instruments on *Skylab*. An x-ray picture of the Sun.

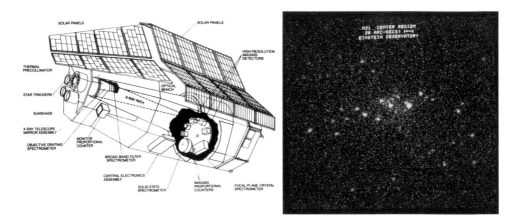

FIGURE 2.5   Cutout view of the *Einstein* spacecraft launched in 1978. An x-ray picture of the center of M31 showing binary x-ray sources and supernova remnants.

METHODOLOGICAL HERITAGE FROM X-RAY ASTRONOMY

While for x-ray astronomy and for many other disciplines in astronomy the need for technological improvements has always been clear, and we have seen the need to go into space and to design new telescopes and detectors, the accompanying changes in methodology were equally important. The significance of the correct methodology is often forgotten, but I am convinced of its importance. My favorite example is that of Tycho Brahe, who more than 400 years ago built the first observatory in the Western world. He had at his disposal instruments and mathematical tools not superior to those available during Hellenistic times, more than 1,000 years before. Yet his observations, which spanned 20 years and were of greater quality than any previously obtained, were the foundation for Kepler's three laws and for Newton's synthesis.

The x-ray astronomers had to adopt new rules in their research approach; while some of these are common to any large-scale scientific enterprise, others were made necessary by the particular requirements of space research and the novelty and uniqueness of the data. We had to learn to apply system engineering not only to the instrumentation and the carriers, but also to the scientific enterprise itself. We called this particular aspect of our work "science system engineering," which included detailed advanced planning for operations and the development of end-to-end automated data systems. These systems permitted pipeline data calibration, reduction, and archiving. They allowed the distribution of high-quality calibrated data to be made available promptly to astronomers in all disciplines, in a form immediately usable for their further analysis.

It turned out that the same requirements were found to be common to some of the largest astronomical projects undertaken after the *Einstein Observatory* both in space and on the ground (they include *HST*, the ESO VLT, and *Chandra*) and that they were imposed by the complexity of the instruments, by the huge quantity of data to be managed, and by the need to carry out operations in real time. I am focusing on these three projects because I had a direct involvement with them, although I know that this changed methodology has been applied very widely in all modern observatories. I will describe briefly these observatories and give some examples of actual discoveries that could not have happened without the application of this new methodology.

HUBBLE SPACE TELESCOPE

HST (Figure 2.6) is perhaps the most famous among the modern telescopes. It has produced an incredible quantity of high-quality data with high angular resolution (0.070 arcsec) and sensitivity that in long exposures reached magnitude 33. Scientifically, it has been extremely productive, with

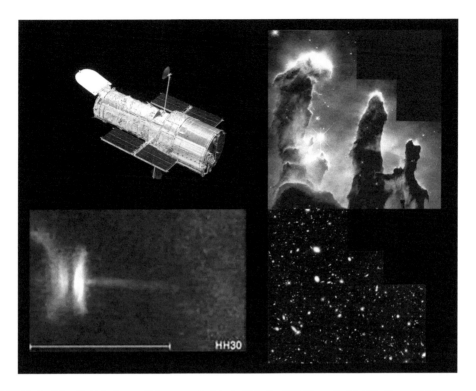

FIGURE 2.6 The *Hubble Space Telescope*, the Eagle Nebula, HH 30, and a Hubble Deep Field.

fundamental discoveries that provided new insights into cosmic phenomena. The insets in Figure 2.6 show some of the almost iconic images it has produced: the Eagle Nebula, a region of star formation where we can distinguish the evaporating gaseous globules (EGGs) where stars are actually forming; a disk and a jet around the forming star HH 30, which is trying to get rid of the remnant angular momentum and magnetic field of the original Nebula from which it formed; and finally, one of the Hubble Deep Fields where, by careful study, we can observe evolution in the types of galaxies that form at different epochs.

The impact of x-ray astronomy experience on the *Hubble* project was not planned by NASA; it was mainly a result of the recommendation by the Horning Committee of the National Academy of Sciences to create an independent institute to direct the scientific operations of the *HST*. The management entity that had won the competition to create the Space Telescope Science Institute (STScI) decided to appoint me as the first director; the early additions of some other x-ray astronomers, such as Rodger Doxsey of the Massachusetts Institute of Technology (MIT) and Ethan Schreier of the Harvard–Smithsonian Center for Astrophysics (HSCfA), to the staff made this transfer of experience almost inevitable. I will mention only some of the changes brought about by the x-ray astronomers in the conduct of the *Hubble* program after the creation of STScI (Giacconi, 2008).

STScI introduced the science system engineering in the *Hubble* program. The best example of its application is that although *Hubble* was supposed to do planetary science, the special software needs required for pointing at planets had not been recognized and had to be developed by the Institute.

The system for correctly pointing the telescope relied on the acquisition of stars as faint as magnitude 15. A catalog of objects to that magnitude did not exist; therefore, STScI had to contract with the Anglo-Australian Observatory for the south and Mount Palomar for the north to obtain Schmidt plates of adequate resolution and sensitivity. Once the photographic material was available, there remained the question of how and when the survey could be digitized. We decided that the only practical solution was to scan all plates before launch and to create a digital catalog instantly available for operations as needed. This ultimately resulted in a Guide Star Catalog containing 435,457,355 stars with magnitude limits between 18.5 and 19.5 and with positional accuracy between 0.3 and 0.75 arcsec (Lasker et al., 1990). The data of this catalog were compressed by using multiwavelet transform techniques and were made available to the entire astronomical community on CDs.

We developed an end-to-end automated data system supporting our work from scientific proposal submission to operations, calibrations, and data reduction, and then to data distribution and archiving. This permitted rapid worldwide dissemination of the data. The need for such an automated system had not been sufficiently understood, nor was it believed to be feasible by most of the optical astronomical community. Yet the planned volume of data acquisition was of the order of 100 gigabits per week and expected to increase as new instruments were brought up in orbit by the astronauts; it has, in fact, increased from launch by a factor of 30. Even more significant were the requirements imposed by data reprocessing and by the use of the archived data for research. The two combined are now at the level of 100 gigabits per day, and they have come to dominate the data volume. Use of archived data played a crucial role in the *Hubble* studies of dark energy, discussed below (also see the chapter by Mark Sullivan in this volume).

STScI unfortunately did not exist during the construction of *Hubble* and could not impose the unified system engineering approach that might have avoided the nasty surprise of finding that *Hubble* was out of focus when it went into orbit. However, as soon as the problem became apparent, it was quickly analyzed by STScI (Burrows et al., 1991), and a scientific and technical solution was developed under STScI scientific leadership.

The result was that as soon as the out-of-focus condition was repaired, *Hubble* became one of the most productive observatories in astronomy. In Figure 2.6 I have shown some of the iconic images that *Hubble* has produced and that have become known around the world. I now want to pay particular attention to one of the most interesting contributions by *Hubble*: the extension of the observations of Type Ia supernovae (SNe Ia) at $z > 1$, that is, in the remote past (Riess et al., 2004).

In Figure 2.7 (this and the following three figures are courtesy of Adam Reiss), I show a composite of the detection of the first SN Ia at $z > 1$. First came the detection with the Advanced Camera for Surveys (ACS), which has a sensitivity of >25 mag. Then the observation was winnowed by noting the characteristic reddening in the UV. The spectrum obtained with the ACS grism spectrometer allowed the measurement of the redshift (never yet done from the ground at this large redshift). Finally, the recovery from the archive of a spectrum obtained with the Near Infrared Camera and Multi-Object Spectrometer (NICMOS) allowed the study in the IR of the shape and peak of the SN Ia emission, yielding the distance.

Figure 2.8 shows the detection of several SNe Ia at large redshifts also obtained with the ACS. The redshifts are $z = 1.39$, 0.46, 0.52, 1.23, and 1.03. The bottom panel shows images of the host galaxies.

Figure 2.9 shows how the *Hubble* observations (*filled red circles*) agree with and extend the ground observations (*open circles*). *Hubble* adds the observations of distant SNe, which are crucial to study the nature of the dark energy, particularly its time variability.

Figure 2.10 shows the time variability of the ratio of dark matter to dark energy. A transition is observed from now, a dark energy era when the Universe accelerates, to a dark matter era in the past when the Universe decelerated. The different curves show the various models one can use to explain these observations: a dust- or chemical-evolution model (*yellow*), an empty Universe (*light blue*), a dark matter–dark energy model (*red*), and finally a matter-only model (*green*). The result is a clear confirmation of the existence of dark energy and of different phases of acceleration and deceleration in the expansion depending on the ratio between dark energy and dark matter.

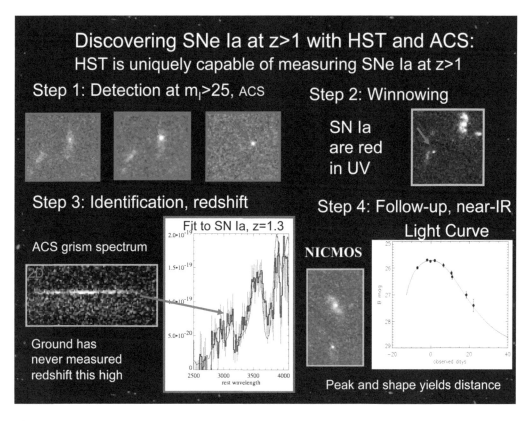

FIGURE 2.7   The detection of a SN Ia at $z = 1.39$ by *Hubble* using the ACS and NICMOS instruments. (Courtesy of Adam Reiss.)

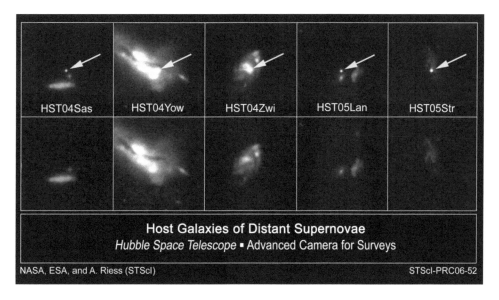

FIGURE 2.8   The detection of several SNe Ia at large z's and their host galaxies. (Courtesy of Adam Reiss.)

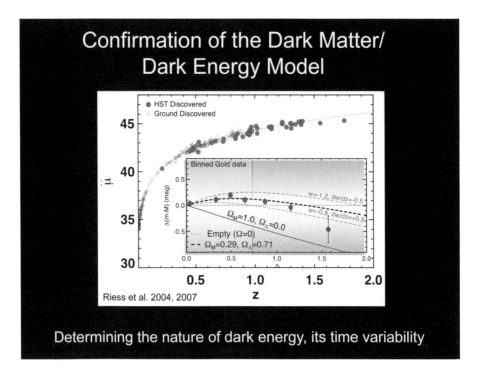

FIGURE 2.9   Confirmation of dark matter/dark energy model. (Courtesy of Adam Reiss.)

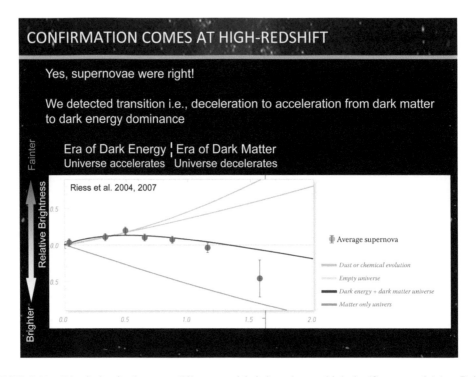

FIGURE 2.10   Discrimination between different models is best done at high z's. (Courtesy of Adam Reiss.)

THE ESO VERY LARGE TELESCOPE (VLT)

I would like now to turn to the impact of the *Einstein Observatory + HST* methodology on ground-based astronomy. This choice is in part due to my familiarity with VLT, but also because VLT embodies this new approach to a larger degree than any other ground-based observatory, both in the construction and in the operation phase. VLT (Figure 2.11) is the largest array of optical telescopes in the world. It consists of four telescopes of 8 m (26 ft) supported by two auxiliary telescopes for wide-field surveys and three auxiliary telescopes for use with the array during interferometric observations. It is located at Cerro Paranal in the Atacama Desert in Chile at an altitude of 2,800 m (9,186 ft), where visibility is perhaps the best in the world.

Figure 2.11 shows VLT on Paranal and some of the VLT icons: Uranus, the Eagle Nebula with the "Pillars of Creation" now transparent in IR, and a spectacular view of the Galactic center region obtained by use of adaptive optics in the IR with NAOS-CONICA ("NACO"—Nasmyth Adaptive Optics System and COudé Near-Infrared CAmera on Yepun, the Chilean Indians' name for Sirius that was given to Unit Telescope 4). I will discuss this observation in greater detail later.

The impact of the new methodology on VLT was even greater than that on *Hubble*. This was due to the fact that the methodology had further developed and that ESO had clear and complete responsibility for both the construction and operation of VLT and VLTI (the VLT Interferometer).*

The transfer of methodology was facilitated by the existence at ESO of the European Coordinating Facility that had cooperated with STScI in producing the *Hubble* data archive. Several of the scientists on the ESO staff had used *Hubble*, and others had been deeply involved in the development of advanced calibration algorithms.

FIGURE 2.11 The *VLT* observatory of ESO at Paranal. A picture of Neptune. An infrared image of the Eagle Nebula, in which the dust is largely transparent. (Compare with Figure 2.6, an optical image, in which the dust is opaque.) A picture of the center of our Galaxy obtained with adaptive optics.

---

* A good description of the developments of these models is given in a number of articles appearing between December 1993 and December 1999 in the *ESO Messenger*, a quarterly publication by ESO, Garching, Germany.

VLT used simulation models for the atmosphere, optical elements, structural dynamics, primary mirror dynamics, wavefront sensors, active optics, field stabilization, and control systems. VLT also developed an end-to-end data system including calibration and data archiving for multiple instruments and telescopes. Its archival capability provides random access to 170 terabits of data and ESO pioneered the development of a calibration model from first principles.

The reason why this intensive modeling was required stemmed from the complexity of the VLT design of an actively controlled thin meniscus never previously realized on such a large mirror. VLT had 150 axial supports that had to be actively controlled, using a Shack–Hartman wavefront sensor with 30×30 lenslets, which would provide information for updates every 30 sec. The corrections followed a modal correction scheme of the primary mirror, while coma and focus corrections occurred on the secondary mirror.

The simulation model took into account the contributions of different elements of the optics, the dynamics of the supporting structure, the primary mirror dynamics, the response of the wavefront sensor, the behavior of the active optics and of the stabilization loops, and the behavior of the control system at different frequencies.

The resulting design has been a great success. The wavefront is stabilized after four updates, which take about 2 min. VLT is then ready to operate at the visual limit and to obtain data continuously until the next pointing, typically several hours. This efficiency decreases the time overheads and increases the time actually spent on observations; thus, VLT provides high efficiency in the conduct of scientific programs.

Figure 2.12 shows a schematic view of the complexity of the infrastructure created to support VLT and VLTI operations. This view shows the four main telescopes and the underground tunnels joining them to the control center on the left from which they are run. It also shows the auxiliary telescopes for interferometry and the interferometric laboratory where the beams of all telescopes are made to interfere after appropriate delays. The interferometric instrumentation is all housed in this underground laboratory. When the beams of all the telescopes are combined, they form an interferometric array of 120 m (394 ft).

Each telescope is provided with instruments at its different foci. The instruments shown in this diagram are those that were provided in the first generation of instruments—they include image detectors and spectrometers with wavelength range from UV to IR. A laser guide star system is provided to support the NAOS-CONICA adaptive optics.

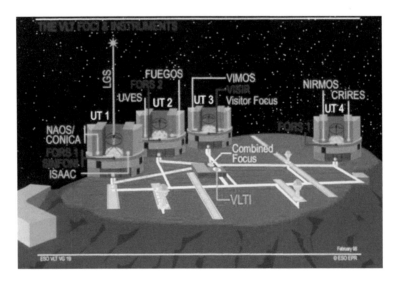

FIGURE 2.12   A schematic view of the four 8 m telescopes, the auxiliary telescopes for interferometry, and the planned first-generation instruments.

All telescopes and instruments are controlled in the single facility already mentioned, and a single data facility provides for the real-time diagnostic and for the reduction and calibration of the data. The data are then transferred to the archive, which after the first 10 years of operations contains 70 terabits of data.

The end-to-end data system of VLT at the time of commissioning was indistinguishable from that of *HST*. It included proposal solicitation and selection support, observing plans preparation and scheduling, observing logs and calibration, engineering data, standard pipeline processing, a science archive, trend analysis, and instrument simulators and handbooks.

The high degree of automation and the careful operations planning have allowed the commissioning of VLT to occur very promptly and the rapid achievement of observational efficiencies considerably greater than Keck.

I have chosen to discuss one particular scientific result to illustrate what has already been obtained—that is, the data on the massive BH at the center of our Galaxy obtained after the commissioning of NAOS and CONICA (Figure 2.13). The high angular resolution of this adaptive optic system (40 mas) has allowed Reinhard Genzel and his group at the Max Planck Institute for Extraterrestrial Physics (MPE) to obtain repeated high-precision observations in the H band of the star designated as S2 in orbit around Sgr A*, completing in 2002 the program they initiated in 1992 (Genzel et al., 2003).

The orbit of the star S2 around Sgr A* is quite remarkable. Its distance of closest approach was measured to be 17 lt-hr in April 2002, with orbital velocity of 8,000 km/sec. This is the predicted Keplerian orbit around a mass of 3 million solar masses.

These new data further constrain the results obtained at the ESO New Technology Telescope (NTT) and at Keck regarding the nature of the central density cusp. Only a boson star or a BH appears compatible with the new data. The mass of the central object is measured to be 2.87 ± 0.15 million solar masses.

The measurement is extremely interesting in itself, but also shows the potential of large optical telescopes on the ground when adaptive optics is used to eliminate atmospheric scattering.

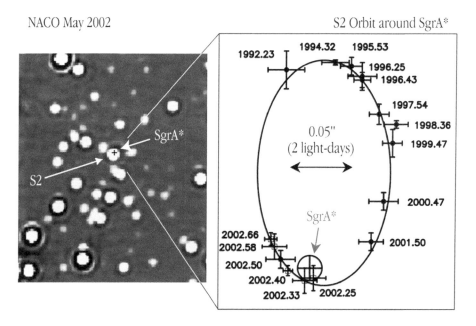

FIGURE 2.13   A picture of the Sgr A region obtained with the adaptive optics NAOS-CONICA instrument and the orbit of S2 around the central black hole. (Courtesy of the European Southern Observatory.)

THE CHANDRA X-RAY OBSERVATORY

Figure 2.14 shows the *Chandra X-ray Observatory* in orbit. It was launched in 1999 and is the most advanced x-ray observatory in the world. It has a mirror of 120 cm (47 in) diameter built of ceramic glass, which was exquisitely polished to yield an angular resolution of 0.5 arcsec. Imaging with charge-coupled device (CCD) detectors in the center field achieves a sensitivity of 10 billion times greater than the 1962 rocket that discovered Sco X-1, an improvement equal to that which occurred in optical astronomy over the last 400 years. The icons in Figure 2.14 display an observation of the SNR in the Crab Nebula, an encounter between two clusters of galaxies showing the shock in the interacting plasmas, and a very deep picture showing supermassive BHs at cosmological distance.

The *Chandra* construction was led by the same group at CfA and Marshall that had produced the *Einstein Observatory* in 1978 and thus incorporated all of the experience accumulated in past x-ray missions, as well as the advances made by *Hubble* in data archiving and distribution (Tananbaum and Weisskopf, 2001).

One of the outstanding scientific contributions produced by the use of *Chandra*'s data is shown in Figure 2.15, an x-ray picture of the Bullet Cluster (Markevitch et al., 2004). While the plasmas show a shock front due to their interaction, the distribution of galaxies and of dark matter (revealed by gravitational lensing) is unperturbed—thus the conclusion that dark matter is interacting weakly.

In Figure 2.16, one of the deepest exposures obtained with *Chandra* (1 million sec) is shown. The x-ray background is resolved in a large number of individual sources with a sky density of 1 per arcmin^2. The sources have all been identified with optical counterparts by use of VLT, *HST*, and *Spitzer* for identification and VLT spectrometers for spectroscopy (Giacconi and the CDFS Team, 2002). More than 90% of them turn out to be supermassive BHs at cosmological distance. This raises interesting questions about the epoch and mechanism of formation of these objects, as well as their dynamic interaction with forming galaxies and clusters. It is interesting to note that this result could be obtained only by using four of the very best telescopes in the world, operating in the x-ray, optical, and IR domains.

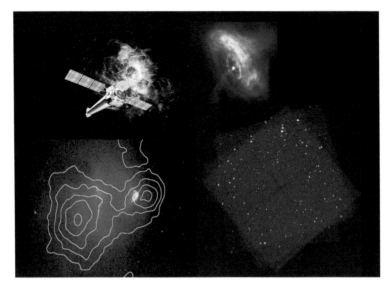

FIGURE 2.14   A view of the *Chandra* spacecraft launched in 1999. An x-ray picture of the Crab Nebula pulsar, the Bullet Cluster, and the Chandra Deep Field South.

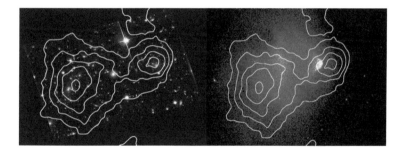

FIGURE 2.15   An x-ray picture of the Bullet Cluster with the dark matter contours superimposed. The dark matter distribution is derived from gravitational lensing.

FIGURE 2.16   The Deep Field South showing hundreds of black holes at cosmological distance.

CONCLUSIONS AND THE FUTURE

The methodological changes I have been discussing are very clear. To make progress in astronomy, we need observations extended over the entire electromagnetic spectrum. All data are now available worldwide. Observatories must now provide not only facilities, but high-quality calibrated data ready for further analysis. Prompt distribution of data contributes to scientific productivity. Astronomical images and online text contribute greatly to outreach and education.

Future research has the formidable task of understanding the nature of dark matter and dark energy, the main constituents of our Universe. This will require advances not only in astronomy, but also in physics, to reach a unified theory that will link the laws of particle physics and cosmology. Only by following such advances will we be in a position to ask questions about the formation of early structures and the future of the Universe. We will continue to pursue an understanding of the formation of stars and planets and to muse on astrobiology.

It seems a very exciting future for astronomical knowledge and for our rational understanding of our place in the cosmos.

REFERENCES

Burrows, C.J., Holtzman, J.A., Faber, S.M., et al. (1991). The imaging performance of the Hubble Space Telescope. *Astrophysical Journal,* 369(2): L21–5.

Elvis, M. (1990). *Imaging X-ray Astronomy, a Decade of Einstein Achievements.* Cambridge: Cambridge University Press.

Genzel, R., Schödel, R., Ott, T. et al. (2003). The stellar cusp around the supermassive black hole in the Galactic center. *The Astrophysical Journal,* 594 (2): 812–32.

Giacconi, R. (2008). *Secrets of the Hoary Deep: A Personal History of Modern Astronomy.* Baltimore: Johns Hopkins University Press.

Giacconi, R. and Rossi, B.B. (1960). A "telescope" for soft x-ray astronomy. *Journal of Geophysical Research,* 65: 773.

Giacconi, R. and the CDFS Team (2002). The Chandra deep field south one million second catalog. *Astrophysical Journal Supplement,* 139: 369–410.

Giacconi, R., Gorenstein, P., Murray, S.S., et al. (1981). The Einstein Observatory and future x-ray telescopes. In: G. Burbidge and A. Hewitt (eds.), *Telescopes for the 1980s.* Palo Alto, CA: Annual Reviews, pp. 195–278.

Giacconi, R., Gursky, H., Paolini, F. et al. (1962). Evidence for x-rays from sources outside the solar system. *Physical Review Letters,* 9: 442.

Giacconi, R., Kellog, E.M., Gorenstein, P. et al. (1971). An x-ray scan of the Galactic plane from Uhuru. *Astrophysical Journal,* 165: L27–5.

Gursky, H., Solinger, A., Kellog, E.M. et al. (1972). X-ray emission from rich clusters of galaxies. *Astrophysical Journal,* 173: L99.

Hirsh, R.F. (1979). Science, technology and public policy: The case of x-ray astronomy, 1959 to 1978. Ph.D. thesis, University of Wisconsin, Madison.

Lasker, B.M., Sturch, C.R., McLean, B.J. et al. (1990). The guide star catalog I: Astronomical foundations and image processing. *Astronomical Journal,* 99: 2019.

Mandelshtam, S.L. and Efremov, A.I. (1958). Research on shortwave solar ultraviolet radiation. In: *Russian Literature of Satellites.* Moscow: Academy of Sciences of the USSR, pp. 47–65. New York: Transactions of the International Physical Index.

Markevitch, M., Gonzales, A.H., Clowe, D. et al. (2004). Direct constraints on the dark matter self-interaction cross section from the merging galaxy cluster 1E 0657-56. *Astrophysical Journal,* 606 (2): 819–24.

Oda, M., Gorenstein, P., Gursky, H., et al. (1971). X-ray pulsations from Cyg X-1 observed from Uhuru. *Astrophysical Journal,* 166: L1.

Riess, Adam G., Strolger, Louis-Gregory, Tonry, John et al. (2004). Type Ia supernova discoveries at $z > 1$ from the Hubble Space Telescope: Evidence of past deceleration and constraints on dark energy evolution. *Astrophysical Journal,* 607 (2): 665–87.

Shreier, E.J., Levinson, R., Gursky, H. et al. (1972). Evidence for the binary nature of Centaurus X-3 from Uhuru x-ray observations. *Astrophysical Journal,* 172: L79.

Tananbaum, H. and Weisskopf, M. (2001). A general description and current status of the Chandra X-ray Observatory. In: H. Inue and H. Krineda (eds.), *Astronomical Society of the Pacific Conference Proceedings,* Vol. 251.

Tananbaum, H., Gursky, H., Kellog, E.M. et al. (1972). Discovery of a periodic pulsating binary source in Hercules from Uhuru. *Astrophysical Journal,* 174: L134.

Vaiana, G.S. and Rosner, R. (1978). Recent advances in coronal physics. *Annual Review of Astronomy and Astrophysics,* 16: 393–428.

3 Searching for Other Earths and Life in the Universe

Geoffrey W. Marcy

CONTENTS

THE GALACTIC ENVIRONMENT FOR LIFE-BEARING PLANETS

Throughout history, philosophers have wondered whether Earth was a special place in the Universe and whether life might exist elsewhere. The great Greek philosopher Aristotle concluded 2,350 years ago that Earth was unique and that life existed only here. Other Greek philosophers disagreed, notably Epicurus and Democritus, both suggesting that the twinkling lights in the night sky might be other "solar systems" containing many planets, and even civilizations that thrive and die with time. The hypothesis of many "Earths" gained considerable weight from our knowledge that 200 billion stars occupy our Galaxy, the Milky Way, many of which are similar to our Sun, within a factor of two in mass, size, chemical composition, and age. If our Sun is a common type of star, perhaps so too are its planets. The laws of physics and chemistry are surely the same everywhere within our Galaxy, suggesting that organic chemistry might proceed, as on Earth, toward ever-more complex compounds. Some of these organics would naturally duplicate themselves chemically, leading to a Darwinian competition among them for atoms, energy, and real estate. Indeed, astronomers have found various amino acids in meteorites, comets, and interstellar clouds showing the biochemical parade already marching down the boulevard of biology.

While the laws of physics and chemistry are the same everywhere in our Galaxy, the "laws" of biology remain poorly known. We do not know whether water is the only key solvent for biochemistry, as it seems to be here on Earth. Could some other liquids, such as methane at low temperatures or magma at high temperatures, serve as the solvent for life-forming chemical reactions? Is DNA the only replicating molecule that contains a suitable computer program for life, or can other molecules serve as the flexible template for successive generations? The greatest biological mystery of all is whether evolution always leads to increasingly intelligent species or whether, instead, intelligence is some lucky occurrence that emerged on the East African savannah 2 million years ago as a result of an unusual confluence of environmental factors. We humans imagine ourselves perched at the top of the evolutionary tree, but both genetic and evolutionary biologists wonder whether we reside on a small branch of the tree of life, having sprouted from a lucky twig.

Our ignorance about the rules of biology in our Universe stems from having only one example to study, life on Earth. In the search for another example, tantalizing destinations such as Mars, Europa, Titan, and Enceladus offer humanity an opportunity for exploration as great as those of the great transoceanic voyages of the 15th century. But reconnaissance imaging of the other planets and moons in our Solar System already shows that large life-forms, including intelligent life, do not exist within our Solar System. The hunt for advanced life forces us to venture into the vast Milky Way, many light-years from Earth at least.

DISCOVERY AND PROPERTIES OF EXOPLANETS

In 1987, my research group began making precise Doppler measurements of more than 100 stars to hunt for planets around them. Other teams, notably those led by Michel Mayor, Bill Cochran, and Bob Noyes, were also surveying many stars. The discovery of the first exoplanets occurred in 1995, with planets found around 51 Pegasi, 70 Virginis, and 47 Ursae Majoris. As of December 2010, more than 500 exoplanets have been discovered. (An updated list of the most accurate masses and orbits of exoplanets is provided at http://exoplanets.org/.) Most exoplanets have been found by precise measurements of the Doppler shift of starlight, revealing the periodic motion of nearby stars that occurs as unseen planets pull gravitationally on them. Astronomers have developed extraordinary techniques for measuring the Doppler effect of the thousands of spectral lines in the spectrum of a star (see Figure 3.1), reaching a current precision of 1 m/sec, human walking speed. The giant planets having masses similar to that of Jupiter can be detected with that Doppler precision, as can smaller planets if they orbit closer to the star. Planets as small as a few times the mass of Earth have been detected by such Doppler measurements, such as for HD 156668 (see Figure 3.2 and Howard et al., 2011).

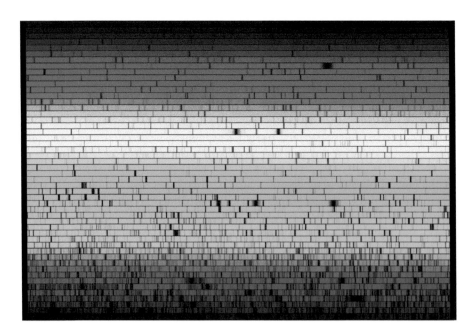

FIGURE 3.1　A typical high-resolution spectrum of a solar-type star used to measure the Doppler shift. The resolution is typically 1.5 km/sec per pixel, and the instrumental smearing of the spectrum and the wavelength calibration of the spectrometer are determined from superimposed iodine absorption lines with a precision of 0.001 pixels. A Jupiter-mass planet orbiting at 5 AU causes a Doppler displacement of about 0.003 pixels, detectable by building a computer model of thousands of absorption lines (seen here) at millipixel sampling.

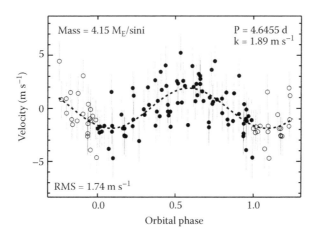

Mass = 4.15 M_E/sini   P = 4.6455 d
                              k = 1.89 m s^{-1}

RMS = 1.74 m s^{-1}

FIGURE 3.2 Measured velocities (Doppler) vs. orbital phase for the K-type star HD 156668 obtained over second years with the Keck telescope. The velocities have a period of 4.6 days and an amplitude of 1.9 m/sec, the second lowest known amplitude. The planet mass is 4.2 $M_\oplus$ (*M*sini), among the lowest mass exoplanets known (see http://exoplanets.org/). The root mean square (RMS) of 1.7 m/sec shows the excellent long-term Doppler precision of the iodine method. (Courtesy of Andrew Howard.)

Astronomers have also found multiple planets orbiting more than 30 different stars, as shown in Figure 3.3, by detecting multiple periodicities in the Doppler shift of the star (Wright et al., 2009). The first multiplanet system discovered was around Upsilon Andromedae with its three planets. That architecture of multiple low-mass objects orbiting a central star established the extrasolar planets as related to the planets in our Solar System, rather than simply large mass–ratio binary stars or a continuation of the brown dwarfs. Figure 3.4 provides the current set of velocities for Upsilon Andromedae, showing that the triple-planet system persists with no obvious perturbations and no additional planets detected.

The record holder among multiplanet systems found by radial velocity measurements is around star 55 Cancri, for which astronomers have detected five planets over a period of 20 years of continuous Doppler measurements (see Figure 3.5 and Fischer et al., 2008). We had no idea in 1988 when we started observing 55 Cancri that our Doppler measurements would reveal any planets at all. Only by going back to the telescope month after month for 20 years, improving our Doppler technique every year, did all five planets emerge from our Doppler measurements. This remarkable planetary system has one planet of at least four times the mass of Jupiter orbiting 6 AU from the star and has four much smaller planets orbiting within 1 AU, an architecture reminiscent of our Solar System, but scaled up in mass (see Figure 3.6). During its formation, 55 Cancri must have been surrounded by a protoplanetary disk considerably more massive than the one out of which our Solar System formed.

Planetary systems such as 55 Cancri and our own Solar System provide astronomers with key evidence about how planets form. Giant clouds of gas and dust in our Galaxy occasionally contract as a result of their own gravity, pulling all the mass inward. Any slight spinning motion in a cloud will cause it to flatten and move like a spinning pizza, with the gas and dust density increasing. The dust particles will collide, stick together, and grow into larger and larger objects, eventually reaching the size of rocks, mountains, and small planets. Those objects eventually collide, growing to the size of Earth or bigger. Any gas remaining in the protoplanetary disk will be gravitationally attracted to those rocky planets, forming a gas giant planet similar to Jupiter or Saturn. Some of these planets may lose their orbital energy to disk material, causing the planets to spiral inward, ending their travel close to the host star. Some

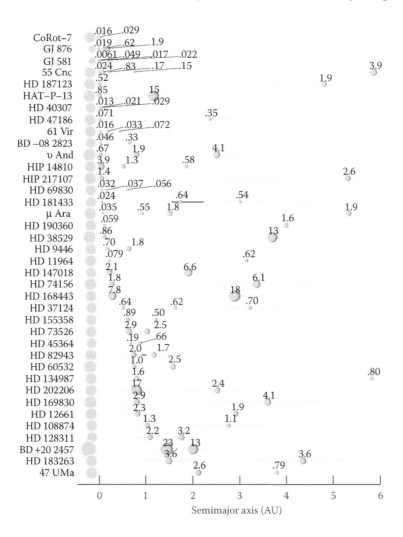

FIGURE 3.3   Schematic sketches of the 38 known multiplanet systems as of February 2010. The star names are on the left, with the yellow dots having sizes proportional to the diameter of the star. Shown on the right in green are the planets located at their average orbital distance (semimajor axis). The turquoise horizontal lines indicate the closest and furthest approach of the planet to the star caused by the orbital eccentricity. The minimum mass, Msini, is indicated in Jupiter masses both numerically and as the size of the dot that scales as $M^{1/3}$. (Courtesy of Wright, J.T., Upadhyay, S., Marcy, G.W. et al. (2009). Ten new and updated multiplanet systems and a survey of exoplanetary systems. *Astrophysical Journal*, 693: 1084–99.)

planets pull gravitationally on other planets, causing their original circular orbits to become elongated ellipses.

THE STATISTICAL PROPERTIES OF OBSERVED EXOPLANETS

The Doppler surveys for planets show that 4% of all solar-type stars have two or more giant planets, not unlike Jupiter and Saturn orbiting the Sun. Surely even more stars have smaller planets, as our Doppler precision of 1 m/sec prevents us from finding planets similar to Neptune, Uranus, and Earth in orbits such as theirs (1 AU or larger). Also, our 20 years of Doppler surveying has allowed us to find only planets with orbital periods of less than 20 years. We can be certain that at least 20% of all stars have planets, and the frequency could be as high as 80%.

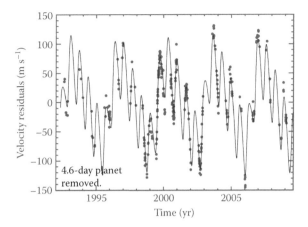

FIGURE 3.4 Residual velocities for Upsilon Andromedae, after subtracting the velocity caused by the innermost planet with its orbital period of 4.6 days and a minimum mass (Msini) of 0.67 M_J. The remaining velocities exhibit two additional periodicities with periods of 241.0 and 1279.6 days, caused by two additional planets having Msini of 1.88 and 4.07 M_J, respectively. This system was the first multiplanet system ever discovered around a main-sequence star, providing the first evidence of a common formation mechanism between extrasolar planets and our Solar System. (Courtesy of Debra Fischer.)

Most giant planets orbit their stars in very noncircular orbits (see Figure 3.7). The orbital eccentricities of the known giant planets span a range from circular (eccentricity of 0) to highly elongated (eccentricity above 0.9), with the average eccentricity being near 0.24 (Marcy et al., 2008; Johnson, 2009). Such elongated orbits are normal, even for those giant planets that orbit their stars as far as Jupiter orbits the Sun, about 5 AU away (see Figure 3.7). It seems likely that multiple planets commonly form, causing the planets to pull gravitationally on one another, often yanking them out of their original circular orbits. Most planetary systems suffer from these gravitational interactions, including some that are so violent that planets are ejected from the system entirely. The

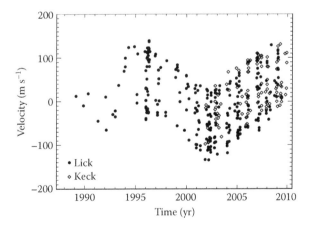

FIGURE 3.5 Doppler shift measurements over 21 years for the solar-type star 55 Cancri. The velocities require five planets to explain them adequately. The five planets have orbital distances (semimajor axes) of 0.038, 0.115, 0.241, 0.78, and 5.8 AU and minimum masses (Msini) of 10.5, 255, 50.2, 45.8, and 1,230 $M_\oplus$, respectively, all in nearly circular orbits. With a giant planet orbiting beyond 5 AU and four smaller planets orbiting closer in, the 55 Cancri system resembles our Solar System, but scaled up in mass. The dots are from the Lick Observatory, and the diamonds are from the Keck Observatory. (Courtesy of Debra Fischer.)

FIGURE 3.6   Artist's rendering of the five planets orbiting 55 Cancri. *Bottom*: The star 55 Cnc is shown near the left. The planet in a 14-day orbit is located just to the left (a dark dot), and the four remaining planets are shown to the right. *Top*: A magnified view of the star and five planets, shown with arbitrary colors; an imaginary moon is included orbiting the outer giant planet. (Artwork by the Lynette Cook, http://extrasolar.spaceart.org/.)

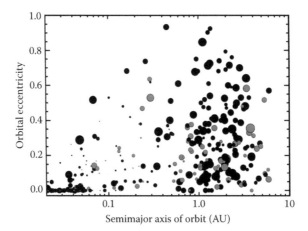

FIGURE 3.7   Orbital eccentricity vs. orbital distance (semimajor axis, on a logarithmic scale) for all 353 known exoplanets having accurately known orbital properties (within 20%; see http://exoplanets.org/). Exoplanets orbiting beyond 0.1 AU exhibit a wide range of eccentricities from e = 0.0 to 0.93, with a median value of 0.24. Planets around other stars reside in much more elongated orbits than do the eight major planets in our Solar System, which are in nearly circular orbits. Indeed, exoplanets beyond 3 AU have a full range of eccentricities, indicating that our Solar System is a rarity, with its giant planets in nearly circular orbits. The dot sizes are proportional to the cube root of minimum mass (*M*sini), showing that exoplanets of all masses exhibit high eccentricity. However, the lower-mass planets have systematically lower eccentricities with a distribution peaking toward circular orbits. The colored dots represent exoplanets in multiplanet systems. Evidently, planets in multiplanet systems exhibit no greater eccentricity than those in single-planet systems and, indeed, show somewhat lower eccentricities by a statistically significant amount. Note that beyond 0.8 AU, the high-eccentricity planets are nearly all in single-planet, not multiplanet, systems.

survivors are usually the more massive planets left in elongated orbits, providing graphic testimony to the gravitational damage done in the past.

Why do the planets in our Solar System have nearly circular orbits? Our Solar System apparently formed with its eight major planets just far enough apart from one another, and with masses just low enough, to avoid the normal damaging chaos. This allowed the small rocky planets, including Earth, to survive and remain in circular orbits. As a result, Earth enjoys a nearly constant distance and illumination from the Sun throughout the year, keeping the temperature within a small range. If it were not in such a circular orbit, the survival and evolution of life might have been impossible. It may be no coincidence that our home planet resides in a circular orbit in one of the rare quiescent planetary systems. About 5% of all giant exoplanets have nearly circular orbits (eccentricity less than 0.05). Our Galaxy surely contains more than 10 billion quiescent planetary systems in which circular orbits prevail and Earth-sized planets survive. Surely some 30% of them, implying 3 billion stars, have a rocky planet in the habitable zone where water remains liquid at the surface.

A rich source of information about the growth and migration of planets can be found in a plot of planet mass versus orbital distance, as shown in Figure 3.8. That plot shows the minimum mass (Msini) as determined from Doppler measurements for the 353 exoplanets that have accurate orbital properties and Msini. The Doppler technique can determine only a minimum possible mass for a planet because of the unknown tilt (inclination) of the orbit. We can measure only the combination of planet mass and the trigonometric sine of the orbital inclination, Msini. The true masses of planets are typically 25% higher than the Msini, assuming randomly oriented orbital planes.

As of February 2010, 353 exoplanets have orbits and Msini values measured to an accuracy of 20% (see updates at http://exoplanets.org/). The plot of planet mass versus orbital distance in Figure 3.8 shows remarkable structure. The planets of highest mass have 10 M_J. Any orbiting objects with masses between 10 and 70 M_J are called "brown dwarfs," too large to be planets, but too small to ignite nuclear reactions in

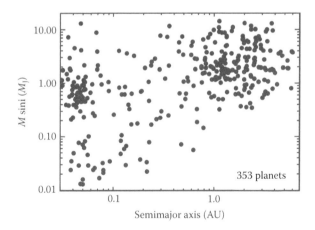

FIGURE 3.8 Minimum planet mass (Msini) vs. orbital distance (semimajor axis on log scale) for the 353 well-measured exoplanets as of February 2010 (see http://exoplanets.org/). This parameter space of planet mass and orbital distance offers rich information about the physics of planet formation. The clump of planets at the upper right shows that planet growth stops above 10 M_J, indicating the maximum size of the feeding zone for even the most massive protoplanetary disks. Another clump at middle left shows a parking mechanism, yet to be identified, for which migration has obviously occurred. But the very few high-mass planets in that inner clump suggest that migration is not common or that parking is not possible for massive planets above a Jupiter mass. The small number of planets between 0.1 and 1.0 AU suggests that migration occurs quickly, on a timescale shorter than the lifetime of the protoplanetary disks. Once a planet begins to migrate, it proceeds quickly from 1 to 0.1 AU, before the disk dissipates. At far left, for planets orbiting within 0.1 AU, very few exoplanets have masses 0.1–0.4 M_J. This cannot be a selection effect. The few planets in that mass range show that runaway gas accretion proceeds above a threshold of 0.1 M_J, making them massive planets.

their centers as stars do. We find very few orbiting objects between 10 and 70 M_J, making a "brown dwarf desert." Very few planets have masses above 1–2 M_J, showing that planets acquire gas gravitationally within their feeding zone of the protoplanetary disk, but rarely acquire more than 10 M_J for even the most massive protoplanetary disks. The other clump at the far left of Figure 3.8 shows the hot "Jupiters" that have somehow migrated and parked at their locations within 0.1 AU. The processes that cause planets to migrate inward and to park close to the star are not well understood. The stars are typically 1–10 billion years old, giving the planets a long time to evolve to their present configuration. Very few high-mass planets (above a Jupiter mass) are within 0.1 AU, suggesting either that migration rarely occurs for the most massive gas giants or that parking is not possible for them, allowing them to be swallowed.

Very few planets are between 0.1 and 1.0 AU, suggesting that migration occurs quickly, on a timescale shorter than the lifetime of the protoplanetary disks. The migration travel time from 5 to 0.1 AU must be shorter than the disk lifetime. Many giant planets have apparently migrated, but they do so quickly and leave few residents between 0.1 and 1 AU when the disk dissipates, analogous to the Hertzsprung gap in the Hertzsprung–Russell (H–R) diagram. Giant planets either stay far from the star or migrate all the way to the star, some parking there and some falling into the star.

Examining the planets that reside inward of 0.1 AU, only a few have masses between 0.1 and 0.4 M_J. This cannot be a selection effect, as less massive planets do exist there (see Figure 3.8) in large numbers: large planets at these close separations are easy to detect. The few planets between 0.1 and 0.4 M_J clearly show the runaway gas accretion that operates quickly to increase the mass of planets once they are above 0.05 M_J (15 $M_\oplus$). This indicates the minimum core mass necessary for that runaway process. When a planet grows to about 15 $M_\oplus$, it can quickly acquire more gas gravitationally to become a gas giant the size of Saturn or Jupiter.

While planets having masses as low as 5 $M_\oplus$ have been detected, very few have been found with Msini much below that. The lowest possible planet mass reported is for Gliese 581e (Mayor et al., 2009), with a Msini of 2 $M_\oplus$, making its likely mass 2.5–3.0 $M_\oplus$. Whether this planet is mostly composed of rock or is a mixture of rock, ice, and gas remains unknown. Residing 30 times closer to its star than Earth is to the Sun, this planet is heated to well more than 230 degrees Celsius (230°C) on the daytime side, so hot that liquid water and clouds seem unlikely. Still, this planet is likely composed largely of rock, as a purely gaseous object of this mass would not be able to gravitationally retain its gases or form in the first place. Other intriguingly small exoplanets that are only slightly larger than Earth include Gliese 1214b, HD 156668b, CoRoT-7b, *Kepler*-10, and 55 Cancri e (see http://exoplanets.org/ for updated references). A possible structure for Gliese 1214b is shown in Figure 3.9. These small planets probably contain large amounts of rock, iron, nickel, and perhaps

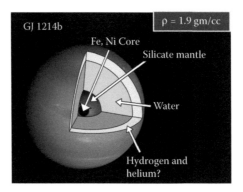

FIGURE 3.9   The interior structure of GJ 1214b. The density of this planet is above that for water (1 g/cm³), but below that for Earth (5.5 g/cm³). Therefore, it is probably composed of a mixture of common substances, notably iron, nickel, silicate rock, and water (and perhaps hydrogen and helium gas), with amounts that yield a final density of 1.9 g/cm³ as observed. (Courtesy of Wright, J.T., Upadhyay, S., Marcy, G.W. et al. (2009). Ten new and updated multiplanet systems and a survey of exoplanetary systems. *Astrophysical Journal*, 693: 1084–99.)

water, offering the suggestion that rocky planets the size of Earth may be common, some with lukewarm temperatures. These planets give support to the estimate that, among the 200 billion stars in the Milky Way, there must be at least 3 billion rocky planets with liquid water on their surfaces.

THE *KEPLER* MISSION: A SEARCH FOR EARTH-SIZE PLANETS

The search for other truly Earth-like planets has begun with NASA's *Kepler* telescope, launched from Kennedy Space Center on March 6, 2009 (Borucki et al., 2010). With its 1 m-diameter collecting area and its 40 charge-coupled device (CCD) light detectors, *Kepler* monitors the brightness of 157,000 normal, hydrogen-burning stars, taking a new measurement every 30 min nonstop for 3.5 years. While you eat and sleep, while relationships blossom and crumble, while wars start and stop, the *Kepler* mission will continue, unfailingly searching for the first Earth-like planets ever discovered. History is in the making.

Kepler can sense a dimming as small as 0.002% for any of those stars over a 6-hour period. If an Earth-sized planet orbits in front of a Sun-like star, it covers 0.01% of the surface of the star, blocking the light and dimming the star by that same amount, easily detectable by *Kepler* (see Figure 3.10). Figure 3.10 shows one of the Jupiter-sized planets discovered by *Kepler*, *Kepler*-8, for which Doppler measurements were made both out of transit (showing the reflex "wobble" of the star giving the mass of the planet) and during transit, showing the Rossiter–McLaughlin effect (Jenkins et al., 2010). The planet apparently blocks first the approaching side of the star (allowing less light with a negative velocity shift) and then the receding side of the star, giving the up-and-down velocity signature seen in the bottom panel of Figure 3.10. Apparently, the planet orbits with prograde motion, but at an angle of 25 degrees between the projected orbital plane and the equator of the star.

In general, *Kepler* verifies the existence of the planets by demanding that the star dim repeatedly at every orbit of the planet, confirming the orbital motion. The photometric variations can be verified as due to planets, rather than eclipsing binaries, by the detailed shape of the photometric variation (fitted well with a planet model and not an eclipsing binary) and by any displacement of the star in the sky during transit. Stars with true planets should show very little or no motion if indeed the dimming is due to a planet, rather than a faint background eclipsing binary star located nearby. In addition, follow-up spectroscopy of the host star will reveal the telltale Doppler periodicity of the host star, confirming the existence of the planet and allowing a measure of the mass and orbit of the planet. Combining the planet's mass (from Doppler measurements) with its diameter (from the amount of dimming) allows us to calculate the density of the planet. The first five planets announced by *Kepler* included four of Jupiter size and another four times the diameter of Earth (see, for example, Figure 3.10).

On February 2, 2011, the *Kepler* mission announced the discovery of 1,235 new planet candidates, most being smaller than three times the diameter of Earth. Remarkably, *Kepler* finds far more planets of 1–3 Earth radii than large Jupiter-size planets of 5–10 Earth radii. Apparently, planets of nearly Earth size greatly outnumber the Jupiter-size planets by a factor of at least 5.

Rocky planets, such as Earth, have a density of 5.5 g/cm^3, while gaseous planets, such as Jupiter or Neptune, have densities between 0.2 and 2.0 g/cm^3. Thus, *Kepler* and Doppler measurements combined will distinguish rocky from gaseous planets. If rocky planets are common, they will be certainly found and verified as solid by *Kepler*. The best stars for this effort are the smallest ones because they have small diameters, small masses, and small luminosities. These small stars allow the Earth-sized planets to block a larger fraction of the light, and their small stellar masses allow them to be pulled more vigorously by the gravitational tug of an Earth-mass planet, helping detection. In addition, planets orbiting very close to these low-luminosity stars will receive only a small amount of starlight, warming them only slightly, so that water on the planet remains in liquid form. Thus, *Kepler* may answer one of humankind's oldest questions about the existence and common occurrence of Earth-like, habitable planets.

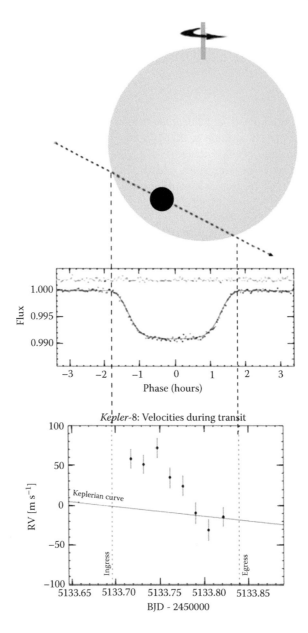

FIGURE 3.10 Schematic drawing and the underlying photometry and Doppler measurements for exoplanet *Kepler*-8b. *Top*: A graphic rendering of the relative size of the planet and star and the path of the orbit across the star. *Middle*: The photometry from the *Kepler* telescope has a precision of 20 μmag over 6 hours (at 12th mag), here showing brightness vs. orbital phase for the 3.54-day period. The star dims by 0.95% with a photometric shape fitted well by a model of a transiting planet in front of an F8 main-sequence star (*solid red line*). At the top of the middle panel is the photometry 180 degrees away from the transit, showing no indication of a secondary transit as would be seen for an eclipsing binary as a false positive. Velocity measurements from the Keck telescope show the usual reflex "wobble" yielding a planet mass of 1.3 M_J. *Bottom*: The measured radial velocities during transit, showing the Rossiter–McLaughlin effect as the planet blocks first the approaching and then the receding portions of the stellar surface, giving a net redshift and then blueshift. The asymmetry in velocities reveals the tilt of the orbital plane relative to the star's equator, as projected onto the sky. (Courtesy of Howard Isaacson.)

Other NASA and European Space Agency (ESA) missions are being considered to detect Earth-like planets. NASA is pursuing the *Space Interferometry Mission* (*SIM*) and the *Terrestrial Planet Finder* (*TPF*). While both are yet to be funded and are at least 5 years away from launch, *SIM* and *TPF* will detect Earth-like planets orbiting nearby Sun-like stars, located within 30 lt-yr, if they are common. These two space-borne telescopes will determine the masses, orbits, and chemical compositions of those rocky worlds. It would be glorious if the major countries in the world could work together cooperatively to jointly fund these giant, space-borne telescopes and to jointly fund the giant telescopes on the ground. Surely it is the rocky planets located within a few light-years of Earth that will be targets for future spectroscopy, revealing their chemistry, atmospheres, continents, oceans, and habitability (see the contributions by Sara Seager and Charles Beichman in this volume).

A statistical sample of terrestrial planets will tell us how commonly bio-friendly worlds occur and will inform us about the stability of Earth's environment. We humans have pursued our origins all over the globe, from sites as diverse as the Afar Region of Ethiopia and the Eocene fissures in Jiangsu Province. Now the digs for our human origins will continue at sites among the stars.

REFERENCES

Borucki, W.J., Kock, D.G., Brown, T.M. et al. (2010). Kepler-4b: Hot Neptune-like planet of a G0 star near main-sequence turnoff. *Astrophysical Journal,* 713: L126–30.

Fischer, D.A., Marcy, G.W., Butler, R.P. et al. (2008). Five planets orbiting 55 Cancri. *Astrophysical Journal*, 675: 790–801.

Howard, A.W., Johnson, J.A., Marcy, G.W. et al. (2011). The NASA-UC Eta-Earth Program: II. A planet orbiting HD 156668 with a minimum mass of four Earth masses. *Astrophysical Journal*, 726: 73–83.

Jenkins, J.M., Borucki, W.J., Koch, D.G. et al. (2010). Discovery and Rossiter–McLaughlin Effect of Exoplanet Kepler-8b. *Astrophysical Journal,* 724: 1108–119.

Johnson, J.A. (2009). International Year of Astronomy Invited Review on Exoplanets. *Publications of the Astronomical Society of the Pacific*, 121: 309–15.

Marcy, G.W., Butler, R.P., Vogt, S.S. et al. (2008). Exoplanet Properties from Lick, Keck and AAT. *Physica Scripta*, vol. T, 130: 014001 (7 pages).

Mayor, M., Bonfils, X., Forveille, T. et al. (2009). The HARPS search for southern extra-solar planets. XVIII. An Earth-mass planet in the GJ 581 planetary system. *Astronomy and Astrophysics*, 507: 487–94.

Wright, J.T., Upadhyay, S., Marcy, G.W. et al. (2009). Ten new and updated multiplet systems and a survey of exoplanetary systems. *Astrophysical Journal*, 693: 1084–99.

Part II

Impact of Telescopes on Our
Knowledge of the Universe

4 The Formation and Evolution of Galaxies

Ben Moore

CONTENTS

INTRODUCTION

The fuzzy band of light in the night sky was speculated to be distant stars by Democritus around 400 BCE; but it was not until 1610, following the invention of the telescope in 1608 by Hans Lipperhey, that this was finally verified by Galileo, who first resolved the Milky Way into millions of stars. Since Hubble determined in the early 20th century that the fuzzy nebulae were distant galaxies, astronomers have carefully attempted to visually classify and catalog galaxies into common sequences. The most famous of these, the Hubble sequence, describes galaxies according to the complexity of their appearance—a sequence that is often mistakenly interpreted as an evolutionary sequence because the scheme smoothly stretches from disks lacking a central bulge, through disk (S0) galaxies that have no apparent structure in the disk component, to elliptical galaxies. The subject of this chapter is to give a modest overview of our current understanding of how galaxies form and evolve from a theoretical perspective and to discuss open questions that will be addressed in future work.

Visual inspection of telescope images reveals a huge diversity in the morphological appearance of galaxies. Furthermore, the appearance of an individual system is very different in various wavelengths. In order to reproduce the wide variety of galaxy morphologies, many transformation mechanisms have been invoked. These are typically gravitational and hydrodynamical interactions, which can move galaxies across the Hubble sequence and create the diversity observed in our Universe. To date, these processes have generally been studied using numerical simulations of preconstructed idealized galaxy models. The ultimate goal is to develop our computational techniques such that environmental effects and morphological evolution can be followed within the cosmological model of a hierarchical Universe.

It has been almost 30 years since the theoretical cosmological framework for the evolution of structure within a cold dark matter (CDM)-dominated Universe was pioneered (Peebles, 1982). More recently, dedicated campaigns of space- and ground-based observations have precisely measured the initial conditions from which structure forms in our Universe—tiny perturbations imprinted on the mass distribution like a network of ocean ripples (e.g., Spergel et al., 2007). The past decade in particular has proved to be an exciting time in cosmology. Astronomers have measured the fundamental parameters that govern the evolution of the Universe. The matter and energy densities, the expansion rate, and the primordial power spectrum are now well constrained; however, only about 1% of the Universe has been physically identified and understood. Thus, although the initial conditions for structure formation are known, the dominant components of matter and energy actually are not. How these fluctuations in the dark matter and baryonic components form galaxies, stars, and planets involves complicated nonlinear processes including gravity, hydrodynamics, radiation, and magnetic fields. Linear theory calculations take us only so far, and we must rely on numerical simulations to follow the detailed structure-formation process.

Until recently, it was difficult to stringently test the predictions of a given cosmological model. Simulations are the ideal means by which to relate theoretical models with observational data, and advances in algorithms and supercomputer technology have provided the platform for increasingly realistic astrophysical modeling. For example, simulations simply could not resolve the central regions of dark matter halos where kinematic and lensing observations constrain the mass distribution (Mao and Schneider, 1998; Simon and Geha, 2007). There has been steady and significant progress in this area—reliable and fundamental predictions of the clustering properties of the dark matter have been made via massively parallel computations (Navarro et al., 1997; Moore et al., 1998; Diemand et al., 2007). The "cusp" and "satellite" problems (see the next section) for the standard CDM represent real observational tests of the properties of a new fundamental particle that makes up most of the mass of the Universe—this is a testament to the achievements of modern numerical simulations. This ability to predict the nonlinear behavior of dark matter clustering has stimulated much work in the observational astronomy and astroparticle physics communities.

Our theoretical understanding of galaxy formation is somewhat behind the stream of quality observational data that comes from ground- and space-based facilities around the world. While theorists are still trying to understand how galaxies assemble themselves from the dark matter and baryons in the Universe, observational astronomers have exquisite data in multiple wavelengths with high-resolution spectral information, element abundances, color maps, and kinematical data. Theorists have not yet succeeded in making a single realistic disk-dominated galaxy via direct simulation—it is still an open question as to how the baryons collect at the centers of galaxies and what causes and regulates star and star-cluster formation.

Here are some observational facts about galaxies:

- All galaxies are observed to sit at the center of an extended "halo" of dark matter.
- Galaxies range in baryonic mass from 10^4 to 10^{13} $M_{\odot}$ (from the smallest satellites of the Milky Way to the giant cD galaxies in clusters).
- No galaxy has been found to contain only gas and no stars.
- No one has observed a dark galaxy (a halo with no baryons), even though CDM halos are predicted to span a mass range from about 10^{-6} to 10^{15} $M_{\odot}$.
- The tight Fisher–Tully and Faber–Jackson relations, relating luminosity to mass, imply that galaxies form in a well-behaved and nonstochastic way.
- Star formation is physically complex, but follows well-defined global scaling laws.
- The baryon fraction of halos decreases from cluster scales, which have the universal baryon fraction, to dwarf galaxies, which have only ~1% of the universal fraction.

- Most of the stars in the Universe are inside luminous spiral and elliptical galaxies, but most galaxies are dwarf spheroidal (dSph) and disk galaxies.
- The morphology of galaxies varies with environment and redshift.
- The rate of star formation of galaxies peaks at a redshift $z = 2$ and then declines.
- The luminosity function of faint galaxies scales as $n(L) \sim L^{-1}$, whereas the mass function of dark matter halos is steep, $n(M) \sim M^{-2}$.
- There are always examples of galaxies that are exceptions to the rule.

COLD DARK MATTER (CDM) HALOS

Given that all galaxies are expected to form and evolve at the centers of dark matter halos, before attempting to follow the complexity of the galaxy formation process, we must first understand the origin and properties of halos (White and Rees, 1978). Gravity is primarily responsible for the development of dark matter halo structure, substructure, and clustering on larger scales. This has allowed theorists to show that the CDM model is remarkably successful at describing the large-scale development of our Universe, from the hot Big Bang to the present day. However, the nature of the dark matter particle is best tested on small scales, where its physical characteristics manifest themselves by modifying the properties of halo structure and substructure. Ultimately, in the coming decade, experimentalists hope to find or rule out the existence of weakly interacting CDM particles at the Large Hadron Collider (LHC) in the laboratory by direct-detection experiments or indirectly by observing the cascade of visible particles that result from the mutual annihilation of the candidate neutralino particles.

During the 1980s, the first simulations of the CDM model were carried out. Large volumes of the Universe were followed from the linear to the nonlinear regimes in an attempt to match the large-scale clustering of galaxies. Indeed, reproducing the filamentary pattern observed in the Harvard–Smithsonian Center for Astrophysics (HSCfA) redshift survey was considered compelling evidence for such models (Davis et al., 1985). At that time, some of the most basic properties of collapsed structures were discovered, for example, the distribution of halo shapes and spin parameters (Frenk et al., 1985).

It was not until the simulations of Dubinski and Carlberg (1991) that individual dark matter halos were simulated at sufficiently high resolution to resolve their inner structure on scales that could be compared with observations. Using a million-particle simulation of a cluster mass halo run on a single workstation for an entire year, these authors found density profiles with a continuously varying slope as a function of radius and central cusps diverging as $1/r$ in their centers. It is still a matter of debate in the literature as to whether the observations support this prediction.

Navarro et al. (1997) published results of simulations of CDM density profiles from scales of galaxies to galaxy clusters. They demonstrated that all halos could be reasonably well fitted by a simple, two-parameter function with a concentration parameter that was related to the halo mass. With only $\sim 10^4$ particles, they could resolve the halo structure to only about 5%–10% of the virial radius, that is, the radius within which the system has reached equilibrium. Shortly afterward, simulations with 1 million particles and high force resolution resolved the overmerging problem—the artificial disruption and loss of substructure by gravitational tides—the resolution was sufficient to resolve cusps in the progenitor halos, enabling us to see that the structures survive the merging hierarchy (Ghigna et al., 1998). The final, surviving substructure population is a relic of the entire merger history of a given CDM halo and provides a unique verification of the hierarchical clustering model on cluster scales (see Figure 4.1).

Recent high-resolution dark matter simulations have demonstrated that CDM halos are remarkably self-similar over a mass range spanning more than 15 orders of magnitude (Diemand et al., 2005). Interestingly, such self-similarity is notably absent in the baryonic component: halos below about $10^7 \, M_\odot$ are completely dark, and detecting them would provide the strongest evidence for a CDM-dominated Universe. The fact that halos are close to self-similar in their radial density profiles and substructure abundances has allowed unique tests of the CDM paradigm (Moore et al., 1999a). The distribution of galaxies on large scales and the detailed properties of galaxy clusters

FIGURE 4.1 Galactic dark matter halo (GHALO): A billion-particle simulation of the dark matter distribution surrounding a galaxy. This simulation took 3 million CPU hours with the Parallel K-D tree GRAVity (PKDGRAV) code on the MareNostrum supercomputer. Tens of thousands of dark matter substructures orbit through the halo and fine streams of dark matter cross the entire system. These simulations show a remarkable self-similar pattern of clustering properties, with entire generations of the merging hierarchy preserved as a series of nested structures and substructures reminiscent of a Russian matryoshka nesting doll. (From Stadel, J., Potter, D., Moore, B., et al., *Monthly Notices of the Royal Astronomical Society*, 398, L21–5, 2009.)

match well with the predictions of our standard cosmological model. On scales that probe the heart of dark matter halos and on dwarf galaxy scales, the evidence for or against the existence of CDM is uncertain (e.g., the missing satellite problem; see Kravtsov, 2010 and references therein). The main difficulty in the comparison between theoretical models and observations, such as rotation curves or satellite abundances, is the uncertainty of the effects of the galaxy-formation process. Baryons can modify the central structure of halos, and radiative and supernova (SN) feedback processes can affect how stars populate low-mass halos.

The recent detection of many faint satellite galaxies orbiting deep within our Galactic halo provides a new and stringent test of the nature of the dark matter particle. Sloan Digital Sky Survey (SDSS) results (Simon and Geha, 2007) may imply as many as 60–70 satellites brighter than 30 mag/arcsec^2 within 400 kpc, and there could be as many as 25 satellites in the inner 50 kpc, a region previously known to host only the Large Magellanic Cloud (LMC) and Sagittarius. The motions of individual stars in these satellites reveal that their central regions are completely dark matter dominated with extreme densities as high as 1 $M_\odot$ per cubic parsec on length scales of ~100 pc. Resolving these regions in numerical simulations has only just been achieved, and indeed the dark matter within the substructure reaches densities this high (Diemand et al., 2008).

These surveys have also revealed numerous streams of stars in the outer "Field of Streams" and inner halo, which are the remnant ghosts of previous Galactic companions that have been tidally destroyed (Ibata et al., 2001; Belokurov et al., 2006). Further evidence for significant dark substructure within other galaxies comes from the anomalous flux ratios in gravitationally lensed quasi-stellar objects (QSOs; Mao and Schneider, 1998). Perturbations to the light path from inner substructure can naturally explain this phenomenon if the projected substructure fraction within 10 kpc is as high as 1%.

GALAXY FORMATION

Here are some things we don't know about galaxy formation:

- How do galaxies acquire their baryons?
- Is the halo mass the main quantity that determines the morphology of a galaxy?
- How do disk galaxies form, in particular, bulgeless Sc/Sd galaxies?
- How do elliptical galaxies and S0 galaxies form?

- The merger history of halos is extremely varied; how do tight scaling laws result?
- What prevents all the low-mass substructures and halos from forming stars?
- How do SN feedback in small halos and active galactic nucleus (AGN) feedback in massive halos affect galaxy formation?
- How did Sc galaxies such as M33 lose most of their baryons?
- How does the galaxy-formation process modify the distribution of dark matter?
- Why does the luminosity function of faint galaxies scale as $n(L) \sim L^{-1}$, whereas the mass function of halos is steep, $n(M) \sim M^{-2}$—i.e., how do galaxies populate dark matter halos?

Our Milky Way (and its local environment) is a "Rosetta stone" for understanding galaxy formation and evolution and for testing cosmological models. It contains several distinct old stellar components that provide a fossil record of its formation—the old stellar halo, globular clusters, and satellite galaxies. We can begin to understand their spatial distribution and kinematics in a hierarchical formation scenario by associating the protogalactic fragments envisaged by Searle and Zinn 30 years ago with the rare peaks able to cool gas that are predicted to form in the CDM density field collapsing at redshifts $z > 10$. Hierarchical structure formation simulations can be used to explore the kinematics and spatial distribution of these early star-forming structures in galaxy halos today (Moore et al., 2006; Madau et al., 2008). Most of the protogalaxies rapidly merge together, their stellar contents and dark matter becoming smoothly distributed and forming the inner galactic stellar and dark halo. The metal-poor globular clusters and old halo stars become tracers of this early evolutionary phase, centrally biased and naturally reproducing the observed steep fall-off with radius. The most outlying peaks fall in late and survive to the present day as satellite galaxies. The observed radial velocity dispersion profile and the local radial velocity anisotropy of Milky Way halo stars are successfully reproduced in this toy model. If this epoch of structure formation coincides with a suppression of further cooling into lower sigma peaks, then the rarity, kinematics, and spatial distribution of satellite galaxies can be produced. Recent numerical work has indicated that the Local Group of galaxies may have been re-ionized from outside, from the nearby-forming Virgo Cluster of galaxies that collapsed before our own Milky Way (Weinmann et al., 2007). This leaves observable effects on the distribution of the old stellar components. However, the above qualitative scenario needs to be tested rigorously using simulations that follow all of the physical processes, not just the dark matter component as all previous studies have used (Bullock et al., 2001; Kravtsov et al., 2004).

Although the theory behind galaxy formation appears well defined, many of these individual assumptions remain untested, and no group has been able to successfully simulate the formation of a Milky Way-like disk galaxy that resembles observed Sb–Sc galaxies. Very recently, there has been a paradigm shift in our understanding of how galaxies obtain their baryons—rather than cooling flows from a hot ionized gaseous halo component, cold inflowing streams of baryons can dominate the accretion (Keres et al., 2005; Dekel and Birnboim, 2006; Dekel et al., 2009). Various groups are coming close to resolving galaxies using computational techniques (e.g., Abadi et al., 2003); however, these simulations resulted in bulge/spheroid-dominated systems with disks that are too small. It is not known whether this is a deficiency in the model or a problem with the numerical simulations (Okamoto et al., 2005). Forming realistic disk galaxies is widely recognized as a major challenge for both numerical simulations and the CDM hierarchical structure formation model. The ultimate goal is to calculate the formation of the Milky Way and the Local Group of galaxies within our concordance cosmological model in exquisite detail, so as to make theoretical predictions for forthcoming ground- and space-based missions such as the Visible and Infrared Survey Telescope for Astronomy (VISTA) Hemisphere Survey (VHS) or *Global Astrometric Interferometer for Astrophysics* (*Gaia*).

MODELING ISSUES I: THE INTERSTELLAR MEDIUM

Modeling the thermodynamics of the interstellar medium (ISM) is an important aspect of galaxy formation and evolution. The ISM has at least three main phases that are in approximate pressure equilibrium: a hot (~10^6 K), low-density (0.01–0.00001 atoms/cm³) ionized space-filling plasma that fills the holes and bubbles within the disk ISM and extends to tens (and possibly hundreds) of kiloparsecs into the galactic halo (Snowden et al., 1998; Wang et al., 2005); a warm (100 K–1,000 K), diffuse phase with densities up to 1 atom/cm³ (Wang et al., 2005); and the dense (>100 atoms/cm³), cold molecular H_2-dominant phase with temperature <50 K (Pohl et al., 2008). The phases are constantly mixed by SN explosions that inject large quantities of thermal energy and momentum in the ISM. The different phases in the ISM coexist as the result of thermal balance between radiative heating and cooling at different densities and, at the same time, thermal instability (perhaps coupled with gravitational instability) that determines the emergence of the dense molecular phase in which star formation takes place. Furthermore, the formation of a galaxy should not be considered within a closed box. There are constant accretion events of satellites and new gas, cooling flows, and recycled material from stripped debris and galactic fountains.

Molecular gas is formed and destroyed via a number of microscopic interactions involving ions, atoms, and catalysis on dust grains. These processes become biased toward the formation, rather than toward the destruction, of molecular hydrogen only at densities >10 atoms/cm³. A great deal of energy in the ISM is nonthermal; this turbulent energy, which is essentially observed as random gas motions, is supersonic and several times larger than the thermal energy at the scale of giant molecular cloud (GMC) complexes. Turbulent kinetic energy is thought to be the main agent that supports the largest molecular clouds (>10^5 $M_\odot$) against global collapse. The partial suppression of gravitational collapse owing to turbulent support also explains the low efficiency of star formation in our galaxy (only a small percentage of the molecular gas mass present in the Milky Way appears to be involved in forming stars).

Modeling the ISM is nontrivial. Recently, Agertz et al. (2007) questioned the ability of smoothed particle hydrodynamics (SPH) codes to follow multiphase gas and basic flow instabilities such as Kelvin–Helmholtz. Galaxies form from a turnaround region that is a megaparsec in size (the angular momentum is generated from a region larger than 10 Mpc). On these scales the gas density is 10^{-7} atoms/cm³, and we need to follow this region as the gas collapses to parsec-scale molecular clouds with densities larger than 100 atoms/cm³. Simultaneously resolving the star-formation process itself inside molecular cloud cores within a cosmological context is a decade away. However, a dynamic range of five decades in length and seven in density necessary to resolve GMC formation has recently been achieved in the highest-resolution adaptive mesh refinement (AMR) simulations (see below).

MODELING ISSUES II: SUBGRID PHYSICAL PROCESSES OF STAR FORMATION AND FEEDBACK

Even in these simulations and for the foreseeable future, modeling the physical processes of star formation, SN feedback, and radiative processes from stars relies on "subgrid" algorithms. These processes cannot be simulated directly as a result of resolution issues and are implemented by hand as realistically as possible, relying on observational scaling laws and theoretical modeling.

Even the thermodynamics of the ISM is partially subgrid, given that current simulations typically lack resolution below 100 pc; this sets a maximum density that can be resolved of the order of 1 atom/cm³, which is close to the density of the warm neutral medium. For this reason cooling processes that are important below $T \sim 10^4$ K are usually neglected. Likewise, radiative heating is partially determined by the thermal and turbulent energy injection from SN explosions, accreting massive black holes (BHs), as well as by the radiation background produced by stars or by cosmic-ray and x-ray heating. These processes are simply included as a constant heating term in the internal energy equation.

Star formation in these simulations is treated in an embarrassingly simplistic way—once the gas reaches some threshold, "star particles" (which individually represent star clusters) are created. This density is taken to be ~0.1–1 atom/cm³ because this is the highest density that can be followed at the current resolution. In essence, the Schmidt–Kennicutt law (the correlation between gas density and star formation rate) is implemented by hand into the simulations.

Once these superstar-sized particles are created, an initial mass function is assumed that can be evolved to determine what fraction of the "stars" explode as supernovae (SNe), returning energy and heavy elements to the ISM. How this energy return is treated ranges from dumping thermal or kinetic energy into the surrounding gas to halting the radiative cooling of gas for a timescale of about 20 million years. The latter approach attempts to account for the energy dissipation timescale of the unresolved turbulence generated by the SNe.

MODELING ISSUES III: MAKING DISK GALAXIES

The angular momentum problem—namely, the fact that disks that form within cosmological simulations do not lie on the Tully–Fisher relation (they are rotating too fast for the amount of stars they have formed)—has received a lot of attention in the literature. However, as resolution increases and subgrid modeling has improved, galaxies are produced that lie quite close to the relation. One of the most puzzling remaining problems is that, until recently, no simulation has managed to produce a pure disk galaxy—all simulated galaxies have a significant bulge/spheroid component that arises in different ways: (i) massive gas clumps form dense star clusters that rapidly sink to the centers of halos, (ii) too many satellites accrete and merge with the forming galaxy, and (iii) gas at the centers of halos continues to form too many stars.

Governato et al. (2010) recently simulated the formation of a small dwarf spiral galaxy that had a negligible bulge component. These SPH simulations have a high spatial resolution as a result of the small mass of the system. By invoking strong SN feedback, star formation was inhibited and the system was violently stirred by the motions of stellar and gaseous clumps. However, it is not clear that SN feedback is the key to understanding the formation of larger disk galaxies such as the Milky Way. It may be the case that star formation efficiency plays the most important role; moreover, this may be a time-dependent process closely linked to H_2 formation (Gnedin and Kravtsov, 2010). Recently, Agertz et al. (2011) were able to simulate the formation of a Milky Way-like Sb disk galaxy within the ΛCDM framework, concluding that a low star-formation efficiency is crucial.

Another origin of all of these problems is possibly the fact that radiative processes are not accurately included. Star clusters that form would quickly evaporate the surrounding gas, leaving the stellar component unbound, which would cause it to disperse before it sinks to the halo center. The protogalaxies that form at high redshift in the simulations have far too many baryons and are so dense that they can sink intact into the central galaxy, creating a spheroid. Those protogalaxies that accrete late form a population of satellites that are too luminous and too numerous. Reducing the baryon fractions in these systems as they form, perhaps by reionization and photo ionization, would greatly reduce the number of stars they could bring into the central galaxy and also make them easier to disrupt in the outer halo. Finally, the central cold neutral hydrogen component at halo center that continues to form stars is rarely observed in galaxies—radiation from OB stars from the bulge or inner disk would be sufficient to keep this material ionized or to prolong the cooling timescale such that star formation in the simulations in the bulge region would be greatly reduced.

The first goal that is achievable in the near future is to resolve the formation of molecular clouds within the gaseous disks that form at the centers of the dark matter halos. This alone would allow us to make many realistic comparisons to existing data and to study how galaxies evolve in different environments. Following this evolution to a redshift $z = 0$ with parsec-scale resolution is likely to be achieved within the next 5 or so years. (Resolving the fragmentation and collapse of individual clouds is perhaps several decades away given existing algorithmic and computational limitations.) Simulations that spatially resolve molecular cloud formation within a cosmological context would allow us to make enormous progress in understanding galaxy formation and the origin of the Hubble sequence.

COLD STREAMS AND COOLING FLOWS: HOW GALAXIES GET THEIR BARYONS

Galaxy formation has two key phases: (1) the early rapid virialization and assembly of the dark matter and old stellar components, a stage when the surviving satellite distribution is established; and (2) the longer-term quiescent stage when secular disk growth proceeds. The complexity in these processes can be followed only with numerical simulations. Semianalytic calculations attempt to incorporate these results to make predictions for the global properties of galaxies for large-scale surveys.

The classic picture of galaxy formation within the CDM scenario assumes that the accreted gas is shock heated to the virial temperature (i.e., a temperature in which the kinetic energy of the gas balances the gravitational potential energy of the mass distribution), cools radiatively, and rains down to form an inner star-forming rotating disk. Recent theoretical studies (Birnboim and Dekel, 2003; Keres et al., 2005) have demonstrated that accretion of fresh gas via cold in-fall can, in fact, be the dominant process for gas accretion for halo masses below 10^{11} $M_{\odot}$. In these halos, the cooling time for gas of temperature $T \sim 10^4$ K is shorter than the timescale of gas compression, and shocks are unable to develop. Cold accretion persists in halos above this mass at $z = 2$, whereas the classical hot mode of gas accretion dominates at lower redshifts. Because of insufficient spatial resolution, these studies could not follow the evolution of the accreting gas and how the cold streams connect to the central galaxies.

Figure 4.2 captures the complex disk-formation process where we observe gas reaching the disk in very distinct ways. This striking image ties together many aspects present in modern theories of galaxy formation and highlights new complexities. Cold streams of gas originating in narrow dark matter filaments effectively penetrate the halo and transport cold metal-poor gas right down to the protogalactic disk to fuel the star-forming region. A comparable amount of metal-enriched material reaches the disk in a process that has previously been unresolved—material that is hydro-dynamically stripped from accreting satellites, themselves small disk systems, through the inter-action with the hot halo and frequent crossings of the cold streams. The cold gas streams into the halo on a highly radial trajectory, eventually forming more orderly rotational motion in an extended disk through two mechanisms. A cold stream can gravitationally swing past the halo center and subsequently collide with a cold stream inflowing from an opposite direction: as the cold stream

FIGURE 4.2   The complex gas flows into a dark matter halo with a forming disk galaxy at redshift $z = 3$. The colors represent the amplitude of temperature (*red*), metallicity (*green*), and density (*blue*). This simulation, performed with the adaptive mesh code RAMSES, represents the state of the art in galaxy formation (Agertz et al., 2009). One can clearly distinguish the cold pristine gas streams in blue connecting directly onto the edge of the disk, the shock-heated gas in red surrounding the disk, and metal-rich gas in green being stripped from smaller galaxies interacting with the hot halo and cold streams of gas. The disk and the interacting satellites stand out because they are cold, dense, and metal rich.

enters the inner halo, it also feels a high confining pressure from the hot halo that has a significant rotational component within about 40 kpc. Shocks from these collisional processes are quickly dissipated because the cooling times are very short, resulting in a denser configuration for the cold gas. As the in-falling stripped material and streams lose their radial energy through these interactions, they connect to the inner disk as extended dense spiraling arms that progressively slow down to match the highly ordered inner disk rotation. The details of the spiral structure and secular instabilities within such cosmological simulations have yet to be explored in detail.

Cold, metal-poor, pristine gas flowing down narrow filaments; metal-enriched gas, stripped from accreting satellites; and cooling flows from the hot halo are all significant sources of baryons. This simulation is the first of its kind to achieve a resolution of GMC formation within a cosmological context (50% resolution in the AMR grid and over 10^7 dark matter particles).

THE IMPORTANCE OF BARYON FRACTION

The baryonic inventory of galaxies of different types and luminosities gives us very important information as to how they form and the processes that affect their formation and evolution. It has been established that large galaxies and galaxy clusters have captured close to the universal baryon fraction available (roughly 6:1 relative to the dark matter component). However, lower-mass galaxies have retained today just a small fraction of this value (e.g., the "baryonic Tully–Fisher relation"; McGaugh et al., 2000; Mayer and Moore, 2004). For example, the Milky Way has captured about 50% of the available baryons (Klypin et al., 2002), whereas nearby M33, the prototypical late-type disk galaxy (close to type Sd), has only ~2% of the universal baryon fraction (Corbelli and Salucci, 2000). Dwarf spheroidal galaxies are even more extremely dark matter dominated (Mateo, 1998); however, additional environmental effects, such as tidal stripping and ram pressure, act on these systems as they orbit with the Galactic halo. There are at least two plausible models for the origin of this relationship between baryonic mass and halo mass: (1) feedback from stars (SNe and the ultraviolet background radiation) may be more efficient at expelling gas in smaller halos, or (2) perhaps reionization preheats the gas, preventing it from cooling efficiently within less massive halos.

For isolated galaxies less massive than the Milky Way, the baryon fraction decreases rapidly, $M_{baryon} \sim V_{vir}^4$, such that the smallest galaxies have captured and cooled just a few percent of the available baryons. Note that if all halos kept hold of the universal value, this relation would scale as $M_{baryon} \sim V_{vir}^3$ (simply resulting from the top-hat collapse model; Gunn and Gott, 1972). Reproducing the observed baryon fractions of galaxies is perhaps the most fundamental goal that should be achieved, for several reasons. In particular, it may help resolve the discrepancy mentioned above between the mass function of halos and the luminosity function of galaxies. If stars form in proportion to the baryon fraction, then we have the fact that the number of stars per halo gives a luminosity $L \sim M_{baryon} \sim V_{vir}^4 \sim (M_{halo})^{4/3}$. Inserting this relation between L and halo mass into the CDM mass function, we find a closer agreement between the faint-end luminosity function and the halo mass spectrum.

Finally, it should be noted that disks that have a lower baryon fraction are considerably more stable against instabilities such as bar formation. Indeed, isolated disk simulations of M33-type galaxies could be reproduced only in models that began with a lower baryon fraction (Kaufmann et al., 2006). Thus, one can also speculate that in order to create pure disk component galaxies, some mechanism for keeping most of the available baryons at large distances from the halo center is required.

THE EFFECTS OF BARYONS ON DARK MATTER HALO STRUCTURE

A lot of work remains to be done to quantify the effects that baryons can have in modifying the distribution of dark matter. Simulations including baryons are more complex and expensive, and, as discussed above, we do not yet have a clear understanding of how galaxies form.

The dark matter density profiles can steepen through adiabatic contraction due to dissipating baryons (Blumenthal et al., 1984; Gnedin et al., 2004). The strength of this effect depends on the baryonic fraction that slowly dissipates via radiative cooling. However, accretion of baryons via cold flows may dominate the growth of many galaxies; thus, it is not yet clear how strongly this changes the inner distribution of dark matter in galaxies. For a halo that cools the cosmologically available baryons into a disk component, the dark matter density at a few percent of the virial radius increases by about a factor of 2, and the final density profile can resemble an isothermal sphere—comparable to observed constraints on elliptical galaxies (Gnedin et al., 2004).

The growth of supermassive BHs or central nuclei can increase or decrease the central dark matter density depending on whether these structures grow adiabatically or through mergers of smaller objects. Gondolo and Silk (1999) explored the effects of slow central BH formation on the CDM cusp in the context of an enhanced annihilation signal. (The leading candidate for CDM is the neutralino, which is its own antiparticle and annihilates into gamma rays in dense regions.) This mechanism can create isothermal cusps on parsec scales, with a boost factor to the gamma-ray flux of several orders of magnitude. On the other hand, if massive BHs grow through merging with other BHs, then binary systems can form that can eject significant amounts of dark matter from the central halo region.

Similar behavior would result from the formation of central stellar nuclei in galaxies. Dissipative growth of nuclear star clusters would increase the central dark matter density, but formation via the dynamical friction acting on sinking star clusters would lead to an inner core (Goerdt et al., 2006). A similar mechanism acts in cluster halos, whereby energy transfer to the dark matter background from dynamical friction acting on massive satellite galaxies gives rise to a constant-density inner region. All of these processes have yet to be studied in a realistic cosmological context, and their effects on dark matter halo structure are unclear.

Feedback from the star-formation process has frequently been invoked to flatten cusps, especially in dwarf galaxies that have challenged the CDM paradigm through observations of rotation curves, stellar velocities, and star-cluster kinematics. A single violent event, which somewhat unrealistically ejects a cosmological baryon fraction from the inner region, can redistribute the dark matter through a central revirialization. However, the most careful study of this process shows the effect to be modest, with a reduction in the central halo density by at most a factor of 2–6 (Gnedin et al., 2004). More realistic SPH simulations in a cosmological context show that SN-driven turbulent gas motions can impart sufficient energy to the dark matter to create a core as large as 400 pc in a Fornax-sized galaxy (Mashchenko et al., 2008; Governato et al., 2010). This effect requires both a significant early central baryon fraction and the Jeans mass to be accurately followed given that bulk motions are driven by starbursts in GMCs. It will be interesting to compare these experiments with higher-resolution adaptive-mesh techniques, including the effects of radiative processes.

More than half of disk galaxies have stellar bars that can transfer angular momentum to dark matter particles through orbital resonances and dynamical friction. The magnitude of this process has been debated in the literature; however, even when a rigid perturber mimicking a bar was placed at the center of a cuspy halo, it affected only the dark matter particles within ~0.1% of the virial radius, or ~300 pc in our Galaxy. The most recent and highest-resolution study of this process demonstrates that the effect of bars on the central dark matter distribution is negligible (Dubinski et al., 2009 and references within).

The shapes of dark matter halos can be dramatically modified as particles respond to the central mass growth by modifying their orbital configurations. The irregular "box orbits" (those that fill all the 3-dimensional space available for a given energy), which support the triaxial nonspherical configurations, are destroyed by the central potential (Kazantzidis et al., 2004). Given that particles move on eccentric orbits with a typical apocentric to pericentric distance of 5:1, halos can be visibly affected out to half the virial radius and become almost spherical close to the galaxy. The change in shape depends on the central baryonic fraction, which is highest for elliptical galaxies and lowest for galaxy clusters and dwarf galaxies, whose halos should barely be affected. The detailed modification of particle orbits within the disk region has yet to be explored, but this could possibly affect the detailed predictions for direct detection experiments.

Galaxy formation also leads to the accretion of gas, stars, and dark matter from satellites into the disk: systems on roughly coplanar orbits suffer dynamical friction against the disk, which brings them into the disk plane where they are disrupted (Read et al., 2008). This process produces a *dark matter disk*, which could contribute a significant fraction of the local dark matter density.

On the smallest scales, substructures orbiting through the galactic disk will lose significant amounts of mass as they suffer disk shocking or heating from individual stars (Zhao et al., 2007; Goerdt et al., 2006). We can estimate a timescale by comparing with similar calculations performed for globular clusters. Disk shocking of globular clusters has a disruption timescale of the order of a Hubble time, implying that these stellar systems have already lost significant mass and some may have been completely disrupted. This period scales as the square of the mean radius of the impacting system, and CDM subhalos are quite extended—at least an order of magnitude larger than globular clusters; thus, the disruption timescale of $\sim 10^9$ years is much shorter than for globular clusters. We therefore expect that the inner 20 kpc of the galactic halo could be smooth in configuration space, but rich in phase space. For the smallest substructures with sizes smaller than a few hundred parsecs, impulsive collisional heating due to encounters with disk stars dominates their mass loss. Over a Hubble time, most of their particles will be lost into small streams, although an inner dense and bound core could survive (Goerdt et al., 2006).

The importance of many of these processes remains to be quantified, and this will be an area of intense activity while simulators attempt to create realistic galaxies from cosmological initial conditions. These studies will also play an important role in precision cosmology, in which future observational missions plan to measure the power spectrum or mass distribution on large scales to percent-level precision, which requires a detailed knowledge of how baryons affect the global properties of their halos.

MORPHOLOGICAL EVOLUTION

During and after its formation, a galaxy can be transformed between morphological types through a variety of physical processes, thus creating the entire Hubble sequence and the observed diversity in galaxy types. The starting point for most scenarios for morphological evolution is a disk galaxy. However, once a disk has formed, a number of mechanisms can transform its morphology.

ELLIPTICAL GALAXIES

Nearby merging galaxies are spectacular as depicted in some examples taken from the SDSS in Figure 4.3. Such images have motivated generations of researchers to investigate the formation of elliptical galaxies via the merger of pairs of disk galaxies. The pioneering mechanical simulations of Holmberg (1941) used disk models, each represented by 37 light bulbs, to show that gravitational interactions could unbind stars and produce tidal tails like the observations. (If he had not fixed the inner seven bulbs in each model, Holmberg would have also discovered the bar instability!)* Holmberg's work was confirmed numerically by Toomre and Toomre (1972); however, it was another 10 years before the binary interactions between pairs of spiral galaxies were completely followed through the merger stage, resulting in elliptical-like systems (Gerhard and Fall, 1983).

It should be pointed out that the frequency of mergers today is rare (see Figure 4.4), and elliptical galaxies are almost as abundant at high redshifts as at low redshifts, suggesting that they formed very early. Investigating the origin of elliptical galaxies using mergers between spiral galaxies modeled on today's spiral galaxies is not re-creating the typical events that led to their formation. These

---

* See John Dubinski's home page for a numerical re-creation of the original experiment: http://www.galaxydynamics.org/gravitas.html.

FIGURE 4.3   SDSS images of a merger among three spiral galaxies (*left*), an elliptical-spiral merger (*middle*), and a triplet of merging elliptical galaxies (*right*). The left image is the UCG 09103 galaxy group, the galaxy centered in the middle image is NGC 2936, and the right image is the cluster CGCG 032-027.

studies need to be carried out in the appropriate cosmological context—from multiple mergers of gas-rich protogalaxies that one might typically observe in the Hubble Deep Field.

Dwarf Spheroidal (dSph) Galaxies

Galaxies within the dense environments of clusters and groups must have undergone some form of transformation given that nearby older systems are observed to be different from their younger, high-redshift counterparts. Understanding the processes that can affect galaxies in different environments is important if we wish to fully answer the question, what determines the structure and appearance of galaxies?

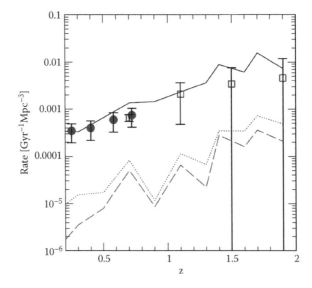

FIGURE 4.4   Total merger rate (major + minor) (*solid line*) and major merger rate only (*dotted line*) of dark matter halos, per unit volume (comoving), as a function of redshift, derived from cosmological simulations (D'Onghia et al., 2008). The rate per unit volume of mergers of Milky Way-sized halos identified at any time is plotted for comparison (*dashed line*). *Filled circles*: rate of mergers and interactions, with mass ratios in the range 1:1 to 1:10, derived by Jogee et al. from the GEMS galaxy evolution survey data. *Open squares*: merger rate from the Hubble Deep Field. (From Conselice, C.J., Bershady, M.A., Dickinson, M., et al., *Astronomical Journal*, 126, 1183–207, 2003.)

Not to be confused with M32-like dwarf elliptical (dE) galaxies, which are very rare, dSph galaxies are also triaxial systems whose shape is supported by the random motions of stars, but they have light profiles similar to those of exponential disks. In fact, dSph galaxies are the most common galaxy type in the Universe; every bright galaxy is surrounded by several dozen. Figure 4.5 shows images of galaxies in the Virgo Cluster taken from the SDSS. They show a possible evolutionary sequence that could take place as disk galaxies orbit in the cluster potential. A bar instability is driven in the existing stellar disk; gravitational interactions from high-speed encounters and from the global cluster potential remove the outer loosely bound stars. A "naked bar" remains, which is heated and becomes more spherical over time. A large fraction of the initial disk is stripped away and adds to the intracluster light component. The kinematics, appearance, light profiles, and abundance of these galaxies are reproduced in these simulations (Mastropietro et al., 2005). An unverified prediction of this scenario is that the dSph galaxies are embedded within very low surface brightness (LSB) streams of tidal debris.

Giant Low Surface Brightness (LSB) Galaxies

There is a small population of extremely LSB disk galaxies with very extended disks, exemplified by Malin 1. The origin of these systems was a puzzle in hierarchical models because the available angular momentum is insufficient to produce such large systems. Furthermore, how stars could form at 100 kpc from the halo center within such low-density gas provided an additional mystery. However, recently a new scenario for the origin of these galaxies has been proposed (Mapelli et al., 2008), which suggests that LSB galaxies are the late stage of the evolution of ring galaxies such as the Cartwheel. Such systems arise from the near head-on collision between a disk and a high-speed satellite, which creates an outward-expanding wave of debris (see Figure 4.6 for a comparison between an observed and simulated giant LSB galaxy). After several billion years, this material reaches distances larger than 100 kpc and has the same kinematical and photometric appearance

FIGURE 4.5 SDSS images of galaxies that I have selected from the Virgo Cluster. *Top left to bottom right*: This illustrates the possible transformation sequence from Sc to dSph due to gravitational interactions (galaxy harassment). (From Moore, B., Katz, N., Lake, G., et al., *Nature*, 379, 613–16, 1996.) The images are all to the same scale such that each galaxy is centered in a square 10 kpc across.

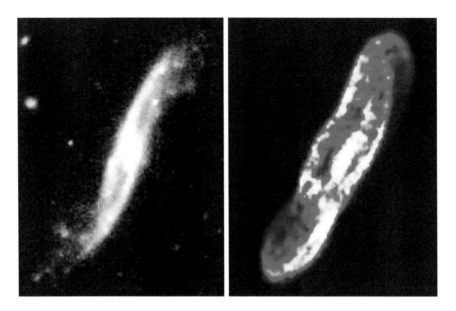

FIGURE 4.6 *Left*: UGC 7069, a giant LSB galaxy with a diameter larger than 120 kpc. *Right*: One of the evolved ring galaxies (shown to the same scale as UGC 7069) that resulted from an off-center collision between a disk galaxy and a satellite. This new model for the formation of Malin 1–type LSB galaxies can be tested by measuring the spatially resolved kinematics of LSB disks at large radii. (From Mapelli, M., Moore, B., Ripamonti, E., et al., *Monthly Notices of the Royal Astronomical Society*, 383, 1223–31, 2008.)

as LSB galaxies. Moreover, the abundance of observed LSB galaxies is consistent with an evolved population of observed ring galaxies.

S0 GALAXIES

An S0 galaxy is a disk galaxy that has no apparent structure in the disk component. It is very difficult to distinguish S0 galaxies from elliptical galaxies unless they are highly inclined—observers look by eye for a break in the projected light profile, a signature of an exponential disk in projection. Classification of S0 galaxies is more art than science, but spatially resolved kinematics could be used to distinguish S0 from elliptical populations.

S0 galaxies constitute a significant fraction of galaxies in clusters, and *Hubble Space Telescope* (*HST*) studies indicate that the population decreases at higher redshifts at the expense of a larger fraction of spiral galaxies. Indeed, the transformation of spiral galaxies to S0 galaxies is the most common proposed mechanism for their formation. One also finds these systems in the lower-density environments of groups and in the field. S0 galaxies in clusters usually contain no gas, but this is not true of more isolated examples. Two basic mechanisms make S0 galaxies, and it is likely that both of these are required: (1) removing the gaseous component from a disk galaxy by ram pressure stripping would leave a featureless disk after star formation is halted (Abadi et al., 1999; Quilis et al., 2000); and (2) heating the dominant stellar component in the disk would suppress the formation of stars and spiral patterns (Moore et al., 1999b). Simulations of the former mechanism show that all the gas can be rapidly removed from a plunging spiral galaxy by the intracluster medium. The latter could be accomplished via galaxy harassment (gravitational interactions) in clusters or by a minor merger with a companion galaxy (see the Figure 4.5 caption). In the field, only the last mechanism can be operating. If S0 galaxies originate from a population of spiral galaxies, then we should observe similar properties in the two populations: bulge-to-disk ratios, luminosity functions,

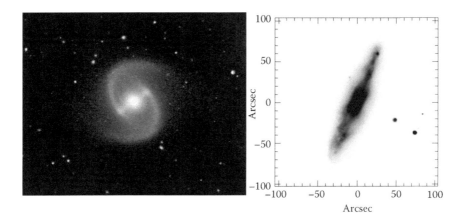

FIGURE 4.7 *Left:* A barred galaxy in the Virgo Cluster. *Right:* NGC 7582, an edge-on galaxy with a box/peanut-shaped bulge that most likely resulted from the buckling instability of an existing bar. (From Quilis, V., Moore, B., and Bower, R. (2000). Gone with the wind: The origin of S0 galaxies in clusters. *Science, 288:* 1617–620.)

and Tully–Fisher kinematical correlations. The literature contains varying degrees of support for these hypotheses.

BARS

About half of all spiral galaxies are barred (Figure 4.7), and the bar fraction is almost constant with redshift. The bar fraction decreases toward late-type spiral galaxies, suggesting that these disks are more stable, likely because of their low baryon fractions (Giordano et al., 2010). Bars can redistribute stars through angular momentum transfer, creating the downward breaks observed in the light profiles of disks at large radius. Bars can funnel gas to the nuclear region, providing fuel for BH accretion. Bars can also buckle, creating pseudo-bulges (Figure 4.7, *right panel*), a mechanism that does not rely on mergers to create a central spheroid. A characteristic of pseudo-bulges is their peanut shape—the Milky Way's bulge appears to have resulted from the secular evolution of a bar (Debattista et al., 2006). This has the interesting implication that the Milky Way formed as a pure disk galaxy even though it is a very massive system.

SUMMARY

The origin of galaxies is a hot topic in astrophysics today, with lots of existing data and much more on the way. In the next decade, ground- and space-based observations are aiming to detect structure formation at very high redshifts, even before the epoch of reionization. Such observations will provide strong constraints on our standard model for structure formation. Unfortunately, perhaps, there is no compelling alternative to the ΛCDM model; therefore, for the time being we can adopt the cosmologist's "standard model," which gives us the initial conditions within which to "create galaxies." The role of simulations and theoretical work is to see whether we can understand structure formation within this model.

Galaxy clusters provide some of the strongest evidence for the fact that we live in a hierarchical Universe. It is remarkable that, beginning with linear fluctuations in the matter component, such massive structures arise, each containing many thousands of galaxies. The galaxies must form before the cluster, merging hierarchically to create the final virialized systems—they certainly cannot form after the cluster because the baryons in the cluster are too hot to accrete into galactic halos and galaxies are moving too fast to merge together. This hierarchy is mimicked exactly within dark matter simulations. Cluster mass halos contain many thousands of galactic-mass halos, each a remnant of the merging hierarchy. The outer regions of these satellites are stripped away, contributing to the smooth mass distribution in the cluster. The central regions survive, and their kinematics and spatial distributions match the observations remarkably well.

Galactic halos themselves preserve the merging hierarchy in a self-similar way, each containing thousands of smaller dark matter substructures. There is some evidence for this from the small observed satellite population of galaxies; however, the number of observed substructures is tiny compared with galaxy clusters. On these scales, astrophysical processes can keep many of the substructures dark. However, if they could be detected, then this would provide the strongest evidence that we live in a CDM-dominated Universe. Their nondetection would be the signature that the power spectrum of fluctuations is truncated on small scales, such as might arise from a warm dark matter component.

Theoretical work is slowly catching up with the existing data, and it is interesting to ask the question, When will we know whether we have a successful theory of galaxy formation? This needs to be more than just a beauty contest. Simulations of galaxy formation in a cosmological context should be able to reproduce the diversity we see in the Universe and the scaling laws that galaxies satisfy as a function of redshift. I believe that such simulations will be carried out within this decade. Accurate numerical simulations are also needed to guide and interpret observational work that attempts to connect astrophysics with fundamental physics. For example, the cross-section between neutralinos and baryons may be much smaller than the experimental searches anticipate, and detecting dark matter in laboratory experiments may prove futile. In this case, astrophysical observations combined with numerical simulations may be the only way to constrain its nature. Likewise, measuring cosmological parameters to a high precision in order to constrain the properties of the dark energy requires equally precise predictions for how galaxies form and how baryons modify halo structure. Such surveys are likely to tell us a great deal about the details of the galaxy-formation process, perhaps of greater interest to astrophysicists than their original goals (White, 2007).

REFERENCES

Abadi, M.G., Moore, B., and Bower, R.G. (1999). Ram pressure stripping of spiral galaxies in clusters. *Monthly Notices of the Royal Astronomical Society*, 308: 947–54.

Abadi, M.G., Navarro, J.F., Steinmetz, M., et al. (2003). Simulations of galaxy formation in a lambda cold dark matter universe. I. Dynamical and photometric properties of a simulated disk galaxy. *Astrophysical Journal*, 591: 499–514.

Agertz, O., Moore, B., Stadel, J., et al. (2007). Fundamental differences between SPH and grid methods. *Monthly Notices of the Royal Astronomical Society*, 380: 963–78.

Agertz, O., Teyssier, R., and Moore, B. (2009). Disk formation and the origin of clumpy galaxies at high redshift. *Monthly Notices of the Royal Astronomical Society*, 397: L64–8.

Agertz, O., Teyssier, R., and Moore, B. (2011). The formation of disk galaxies in a ΛCDM universe. *Monthly Notices of the Royal Astronomical Society*, 410: 1391–1408.

Belokurov, V., et al. (2006). The Field of Streams: Sagittarius and its siblings. *Astrophysical Journal*, 642: L137–40.

Birnboim, Y. and Dekel, A. (2003). Virial shocks in galactic halos? *Monthly Notices of the Royal Astronomical Society*, 345: 349–64.

Blumenthal, G.R., Faber, S.M., Primack, J.R., et al. (1984). Formation of galaxies and large-scale structure with cold dark matter. *Nature*, 311: 517–25.

Bullock, J.S., Kolatt, R.S., Sigad, Y., et al. (2001). Profiles of dark halos: Evolution, scatter and environment. *Monthly Notices of the Royal Astronomical Society*, 321: 559–75.

Conselice, C.J., Bershady, M.A., Dickinson, M., et al. (2003). A direct measurement of major galaxy mergers at $z<3$. *Astronomical Journal*, 126: 1183–207.

Corbelli, E. and Salucci, P. (2000). The extended rotation curve and the dark matter halo of M33. *Monthly Notices of the Royal Astronomical Society*, 311: 441–47.

Davis, M., Efstathiou, G., Frenk, C.S., et al. (1985). The evolution of large-scale structure in a universe dominated by cold dark matter. *Astrophysical Journal*, 292: 371–94.

Debattista, V.P., Mayer, L., Carollo, C.M., et al. (2006). The secular evolution of disk structural parameters. *Astrophysical Journal*, 645: 209–27.

Dekel, A., and Birnboim, Y. (2006). Galaxy bimodality due to cold flows and shock heating. *Monthly Notices of the Royal Astronomical Society,* 368: 2–20.

Dekel, A., Birnboim, Y., Engel, G., et al. (2009). Cold streams in early massive hot halos as the main mode of galaxy formation. *Nature,* 457: 451–54.

Diemand, J., Kuhlen, M., and Madau, P. (2007). Formation and evolution of galaxy dark matter halos and their substructure. *Astrophysical Journal,* 667: 859–77.

Diemand, J., Kuhlen, M., Madau, P., et al. (2008). Clumps and streams in the local dark matter distribution. *Nature,* 454: 735–38

Diemand, J., Moore, B., and Stadel, J. (2005). Earth-mass dark-matter halos as the first structures in the early Universe. *Nature,* 433: 389–91.

D'Onghia, E., Mapelli, M., and Moore, B. (2008). Merger and ring galaxy formation rates at $z<2$. *Monthly Notices of the Royal Astronomical Society,* 389: 1275–283.

Dubinski, J., Berentzen, I., and Shlosman, I. (2009). Anatomy of the bar instability in cuspy dark matter halos. *Astrophysical Journal,* 697: 293–310.

Dubinski, J. and Carlberg, R.G. (1991). The structure of cold dark matter halos. *Astrophysical Journal,* 378: 496–503.

Frenk, C.S., White, S.D.M., Efstathiou, G., et al. (1985). Cold dark matter, the structure of galactic halos and the origin of the Hubble sequence. *Nature,* 317: 595–97.

Gerhard, O.E. and Fall, S.M. (1983). Tidal interactions of disc galaxies. *Monthly Notices of the Royal Astronomical Society,* 203: 1253–268.

Ghigna, S., Moore, B., Governato, F., et al. (1998). Dark matter halos within clusters. *Monthly Notices of the Royal Astronomical Society,* 300: 146–62.

Giordano, L., Tran, K.-V.H., Moore, B., et al. (2010). Multi-wavelength properties of barred galaxies in the local universe. I:Virgo Cluster. *Astrophysical Journal,* submitted (arXiv:1002.3167v1 [astro-ph.CO]).

Gnedin, N.Y. and Kravtsov, A.V. (2010). On the Kennicutt–Schmidt relation of low-metallicity high-redshift galaxies. *Astrophysical Journal,* 714: 287–95.

Gnedin, O.Y., Kravtsov, A.V., Klypin, A.A., et al. (2004). Response of dark matter halos to condensation of baryons: Cosmological simulations and improved adiabatic contraction model. *Astrophysical Journal,* 616: 16–26.

Goerdt, T., Moore, B., Read, J.I., et al. (2006). Does the Fornax dwarf spheroidal have a central cusp or core? *Monthly Notices of the Royal Astronomical Society,* 368: 1073–77.

Gondolo, P. and Silk, J. (1999). Dark matter annihilation at the Galactic center. *Physical Review Letters,* 83: 1719–722.

Governato, F., Brook, C., Mayer, L., et al. (2010). Bulgeless dwarf galaxies and dark matter cores from supernova-driven outflows. *Nature,* 463: 203–6.

Gunn, J.E. and Gott, J.R., III (1972). On the infall of matter into clusters of galaxies and some effects on their evolution. *Astrophysical Journal,* 176.

Holmberg, E. (1941). On the clustering tendencies among the nebulae. II. A study of encounters between laboratory models of stellar systems by a new integration procedure. *Astrophysical Journal,* 94: 385–95.

Ibata, R., Lewis, G.F., Irwin, M., et al. (2001). Great circle tidal streams: Evidence for a nearly spherical massive dark halo around the Milky Way. *Astrophysical Journal,* 551: 294–311.

Kaufmann, T., Mayer, L., Wadsley, J., et al. (2006). Cooling flows within galactic halos: The kinematics and properties of infalling multiphase gas. *Monthly Notices of the Royal Astronomical Society,* 370: 1612–622.

Kazantzidis, S., Kravtsov, A.V., Zentner, A.R., et al. (2004). The effect of gas cooling on the shapes of dark matter halos. *Astrophysical Journal,* 611: L73–6.

Keres, D., Katz, N., Weinberg, D.H., et al. (2005). How do galaxies get their gas? *Monthly Notices of the Royal Astronomical Society,* 363: 2–28.

Klypin, A., Zhao, H., and Somerville, R.S. (2002). Lambda CDM-based models for the Milky Way and M31. I. Dynamical models. *Astrophysical Journal,* 573: 597–613.

Kravtsov, A.V. (2010). Dark matter substructure and dwarf galactic satellites. *Advances in Astronomy,* 2010: 1–22.

Kravtsov, A.V., Gnedin, O.Y., and Klypin, A.A. (2004). The tumultuous lives of galactic dwarfs and the missing satellites problem. *Astrophysical Journal,* 609: 482–97.

Madau, P., Kuhlen, M., Diemand, J., et al. (2008). Fossil remnants of reionization in the halo of the Milky Way. *Astrophysical Journal,* 689: L41–4.

Mao, S. and Schneider, P. (1998). Evidence for substructure in lens galaxies? *Monthly Notices of the Royal Astronomical Society,* 295: 587–94.

Mapelli, M., Moore, B., Ripamonti, E., et al. (2008). Are ring galaxies the ancestors of giant low surface brightness galaxies? *Monthly Notices of the Royal Astronomical Society,* 383: 1223–231.

Mashchenko, S., Wadsley, J., and Couchman, H.M.P. (2008). Stellar feedback in dwarf galaxy formation. *Science,* 319: 174–7.

Mastropietro, C., Moore, B., Mayer, L., et al. (2005). Morphological evolution of discs in clusters. *Monthly Notices of the Royal Astronomical Society,* 364: 607–19.

Mateo, M.L. (1998). Dwarf galaxies of the Local Group. *Annual Review of Astronomy and Astrophysics,* 36: 435–506.

Mayer, L. and Moore, B. (2004). The baryonic mass-velocity relation: Clues to feedback processes during structure formation and the cosmic baryon inventory. *Monthly Notices of the Royal Astronomical Society,* 354: 477–84.

McGaugh, S.S., Schombert, J.M., Bothun, G.D., et al. (2000). The baryonic Tully–Fisher relation. *Astrophysical Journal,* 533: L99–102.

Moore, B., Diemand, J., Madau, P., et al. (2006). Globular clusters, satellite galaxies and stellar halos from early dark matter peaks. *Monthly Notices of the Royal Astronomical Society,* 368: 563–70.

Moore, B., Ghigna, S., Governato, F., et al. (1999a). Dark matter substructure within galactic halos. *Astrophysical Journal,* 524: L19–22.

Moore, B., Lake, G., Quinn, T., and Stadel, J. (1999b). On the survival and destruction of spiral galaxies in clusters. *Monthly Notices of the Royal Astronomical Society,* 304: 465–74.

Moore, B., Governato, F., Quinn, T., et al. (1998). Resolving the structure of cold dark matter halos. *Astrophysical Journal,* 499: L5.

Moore, B., Katz, N., Lake, G., et al. (1996). Galaxy harassment and the evolution of clusters of galaxies. *Nature,* 379: 613–16.

Navarro, J.F., Frenk, C.S., and White, S.D.M. (1997). A universal density profile from hierarchical clustering. *Astrophysical Journal,* 490: 493–508.

Okamoto, T., Eke, V.R., Frenk, C.S., et al. (2005). Effects of feedback on the morphology of galaxy discs. *Monthly Notices of the Royal Astronomical Society,* 363: 1299–314.

Peebles, P.J.E. (1982). Large-scale background temperature and mass fluctuations due to scale-invariant primeval perturbations. *Astrophysical Journal,* 263: L1–5.

Pohl, M., Englmaier, P., and Bissantz, N. (2008). Three-dimensional distribution of molecular gas in the barred Milky Way. *Astrophysical Journal,* 677: 283–91.

Quilis, V., Moore, B., and Bower, R. (2000). Gone with the wind: The origin of S0 galaxies in clusters. *Science,* 288: 1617–620.

Read, J.I., Lake, G., Agertz, O., et al. (2008). A dark disc in the Milky Way. *Astronomische Nachrichten,* 329: 1022–24.

Simon, J.D. and Geha, M. (2007). The kinematics of the ultra-faint Milky Way satellites: Solving the missing satellite problem. *Astrophysical Journal,* 670: 313–31.

Snowden, S.L., Egger, R., Finkbeiner, D.P., et al. (1998). Progress on establishing the spatial distribution of material responsible for the 1/4 keV soft x-ray diffuse background local and halo components. *Astrophysical Journal,* 493: 715–29.

Spergel, D.N., Bean, Doré, O., et al. (2007). *Wilkinson Microwave Anisotropy Probe (WMAP)* three year results: Implications for cosmology. *Astrophysical Journal Supplement Series,* 170: 377 (91 pages).

Stadel, J., Potter, D., Moore, B., et al. (2009). Quantifying the heart of darkness with GHALO—a multibillion particle simulation of a galactic halo. *Monthly Notices of the Royal Astronomical Society,* 398: L21–5.

Toomre, A. and Toomre, J. (1972). Galactic bridges and tails. *Astrophysical Journal,* 178: 623–66.

Wang, Q.D., Yao, Y., Tripp, T. M., et al. (2005). Warm-hot gas in and around the Milky Way: Detection and implications of O VII absorption toward LMC X-3. *Astrophysical Journal* 635: 386–95.

Weinmann, S.M., Macciò, A.V., Iliev, I.T., et al. (2007). Dependence of the local reionization history on halo mass and environment: Did Virgo reionize the Local Group? *Monthly Notices of the Royal Astronomical Society,* 381: 367–76.

White, S.D.M. (2007). Fundamentalist physics: Why dark energy is bad for astronomy. *Reports on Progress in Physics,* 70: 883–97.

White, S.D.M. and Rees, M.J. (1978). Core condensation in heavy halos—a two-stage theory for galaxy formation and clustering. *Monthly Notices of the Royal Astronomical Society,* 183: 341–58.

Zhao, H., Hooper, D., Angus, G.W., et al. (2007). Tidal disruption of the first dark matter haloes. *Astrophysical Journal,* 654: 697–701.

5 Structure Formation in the Universe: From the Dark Side to First Light

Naoki Yoshida

CONTENTS

Rich structures in the Universe such as stars, galaxies, and galaxy clusters have developed over a very long time. It is thought that our Universe is now 13.7 billion years old and that no such structures existed when it began with the Big Bang. Recent observations utilizing large, ground-based telescopes include distant galaxies and quasars that existed when the Universe was less than 1 billion years old. We can track the evolution of the cosmic structure from the present day all the way back to such an early epoch. Understanding how and when the cosmic structure grows is a major goal in modern cosmology and astronomy.

THE LARGE-SCALE STRUCTURE

THE EXPANDING UNIVERSE

Observations of the distribution of distant galaxies suggest that the Universe is approximately homogeneous and isotropic at very large length scales. The almost perfectly isotropic nature of the cosmic microwave background (CMB) radiation also indicates that the Universe *was* homogeneous and isotropic, although a variety of clumpy structures, such as galaxies, are seen in the local Universe. It is also found, for instance in the catalogs of galaxy redshift surveys, that there are some patterns or prominent "structures" that extend over millions of light-years (see Figure 5.1). These structures compose the large-scale structure of the Universe. They exist not only in the local Universe,

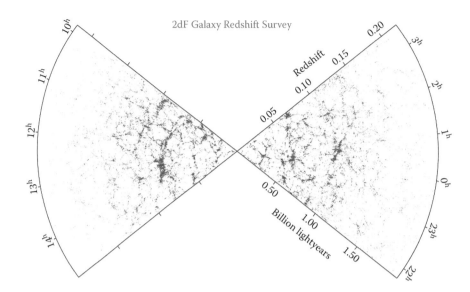

FIGURE 5.1   The distribution of galaxies probed by the 2-degree Field Galaxy Redshift Survey. (Adapted from Colless, M., Dalton, G., Maddox, S. et al., *Monthly Notices of the Royal Astronomical Society*, 328, 1039–63, 2001. With permission.)

but also essentially everywhere. Furthermore, recent observations of distant galaxies revealed that large-scale structure already existed when the age of the Universe was just one-tenth of what it is today. The Universe appears to have undergone the transition from a smoother state to the currently observed clumpy state quite rapidly, but details of this transition remain largely unknown.

The so-called standard theory of structure formation posits that the present-day clumpy appearance of the Universe developed through gravitational amplification of the initial matter density fluctuations. This basic picture is now supported by an array of observations, including the measurement of the CMB anisotropies. The CMB observation by NASA's *Wilkinson Microwave Anisotropy Probe (WMAP)* also confirmed that tiny density fluctuations were seeded at a very early epoch when the Universe went through a rapid expansion phase called "inflation" (see the chapter by Kallosh and Linde in this volume). These two elements—the inflationary Universe and gravitational evolution—form the basic framework of structure formation.

Because the Universe is *approximately* homogeneous, its overall expansion history can be described by a set of simple equations, which can be derived from Einstein's theory of general relativity. The equations contain just a handful of parameters, and, surprisingly, these cosmological parameters are now known with relative certainty.

The cosmological parameters describe the overall geometry of the Universe, the energy content, and the current rate of cosmic expansion. Recent observations are summarized as follows: Our Universe is dominated by two mysterious components called dark energy and dark matter. The former constitutes about three-fourth of the Universe's total energy, a large majority. The latter contributes about one-fourth of the energy and is largely responsible for structure formation. Ordinary elements such as hydrogen, carbon, and magnesium that we encounter in our daily lives constitute only 4% of the total energy density. These elements, however, compose our human body, and thus we are made of a very special "minority" of elements in the Universe. Thus, ordinary matter with which we are acquainted makes up only about 16% of all the matter in the Universe. We use the term "matter" to refer to the sum of these two components.

Let us now discuss the dynamic of an expanding universe by using an illustrative example. For a homogeneous and isotropic universe, the rate of its expansion is governed by a very simple equation:

$$H^2 = H_0^2 \, (a^{-3}B + C). \qquad (5.1)$$

Here, a is a factor that quantifies the extension of the Universe, called the cosmic expansion factor; $H = \dot{a}/a$ is the rate of change of a, with H_0 being the present-day value; B is the present-day matter density; and C is that of dark energy. Simply, the right-hand side implies that the overall expansion is determined by a balance between matter and dark energy. The reason why the term B appears with a factor a^{-3} can easily be understood by noting that matter density has a dimension of mass over length to the power of 3; matter density changes at a rate inversely proportional to volume and thus decreases rapidly as the Universe expands. In contrast, the dark energy term C does not contain a factor of a, which means that the density of dark energy remains constant. It is this feature that is responsible for dark energy's dominance in the later evolution of the Universe.

The dynamics of cosmic expansion can be understood easily by looking at a few simplified models. First, let us ignore dark energy and imagine a matter-dominated universe. By setting $B = 1$, $C = 0$, we obtain

$$\frac{\mathrm{d}a}{\mathrm{d}t} = \frac{1}{a^{1/2}}, \qquad (5.2)$$

which can be readily integrated to yield a solution,

$$a = t^{2/3}. \qquad (5.3)$$

Here, and in the following, we ignore constant coefficients in equations. A matter-dominated universe expands in time at a rate proportional to $t^{2/3}$. The expansion factor grows to 1.6 times the original value when t gets 2 times larger, becomes 4.6 times when t gets 10 times larger, and so on. This is not a surprisingly rapid rate. Next, let us consider a universe dominated by dark energy. By setting $B = 0$, $C = 1$, we obtain

$$\frac{\mathrm{d}a}{\mathrm{d}t} = a, \qquad (5.4)$$

which has a solution

$$a = \exp(t). \qquad (5.5)$$

This shows that a dark energy-dominated universe expands exponentially in time. How rapid is it? The expansion factor grows 7.4 times when t gets 2 times larger; 22,000 times when t gets 10 times larger; and, when t gets 1,000 times larger, a becomes so large that the number does not fit on this page (10^{434}). This is a defining characteristic of a dark energy-dominated universe: it expands exponentially. The actual evolution of a universe that contains both dark matter and dark energy is more complicated, but it is qualitatively explained by a combination of the above two simple cases. As mentioned earlier, various observations suggest that our Universe has $B \sim 1/4$ and $C \sim 3/4$, and so it is dominated by dark energy. Our Universe is currently in a rapid expansion phase.

The model incorporating both dark energy and dark matter is now credibly applied to the evolution of our Universe. Structure formation is seeded by the initial density fluctuations and is driven by gravitational instability. Given these starting points, and aided by large-scale computer simulations, theoretical models are now able to make accurate predictions for a variety of phenomena, from galaxy clustering to gravitational lensing statistics.

PROBING THE LARGE-SCALE STRUCTURE

The distribution of galaxies in the sky has been used for studies of the large-scale structure of our Universe for many years. Prominent clustering features were found in the projected galaxy distribution in the Lick catalog compiled in the 1960s. Galaxy redshift surveys added the 3rd dimension, in terms of redshift, by which one can make a full 3-dimensional map of the galaxy distribution. Statistical methods such as 2-point correlation functions and the power spectrum are most often used to quantify the clustering of galaxies (Springel et al., 2006). The 2-point correlation function is a measure of the number of pairs (of galaxies) at a certain separation distance and describes how "crowded" the distribution of galaxies is. This quantity as a function of relative distance gives a complete description of galaxy distribution, against which predictions from theoretical models are tested.

Nowadays, the basic statistics are measured precisely, and thus they can be used to determine cosmological parameters because different cosmological models predict slightly different shapes and amplitudes for the correlation function. The two most recent galaxy redshift surveys, the 2-degree Field (2dF) Survey and the Sloan Digital Sky Survey (SDSS), are providing unprecedented data in terms of both quality and quantity.

Detailed analyses of these data suggest consistently that dark matter is "cold," that is, that dark matter particles were almost at rest at early epochs (see the chapter by Bloom in this volume). The cold dark matter (CDM) model can be used to predict the formation of large-scale structure. An illustrative example is given in Figures 5.2 and 5.3. These figures compare the large-scale structure in a virtual universe with one in the real Universe. Figure 5.3 shows the galaxy-to-galaxy, two-point correlation functions derived from Virgo Consortium's Millennium Simulation, a large cosmological simulation in which a sophisticated galaxy-formation model is implemented. The result is compared with the observed correlation functions from the 2dF Survey. The close agreement is a triumph of the standard CDM model.

Extremely large-scale structures of size greater than ~100 Mpc* are rarely formed in the standard model. The Harvard–Smithsonian Center for Astrophysics (HSCfA) "Great Wall" extends over ~200 Mpc; a feature found in the Sloan Digital Sky Survey extends over twice this length (see Figure 5.2). In computer simulations, only a few such large-scale structures are found in a large volume. Thus, existence of the largest-scale structure may challenge the standard model, if commonly observed in the local and distant Universe. Planned future observations will probe a larger area of space and serve to clarify whether a larger-scale structure exists in our Universe.

While the distribution of galaxies provides an overall picture of matter distribution, there remains the complex issue of "bias." One usually assumes that galaxies accurately trace underlying mass. However, galaxies may be located in only high-density regions, hence the distribution may be somewhat biased. Estimating bias with respect to the underlying mass is nontrivial. Bias could depend on length, scale, and time, and could be nonlinear with respect to the local mass density. Hence, it would be ideal if one could directly map the matter distribution in some observational way.

A magnificent method known as gravitational lensing provides a way of estimating this mass distribution. In the next section, we review recent progress in observations of gravitational lensing and discuss some future prospects.

DARK MATTER DISTRIBUTION

Gravitational lensing provides a unique and powerful method to map the distribution of dark matter. Light from distant galaxies is deflected by foreground matter that lies between the galaxies and us. This results in the shapes of the galaxies appearing significantly deformed (see Figure 5.4). By

---

* 1 parsec (pc) is about 3.24 light-years (lt-yr); 1 megaparsec (Mpc) is 1 million pc.

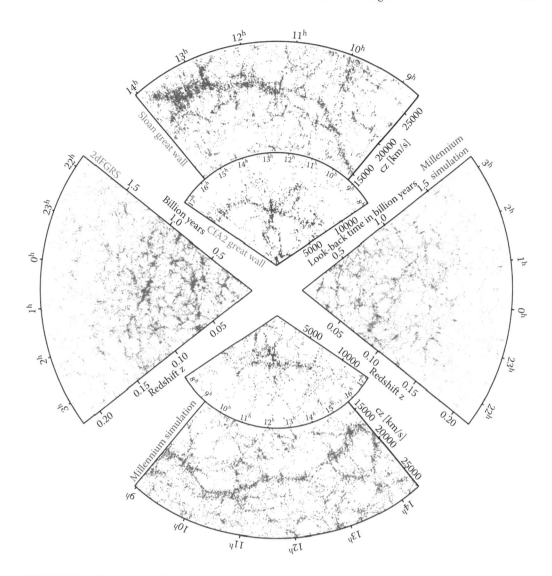

FIGURE 5.2 Simulations of large-scale structure formation. The "pie" diagrams compare the distribution of galaxies in the real Universe (left, upper "pies") and in a simulated universe (right, lower "pies"). (From Springel, V., Frenk, C.S., and White, S.D.M., *Nature*, 440, 1137–44, 2006. With permission.)

knowing that it is a geometrical effect, one can construct a reverse procedure. The foreground mass distribution can be, in principle, inferred from the degree of deformation of galaxy shapes. In practice, one needs to observe a number of background galaxies using a large, ground-based telescope or a space telescope.

Weak gravitational lensing is a technique that evaluates the alignment of images of background galaxies to map the large-scale mass distribution. A number of weak-lensing surveys have already been conducted, and larger area surveys are currently occurring. The statistics of the matter distribution can be used to determine cosmological parameters in the same way as galaxy distributions are used. Indeed, the next-generation weak-lensing surveys are expected to provide the most precise measurements of the matter power spectrum.

Recently, it has been proven that it is feasible to locate clusters of galaxies as matter density enhancements using weak gravitational lensing. Conventionally, galaxy clusters are found either by x-ray observations or by optical surveys. An advantage of gravitational lensing is that the obtained

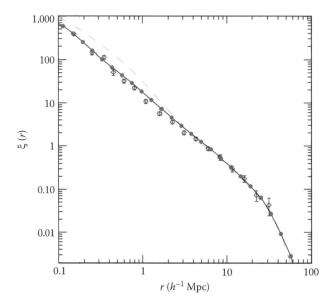

FIGURE 5.3 The 2-point correlation functions showing how "crowded" the galaxies' distribution is. The "crowdedness" is shown as a function of separation length for a pair of galaxies. The solid line is the result from a simulated model universe, whereas the open circles with small error bars are from the real galaxy distribution. The dashed line is the correlation function for matter. There is a clear difference between the matter distribution and the galaxy distribution at small separation lengths. (From Springel, V., White, S. D. M., Jenkins, A. et al., *Nature*, 435, 629–36, 2005. With permission.)

FIGURE 5.4 Galaxy cluster Abell 1689 observed by the *Hubble Space Telescope*. (From *HST* website http://hubblesite.org/. See http://hubblesite.org/gallery/album/entire/pr2003001a/.)

galaxy cluster sample is not biased toward luminous systems. Optical- or x-ray-selected catalogs suffer from the unavoidable bias effect.

Figure 5.5 shows the weak-lensing mass map obtained by the Subaru Suprime 2 Square Degree Survey. It displays the projected mass density in a small patch of the sky. This map contains 14 density peaks (dense regions), among which 11 peaks are confirmed to be galaxy clusters by subsequent optical observations. Of these, seven peaks are newly discovered clusters, demonstrating the ability of weak-lensing surveys to locate new galaxy clusters. In principle, the cluster number counts can be directly compared with model predictions or results obtained from numerical simulations and, thus, can be used to put strong constraints on cosmological models. While the number of samples in the Subaru survey is still too small, it is consistent with the prediction from the standard Lambda-CDM (ΛCDM) model. A significantly larger sample will enable us to estimate how rapidly structure grew in the past billion years. The simple statistics of the number-density evolution can be used to infer the nature of dark energy.

It is also possible to map the 3-dimensional distribution of matter by gravitational lensing observations. Foreground matter deflects light from galaxies at different distances. When selecting source galaxies by their distances from us, one can obtain the mass distribution in a certain range of distance. This technique is called lensing tomography. Recently, a group of astronomers used the data from the COSMOS survey to conduct lensing tomography. The 3-dimensional mass distribution obtained from this displayed complex morphology. This confirmed that light traces matter, meaning that the distribution of galaxies and that of mass are indeed similar.

Planned lensing observations will utilize the Subaru-HyperSprimeCam (HSC), Panoramic Survey Telescope and Rapid Response System (Pan-STARRS), and the Large Synoptic Survey Telescope (LSST) to conduct wide-field lensing tomography surveys. These will work as ultimate

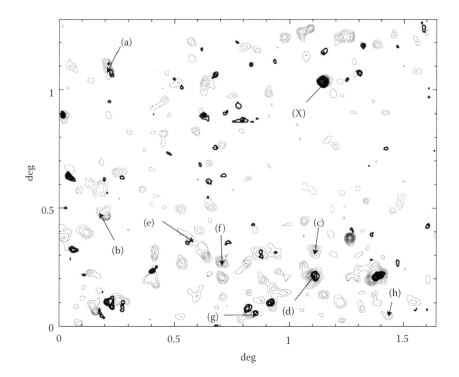

FIGURE 5.5 The weak-lensing mass map obtained by the Subaru Suprime-Cam 2 Square-Degree Field survey. The area covers a 2.1 deg^2 field, in which 14 significant peaks (galaxy clusters) are found. (From Miyazaki, S., Hamana, T., Shimasaku, K. et al., *Astrophysical Journal*, 580, L97–100, 2002.)

"dark matter telescopes," which can probe large-scale distributions of matter in the Universe in an unbiased manner.

THE FIRST STARS

The preceding section introduced the large-scale structure of the Universe and explained how the structure formed and evolved in an expanding universe. Following is a discussion of how smaller structures—the first stars in the Universe—arose.

THE DARK AGES

Current-generation telescopes discovered galaxies, quasars, and gamma-ray bursts (GRBs) at very large distances, i.e., those in the distant past. The Universe at a much younger age can be seen almost directly as the (CMB) radiation. Between these two epochs lies the remaining frontier of astronomy, when the Universe was about several hundred million years old. The epoch is called the cosmic Dark Ages.

Shortly after the cosmological recombination epoch when hydrogen atoms were formed and photons were released, the optical photons disappeared. At this point, the Universe would have appeared completely dark to human eyes. A long time had to pass until the first stars were born, which then illuminated the Universe once again and terminated the Dark Ages. The first stars are thought to be the first sources of light, as well as the first sources of heavy elements that enabled ordinary stellar populations, planets, and, ultimately, life to emerge (Bromm et al., 2009).

In recent years, there have been a number of theoretical studies on this unrevealed era in cosmic history. Perhaps the current fascination with such studies is due to recent progress in observational astronomy. Existing telescopes are already probing early epochs close to the end of the Dark Ages, and planned observational programs are aimed at direct detection of light from stars and galaxies farther away.

For now, theoretical studies can potentially describe the detailed process of the formation of the first objects for two main reasons: (1) the initial conditions, as determined cosmologically, are well established, so that statistically equivalent realizations of a standard model universe can be accurately generated; and (2) the requisite theories of physics, such as gravitation, hydrodynamics, and atomic and molecular processes in a hydrogen-helium gas, are understood. In principle, therefore, it is possible to make solid theoretical predictions for the formation of early structure and of the first stars in the standard model universe. Computer simulations are often used to tackle the nonlinear problems of structure formation.

Recent progress in observations and some key physical processes are discussed below, followed by the results from state-of-the-art simulations of early cosmic structure formation.

RECENT PROGRESS IN OBSERVATIONS

Large, ground-based telescopes have discovered very luminous objects—quasars—that existed when the Universe was less than 1 billion years old (about 5% of its current age). Moreover, further observations have confirmed that some stars must have been present. For example, the most distant known quasar, SDSS J1148+5251, contains substantial amounts of heavy elements such as carbon, oxygen, and iron, as well as dust grains. These heavy elements are not of cosmic origin, but they must have been formed earlier in massive stars before being expelled by supernovae (SNe) and stellar winds and then incorporated into the material that later condensed to produce this quasar. This is a significant piece of evidence that indicates the early formation of stars.

Astronomers have been making a continuous effort to find the first-generation stars in our Galaxy. It has been only recently that stars with little heavy-element content were discovered within the Milky Way (Frebel et al., 2005). The current record-holder contains less than 0.00001% of the iron in the Sun (Figure 5.6). The observed elemental abundance patterns indicate several possibilities for

FIGURE 5.6 An "oldest" star in constellation Hydra. The star contains an extremely small amount of iron and hence mostly consists of hydrogen and helium, indicating that it is one of the early generation of stars. (From National Astronomical Observatory of Japan website http://www.naoj.org/Pressrelease/2005/04/13/ HE0 107.jpg. With permission.)

the nature of their *ancestors*, including an interesting scenario in which supernova (SN) explosions of massive primordial stars enriched the parent gas clouds from which these halo stars were born. Assuming this to be the case, we can derive important properties of the first generation of stars in the Universe.

There is another potential source of information about early star formation in the Universe. The CMB comes almost directly from the distant region called the last scattering surface. However, a small fraction of CMB photons are scattered again by electrons in the intergalactic space between it and us. The portion of photons that are "rescattered" is roughly proportional to the integrated number of electrons in intergalactic space along a line of sight. Because the Universe was almost neutral, electrons in intergalactic space must have been ejected from atoms by some physical mechanisms. A plausible explanation is that an early generation of stars emitted ultraviolet (UV) photons, and these high-energy photons ionized hydrogen and helium in the intergalactic medium. According to a recent measurement of CMB polarization by the *WMAP* satellite, this "rescattered" portion of photons is about 10% of the CMB, which means that ionization of the intergalactic medium occurred a few hundred million years into the Universe's development.

Altogether, these observations suggest an early episode of star formation. It is the time for theorists to propose a viable scenario for early star formation within the standard cosmological model.

HIERARCHICAL STRUCTURE FORMATION

The generic feature of structure formation in the standard model is based on weakly interacting CDM. The primordial density fluctuations predicted by popular inflationary universe models have very simple characteristics. The fluctuations are described by a random field (just like noise) and have a nearly scale-invariant power spectrum, which means that no particular size is specified for structures to form.

The power spectrum is a measure of fluctuation amplitudes as a function of wavelength. Let us imagine that the distribution of matter is described by a superposition of many sinusoidal waves with different wavelengths. If the amplitude is larger for larger wavelength modes, the resulting matter distribution would look very smooth, whereas in the opposite case, the matter distribution would be clumpy. Hence, the power spectrum essentially describes the relative "power" of modulations of different wavelengths.

The power spectrum predicted by the standard CDM model has progressively larger amplitudes on smaller length scales. Hence, structure formation is expected to proceed in a "bottom-up" manner, with smaller objects forming first.

It is useful to work with a properly defined mass variance to obtain the essence of hierarchical structure formation. To this end, we need to use a few mathematical formulae, in order to determine the amplitude of density fluctuations of a given size. The mass variance is defined by a weighted integral of the power spectrum $P(k)$ as

$$\sigma^2(M) = \frac{1}{2\pi^2} \int P(k) \left[W(kR) \right]^2 k^2 dk, \tag{5.6}$$

where the window function $W(kR)$ describes an appropriate weight for a given mass M. Essentially, $\sigma(M)$ describes if and when objects with M can form. There is a threshold over density for gravitational collapse. When $\sigma(M)$ exceeds the threshold, fluctuations corresponding to the mass M can grow nonlinearly to collapse. Figure 5.7 shows the mass variance and the collapse threshold at $z = 0, 5$, and 20. At $z = 20$ (when the Universe was about 200 million years old), the mass of a clump at a rare 3-sigma (3σ) density peak is just about 10^6 solar masses, or $10^6 M_\odot$. Clumps with this mass will start to collapse at this point. As will be shown later, this is the characteristic mass of the first objects in which the cosmic primordial gas could cool and condense by molecular hydrogen cooling.

The mass variance is sensitive to the shape of the initial power spectrum. For instance, in cosmological models that assume that dark matter particles move fast, the power spectrum has an exponential cutoff at the free-streaming scale of the particles, and the mass variance at the corresponding small mass scale is significantly reduced. In such models, early structure formation is effectively delayed; hence, small nonlinear objects form later than in the standard CDM model. Thus, the formation epoch of the first objects has a direct link to the nature of dark matter and the physics of inflation.

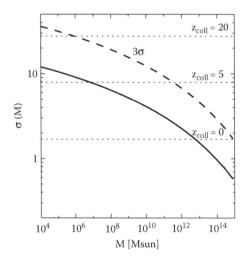

FIGURE 5.7 Mass variance and collapse thresholds for the standard cosmological model. The cross points of the two curves and three horizontal lines (for $z = 0, 5, 20$) indicate the characteristic masses that collapse to form gravitationally bound objects at the respective epochs. For example, a high-density ("3σ") peak will collapse to form a $10^6 M_\odot$ object at $z = 20$.

Formation of the First Cosmological Objects

The basics of the formation of nonlinear dark matter clumps are easily understood; clumps form by gravitational instability from initial density fluctuations. Because of its hierarchical nature, dark matter forms clumps in essentially the same way regardless of mass and the formation epoch. The first "dark" objects are thus well defined and are, indeed, clumps of a very small mass set by the initial random motion of dark matter particles.

The formation of the first "luminous" objects involved a number of physical processes, and this is much more complicated. The study of the evolution of primordial gas in the early Universe and the origin of the first baryonic objects has a long history. Recent development in theoretical astrophysics and the emergence of the standard cosmological model have enabled us to ask more specific questions, such as "When did the first objects form?" and "What is the characteristic mass?"

Sophisticated computer simulations of early structure formation show that this process likely began within the first 100 million years of the Universe's development. In these simulations, dense, cold clouds of self-gravitating molecular gas develop in the center of small dark matter clumps and contract into protostellar clouds with masses in the range ~100–1,000 $M_\odot$. Figure 5.8 shows the projected gas distribution in a cosmological simulation that includes hydrodynamics and primordial gas chemistry. Star-forming gas clouds are found at the knots of filaments, resembling the large-scale structure of the Universe (see the first section, above), although much smaller in length and mass. This results in the hierarchical nature of structure in the CDM Universe.

Unlike the formation of dark matter clumps, which is solely driven by gravity, star formation involves several major processes schematically shown in Figure 5.9 and described as follows: For

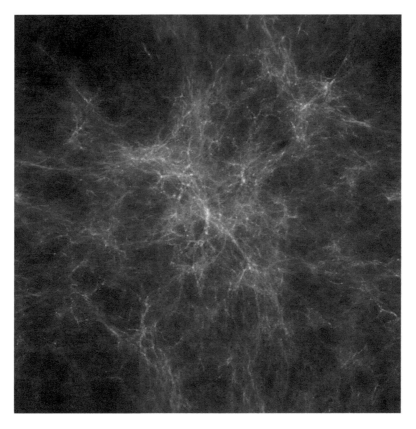

FIGURE 5.8   The projected gas distribution at $z = 17$ in a cubic volume of 100,000 lt-yr on a side. The cooled dense gas clouds appear as bright spots at the intersections of the filamentary structures.

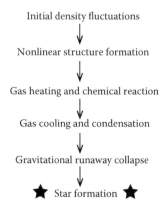

FIGURE 5.9   Five steps for star formation in the early Universe.

star formation to begin in the early Universe, a sufficient amount of cold dense gas must accumulate in a dark matter clump. The primordial gas in its original composition cannot efficiently cool radiatively. Trace amounts of molecular hydrogen (H_2) can form via a sequence of chemical reactions:

$$\text{Hydrogen} + \text{Electron} \rightarrow \text{Hydrogen ion} + \text{Photon}, \qquad (5.7)$$

$$\text{Hydrogen ion} + \text{Hydrogen} \rightarrow \text{Hydrogen molecule} + \text{Electron}. \qquad (5.8)$$

Here, a hydrogen ion is a negatively charged ion, not a proton. H_2 molecules, once formed, can change their quantum rotational and vibrational levels by emitting photons. This allows the gas to lose energy (and, hence, to become cooler), causing it to condense into a dense gas cloud.

The critical temperature for this to occur in a gas cloud is ~1,000 K, several times room temperature. This critical temperature is determined by the condition that a sufficient number of molecules are formed before electrons are depleted in the gas by recombining with protons. Remarkably, this simple argument provides an accurate estimate. There is an important dynamical effect, however. The gas in clumps that grow rapidly (primarily by mergers) is unable to cool efficiently because of the effects of gravitational and gas dynamical heating. In this interplay, the formation process is significantly affected by the dynamics of the gravitational collapse of dark matter.

THE ROLE OF DARK MATTER AND DARK ENERGY

When and how primordial gas clouds are formed is critically affected by the particle properties of dark matter, by the shape and the amplitude of the initial density perturbations, and by the overall expansion history of the Universe. Here, we introduce two illustrative examples: a cosmological model in which dark matter is assumed to be "warm" and another model in which dark energy properties change as a function of time.

If dark matter particles are light, they move with substantial random velocities (hence, they are "warm," rather than "cold"). In this case, the matter power spectrum has a sudden cutoff at a particular length scale. This results in the corresponding mass variance at small mass being significantly reduced (see Equation 5.6), indicating that there is little density perturbation at small wavelengths. The effect is clearly seen in Figure 5.10. The gas distribution is much smoother in the warm dark matter (WDM) model. Dense gas clouds (tantamount to star nurseries) are formed in filamentary shapes, rather than being embedded in dark matter clumps. It is expected that stars are formed along the filaments. Filamentary gas clouds are intrinsically unstable and can lead to the formation

Cold dark matter   Warm dark matter

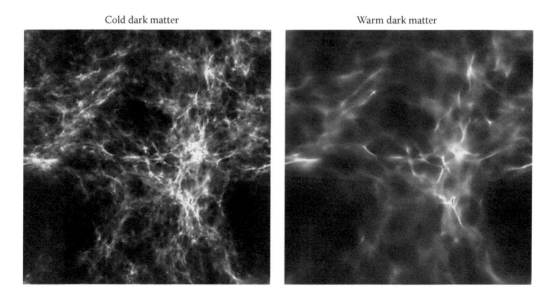

FIGURE 5.10   The projected gas distribution in the early Universe for the standard cold dark matter (CDM) model (left) and for a warm dark matter (WDM) model (right). The region shown has a side length of 100,000 lt-yr on a side. We see much smoother matter distribution in the WDM model, in which stars are formed in the prominent filamentary structure.

of multiple stars at once. Currently, it is not possible to determine which model is correct (CDM or WDM), but future observations of the first stars and their signatures might help us to better understand the nature of dark matter.

Dark matter particles might affect primordial star formation in a very different way. A popular candidate for describing the constitution of dark matter are supersymmetric particles—neutralinos, for instance. Neutralinos are predicted to have a large cross section for pair annihilation; when two neutralinos collide, they will eventually be converted to high-energy photons, electrons, and neutrinos. When the particle annihilation takes place in very dense gas clouds, the produced photons and electrons can be captured by the gas in a variety of ways, resulting in an effective heating of the gas clouds. Because primordial gas clouds are formed at the center of dark matter clumps, where the dark matter density is very large, the dark matter annihilation rate and resulting energy input can be significant. The additional heat supplied by dark matter can even halt gravitational collapse of the gas cloud. In this case, it is unlikely that stars will be formed. Although the net effect of dark matter annihilation remains highly uncertain, it would be necessary to include the gas heating mechanism if neutralinos are detected in laboratories and are proved to be the bulk of dark matter.

Dark energy affects the formation of the first stars in an indirect way. The growth rate of density perturbations is a function of the cosmic expansion parameter, which is determined by the energy content of the Universe. Most generally, the energy density of dark energy can be written as

$$\rho_{DE}(a) \propto \exp\left[\int^{a} -3\frac{da'}{a'}\left(1+\omega(a')\right)\right] \tag{5.9}$$

where a is the cosmic expansion parameter, and $w(a)$ defines the effective equation of state of dark energy via $P_{DE} = w\rho_{DE}$; the equation of state describes how pressure is related to density. For a gas consisting of ordinary matter (e.g., air), the pressure is proportional to the density such that $P \propto \rho$. Dark energy has a strange property in that it has a negative pressure with $w < 0$. It effectively causes a repulsive force throughout the Universe. Hence, it *accelerates* the cosmic expansion. For the simplest model of dark energy (i.e., Einstein's cosmological constant with $w = -1$), cosmic expansion is accelerated only at late epochs, say, in the most recent several billion years, which seems unimportant for early structure formation. However, some dark energy models predict a time-dependent equation of state, which effectively shifts the formation epoch of early-generation stars to early or later epochs. The number of star-forming gas clouds at a given time varies significantly for an evolving dark energy model, even when the models are constructed such that the present-day large-scale structure is the same.

It is, in principle, possible to infer how early cosmic reionization began from direct observations, but currently available data provide only an integral constraint. We will need to wait for a long time until future radio observations map out the distribution of neutral hydrogen in the early Universe by detecting redshifted 21 cm emission. Such observations will ultimately determine the abundance of the sources of re-ionization and will tell us about how rapidly structures grew during the first billion years.

FORMATION OF THE FIRST STARS

This section provides more details about the formation process of the first-generation stars. The dynamics of primordial gas cloud collapse have been studied extensively over the past few decades. According to the theory, primordial stars are formed in an "inside-out" manner, where a tiny protostar forms first and then grows by accreting the surrounding gas.

Impressive progress in computing power in the past few decades, coupled with the development of sophisticated algorithms, have made it possible to perform an *ab initio* simulation of primordial star formation (Yoshida et al., 2008). Recently, the formation of a primordial protostar* has been described in a full cosmological context (see Figure 5.11). The simulation has an extraordinary spatial resolution, so that the highest gas densities reach "stellar" density (~0.01 g/cm³), and thus it offers a complete picture of how the first protostars form from primeval density fluctuations left over from the Big Bang. Unlike most simulations of star formation, this particular simulation does not assume any *a priori* equation of state for the gas. The thermal and chemical evolution is fully determined by molecular and atomic processes, which are treated in a direct, self-consistent manner. The simulation is more like an early universe "experiment," rather than a simulation of star formation. The formation of a star is not designated, but it is the natural outcome of many physical processes.

The entire formation process is described as follows: Small dark clumps of about 1 million solar masses assembled when the Universe was a few million years old. Through the action of gravity and

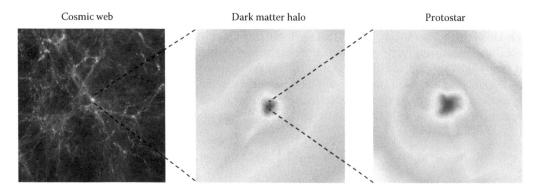

FIGURE 5.11 Projected matter distribution around the protostar. Shown regions are, from left to right, the large-scale gas distribution (45,000 lt-yr on a side), the dark matter halo (1,500 lt-yr on a side), and the newborn protostar (17,000,000 km on a side).

---

* A protostar is the beginning phase of stellar evolution. It is typically much smaller than an ordinary star.

radiative cooling, star-forming gas clouds collected in their host dark clump. Figure 5.11 shows the projected gas density in and around the prestellar gas cloud. Note that the figures were made from a single simulation that covers a very large dynamic range of ~10^{13} in length scale. It is equivalent to resolving a centimeter-sized object in the Solar System. In Figure 5.11, we see substantial variations in density and temperature even in the innermost 10 solar-radii region around the newly formed protostar, or 6.95 million kilometers.

Through a series of chemical reactions and cooling, the cloud contracted roughly isothermally, so that the density increased over 20 orders of magnitude; but the temperature increased by only a factor of 10. The strong thermal pressure at the center finally stopped the gravitational collapse, and hydrodynamic shock waves were generated (see the right panel in Figure 5.11). This resulted in the birth of the first protostar, having a mass equivalent to 1% of the Sun's mass. This was just the beginning, however. The protostar would quickly grow to be a big star.

On the assumption that there is only one stellar seed (protostar) at the center of the parent gas cloud, the subsequent protostellar evolution can be calculated using the standard theory of star formation. A critical quantity here is gas mass accretion rate—how rapidly the star can grow in mass. A primordial protostar has a very large accretion rate, greater than one-hundredth of a solar mass per year. It is easy to infer that, with this large accretion rate, a protostar can grow quickly to become a massive star. The resulting mass can be as large as 100 times that of the Sun.

Overall, the lack of vigorous fragmentation, the large gas mass accretion rate (the star's growth rate), and the absence of dust grains in the accreting gas provide favorable conditions for the formation of massive, even very massive, stars in the early Universe.

THE FIRST LIGHT

The birth and death of the first generation of stars have important implications for the evolution of the intergalactic gas in the early Universe. At the end of the Dark Ages, the neutral, chemically pristine gas was reionized* by the first stars, but it was also enriched with heavy elements when these stars ended their lives as energetic SNe. The importance of SN explosions, for instance, can be easily appreciated by noting that only light elements (hydrogen, helium, and a very small amount of lithium) were produced during the nucleosynthesis phase in the early Universe. Chemical elements heavier than lithium are thus thought to be produced exclusively through stellar nucleosynthesis, and they must have been expelled by SNe or by stellar winds if we are to account for various observations of galaxies and of the intergalactic medium that contains heavy elements.

Altogether, the various influences that the first stars might have caused are called "feedback effects." Feedback may have played a crucial role in the formation of primeval galaxies. Here, two important effects are introduced, and a few unsolved problems are highlighted.

The first feedback effect is caused by radiation from the first stars. Stars emit radiation in a wide range of energy, from infrared (IR) to UV, and even to x-ray. Because photons with different energies interact with a gas differently, both negative and positive effects, in terms of star formation, can be caused by radiation from the first stars.

UV photons are capable of ionizing hydrogen by a reaction

$$\text{Hydrogen atom} + \text{Ultraviolet photon} \rightarrow \text{Proton} + \text{Electron}. \tag{5.10}$$

Individual stars emit a number of UV photons and ionize a small volume of gas surrounding them. This is the initial stage of cosmic reionization. The ionized patches around stars are called

---

* The cosmic gas was initially fully ionized, being in a plasma state, but became neutral at the recombination epoch when the age of the Universe was 380,000 years old. Some time later, the gas was ionized once again by the first stars; hence, the use of the term "reionization" to describe the event.

H II regions.* Early H II regions are different from present-day H II regions in two ways: (1) the first stars and their parent gas cloud are hosted by a dark matter clump, and the gravitational force exerted by dark matter keeps the surrounding gas dense and physically compact; (2) the gas density profile around the first star is concentrated, meaning that the gas density is very large near the star, but is much lower in the outskirts. These two conditions make the evolution different from that of present-day local H II regions.

Figure 5.12 shows the structure of an early H II region in a cosmological simulation. The star-forming region is a dense molecular gas cloud within a small mass (~$10^6 \, M_\odot$ dark matter clump). A single massive primordial star with $M = 100M_\odot$ is embedded at the center. The formation of the H II region is characterized by slow expansion of an ionization front (I-front) near the center initially, followed by rapid propagation of the I-front throughout the outer gas envelope. Photoionization effectively heats the gas. Hence, reionization may be said to be "reheating." The heated gas rapidly escapes out of a small dark matter clump. It takes a very long time for the gas to come back, cool, and condense once again to form a star-forming gas cloud.

A second feedback effect comes from the fact that UV photons with smaller energies might have played a very different role. If the formation of H_2 is strongly suppressed by a far-UV background radiation, star formation proceeds in a quite different manner, possibly leading to black hole (BH) formation. Under a strong far-UV radiation field, a primordial gas cloud cools and condenses nearly isothermally by atomic cooling, which operates at higher temperatures (~8,000 K) than molecular cooling does. Gas cloud collapse under these conditions can be very rapid, once it begins, and the entire cloud can collapse to be an intermediate-mass BH. The origin of supermassive BHs in the early Universe is an ongoing mystery. The first star remnants could be candidates for the seeds of BHs.

SUPERNOVA (SN) EXPLOSIONS

Massive stars end their lives as SNe. Such energetic explosions in the early Universe were violently destructive; they blew the ambient gas out of the gravitational well of small-mass dark matter clumps, causing an almost complete gas evacuation. Because massive stars process a substantial portion of their mass into heavy elements, SN explosions can cause prompt chemical enrichment, at least locally. This may even provide an efficient mechanism to pollute the surrounding intergalactic medium with heavy elements to an appreciable degree.

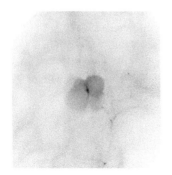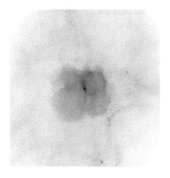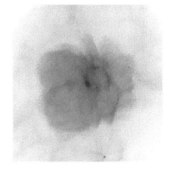

FIGURE 5.12   The first star illuminates the dark Universe. The panels show the evolution of the gas temperature around the formed first star at the center. From left to right, the depicted times correspond to 1, 2, and 3 Myr after the star formed. The shown region is 15,000 lt-yr on a side.

* Neutral hydrogen is termed H I, whereas ionized hydrogen (proton) is termed H II.

Stars with mass greater than 100 times that of the Sun are thought to trigger a particular type of SN. Such a massive star is so heavy that, once the core starts collapsing, rapid nuclear reactions take place in the core, which produce an extremely large energy (see the chapter by Fillipenko in this volume). Then the star is disrupted completely, expelling all the heavy elements synthesized during, and at the end of, its life. A recently discovered luminous SN (Figure 5.13) is thought to be caused by this mechanism. First stars probably died similarly to this energetic SN.

SN explosions could also act as a trigger of star formation. Blast waves may sweep up the surrounding gas and form a dense shell, in which dense gas clouds can be born. For this mechanism to work, the density and its radial profile around the SN site are of particular importance. SN remnants cool both adiabatically (by expansion) and by radiative cooling. The efficiency of cooling is critically determined by the gas distribution inside the blast wave. If the gas has already been evacuated by radiative feedback prior to explosion (see the previous section), blast waves propagate over a long distance, distributing the mechanical energy of the event over a large volume of space.

Detailed calculations show that a large portion of the explosion energy is lost in a few million years by atomic cooling. For even greater explosion energies, SN remnants cool by another mechanism. These are clearly different from local Galactic SNe. In the early Universe, the number of CMB photons was much greater than it is today because of cosmic expansion and the redshift of radiation. Thus, electrons in a gas could have met many CMB photons and transmitted their thermal energy to CMB photons via the so-called inverse Compton process. This is the most efficient cooling process for a diffuse, low-density gas. In principle, this process imprints a distinct signature in the fluctuations of CMB, but the signal is thought to be very small.

SN explosions prevent subsequent star formation for some time in the area where they originated. Although heavy elements expelled by early SNe could greatly enhance the gas cooling efficiency, the onset of this "second-generation star formation" may be delayed because of gas evacuation, particularly in low-mass systems. This again supports the notion that early star formation is likely episodic. If the first stars are massive, only one period of star formation is possible for the small host dark mass

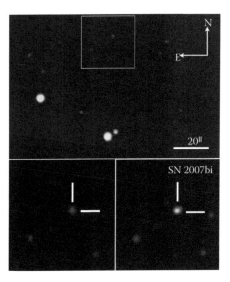

FIGURE 5.13 Supernova 2007bi, a candidate pair-instability supernova (SN). The top panel shows the region of the sky where the SN appeared. The rectangular box indicates the SN site. The bottom-left panel shows the zoom image around the host galaxy before the SN explosion, whereas the bottom-right panel shows the same region when the bright SN appeared. A theoretical model suggests that a very massive star with mass greater than 100 times that of the Sun triggered the energetic explosion. (Courtesy of Dr. Stephen Smartt; Reproduced from Young, D.R., Smartt, S. J., Valenti, S. et al., *Astronomy and Astrophysics*, 512, 70–88, 2010; see: http://star.pst.qub.ac.uk/supernovae/supernova_followup/. With permission.)

clump and its descendants within about 10 million years. The global cosmic star formation activity increased only after a number of large-mass ($>10^{8-9} M_{\odot}$) systems—the first galaxies—were assembled. These first galaxies will be detected by next-generation telescopes. The *James Webb Space Telescope* (Figure 5.14) is expected to discover the brightest of these early galaxies.

Primeval Galaxies

The hierarchical nature of cosmic structure formation naturally predicts that stars or stellar-size objects form first, preceding the formation of galaxies. Through the feedback processes described in the previous section, the first generation of stars set the scene for the subsequent galaxy formation. In this section, we call a large-mass system that is assembled after some influence of the first star "a first galaxy." We adopt this crude, and somewhat ambiguous, definition for the sake of simplicity, although the size and stellar content of a first galaxy might be similar to those of present-day star clusters. The characteristic minimum mass of a first galaxy (including dark matter) is perhaps $\sim 10^{8} M_{\odot}$, within which supernovae ejecta can be retained.

The first galaxies were assembled through a number of large and small mergers, generating turbulence dynamically. The star formation process changed from a quiescent one (as for the first stars) to a highly complicated, but organized one. Recently, there were a few attempts to simulate the formation of first galaxies in a cosmological context. The results generally indicate that star formation in the large-mass system is still an inefficient process overall. However, a remarkable difference is that, in the first galaxies, the interstellar medium was already metal enriched. Heavy atoms can emit line photons efficiently, and dust grains radiate energy constantly. Theoretical calculations show that cooling by heavy elements and by dust grains can substantially change the thermal evolution of gas clouds. Altogether, the necessary conditions may be set for the formation of small-mass stars for the first time in the history of the Universe. The combined effects of strong turbulence and

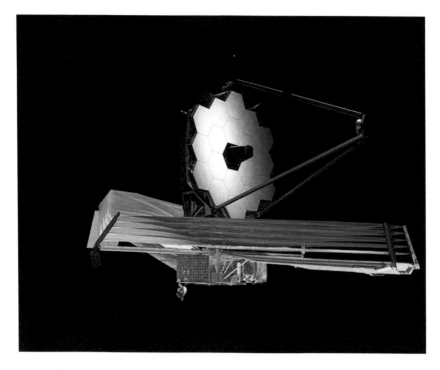

FIGURE 5.14 NASA's *James Webb Space Telescope*, scheduled to be launched in 2013. (From NASA website http://www.jwst.nasa.gov/images2/013526.jpg.)

metal enrichment might make the relative number of stars of different masses close to that in the present-day star-forming regions.

Understanding the formation of the first galaxies is very challenging, because of the complexities described above. Nevertheless, it is definitely a subject where theoretical models can be tested against direct observations in the near future.

PROSPECTS FOR FUTURE OBSERVATIONS

A number of observational programs are planned to detect the first stars and galaxies, both directly and indirectly. In closing this chapter, the following discussion addresses prospects for future observations.

The first galaxies will be the main target of next-generation optical and IR telescopes, while rich information on the first stars will be obtained from observations of the CMB polarization, the near-IR background, high-redshift SNe and GRBs, and the so-called Galactic archeology.

One of the most important constraints on the observation of early star formation is obtained from the CMB optical depth to Thomson scattering. Data obtained over the 5-year operation of the *WMAP* satellite suggest that roughly 10% of CMB photons are scattered by electrons somewhere between the surface of the last scattering and us. The measurement provides an integral constraint on the total ionizing photon production at $z > 6$. More accurate polarization measurements by the European *Planck* satellite will further reveal the reionization history of the Universe, namely the variation of electron fraction in the cosmic gas as a function of time.

Perhaps the most promising *direct* method is the observation of SN explosions of the first-generation massive stars. Once photometrically detected, core-collapse and pair-instability SNe that arise from stars with different masses may be discriminated by their light-curve rise times and emission-line signatures. The late-time luminosities of core-collapse SNe are largely driven by circumstellar interaction. Brightness variations, the so-called light curves, can be utilized to identify these high-z SNe.

GRBs are intrinsically very bright and, thus, are detectable out to redshifts $z > 10$. Recent evidence indicates that GRBs trace the formation of massive stars. Very likely, GRBs are triggered when massive stars end their lives. As already discussed, the first stars in the Universe are predicted to be massive, and so they are progenitors of energetic SNe and associated GRBs at high redshifts. Recently, NASA's *Swift* satellite has detected a GRB originating at $z > 8$, thus demonstrating the promise of GRBs as probes of the early Universe. With higher energy sensors than *Swift*, the new *Fermi* satellite has also detected a GRB at $z > 4$, opening a new window to the early Universe at high energy. It might come with (or without) surprise in the future that very high-z GRBs are detected by these satellites.

There is yet another way of obtaining information on early star formation from our Galaxy. Very metal-poor stars—the stellar relics—indicate the conditions under which these low-mass stars formed. While conventional searches for metal-poor stars are done for halo stars orbiting near the Sun, the Apache Point Observatory Galactic Evolution Experiment (APOGEE) project is aimed at observing ~100,000 stars in the bulge of the Milky Way. It is expected that early generations of stars and their remnants are located near the Galactic center at the present epoch. The nature of early metal enrichment must be imprinted in the abundance patterns in many of the bulge stars.

In the longer term, radio observations using Square Kilometer Array (SKA) (Figure 5.15) will map the topology of reionization, revealing how and by what sources the Universe was reionized. Abundant neutral hydrogen in the intergalactic medium radiates at a particular wavelength of 21 cm. The radiation is redshifted by cosmic expansion to become meter-length radio waves. A large array of radio telescopes can detect the signals. By using the simple relation of the wavelength and redshift [$\lambda_{detected} = 21$ cm $\times (1 + z)$], radio observations can make a full 3-dimensional map of the distribution of neutral hydrogen in the early Universe. High angular and frequency resolution

FIGURE 5.15 Square Kilometer Array (SKA), an extremely large number of radio telescopes to be in operation in the late 2010s. (From the SKA website http://www.skatelescope.org/photo/Dishes_overview_compressed.jpg. With permission.)

observations allow us to probe matter density fluctuations at small length scales, which are inaccessible by CMB observations.

Altogether, these future observations will finally fill the gap of the "Dark Age" in our knowledge of the history of the Universe.

REFERENCES

Bromm, V., Yoshida, N., Hernquist, L. et al. (2009). The formation of the first stars and galaxies. *Nature,* 459: 49–54.

Colless, M., Dalton, G., Maddox, S. et al. (2001). The 2dF Galaxy Redshift Survey: Spectra and redshifts, *Monthly Notices of the Royal Astronomical Society,* 328: 1039–63.

Frebel, A., Aoki, W., Christlieb, N. et al. (2005). Nucleosynthetic signatures of the first stars. *Nature,* 434: 871–73.

Miyazaki, S., Hamana, T., Shimasaku, K. et al. (2002). Searching for dark matter halos in the Suprime-Cam 2 Square Degree Field. *Astrophysical Journal,* 580: L97–100.

Springel, V., Frenk, C.S., and White, S.D.M. (2006). The large-scale structure of the universe. *Nature,* 440: 1137–144.

Springel, V., White, S. D. M., Jenkins, A. et al. (2005). Simulations of the formation, evolution and clustering of galaxies and quasars. *Nature,* 435: 629–36.

Yoshida, N., Omukai, K., and Hernquist, L. (2008). Protostar formation in the early universe. *Science,* 321: 669–71.

Young, D.R., Smartt, S. J., Valenti, S. et al. (2010). Two type Ic supernovae in low-metallic diversity of explosions. *Astronomy and Astrophysics,* 512: 70–88.

6 An Overview of Supernovae, the Explosive Deaths of Stars

Alexei V. Filippenko

CONTENTS

INTRODUCTION

Our own Sun, a typical main-sequence star, will die rather quietly, without exploding violently (e.g., Iben, 1974). It is currently fusing hydrogen into helium in its core, where temperatures are about 15 million kelvin (K), at a steady rate via the proton-proton chain, maintaining a luminosity that increases extremely slowly over its 10-billion-year main-sequence lifetime.

In about 5 billion years, the Sun's core will be nearly pure helium, but the temperature will be too low for fusion to carbon and oxygen. Losing energy to its surroundings, the helium core will gravitationally contract, heating the hydrogen-burning layer around it and causing much more rapid fusion. This will cause the Sun to become much brighter and larger on a relatively short timescale (hundreds of millions of years), producing a red giant having about $100\,L_\odot$ and extending nearly to the orbit of Mercury. Everything on Earth will be fried—we had better move to another planet or even another Solar System long before that happens!

The temperature of the contracting helium core will rise, and eventually it will become so hot that the triple-alpha process, where three helium nuclei fuse to form carbon, will commence. Some of the carbon nuclei will fuse with helium to form oxygen. This "horizontal branch" phase will last much less time than the main-sequence phase, because the Sun will be much more luminous at this stage, and the energy generated per nuclear reaction will be less than in hydrogen fusion. A carbon-oxygen core will form, and then it, too, will begin to gravitationally contract, unable to fuse into heavier elements. This will cause the surrounding helium-burning and hydrogen-burning shells to fuse faster, producing an even more luminous and larger red giant—the "asymptotic giant branch" phase.

Subsequently, it is thought that the Sun will gently blow off its outer atmosphere of gases, causing them to expand outward. Photoionized by the ultraviolet radiation emanating from the partially denuded core of the dying star, this gas will glow; the result will be a beautiful object called a "planetary nebula" because, when viewed through small telescopes, such objects can resemble the disks of planets (for a review, see Kaler, 1985). Some examples of these wonderful nebulae are shown in Figure 6.1; one can see the dying star in the center and the shells of ionized gases slowly expanding away.

Our dying Sun, at this time having only 50%–60% of its initial mass, will continue to contract and radiate the energy stored in its interior and eventually become a "white dwarf," a very small star about the size of the Earth (instead of being about 100 times larger than the Earth as the Sun is now). Its density will be so high that it will be supported not by thermal pressure, but rather primarily by electron degeneracy pressure. It will radiate the thermal energy stored in its interior, but not generate any new energy through nuclear reactions. In this sense, it can be thought of as a "retired star," spending its life savings of energy. It will continue in this state essentially forever, becoming progressively dimmer until it will be difficult to detect—a "black dwarf."

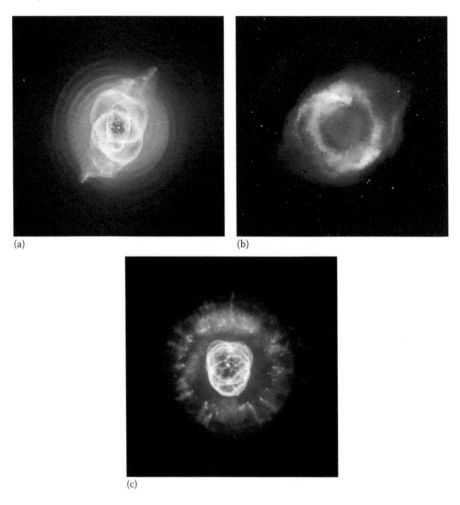

(a)   (b)

(c)

FIGURE 6.1a,b,c *Hubble Space Telescope* (*HST*) optical images of planetary nebulae. (a) NGC 6543, the "Cat's Eye Nebula"; credit: NASA, ESA, The *Hubble* European Space Agency Information Centre, and The *Hubble* Heritage Team (STScI/AURA) (http://imgsrc.hubblesite.org/hu/db/images/hs-2004-27-a-full_jpg.jpg). (b) NGC 7293, the "Helix Nebula"; credit: NASA, NOAO, ESA, The *Hubble* Helix Nebula Team—M. Meixner (STScI) and T.A. Rector (NRAO) (http://imgsrc.hubblesite.org/hu/db/images/hs-2003-11-a-full_jpg.jpg). (c) NGC 2392, the "Eskimo Nebula"; credit: NASA, Andrew Fruchter, and the ERO Team—Sylvia Baggett (STScI), Richard Hook (ST-ECF), and Zoltan Levay (STScI) (http://imgsrc.hubble-site.org/hu/db/images/hs-2000-07-a-full_jpg.jpg).

SUPERNOVAE—CELESTIAL FIREWORKS!

Some stars literally explode at the end of their lives, either completely obliterating themselves or leaving only an exceedingly compact remnant (e.g., Trimble, 1982, 1983). These "supernovae" (SNe; see Figure 6.2 for an example) are among the most energetic phenomena known in the Universe. They can become a few billion times as powerful as the Sun. If you like fireworks shows, you can't do any better than watching stars explode!

SNe are very important for many reasons. They heat the interstellar medium, in some cases creating galactic fountains and winds. Some of them produce neutron stars and black holes (BHs), extreme environments in which we can further explore the laws of physics. They produce highly energetically charged particles—cosmic rays. As I will discuss later in this chapter, their huge luminosities and (generally) almost uniform properties make them attractive yardsticks for cosmological distance measurements.

Most important from the human perspective, SNe eject heavy elements—both the elements synthesized during the normal lives of the progenitor stars and new elements produced through nuclear reactions during the explosions themselves. Elements such as the carbon in our cells, the oxygen that we breathe, the calcium in our bones, and the iron in our red blood cells are produced within stars and ejected into space, making them available as the raw material for new stars, new planets, and ultimately new life.

Astronomers have analyzed the expanding gases in supernova (SN) remnants such as the Crab Nebula (Figure 6.3; see Hester, 2008 for a review), which resulted from an explosion nearly 1,000 years ago; heavy elements have been found in the ejected gases that could not have been present when the star was born.

These gases continue to expand for thousands of years, but eventually they merge with other clouds of gas and become gravitationally bound in gigantic nebulae, such as the famous "Pillars of Creation" (Figure 6.4a). Within the dense clouds of gas, gravitational collapse occurs and new stars are formed. Around some of those new stars there are disks of gas and other debris (Figure 6.4b),

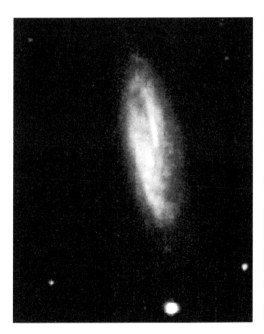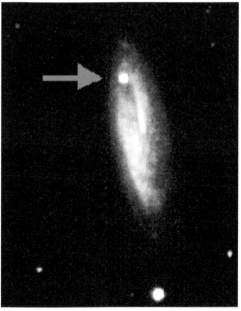

FIGURE 6.2   Optical images of NGC 7541 before and after the appearance of SN 1998dh, obtained with the 0.76-m Katzman Automatic Imaging Telescope (KAIT) at Lick Observatory. (Credit: Alex Filippenko and Weidong Li.)

FIGURE 6.3 *HST* optical image of the Crab Nebula, the remnant of the bright Milky Way SN of 1054 CE. Credit: NASA, ESA, J. Hester, and A. Loll (Arizona State University) (http://imgsrc.hubblesite.org/hu/db/images/hs-2005-37-a-full_jpg.jpg).

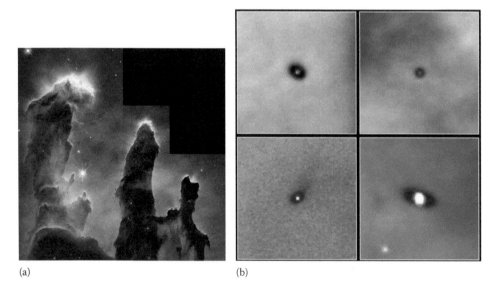

(a)   (b)

FIGURE 6.4a,b (a) *HST* optical image of the Pillars of Creation (M16, the Eagle Nebula); credit: NASA, ESA, STScI, J. Hester, and P. Scowen (Arizona State University) (http://imgsrc.hubblesite.org/hu/db/images/hs-1995-44-a-full_jpg.jpg or http://imgsrc.hubblesite.org/hu/db/images/hs-1995-44-a-full_tif.tif). (b) Protoplanetary disks around young stars in the Orion Nebula; credit: Mark McCaughrean (Max-Planck-Institute for Astronomy), C. Robert O'Dell (Rice University), and NASA (http://hubblesite.org/newscenter/archive/releases/1995/45/image/b/format/web_print).

which can contract to form new planets. Some of those planets may be rocky, Earth-like planets. And in some cases, complex molecules such as DNA can form—the basis of life. The process of stellar death and birth throughout the electromagnetic spectrum is observed to obtain a more detailed understanding of the physical principles.

My research group at the University of California, Berkeley, is studying the process by which some massive stars explode, the rates of different kinds of SNe, their progenitors, and their nucleosynthetic products. But stars explode only very rarely, just a few times per typical large galaxy per century. So to find large numbers of SNe that we can study, we have developed a robotic telescope at Lick Observatory, roughly a 2-hour drive from Berkeley. Expertly programmed by my close collaborator, Weidong Li, the Katzman Automatic Imaging Telescope (KAIT; Filippenko et al., 2001) takes charge-coupled device images of more than 1,000 relatively nearby galaxies each night, close to 10,000 each week. It immediately compares the new images with the old images, and usually these look the same; but sometimes a new object appears—and this will make the observed star a SN candidate (Figure 6.5).

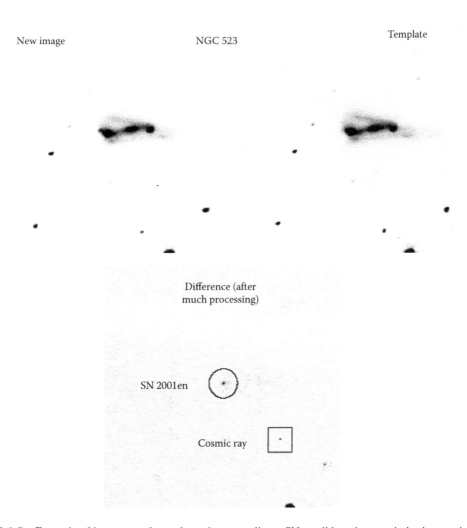

FIGURE 6.5 Example of image template subtraction, revealing a SN candidate that was indeed a genuine SN (SN 2001en). A cosmic-ray hit and two poorly subtracted stars are also visible. Credit: Alex Filippenko and Weidong Li.

I have a team of undergraduate students who examine the new images to determine whether the object is indeed a SN. Over the past decade, KAIT and my team of assistants have been the world leaders in finding nearby SNe. We have found about 40% of the world's supply of relatively nearby SNe and an even greater fraction of those discovered early, shortly after the explosion, when one can learn much about the phenomenon.

We also conduct detailed observations of the SN brightness as compared with time—that is, we perform photometry on filtered images and obtain light curves, typically in the *BVRI* bands.* This is done mostly automatically with a program developed by one of my graduate students, Mohan Ganeshalingam, together with Weidong Li. In addition, we obtain optical spectra of as many SNe as possible, typically with the 3 m (10 ft) Shane reflector at Lick Observatory and with other facilities when possible.

The spectra of SNe within a few weeks past peak brightness can be grouped into several distinct categories (Figure 6.6; see Filippenko, 1997 for a review). Those with obvious hydrogen lines are called Type II, while SNe lacking clear evidence of H are dubbed Type I. The classical Type I SNe are now known as Type Ia; they are distinguished by the presence of a strong absorption line due to Si II at an observed wavelength of about 6,150 Å. If the spectra lack H and obvious Si, but exhibit lines of He I, the SNe are dubbed Type Ib, while if both H and He (and obvious Si) are missing, we call them Type Ic. If a SN II exhibits relatively narrow (width of about 1,000 km/s) emission lines in its spectrum, it is called a Type IIn (for "narrow") SN. If the spectrum of a SN II exhibits H at early times, but resembles a SN Ib or SN Ic (with no H, but strong emission lines of intermediate-mass elements such as O and Ca) at late times, it is a Type IIb SN.

Type II SNe are also subdivided according to their light-curve shapes (Barbon et al., 1979; Doggett and Branch, 1985): those with an extended interval (sometimes up to a few months) of nearly constant brightness are called Type II plateau SNe, whereas those that decline linearly (in magnitudes) are known as Type II linear SNe.

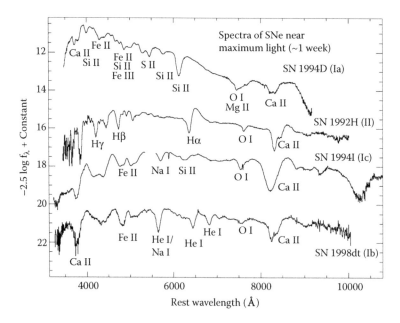

FIGURE 6.6 Spectra of different types of SNe, all about 1 week past maximum brightness. (Credit: Thomas Matheson and Alex Filippenko; reproduced with permission of Thomas Matheson.)

---

* See the discussion of these broadband filters in Ganeshalingam et al. (2010).

TYPE IA SUPERNOVAE

Physically, a SN Ia is thought to result from the thermonuclear runaway of a carbon-oxygen (C/O) white dwarf star that approaches the limiting mass of a body held up against gravity by electron degeneracy pressure—the Chandrasekhar limit (Chandrasekhar, 1931) of about 1.4 $M_\odot$ (e.g., Nomoto et al., 1984; Woosley and Weaver, 1986; Khokhlov, 1991; Hillebrandt and Niemeyer, 2000). As the white dwarf gains mass from a relatively normal companion star, its diameter decreases (a consequence of the physical state of gravitationally bound degenerate matter), and its constituent nuclei get closer and closer together. Nuclear fusion can start to occur, but initially the cooling rate exceeds the heating rate. When the heating rate exceeds the cooling rate, a thermonuclear runaway is initiated; the energy generated by nuclear reactions is not able to significantly expand the star (because the equation of state of a degenerate gas is essentially independent of its temperature). Instead, the nuclei become more energetic and, thus, more likely to fuse.

The thermonuclear runaway produces a large amount ($\sim 0.6~M_\odot$) of radioactive ^{56}Ni, which decays to radioactive ^{56}Co with an *e*-folding time of about 1 week and subsequently into stable ^{56}Fe with an *e*-folding time of about 2.5 months. This decay chain provides the observed optical light of a SN Ia; the nuclear energy emitted during the runaway explosion is quickly used up through adiabatic expansion of the small white dwarf.

It is not yet known whether the companion (mass donor) star is a red giant, subgiant, or main-sequence star (for a review, see Branch et al., 1995). Moreover, it turns out to be difficult to avoid surface nova eruptions that eject the accreted material (which happens at very low accretion rates) and also to not build up a H envelope (which happens at high accretion rates); the mass accretion rate needs to be within a narrow range of acceptable values, and is thus unlikely. The Chandrasekhar limit is difficult to approach, yet sub-Chandrasekhar models have serious problems, thus decreasing the number of likely progenitors. In addition, no H lines have ever been convincingly seen in the spectrum of a SN Ia, yet there should be some H ablated from the donor star. Another problem with the "single-degenerate model" is that it may have trouble explaining the few cases of SNe Ia resulting from apparently "super-Chandrasekhar mass" white dwarfs (e.g., Howell et al., 2006).

Thus, an alternative hypothesis is that at least some SNe Ia arise from binary white dwarf systems that spiral toward one another and eventually merge as a result of the emission of gravitational waves. The white dwarf having the lower mass is physically larger and is tidally disrupted by its more massive companion. The resulting disk of material subsequently accretes onto this remaining white dwarf, thereby increasing its mass and causing it to undergo a thermonuclear runaway as described above. Whether most SNe Ia come from single-degenerate or double-degenerate systems is one of the outstanding unsolved questions in SN research.

In all cases, the white dwarf should explode when its mass is close to the Chandrasekhar limit; thus, one might expect that the peak luminosity is the same among SNe Ia. This is observed to be approximately true for many objects, but there is still a considerable range in peak luminosity; SNe Ia appear to produce a range of ^{56}Ni masses, perhaps because of differences in the white dwarf metallicity, C/O mass ratio, rotation, initial white dwarf mass, or some other variable. The SN 1991bg-like objects (e.g., Filippenko et al., 1992a) tend to be quite underluminous, have faster brightness decline rates, exhibit redder early-time spectra with lines of lower-ionization species, and occur mostly in early-time host galaxies. Conversely, the SN 1991T-like objects (e.g., Filippenko et al., 1992b) tend to be somewhat overluminous, have slower brightness decline rates, exhibit bluer early-time spectra with lines of relatively high-ionization species, and occur in late-type host galaxies. Phillips (1993) was the first to conclusively demonstrate that the peak luminosity is correlated with the rate of brightness decline, although the relationship had previously been suspected by Pskovskii (1977).

TYPE IA SUPERNOVAE AND THE ACCELERATING UNIVERSE

The enormous peak luminosity of SNe Ia, their relative homogeneity, and the fact that we can individually calibrate their peak luminosities by measuring the rate of brightness decline (the Phillips relation) led to the emergence of these objects as superb "custom yardsticks" with which to measure cosmological distances (see Filippenko, 2005 for a review). Riess et al. (1995), Hamuy et al. (1996), and other later studies used relatively nearby SNe Ia to refine the Phillips (1993) relation; Riess et al. (1996) also showed that measurements through several filters could be used to remove the effects of interstellar reddening, allowing galaxy distances to be determined with a precision of ~8% (Figure 6.7). Thus, it was natural to also search for very distant SNe Ia and to trace the expansion history of the Universe, thereby predicting its future—presumed to be either eternal expansion (although always decelerating) or eventual collapse.

In the early 1990s, the Supernova Cosmology Project (SCP), led by Saul Perlmutter of the Lawrence Berkeley Laboratory, started finding high-redshift SNe Ia. The High-*z* Supernova Search Team (HZT), led by Brian P. Schmidt of the Australian National University, soon followed suit; they began looking for distant SNe only after spending considerable time studying nearby SNe and

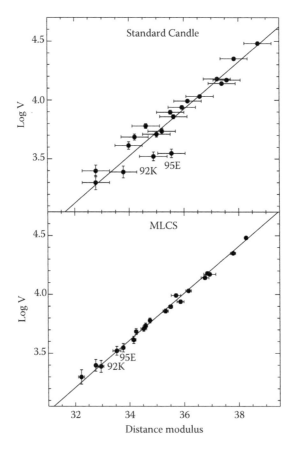

FIGURE 6.7 Hubble diagram with Type Ia SNe, before (*top*) and after (*bottom*) correction for reddening and intrinsic differences in luminosity (A. G. Riess, 2001, private communication). The ordinate is the distance modulus (mag; Cepheid distance scale), and the abscissa is galaxy recession velocity (km/s) in the rest frame of the cosmic background radiation. The multicolor light-curve shape method (MLCS; From Riess, A.G., Press, W., and Kirshner, R.P., *Astrophysical Journal*, 473, 88–109, 1996. With permission.) was used in the bottom diagram for luminosity and reddening corrections. The dispersion drops from 0.42 mag (*top*) to only 0.15 mag (*bottom*) after applying the MLCS method. Credit: Adam G. Riess; reproduced with permission of Adam G. Riess.

demonstrating that they could be used as accurate cosmological tools. Both teams employed a wide-field camera on the 4 m (13 ft) Blanco telescope at the Cerro Tololo Inter-American Observatory (and sometimes other telescopes) to take deep images of selected regions of the sky. By obtaining a new set of images of the same regions a few weeks later, they could search for new SN candidates (Figure 6.8). These were spectroscopically confirmed (and redshifts were measured; typically $z \approx 0.3$–0.8) with the Keck 10 m (33 ft) telescopes in Hawaii, the 4.5 m (15 ft) Multiple Mirror Telescope in Arizona, and the European Southern Observatory (ESO) 3.6 m (12 ft) telescope in Chile. Follow-up images of the confirmed SNe Ia with ground-based telescopes and with the *Hubble Space Telescope* (*HST*) were used to determine the peak observed brightness, the rate of decline, and (by the HZT) the likely interstellar reddening. From these measurements, the distance of each SN (and hence that of its host galaxy) could be derived.

The resulting Hubble diagram of distance as compared with redshift revealed a stunning result: for a given redshift, the measured distances were larger than expected in a decelerating Universe, or even in a Universe with constant expansion speed. Unless something else was corrupting the data, the most reasonable conclusion was that the expansion of the Universe has been accelerating during the past 4–5 billion years! The HZT was the first to announce clear evidence for this conclusion, at a meeting in Marina del Rey in February 1998 (Filippenko and Riess, 1998); it also was the first to publish the result in a refereed journal (Riess et al., 1998; submitted in March 1998 and published in September 1998). The SCP submitted its paper in September 1998, and it was published in June 1999 (Perlmutter et al., 1999).

Although the discovery was initially viewed with considerable skepticism, by December 1998, there had been no clear errors found in the data or the analysis methods. The editors of *Science* magazine thus proclaimed the discovery as the top breakthrough in 1998 in all fields of science. The cause of the acceleration was not yet known (as is still the case); either there exists gravitationally repulsive "dark energy" of unknown origin, or Einstein's general theory of relativity is incorrect.

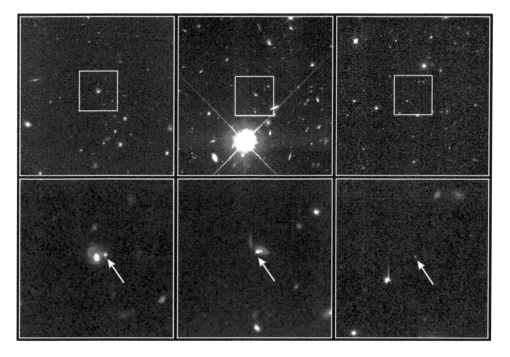

FIGURE 6.8   *HST* image of three distant galaxies with Type Ia supernovae (marked with arrows) in them. Panels in the bottom row show only the insets marked in the top-row panels. Credit: P. Garnavich (Harvard–Smithsonian Center for Astrophysics) and the High-z Supernova Search Team and NASA (http://imgsrc.hubblesite.org/hu/db/images/hs-1998-02-a-print.jpg).

The origin and physical nature of dark energy is often considered to be the most important observationally motivated problem in all of contemporary physics.

Within the next decade after 1998, many hundreds of additional high-redshift SNe Ia were found, and systematic errors were evaluated in greater detail, confirming and refining the initial results (e.g., Astier et al., 2006; Wood-Vasey et al., 2007; Kowalski et al., 2008; and references therein). Moreover, SNe Ia at $z \gtrsim 1$ were used to show that the Universe went through an early phase of deceleration during the first ~9 Gyr of its existence (Riess et al., 2004, 2007). Additional techniques, such as measurements of the angular power spectrum of temperature fluctuations in the cosmic microwave background radiation (CMBR), the growth of large-scale structure in the Universe, x-ray observations of clusters of galaxies, baryon acoustic oscillations, the integrated Sachs–Wolfe effect, weak gravitational lensing, and others confirm the SN results and lead to a "concordance cosmology" in which dark energy constitutes ~75% of the matter-energy content of the Universe (see Frieman et al., 2008 for a recent review, as well as the contribution by M. Sullivan to this volume).

SNe Ia have also been used to make the most precise direct measurement of the Hubble constant, H_0 (Riess et al., 2009). By using *HST* observations of Cepheid variables in the host galaxies of nearby SNe Ia and the "Maser Galaxy" NGC 4258, the peak luminosities of the SNe Ia were calibrated. The average value was then used to determine the zero point of the Hubble diagram based on a set of 240 SNe Ia at $z < 0.1$, yielding $H_0 = 74.2 \pm 3.6$ (km/s)/Mpc, where the uncertainty of only 4.8% includes both statistical and systematic errors. This is in excellent agreement with the result derived from the *Wilkinson Microwave Anisotropy Probe* (*WMAP*) 3-year data set under the *assumption* of a flat Universe: $H_0 = 73.2 \pm 3.2$ (km/s)/Mpc (Spergel et al., 2007).

CORE-COLLAPSE SUPERNOVAE

TYPE II SUPERNOVAE

Hydrogen-rich SNe (Type II) tend to occur in spiral galaxies, often in or close to spiral arms, where massive stars are born and die. Thus, they have long been attributed to explosions of massive stars ($\gtrsim 8\ M_\odot$) at the end of their lives, generally while in the red supergiant stage (e.g., Woosley and Weaver, 1986; Arnett, 1987; Bethe, 1990; Herant et al., 1994; Burrows et al., 1995; Janka and Mueller, 1996; Burrows et al., 2006). Recently, red supergiant progenitor stars have been found at the sites of SNe II in archival images of the host galaxies (Smartt, 2009, and references therein). For example, Figure 6.9 shows *HST* images of the site of SN 2005cs before and after the explosion (Li et al., 2006), revealing the massive progenitor.

During the normal evolution of a massive star, the ashes of one set of nuclear reactions become the fuel for the next set, eventually giving rise to an onion-like structure consisting of layers of progressively heavier elements (H, He, C + O, O + Ne + Mg, Si + S, and finally Fe in stars having initial masses of at least 9 or 10 $M_\odot$; Figure 6.10). But the Fe nuclei in the core cannot fuse together to form heavier elements; the binding energy per nucleon is highest for the Fe-group elements, so fusion is an endothermic (rather than exothermic) process. Thus, the mass of the Fe core builds up as the surrounding shells of lighter elements continue to fuse, and eventually it reaches a value close to the Chandrasekhar limit. Collapse ensues, and the electrons combine with protons to form neutrons and neutrinos.

The proto-neutron star reaches supernuclear densities and rebounds, thus creating a shock wave that pummels its way outward, approaching the explosion of the star. A vivid analogy for such a "mechanical bounce" can be demonstrated by placing a tennis ball on top of a basketball and dropping the two of them simultaneously; the tennis ball bounces to a height much greater than its original height. However, in the case of a collapsing Fe core, most of the energy of the outward-moving shock wave gets consumed through the dissociation of the outer-core Fe into its constituent He nuclei; thus, the shock stalls, and the rest of the star implodes. Something else must force the explosion, or it will fail.

FIGURE 6.9 *HST* images of the Whirlpool Galaxy (M51; *left*) and the site of the Type II SN 2005cs before (*top right*) and after (*bottom right*) the appearance of the SN. Credit: NASA, ESA, W. Li, and A. Filippenko (University of California, Berkeley), S. Beckwith (STScI), and the *Hubble* Heritage Team (STScI/AURA) (http://hubblesite.org/newscenter/archive/releases/star/supernova/2005/21/image/a/format/web_print/).

Many researchers think that one key is neutrinos. A tremendous amount of gravitational binding energy is released during the collapse of the Fe core, heating the resulting neutron star to temperatures of over 100 billion K. At such high temperatures, by far the dominant cooling mechanism is the production and nearly immediate escape of neutrinos and antineutrinos. About 99% of the core implosion energy is carried away by these elusive particles, and only 1% couples with the surrounding material, causing it to explode. The visual display is a minor sideshow, amounting to just 0.01% of the total released energy. However, the details of the core-collapse SN explosion mechanism are still not entirely clear; multidimensional numerical simulations incorporating all known physics (e.g., neutrino-particle interactions, magnetohydrodynamic jets, convection, standing accretion shock instability, acoustic waves) still do not convincingly produce successful SNe. The future detection of gravitational waves may provide important clues to the core-collapse explosion mechanism (Murphy et al., 2009).

During the explosion, a large number of free neutrons are available in the layers containing heavy elements. Rapid neutron capture ensues, building up a wide range of elements and isotopes. This "explosive nucleosynthesis" accounts for many of the elements in the periodic table heavier than Fe.

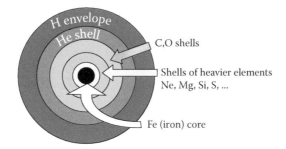

FIGURE 6.10 Cross section of a red supergiant star immediately prior to core collapse, showing the onion-like structure of layers of progressively heavier elements. Credit: Alex Filippenko and Frank Serduke.

At or near the center of an expanding, chemically enriched SN remnant, we expect there to be a compact (radius about 10 km), very dense neutron star having about 1.4 $M_\odot$. Indeed, such an object is found in the Crab Nebula (Figure 6.3); it manifests itself as a "pulsar" with a period of just 0.033 sec. Pulsars are rapidly rotating, highly magnetized neutron stars whose magnetic axis is offset from the rotation axis. Energetic charged particles are accelerated along the magnetic axis, producing narrow, oppositely directed beams of electromagnetic radiation. If at least one of these beams sweeps across the observer's line of sight during each full rotation of the neutron star, a brief flash of light will be seen. Although pulsars were first discovered in the 1960s with radio telescopes, the youngest ones, such as that associated with the Crab Nebula, also shine at shorter wavelengths such as visible light and x-rays.

THE PECULIAR TYPE II SUPERNOVA 1987A

Thus far, the best-studied Type II SN is SN 1987A in the Large Magellanic Cloud (LMC), at a distance of only about 170,000 light-years. This was the brightest SN since Kepler's SN 1604 and Tycho Brahe's SN 1572, and it was studied with modern instruments throughout the electromagnetic spectrum (for reviews, see Arnett et al., 1989; McCray, 1993). It occurred near the Tarantula Nebula, the largest star-forming region in the local group of galaxies, a site where many massive stars have recently formed.

Existing presupernova images reveal the SN progenitor star, the B3 supergiant Sanduleak −69°202, whose initial mass was probably in the range 14–20 $M_\odot$ (see Smartt, 2009, and references therein). This was the first direct evidence that SNe II come from massive, evolved stars. However, it is not yet entirely clear why the progenitor was a blue supergiant rather than a red one; possible causes include the low metallicity of the LMC, rapid rotation, or a previous merger event with a binary companion star. But given the nature of the progenitor, it is easy to understand some initially surprising aspects of SN 1987A, such as its low luminosity at early times and its peculiar light curve; much of the explosion energy was lost to adiabatic expansion of the relatively small star (e.g., Woosley et al., 1987).

A tremendously exciting discovery was a burst of antineutrinos (and presumably neutrinos) associated with SN 1987A, using two large underground tanks of water originally designed to search for proton decay (Hirata et al., 1987; Bionta et al., 1987). Antineutrinos from SN 1987A combined with protons to form neutrons and high-speed positrons, which traveled faster than the local speed of light in water and thus emitted Cerenkov light; each of the Kamioka and Irvine–Michigan–Brookhaven experiments detected about 10 such bursts. (Possibly one or two of the flashes were caused by the superluminal motion of electrons accelerated by a neutrino collision, but the antineutrino-proton interaction cross section is much larger.) This detection marked the birth of extrasolar neutrino astronomy, and it signaled at least the temporary formation of a neutron star.

Another very important aspect of SN 1987A is that satellites detected gamma rays from the SN having energies that are consistent only with certain short-lived isotopes of cobalt. This provides compelling evidence that heavy elements were created through the process of explosive nucleosynthesis in the SN, as had been expected. Other elements were also created by the SN and ejected into space. Maybe someday, far in the future, they will lead to the formation of planets and new life-forms.

The presence of several rings of gas surrounding SN 1987A, as seen in Figure 6.11a, indicates that the progenitor star had experienced some amount of mass loss in the few tens-of-thousands of years prior to exploding. More recent *HST* images reveal that the inner ring has been lit up at roughly 20 locations ("hot spots") because of a collision with the most rapidly moving SN ejecta, as shown in Figure 6.11b. This figure also shows that the main part of the SN 1987A ejecta are spatially resolved, and the largest amount of expansion seems to have occurred in a direction roughly perpendicular to the plane of the inner ring. Moreover, there is no clear evidence for a neutron star

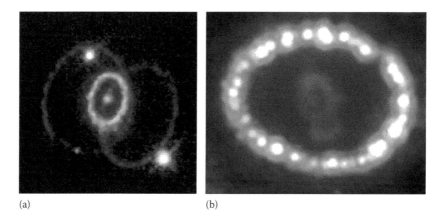

(a)   (b)

FIGURE 6.11a,b (a) *HST* image of SN 1987A and several rings, obtained in February 1994; credit: Christopher Burrows, ESA/STScI, and NASA (http://hubblesite.org/newcenter/archive/releases/star/supernova/1994/22/image/a/). (b) *HST* image of SN 1987A and the inner ring obtained in December 2006, roughly the 20-year anniversary of SN 1987A. Many "hot spots" have lit up, and the extended ejecta are clearly aspherical; credit: NASA, ESA, P. Challis, and R. Kirshner (Harvard–Smithsonian Center for Astrophysics) (http://hubblesite.org/newcenter/archive/releases/star/supernova/2007/10/image/h/).

within the ejecta (e.g., Graves et al., 2005); perhaps the initial neutron star subsequently collapsed to form a BH, but it might also be hidden by newly formed dust.

Stripped-Envelope Core-Collapse Supernovae

Massive stars that explode via the mechanism described above for SNe II, but bereft of most or all of their outer envelope, are known as "stripped-envelope core-collapse SNe." The envelopes are eliminated through winds or transfer to a binary companion star prior to the explosion (e.g., Wheeler and Harkness, 1990; Woosley et al., 1995).

Type Ib SNe, for example, whose optical spectra exhibit He I lines instead of H, are almost certainly massive stars that have lost their outer H envelope (e.g., Filippenko, 1997, and references therein). They tend to occur in late-type galaxies, near regions of active star formation, and analysis of their spectra is also consistent with the explosion of a massive, H-poor progenitor. A schematic cross section of the progenitor of a SN Ib immediately prior to the explosion is shown in Figure 6.12 in comparison with that of a SN II.

In some cases, not all of the progenitor star's hydrogen envelope is removed prior to the explosion, giving rise to a Type IIb SN (Woosley et al., 1987). The first known example was SN 1987K

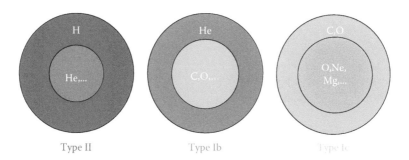

Type II   Type Ib   Type Ic

FIGURE 6.12 Idealized cross sections of the progenitor stars of Type II, Ib, and Ic SNe, immediately prior to the explosion. Only the outer few layers are shown, but the layers interior to them may be seen in Figure 6.10. (Credit: Alex Filippenko and Frank Serduke.)

(Filippenko, 1988), but a much more extensively studied case was SN 1993J (e.g., Filippenko et al., 1993, 1994; Matheson et al., 2000a, b). Maund et al. (2004) illustrate a late-time optical spectrum of SN 1993J that appears to reveal the companion star, a luminous B-type star that probably gained some of the mass lost by the progenitor of SN 1993J. However, the plentiful circumstellar gas evident from optical, radio, and x-ray studies indicates that part of the material was not captured by the companion.

In the case of SNe Ic, the massive progenitor star lost its H envelope and most, if not all, of the He layer as well—again, either through winds, mass transfer, or some combination of these factors. The degree of stripping probably depends on many variables, such as the star's mass, metallicity, rotation rate, and (if in a binary system) the separation from the companion star.

Spectropolarimetry of core-collapse SNe may reveal asymmetries in the explosion; the electron-scattering atmosphere can produce a net polarization in the received light if it is not perfectly spherical (for reviews, see Leonard and Filippenko, 2005; Wang and Wheeler, 2008). A thorough study of the Type II-P SN 2004dj (Leonard et al., 2006) showed that the polarization was nearly zero in the first few months after the explosion, but then jumped to a substantial value, subsequently declining gradually with time. This indicates that although the H envelope was spherical, the He core and deeper layers were progressively more aspherical (Figure 6.13).

The above trend is confirmed by spectropolarimetry of Type IIb, Ib, and Ic SNe: the more highly stripped progenitors exhibit greater asymmetries, suggesting that the inner parts of massive exploding stars are less spherical than the outer parts. Axial ratios of 1.2–1.3 are commonly inferred from the data (Wang and Wheeler, 2008, and references therein). Exactly how the observed asymmetries are produced, and what their link is with the explosion mechanisms, is not yet known; indeed, successful explosions of core-collapse SNe are still difficult to produce in numerical calculations.

Probing supernova geometry with spectropolarimetry

Depth into atmosphere probed by data:

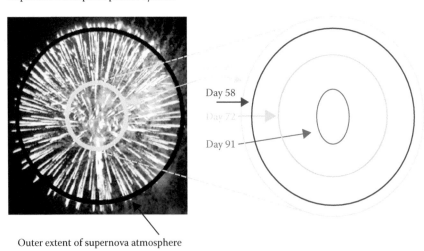

Outer extent of supernova atmosphere

FIGURE 6.13 Multi-epoch spectropolarimetry of the Type II-P SN 2004dj reveals that the percentage polarization increases with time, implying that the degree of asphericity increases with depth in the ejecta. (Credit: Douglas C. Leonard and Alex Filippenko; reproduced with permission of Douglas C. Leonard.)

GAMMA-RAY BURSTS AND ASSOCIATED SUPERNOVAE

In the 1960s, the *Vela* spy satellites detected occasional bursts of gamma rays coming from seemingly random parts of the sky. These were not the possible violations of the partial nuclear-test-ban treaty they were designed to find, but rather events of cosmic origin (Klebesadel et al., 1973). Over the years, many additional "gamma-ray bursts" (GRBs) were found. (For a recent review of GRBs, see Gehrels et al., 2009.) Of these, two subclasses emerged: "long-duration GRBs" and "short-duration GRBs," with an approximate dividing line of ~2 sec for the duration of the gamma-ray outburst. Also, the short-duration bursts are generally characterized by "harder" spectra having relatively more high-energy radiation than the long-duration bursts.

It gradually became increasingly clear that the distribution of GRBs in the sky was isotropic, or nearly so; they definitely were not distributed in a manner consistent with being a population of objects in the disk of the Milky Way. A highly local origin (e.g., part of the Oort cloud of rocky iceballs) was unlikely and could be ruled out through several arguments. The other possibilities were that the origin was in the outer Galactic halo or that most GRBs are at cosmological distances. If exceedingly distant, however, the implied isotropic energies of GRBs were truly stupendous, the equivalent of roughly 1 $M_\odot$ of material converted to energy. Thus, quite a few astronomers favored the extended-halo hypothesis.

The launch of the *Compton Gamma-Ray Observatory* (*CGRO*) in April 1991 led to compelling evidence for a cosmological origin. Its Burst and Transient Source Experiment (BATSE) detected hundreds (and eventually thousands) of GRBs, revealing such a highly isotropic distribution that the halo model was essentially ruled out (Figure 6.14); there was no large-scale anisotropy visible, and there was no concentration of GRBs toward the Andromeda Galaxy (which would have been expected to contain a similar halo of GRBs).

The definitive proof that at least some (and probably most) GRBs are extremely distant objects came in the mid-1990s from the identification of x-ray, radio, and optical counterparts of GRBs (for a review, see van Paradijs et al., 2000). Specifically, the *BeppoSAX* ("Satellite per Astronomia X" [Italian for "X-ray Astronomy Satellite"] named for Giuseppe Occhialini) satellite was able to obtain x-ray images of the afterglow of some GRBs, allowing accurate positions to be measured,

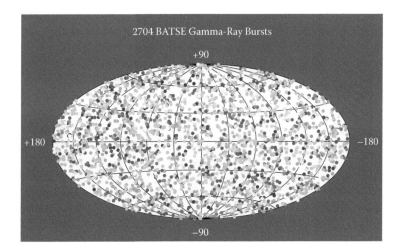

FIGURE 6.14 Observed distribution in the sky of 2704 GRBs detected by BATSE onboard *CGRO* during its 9-year mission. There is no clear deviation from isotropy in this equal-area all-sky projection; in particular, the Galactic plane would appear as a horizontal line in the middle of the diagram. Red points correspond to bright long-duration GRBs, purple points to weak ones; there is incomplete data for the grey points. Credit: NASA/Marshall Space Flight Center, Space Sciences Laboratory (http://gammaray.msfc.nasa.gov/batse/grb/skymap/).

and detections at radio and optical wavelengths soon followed. These afterglows appeared to be in or close to galaxies whose redshifts were subsequently measured to be high (e.g., $z \approx 1$). Moreover, spectra of the optical afterglows revealed absorption lines from intervening high-redshift clouds of gas. The first example was GRB 970508, whose optical afterglow spectrum exhibited an absorption-line system at $z = 0.835$ (Metzger et al., 1997).

The implied amount of gamma-ray energy emitted by GRBs was staggering, under the assumption of isotropic emission. Even at other wavelengths, these objects can be enormously luminous. The optical counterpart of GRB 990123 ($z = 1.6$) could be seen with a good pair of binoculars, and the optical counterpart of GRB 080319B ($z = 0.93$) reached naked-eye visibility (e.g., Bloom et al., 2009)! Perhaps, however, the emission was not isotropic. Indeed, various lines of evidence involving GRB light curves and spectra were used to argue that the radiation from GRBs is produced by highly relativistic particles and collimated into two oppositely directed, very narrow (a few degrees wide) beams—and we see only those GRBs whose beams are pointing at us.

Specifically, long-duration GRBs were generally found in galaxies having an anomalously high rate of star formation, with evidence for the presence of many very massive stars. The most likely scenario is that a highly collimated "fireball" of matter from a dying, massive star leads to relativistic shocks and the production of synchrotron radiation (see Mészáros, 2002 for a review). Although gamma rays are generated primarily by internal shocks within the jets, most of the radiation at lower energies is produced by the collision of the jets with circumstellar material. In particular, the "collapsar" model (Woosley, 1993; MacFadyen and Woosley, 1999) postulates that two relativistic beams of particles propagate along the rotation axis of an accretion disk around a BH that was formed when the rapidly rotating core of a very massive star collapses. The progenitor star should generally have been stripped of its outer envelope of at least H, and possibly even He, to decrease the amount of material through which the high-speed jets must propagate on their way out from the core.

Tentative evidence for the collapsar model was provided in 1998 by the discovery that the peculiar Type Ic SN 1998bw may have been temporally and spatially coincident with GRB 980425 (Galama et al., 1998). SN 1998bw was much more luminous than typical SNe Ic, had a broader light curve, and exhibited substantially higher ejecta velocities (roughly $0.1c$). These distinguishing characteristics increase the likelihood that SN 1998bw, dubbed a "hypernova" by a few researchers (e.g., Iwamoto et al., 1998), was indeed associated with GRB 980425, but some doubts remained; moreover, the GRB itself was very subluminous, so perhaps its possible link to SNe Ic was not representative of GRBs.

The light curves of some typical long-duration GRBs were also consistent with the collapsar model, in that the declining optical afterglow exhibited a "bump" in brightness a few weeks after the GRB event that could be attributed to light from an associated SN similar to SN 1998bw (e.g., Bloom et al., 1999). However, without a confirming spectrum, the true nature of the "bump" was not certain; it might, for example, be caused by collision of material in the jets with additional circumstellar gas.

Much stronger evidence for a link between long-duration GRBs and peculiar, overluminous, broad-lined SNe Ic came from GRB 030329. After the optical afterglow of the GRB had faded sufficiently, optical spectra of GRB 030329 during the time a bump was present in the afterglow light curve closely resembled those of SN 1998bw at comparable epochs (Matheson et al., 2003; Stanek et al., 2003; and others); this can be seen in Figure 6.15. The associated SN was dubbed SN 2003dh. Given that GRB 030329 was a normal-luminosity GRB, unlike GRB 980425, it was more difficult to argue against a probable link. That same year, additional evidence for a connection between broad-lined SNe Ic and long-duration GRBs was provided by GRB 031203 and SN 2003lw (Malesani et al., 2004).

However, evidence for a SN component, either in light curves or spectra, has not been found in the case of some long-duration GRBs. Perhaps a fraction of these, especially the GRBs, whose optical afterglows are also very faint or nonexistent, are hidden by dust. Or, perhaps the SNe failed to succeed, with the ejecta actually collapsing into the newly formed BH. (For a thorough discussion of the collapsar model and observational evidence for SNe associated with GRBs, see Woosley and Bloom, 2006.)

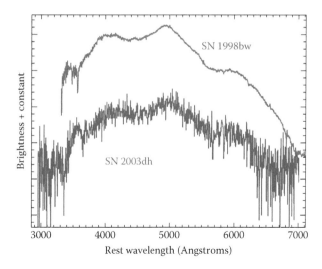

FIGURE 6.15 The optical spectrum of SN 2003dh (associated with GRB 030329), though noisy, is very similar to that of SN 1998bw (associated with GRB 980425), a peculiar broad-lined Type Ic SN. (Credit: Thomas Matheson; reproduced with permission.)

The above discussion of the physical nature of GRBs was limited to the long-duration GRBs because, for many years, no afterglows of short-duration GRBs had been detected, and their positions in the sky were in most cases not accurately known. This changed, however, with the launch of the *Swift* satellite in November 2004; after GRB detection with the Burst and Transient (BAT) instrument, an x-ray telescope (XRT) aboard *Swift* was sometimes able to reveal an associated fading x-ray counterpart. It was found that short-duration GRBs are generally not located in galaxies with active formation of massive stars; indeed, some objects (e.g., GRB 050509B; Bloom et al., 2006) seemed to be located in early-type galaxies having old stellar populations.

These data supported previous suspicions that many, if not most, short-duration GRBs are at cosmological distances and produced by the merging of two neutron stars to form a BH, or perhaps by the merging of a neutron star and a BH. (Some short-duration GRBs are probably associated with relatively nearby magnetars—neutron stars with exceedingly strong magnetic fields.) In both cases, given a sufficiently closely spaced binary system, the emission of gravitational waves can lead to an in-spiral and merging within a time substantially less than the Hubble time, although not on such short time scales as to be associated with a young stellar population. Details of the formation of the relativistic jet are still not understood, but the mechanism is probably related to the accretion disk formed when one of the in-spiraling objects is tidally disrupted by the other. In this case, an associated optical SN should not be seen, and indeed, no such objects have ever been detected at the locations of short-duration GRBs.

Let me end this discussion of GRBs by mentioning their potential for extending the Hubble diagram to substantially higher redshifts than those possible with SNe Ia. Although one might not expect long-duration GRBs to be good standard candles given that the relevant physical conditions should span a wide range and may evolve with cosmic time (e.g., Friedman and Bloom, 2005), some tantalizing correlations have been identified (e.g., Amati et al., 2002; Ghirlanda et al., 2004; Schaefer, 2007). A major problem is the dearth of low-redshift examples with which to compare the high-redshift GRBs. However, it is possible that SNe Ia may be used to calibrate GRBs at $z \leq 1$; the distance modulus (observed magnitude minus absolute magnitude) of a SN Ia is independent of the assumed cosmological model, and GRBs at the same redshifts as SNe Ia should have the same luminosity distances (Liang et al., 2008). On the other hand, low-redshift GRBs are generally observed to have a lower isotropic energy release than those at high redshift, a selection bias probably arising from volume effects and the intrinsic luminosity function; this casts some doubt on the validity of calibrating high-redshift GRBs with low-redshift events. At present, long-duration GRBs are not competitive with SNe Ia and some other techniques such as probes of dark energy, but the situation might improve in the future.

FUTURE RESEARCH

There are many interesting and important questions concerning SNe and GRBs that remain to be answered. I briefly mention some of them here to illustrate the opportunities for future research.

1. SNe Ia are the explosions of white dwarfs, but how exactly do these stars get close to the Chandrasekhar limit? Are they single degenerates or double degenerates, or do both types of systems exist?
2. Can some SNe Ia be produced by super-Chandrasekhar or sub-Chandrasekhar mass white dwarfs?
3. What is the nature of the burning front in a SN Ia? Is it initially subsonic and then supersonic, and what determines the transition? Does it begin somewhat away from the center and propagate out in a main direction or in a more spherically symmetric manner?
4. Are there "super nova" (as opposed to "supernova") explosions that involve only the outer layers of a white dwarf, rather than the thermonuclear disruption of the entire white dwarf?
5. Do SNe Ia evolve with time? That is, was the peak power of SNe Ia billions of years ago the same as it is now? This is crucial to know if we are to continue using these objects to constrain the properties of the mysterious "dark energy" that is currently accelerating the expansion of the Universe.
6. How exactly does a massive star blow up at the end of its life? We know that the Fe core collapses to form a neutron star (or, in some cases, a BH), but what leads to a successful explosion of the surrounding layers of gas?
7. We know that the core of SN 1987A initially collapsed to form a neutron star, given the detected neutrinos. But did the neutron star subsequently collapse to form a BH?
8. Did the progenitors of Type Ib and Ic SNe lose their mass through winds or via mass transfer to a companion star in a binary system? What are the initial mass ranges of the progenitors?
9. Do all long-duration GRBs have SNe associated with them, and are they always broad-lined SNe Ic? Does the core of a massive star necessarily collapse to form a BH rather than a neutron star?
10. What are the details of the explosion mechanism of long-duration GRBs, and how are the relativistic jets formed?
11. Can long-duration GRBs be calibrated sufficiently well to be of substantial utility in the determination of cosmological parameters?
12. Do short-duration GRBs result from the merging of two neutron stars, or a neutron star merging with a BH? Are there any other mechanisms, such as magnetars?
13. Do exceedingly massive stars ($>150\ M_\odot$) explode as "pair-instability SNe?" In these theoretically predicted objects, the massive oxygen core becomes so hot that the highest-energy gamma rays spontaneously turn into electron-positron pairs, robbing the star's core of pressure and causing a contraction that subsequently leads to a gigantic thermonuclear runaway explosion.
14. What kind of pre-explosion mass loss do massive stars go through? In some cases, such as η Carinae, there are violent outbursts that do not completely destroy the star. What causes these gigantic "burps?"
15. What are the explosion mechanisms of the many peculiar varieties of SNe that have been found in the past few years?

CONCLUSION

I have provided a brief review of all types of SNe and the probable link between some GRBs and peculiar SNe. SNe Ia arise from the thermonuclear runaway of a C/O white dwarf whose mass grows to a value close to the Chandrasekhar limit through accretion from a companion star. Many details, however, are still unknown. The fact that their peak optical luminosities are very high, nearly uniform, and able to be calibrated has made SNe Ia enormously useful as distance indicators and led to the discovery that the expansion of the Universe is currently accelerating after initially decelerating.

Stars having initial masses above about 8 $M_\odot$ explode as core-collapse SNe: Type II (hydrogen rich), IIb (low-mass H envelope), Ib (no H; He outer envelope), and Ic (no H or He). SN 1987A, the brightest SN since the time of Galileo, has been extensively observed and has confirmed several key aspects of SN theory, but it also provides some interesting surprises. Spectropolarimetry shows that asymmetries are important in core-collapse SNe, although it is not yet known how some of them arise.

Cosmic GRBs, which for decades were an enigma, are now known to generally reside at cosmological distances. Moreover, there is strong evidence linking at least some of the long-duration GRBs with especially luminous, energetic SNe Ic. Long-duration GRBs may be used to extend the Hubble diagram to higher redshifts than those achievable with SNe Ia.

The research fields of SNe and GRBs are among the most exciting and important in all of astrophysics. The past few decades have led to an explosion of data throughout the electromagnetic spectrum, and even the detection of neutrinos from SN 1987A. Another quantum leap should occur when gravitational waves are detected from merging neutron stars or perhaps from highly asymmetric core-collapse SNe and GRBs.

Humans have made tremendous breakthroughs in their study of celestial phenomena, as is well illustrated in the case of high-energy transients such as SNe and GRBs. Galileo himself would surely have been overjoyed had he known what was in store during the 400 years since his first use of simple, humble telescopes to make monumental discoveries concerning our Solar System and the Milky Way. The future of investigations of SNe and GRBs appears equally bright, with wonderful new visions and insights surely lurking just beyond our current horizons.

ACKNOWLEDGMENTS

I would like to thank the organizers of the *New Vision 400* conference in 2008 for creating a very stimulating and informative program and our local hosts in Beijing for their hospitality. My group's research on SNe and GRBs has been supported by the US National Science Foundation (most recently with grant AST-0908886), NASA, the Sylvia & Jim Katzman Foundation, the Richard & Rhoda Goldman Fund, Gary and Cynthia Bengier, and the TABASGO Foundation.

REFERENCES

Amati, L., Frontera, F., Tavani, M., et al. (2002). Intrinsic spectra and energetics of BeppoSAX gamma-ray bursts with known redshifts. *Astronomy & Astrophysics,* 390: 81–9.

Arnett, W.D. (1987). Supernova theory and supernova 1987A. *Astrophysical Journal,* 319: 136–42.

Arnett, W.D., Bahcall, J.N., Kirshner, R.P., et al. (1989). Supernova 1987A. *Annual Review of Astronomy and Astrophysics,* 27: 629–700.

Astier, P., Guy, J., Regnault, N., et al. (2006). The Supernova Legacy Survey: Measurement of Ω_M, Ω, and w from the first year data set. *Astronomy & Astrophysics,* 447: 31–48.

Barbon, R., Ciatti, F., and Rosino, L. (1979). Photometric properties of Type II supernovae. *Astronomy & Astrophysics,* 72: 287–92.

Bethe, H.A. (1990). Supernova mechanisms. *Reviews of Modern Physics,* 62: 801–66.

Bionta, R.M., Blewitt, G., Bratton, C.B., et al. (1987). Observation of a neutrino burst in coincidence with Supernova 1987A in the Large Magellanic Cloud. *Physical Review Letters,* 58: 1494–496.

Bloom, J.S., Kulkarni, S.R., Djorgovski, S.G., et al. (1999). The unusual afterglow of the gamma-ray burst of 26 March 1998 as evidence for a supernova connection. *Nature,* 401: 453–56.

Bloom, J.S., Perley, D.A., Li, W., et al. (2009). Observations of the naked-eye GRB 080319B: Implications of nature's brightest explosion. *Astrophysical Journal,* 691: 723–37.

Bloom, J.S., Prochaska, J.X., Pooley, D., et al. (2006). Closing in on a short-hard burst progenitor: Constraints from early-time optical imaging and spectroscopy of a possible host galaxy of GRB 050509b. *Astrophysical Journal,* 638: 354–68.

Branch, D., Livio, M., Yungelson, L.R., et al. (1995). In search of the progenitors of Type Ia supernovae. *Publications of the Astronomical Society of the Pacific,* 107: 1019–29.

Burrows, A., Hayes, J., and Fryxell, B.A. (1995). On the nature of core-collapse supernova explosions. *Astrophysical Journal,* 450: 830–50.

Burrows, A., Livne, E., Dessart, L., et al. (2006). A new mechanism for core-collapse supernova explosions. *Astrophysical Journal,* 640: 878–90.

Chandrasekhar, S. (1931). The maximum mass of ideal white dwarfs. *Astrophysical Journal,* 74: 81–2.

Doggett, J.B. and Branch, D. (1985). A comparative study of supernova light curves. *Astronomical Journal,* 90: 2303–311.

Filippenko, A.V. (1988). Supernova 1987K—Type II in youth, Type Ib in old age. *Astronomical Journal,* 96: 1941–948.

Filippenko, A.V. (1997). Optical spectra of supernovae. *Annual Review of Astronomy and Astrophysics,* 35: 309–55.

Filippenko, A.V. (2005). Type Ia supernovae and cosmology. In: E.M. Sion, S. Vennes, and H.L. Shipman (eds.), *White Dwarfs: Cosmological and Galactic Probes,* 97–133. Dordrecht: Springer.

Filippenko, A.V. and Riess, A.G. (1998). Results from the High-Z Supernova Search Team. *Physics Reports,* 307: 31–44.

Filippenko, A.V., Li, W.D., Treffers, R.R., et al. (2001). The Lick Observatory supernova search with the Katzman Automatic Imaging Telescope. In: W.P. Chen, C. Lemme, and B. Paczyński (eds.), *Small-Telescope Astronomy on Global Scales,* 121–30. San Francisco: ASP.

Filippenko, A.V., Matheson, T., and Barth, A.J. (1994). The peculiar Type II Supernova 1993J in M8: Transition to the nebular phase. *Astronomical Journal,* 108: 2220–225.

Filippenko, A.V., Matheson, T., and Ho, L.C. (1993). The "Type IIb" Supernova 1993J in M8: A close relative of Type Ib Supernovae. *Astrophysical Journal,* 415: L103–6.

Filippenko, A.V., Richmond, M.W., Branch, D., et al. (1992a). The subluminous, spectroscopically peculiar Type Ia supernova 1991bg in the elliptical galaxy NGC 4374. *Astronomical Journal,* 104: 1543–556.

Filippenko, A.V., Richmond, M.W., Matheson, T., et al. (1992b). The peculiar Type Ia SN 1991T: Detonation of a white dwarf? *Astrophysical Journal,* 384: L15–18.

Friedman, A.S. and Bloom, J.S. (2005). Toward a more standardized candle using gamma-ray burst energetics and spectra. *Astrophysical Journal,* 627: 1–25.

Frieman, J.A., Turner, M.S., and Huterer, D. (2008). Dark energy and the accelerating universe. *Annual Review of Astronomy and Astrophysics,* 46: 385–432.

Galama, T., Vreeswijk, P. M., van Paradijs, J., et al. (1998). An unusual supernova in the error box of the gamma-ray burst of 25 April 1998. *Nature,* 395: 670–72.

Ganeshalingam, M., Li, W., Filippenko, A.V., et al. (2010). Results of the Lick Observatory Supernova Search Follow-up Photometry Program: BVRI Light Curves of 165 Type Ia Supernovae. *Astrophysical Journal Supplement Series,* 190: 418–48.

Gehrels, N., Ramirez-Ruiz, E., and Fox, D.B. (2009). Gamma-ray bursts in the *Swift* Era. *Annual Review of Astronomy and Astrophysics,* 47: 567–617.

Ghirlanda, G., Ghisellini, G., and Lazzati, D. (2004). The collimation-corrected gamma-ray burst energies correlate with the peak energy of their vF_v spectrum. *Astrophysical Journal,* 616: 331–38.

Graves, G.J.M., Challis, P.M., Chevalier, R.M., et al. (2005). Limits from the *Hubble Space Telescope* on a Point Source in SN 1987A. *Astrophysical Journal,* 629: 944–59.

Hamuy, M., Phillips, M.M., Suntzeff, N.B., et al. (1996). The absolute luminosities of the Calán/Tololo supernovae. *Astronomical Journal,* 112: 2391–397.

Herant, M., Benz, W., Hix, W.R., et al. (1994). Inside the supernova—A powerful convective engine. *Astrophysical Journal,* 435: 339–61.

Hester, J.J. (2008). The Crab Nebula: An astrophysical chimera. *Annual Review of Astronomy and Astrophysics,* 46: 127–55.

Hillebrandt, W. and Niemeyer, J.C. (2000). Type Ia supernova explosion models. *Annual Review of Astronomy and Astrophysics,* 38: 191–230.

Hirata, K., Kajita, T., Koshiba, M., et al. (1987). Observation of a neutrino burst from the Supernova SN1987A. *Physical Review Letters,* 58: 1490–493.

Howell, D.A., Sullivan, M., Nugent, P.E., et al. (2006) The Type Ia supernova SNLS-03D3bb from a super-Chandrasekhar-mass white dwarf star. *Nature,* 443: 308–11.

Iben, I., Jr. (1974). Post main sequence evolution of single stars. *Annual Review of Astronomy and Astrophysics,* 12: 215–56.

Iwamoto, K., Mazzali, P.A., Nomoto, K., et al. (1998). A hypernova model for the supernova associated with the gamma-ray burst of 25 April 1998. *Nature,* 395: 672–74.

Janka, H-T. and Mueller, E. (1996). Neutrino heating, convection, and the mechanism of Type-II supernova explosions. *Astronomy & Astrophysics,* 306: 167–98.

Kaler, J.B. (1985). Planetary nebulae and their central stars. *Annual Review of Astronomy and Astrophysics,* 23: 89–117.

Khokhlov, A.M. (1991). Delayed detonation model for Type Ia supernovae. *Astronomy & Astrophysics,* 245: 114–28.

Klebesadel, R.W., Strong, I.B., and Olson, R.A. (1973). Observations of gamma-ray bursts of cosmic origin. *Astrophysical Journal,* 182: L85–8.

Kowalski, M., Rubin, D., Aldering, G., et al. (2008). Improved cosmological constraints from new, old, and combined supernova data sets. *Astrophysical Journal,* 686: 749–78.

Leonard, D.C. and Filippenko, A.V. (2005). Spectropolarimetry of core-collapse supernovae. In: M. Turatto, S. Benetti, L. Zampieri, et al. (eds.), *1604–2004, Supernovae as Cosmological Lighthouses,* 330–36. San Francisco: ASP.

Leonard, D.C., Filippenko, A.V., Ganeshalingam, M., et al. (2006). A non-spherical core in the explosion of Supernova SN 2004dj. *Nature,* 440: 505–7.

Li, W., Van Dyk, S.D., Filippenko, A.V., et al. (2006). Identification of the Red Supergiant Progenitor of Supernova 2005cs: Do the Progenitors of Type II-P Supernovae Have Low Mass? *Astrophysical Journal* 641: 1060–70.

Liang, N., Xiao, W.K., Liu, Y., et al. (2008). A cosmology-independent Calibration of gamma-ray burst luminosity relations and the Hubble diagram. *Astrophysical Journal,* 685: 354–60.

Malesani, D., Tagliaferri, G., Chincarini, G., et al. (2004). SN 2003lw and GRB 031203: A bright supernova for a faint gamma-ray burst. *Astrophysical Journal,* 609: L5–8.

Matheson, T., Filippenko, A.V., Barth, A.J., et al. (2000a). Optical spectroscopy of Supernova 1993J during its first 2500 days. *Astronomical Journal,* 120: 1487–98.

Matheson, T., Filippenko, A.V., Ho, L.C., et al. (2000b). Detailed analysis of early to late-time spectra of Supernova 1993J. *Astronomical Journal,* 120: 1499–515.

Matheson, T., Garnavich, P.M., Stanek, K.Z., et al. (2003). Photometry and spectroscopy of GRB 030329 and its associated Supernova 2003dh: The first two months. *Astrophysical Journal,* 599: 394–407.

MacFadyen, A.I. and Woosley, S.E. (1999). Collapsars: gamma-ray bursts and explosions in "failed supernovae." *Astrophysical Journal,* 524: 262–89.

Maund, J.R., Smartt, S.J., Kudritzki, R.P., et al. (2004). The massive binary companion star to the progenitor of Supernova 1993J. *Nature,* 427: 129–31.

McCray, R. (1993). Supernova 1987A revisited. *Annual Review of Astronomy and Astrophysics,* 31: 175–216.

Mészáros, P. (2002). Theories of gamma-ray bursts. *Annual Review of Astronomy and Astrophysics,* 40: 137–69.

Metzger, M.A., Djorgovski, S.G., Kulkarni, S.R., et al. (1997). Spectral constraints on the redshift of the optical counterpart to the gamma-ray burst of 8 May 1997. *Nature,* 387: 878–80.

Murphy, J.W., Ott, C.D., and Burrows, A. (2009). A model for gravitational wave emission from neutrino-driven core-collapse supernovae. *Astrophysical Journal,* 707: 1173–90.

Nomoto, K., Thielemann, F.-K., and Yokoi K. (1984). Accreting white dwarf models of Type I supernovae. III—carbon deflagration supernovae. *Astrophysical Journal,* 286: 644–58.

Perlmutter, S., Aldering, G., Goldhaber, G., et al. (1999). Measurements of Ω and Λ from 42 high-redshift supernovae. *Astrophysical Journal,* 517: 565–86.

Phillips, M.M. (1993). The absolute magnitudes of Type Ia supernovae. *Astrophysical Journal,* 413: L105–08.

Pskovskii, Yu.P. (1977). Light curves, color curves, and expansion velocity of Type I supernovae as functions of the rate of brightness decline. *Soviet Astronomy,* 21: 675–82.

Riess, A.G., Filippenko, A.V., Challis, P., et al. (1998). Observational evidence from supernovae for an accelerating universe and a cosmological constant. *Astronomical Journal,* 116: 1009–38.

Riess, A.G., Macri, L., Casertano, S., et al. (2009). A Redetermination of the Hubble Constant with the *Hubble Space Telescope* from a Differential Distance Ladder. *Astrophysical Journal* 699: 539–63.

Riess, A.G., Press, W., and Kirshner, R.P. (1995). Using Type Ia supernova light curve shapes to measure the Hubble constant. *Astrophysical Journal,* 438: L17–20.

Riess, A.G., Press, W., and Kirshner, R.P. (1996). A precise distance indicator: Type Ia supernova multicolor light-curve shapes. *Astrophysical Journal,* 473: 88–109.

Riess, A.G., Strolger, L-G., Casertano, S., et al. (2007). New *Hubble Space Telescope* discoveries of Type Ia supernovae at $z \geq 1$: Narrowing constraints on the early behavior of dark energy. *Astrophysical Journal,* 659: 98–121.

Riess, A.G., Strolger, L-G., Tonry, J., et al. (2004). Type Ia supernova discoveries at $z > 1$ from the *Hubble Space Telescope*: Evidence for past deceleration and constraints on dark energy evolution. *Astrophysical Journal,* 607: 665–87.

Schaefer, B.E. (2007). The Hubble diagram to redshift >6 from 69 gamma-ray bursts. *Astrophysical Journal,* 660: 16–46.

Smartt, S.A. (2009). Progenitors of core-collapse supernovae. *Annual Review of Astronomy and Astrophysics,* 47: 63–106.

Spergel, D.N., Bean, R., Doré, O., et al. (2007). Three-Year *Wilkinson Microwave Anisotropy Probe* (*WMAP*) observations: Implications for cosmology. *Astrophysical Journal Supplement Series,* 170: 377–408.

Stanek, K.Z., Matheson, T., Garnavich, P.M., et al. (2003). Spectroscopic discovery of the Supernova 2003dh associated with GRB 030329. *Astrophysical Journal,* 591: L17–20.

Trimble, V. (1982). Supernovae. Part I: The events. *Reviews of Modern Physics,* 54: 1183–224.

Trimble, V. (1983). Supernovae. Part II: The aftermath. *Reviews of Modern Physics,* 55: 511–63.

van Paradijs, J., Kouveliotou, C., and Wijers, R.A.M.J. (2000). Gamma-ray burst afterglows. *Annual Review of Astronomy and Astrophysics,* 38: 379–425.

Wang, L. and Wheeler, J.C. (2008). Spectropolarimetry of supernovae. *Annual Review of Astronomy and Astrophysics,* 46: 433–74.

Wheeler, J.C. and Harkness, R.P. (1990). Type I Supernovae. *Reports on Progress in Physics,* 53: 1467–557.

Wood-Vasey, W.M., Miknaitis, G., Stubbs, C.W., et al. (2007). Observational constraints on the nature of dark energy: First cosmological results from the ESSENCE supernova survey. *Astrophysical Journal,* 666: 694–715.

Woosley, S.E. (1993). Gamma-ray bursts from stellar mass accretion disks around black holes. *Astrophysical Journal,* 405: 273–77.

Woosley, S.E. and Bloom, J.S. (2006). The supernova gamma-ray burst connection. *Annual Review of Astronomy and Astrophysics,* 44: 507–56.

Woosley, S.E. and Weaver, T. (1986). The physics of supernova explosions. *Annual Review of Astronomy and Astrophysics,* 24: 205–53.

Woosley, S.E., Langer, N., and Weaver, T.A. (1995). The presupernova evolution and explosion of helium stars that experience mass loss. *Astrophysical Journal,* 448: 315–38.

Woosley, S.E., Pinto, P.A., Martin, P.G., et al. (1987). Supernova 1987A in the Large Magellanic Cloud—the explosion of an ~20 $M_\odot$ star which has experienced mass loss? *Astrophysical Journal,* 318: 664–73.

7 The Dark Secrets of Gaseous Nebulae: Highlights from Deep Spectroscopy

Xiao-Wei Liu

CONTENTS

EMISSION-LINE NEBULAE

The existence and distribution of the chemical elements and their isotopes are a consequence of nuclear processes that have taken place in the past during the Big Bang and subsequently in stars and in the interstellar medium (ISM) where they are still ongoing (Pagel, 1997). A large body of our knowledge of the distribution and production of elements in the Universe rests on observations and analyses of photoionized gaseous nebulae. Ionized and heated by strong ultraviolet (UV) radiation fields, photoionized gaseous nebulae glow by emitting strong emission lines (Osterbrock and Ferland, 2005). They are therefore also commonly named emission-line nebulae.

Examples of emission-line nebulae include H II regions, planetary nebulae (PNe), and the broad and narrow emission-line regions found in active galactic nuclei (Figure 7.1). H II regions are diffuse nebulae found around newly formed young, massive stars and trace the current status of the ISM. Giant extra-Galactic H II regions, signposts of massive star formation activities, are among the most prominent features seen in a gas-rich, star-forming galaxy. In some galaxies, the star-forming activities are so intense that the whole galaxy becomes a giant H II region. Such galaxies are called H II or starburst galaxies and are observable to large cosmic distances.

PNe are among the most beautiful objects in the sky and, arguably, the queens of the night. They were given the name by William Herschel (Herschel, 1785) based on their distinct structures and, for some of them, nearly circular and overall uniform appearances resembling the greenish disk of a planet. They have, however, nothing to do with a planet (but see later) and are, in fact, expanding gaseous envelopes expelled by low- and intermediate-mass stars in late evolutionary stage after the exhaustion of central nuclear fuel at the end of the asymptotic giant branch (AGB) phase. Because of their relatively simple geometric structure, a nearly symmetric shell of gas ionized by a single, centrally located white dwarf, PNe are ideal cosmic laboratories to study the atomic and radiative

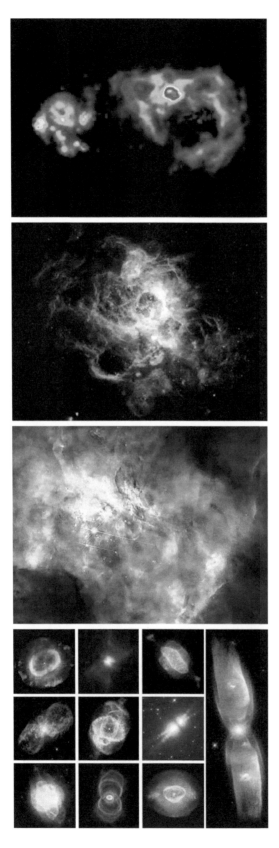

FIGURE 7.1   Examples of emission-line nebulae. *Left to right:* PNe (*HST* images obtained by B. Balick and collaborators; see http://www.astro.washington.edu/users/balick/WFPC2/index.html). The H ɪɪ region M42 (the Orion Nebula; *Hubble* Heritage image obtained by C. R. O'Dell and S. K. Wong, see http://hubblesite.org/newscenter/archive/releases/1995/45); NGC 604, a giant extra-Galactic H ɪɪ region in the outskirts of the Local Group spiral galaxy M33 (*Hubble* Heritage image obtained by H. Yang, see http://heritage.stsci.edu/2003/30/supplemental.html; The starburst galaxy I Zw 18 (based on *HST*/WFPC2 data obtained by E. Skillman). The linear sizes of these objects differ vastly, ranging from about a tenth of a parsec (1 pc = 3.262 lt-yr = 3.086×10^{16} m) for PNe and M 42 to several hundred parsecs for NGC 604 and I Zw 18.

processes governing cosmic low-density plasmas. PNe have played and continue to play a central role in formulating the theory of photoionized gaseous nebulae. Representing a pivotal, albeit transient, evolutionary phase of low- and intermediate-mass stars (the overwhelming majority in a galaxy), PNe play a major role in the Galactic ecosystem—in the constant enrichment of metals, in the formation and destruction of molecules and dust grains, and in the recycling of gas in the ISM. Today, studies of PNe have gone far beyond the objects themselves. PNe are widely used to trace the kinematics of the host galaxies and the intracluster stellar populations. They have even been successfully utilized to measure the Hubble constant of cosmic expansion.

Recent progress in observational techniques, atomic data, and high-performance computation have enabled reliable measurements and analyses of lines as faint as one-millionth of $H\beta$, including weak optical recombination lines (ORLs) from abundant heavy elements (C, N, O, Ne, and Mg), and collisionally excited lines (CELs) from rare elements, such as fluorine and s- and r-process elements. This allows one to address some of the longstanding problems in nebular astrophysics and opens up new windows of opportunity.

In this chapter, I will briefly review the development of the theory of photoionized gaseous nebulae, highlighting some of the key events. I will then present some recent developments of deep spectroscopy of PNe and H II regions, concentrating on observations of faint heavy-element ORLs. I will show that there is strong evidence that nebulae contain another previously unknown component of cold (about 1,000 K), high-metallicity plasma, probably in the form of H-deficient inclusions embedded in the warm (about 10,000 K) diffuse nebula of "normal (i.e., near solar) composition." This cold gas emits essentially all the observed flux of heavy-element ORLs, but is too cool to excite any significant optical or UV CELs and thus is invisible via the latter. The existence of H-deficient gas in PNe and probably also in H II regions, not predicted by current stellar-evolution theory, provides a natural solution to the longstanding dichotomy between nebular plasma diagnostics and abundance determinations using CELs on the one hand and ORLs on the other, a discrepancy that is ubiquitously observed in Galactic and extra-Galactic PNe as well as H II regions.

THE FOUNDING OF THE THEORY OF PHOTOIONIZED GASEOUS NEBULAE

The applications of the principle of "chemical analysis by observations of spectra" (expounded by G. Kirchhoff and R. W. Bunsen, 1860) by W. Huggins and W. A. Miller to the analyses of stellar and nebular spectra in the 1860s heralded the rise of astrophysics. In an accompanying paper to their monumental work on stellar spectra, Huggins and Miller presented their first visual spectroscopic observations of eight PNe, including the Cat's Eye Nebula NGC 6543 in Draco (Huggins and Miller, 1864). Instead of the dark (absorption) Fraunhofer lines observed in the spectra of the Sun and other stars, they found bright emission lines in the nebular spectra and concluded that the nebulae must consist of "enormous masses of luminous gas or vapor." While the Fraunhofer F line ($H\beta$ at 4861 Å) did appear in emission, the two bright, nearby lines λλ4959, 5007 were not Fraunhofer lines at all. The availability of dry photographic plates (light-sensitive silver bromide salts held in a gelatin emulsion on the glass), replacing the old wet collodion plates, made it possible to record long exposures of nebular spectra, and more nebular emission lines were revealed in the blue and UV wavelength regions. In 1881, Huggins successfully photographed a UV spectrum of the nearest bright H II region, the Orion Nebula, and detected a strong emission line near 3728 Å (Huggins, 1881).

By the late 1920s, dozens of nebular lines had been detected and their wavelengths accurately measured (Wright, 1918), yet most of them remained unidentified and were attributed to some hypothetical element "nebulium." The big breakthrough in understanding nebular spectra came in 1927 from Ira Bowen, who, stimulated by a heuristic speculation by H. N. Russell that the nebulium lines must be caused by abundant "atoms of known kinds shining under unfamiliar conditions" such as in gas of very low density (Russell et al., 1927), discovered that eight of the strongest nebular lines were caused by the forbidden transitions from the low-excitation metastable states of the ground electron configurations of singly ionized oxygen, singly ionized nitrogen, and doubly ionized oxygen (Bowen, 1927a,b,c). For the first time, the physical processes in gaseous nebulae could be understood. This important discovery paved the way for future studies of nebular structure and chemical composition.

Great strides forward in nebular spectral observations were made starting in 1910 and continued through the 1930s, powered by a number of technological inventions, such as the Schmidt camera (Schmidt, 1938), the high-efficiency blazed diffraction grating (Wood, 1910), and the image slicer (Bowen, 1938). Deep exposures revealed faint lines from the refractory elements such as potassium, calcium, silicon, magnesium, and iron, demonstrating that nebulae are qualitatively made from similar material to stars (Bowen and Wyse, 1939). During the next 30 years, the structures and the underlying physics governing photoionized gaseous nebulae were worked out quantitatively, including photoionization and recombination (Zanstra, 1927; Strömgren, 1939), heating and cooling (Zanstra, 1929; Baker et al., 1938; Aller et al., 1939; Spitzer, 1948), recombination and collisional excitation (Baker and Menzel, 1938; Menzel et al., 1941; Seaton, 1954; Burgess, 1958; Seaton, 1959a,b; Seaton and Osterbrock, 1957), and elemental abundance determinations (Bowen and Wyse, 1939; Menzel and Aller, 1941; Wyse, 1942; Aller and Menzel, 1945). The origin of PNe, as descendants of red giant stars, was also understood (Shklovskii, 1957). Photoionization models incorporating all known physics were constructed (Hjellming, 1966; Goodson, 1967; Harrington, 1968; Rubin, 1968; Flower, 1969a,b), and the models reproduced observations well. The theory of photoionized gaseous nebulae as it stood in late 1960s was nicely summarized in Osterbrock (1974).

CELS AND ORLS: THE DICHOTOMY

While the theory seemed well established and solid, there were dark clouds hovering on the horizon. One concerned the measurement and interpretation of weak nebular emission lines, and the other concerned the possible presence of significant temperature inhomogeneities in nebulae and their effects on nebular abundance determinations.

Except for a few lines excited under specific environments, such as the Bowen fluorescence lines (e.g., O III λλ3133, 3341, 3444; Bowen, 1934, 1935), strong lines radiated by photoionized gaseous nebulae fall into two categories: recombination lines (RLs) and CELs. Hydrogen and helium ions, the main baryonic components of an ionized gaseous nebula, capture free electrons and recombine, followed by cascades to the ground state. During this process, a series of RLs are radiated (e.g., H α λ6563; H β λ4861; He I λλ4472, 5876, 6678; and He II λ4686). The ground electron configuration of multi-electron ions of heavy elements yields some low-excitation energy levels (within a few eV from the ground state, such as O^{++} $2p^2$ $^3P^o_{0,1,2}$, $^1D^o_2$, $^1S^o_0$), which can be excited by impacts of thermal electrons, typically having energies of ~1 eV in photoionized gaseous nebulae of solar composition. Follow-up de-excitation by spontaneous emission yields the so-called CELs. Quite often, those lines are electron dipole forbidden transitions, so they are commonly called forbidden lines (e.g., [O II] λλ3726, 3729; [O III] λλ4959, 5007; [N II] λλ6548, 6584; and [Ne III] λλ3868, 3967). Recombination of free electrons to bound states of hydrogen and helium ions also yields weak nebular continuum emission. For example, recombination to the H I $n = 2$ state yields the near-UV nebular continuum Balmer discontinuity at λ < 3646 Å. Recombination of heavy-element ions, followed by cascading to the ground state, also produces RLs. However, for typical cosmic composition, i.e., approximately solar, even the most abundant heavy element oxygen has a number density less than one-thousandth of hydrogen. Thus, heavy-element RLs are much weaker, at the level of a few-thousandths of H β or less, and are observable generally only in the optical part of the spectrum. Those lines are therefore often called ORLs. Sample ORLs from abundant heavy-element ions that have been well studied and that are discussed in the current contribution include C II λ4267, C III λ4649, N II λ4041, O I λ7773, O II λ4089, O III λ3265, Ne II λ4392, and Mg II λ4481.

Except for IR fine-structure lines arising from ground spectral terms, emissivities of CELs have an exponential dependence on electron temperature, T_e, $\varepsilon(X^{+i}, \lambda) \propto T_e^{-1/2} \exp(-E_{ex}/kT_e)$ (the Boltzmann factor), where E_{ex} is excitation energy of the upper level of transition emitting wavelength

λ by ion X^{+i} following collisional excitation by electron impacts (Figure 7.2). For any energy level above the ground, a critical density N_c can be defined such that collisional de-excitation dominates over spontaneous radiative decay while depopulating the level; see Osterbrock and Ferland, 2005. At low densities, $N_e \ll N_c$, i.e., electron density N_e much lower than the upper level's critical density N_c, we have $\varepsilon(X^{+i}, \lambda) \propto N(X^{+i})N_e$, where $N(X^{+i})$ is number density of ion X^{+i}. At high densities, $N_e \gg N_c$, emission of CELs is suppressed by collisional de-excitation and $\varepsilon(X^{+i}, \lambda) \propto N(X^{+i})$. Unlike CELs, emissivities of RLs *increase* with *decreasing* T_e by a power law. Under typical nebular conditions ($N_e \ll 10^8\,\mathrm{cm}^{-3}$), $\varepsilon(X^{+i}, \lambda) \propto T_e^{-\alpha}N(X^{+i+1})N_e$, where $\alpha \sim 1$ and $N(X^{+i+1})$ is number density of the *recombining ion* X^{+i+1}.

In a groundbreaking work, Arthur B. Wyse published very deep spectra that he obtained with I. S. Bowen using the Lick 36 in (91 cm) reflector for a number of bright PNe (Wyse, 1942). About 270 spectral lines were detected. Many were weak permitted transitions from abundant C, N, and O ions. In the Saturn Nebula NGC 7009 alone,

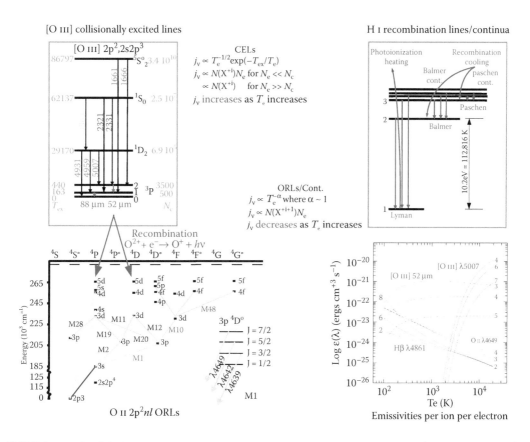

FIGURE 7.2 *Top left*: Schematic diagram showing the four lowest spectral terms of O^{2+} formed by the $2p^2$ ground and $2p^3$ electron configurations. The levels are labeled with excitation energy T_{ex} (in kelvin) and critical density N_c (/cm³). *Bottom left*: Schematic Grotrian diagram of O II, illustrating the O II recombination line spectrum produced by recombination of O^{2+}, with different multiplets, M1, M2,…, M48, labeled. *Top right*: Photoionization and recombination of hydrogen. Photoionization heats the gas, whereas recombination cools the gas. *Bottom right*: Emissivities of the collisionally excited [O III] $\lambda5007$ optical forbidden line and 52 µm far-IR fine-structure line, and of the recombination lines H β $\lambda4861$ and O II $\lambda4649$, as a function of electron temperature T_e for selected electron densities. The curves are labeled with logarithms of electron densities. Emissivities of recombination lines depend only weakly on electron density under typical (low density) nebular conditions.

. . . more than 20 lines or blends of O II have been observed, free of blending with lines of other elements. There are good grounds for the assumption, previously made, that these lines originate in electron captures by O III ions, just as the Balmer lines originate in electron captures by H II ions, and that therefore the relative intensities of the oxygen and hydrogen lines give a measure of the rates of recombination, and therefore of the relative abundance, of the two kinds of ions ... The important thing is that for this nebula a sufficient number of oxygen lines has been observed to eliminate all doubt of the correctness of their identifications, even though they are very faint; and also that they permit the direct estimate of the relative abundance of H II and O III ions,..., by a method that is relatively independent of assumptions regarding electron density and velocity distribution or their variations throughout the nebula.

From the O II ORL strengths, Wyse deduced O/H abundances, which were much higher than the values obtained by Menzel and Aller (1941), from analyses of the [O III] forbidden lines. He concluded, "The discrepancy, then, between the relative abundance found by Menzel and Aller, on the one hand, and from the recombination spectra, on the other, is of the order of 50 for NGC 7027 and of 500 for NGC 7009." Shortly after the publication of this prophetic article, Wyse was called to serve in World War II and died tragically on duty the night of June 8, 1942, in a disastrous accident over the Atlantic Ocean, off the New Jersey coast, 17 days shy of his 33rd birthday.

As stressed by Wyse, ionic abundances deduced from intensities of heavy-element ORLs relative to H β, a method based on comparing lines excited by similar mechanisms, have the advantage that they are almost independent of the nebular thermal and density structures. In contrast, ionic abundances deduced from the intensity ratio of the collisionally excited, much stronger [O III] $\lambda\lambda4959$, 5007 forbidden lines relative to H β have an exponential dependence on the adopted nebular electron temperature. Theoretically, ionic abundances deduced from heavy-element ORLs should thus be more reliable, provided that the lines can be measured accurately. Unfortunately, later development showed that accurate flux measurements for faint nebular emission lines were *not* possible after all with the technique available then, i.e., spectrophotography, as a result of the nonlinearity of photographic plates. Via detailed comparisons between the observed fluxes of H I and He II RLs and continua, and those predicted by the recombination theory, it became clear that spectrophotographic observations systematically overestimated intensities of faint lines by as much as over a factor of 10 (Seaton, 1960; Kaler, 1966; Miller, 1971; Miller and Mathews, 1972). That spectrophotographic measurements of faint lines cannot be trusted seemed to be further supported by work in the 1980s that contrasted C^{2+}/H^+ ionic abundances deduced from the collisionally excited C III] $\lambda\lambda1907$, 1909 intercombination lines and from the faint C II $\lambda4267$ ORL (see Barker, 1991 and references therein), suggesting that the intensity of the faint C II $\lambda4267$ line had either not been interpreted correctly or had been grossly overestimated (Rola and Stasińska, 1994).

In another worrisome development, Manuel Peimbert showed that if nebulae are nonisothermal and have (localized and random) temperature fluctuations, then T_e deduced from the [O III] nebular to auroral line intensity ratio ($\lambda4959 + \lambda5007$)/$\lambda4363$ will overestimate the average emission temperature of the $\lambda\lambda4959$, 5007 nebular lines and of H β at 4861 Å, leading to an underestimated O^{2+}/H^+ ionic abundance ratio deduced from the ($\lambda4959 + \lambda5007$)/H β ratio (Peimbert, 1967). He presented evidence pointing to the presence of large temperature fluctuations in PNe and H II regions by finding that T_e's derived from the Balmer jump, T_e(BJ) are systematically lower than those derived from the [O III] forbidden line ratio, T_e([O III]) (Peimbert 1967, 1971). Given the weakness of nebular continuum emission, measuring the BJ accurately was no easy task, and his results were disputed (Barker, 1978). Theoretically, while some systematic spatial temperature variations undoubtedly occur within a nebula resulting from changes in ionization structure and cooling rates as a function of position induced by varying ionization radiation field and density distribution, no known mechanisms are capable of generating large, localized temperature fluctuations, certainly nothing of the magnitude implied by Peimbert's measurements, which yield a typical value of 0.055 for the temperature fluctuation parameter, t^2, or fluctuations of an amplitude of 23%.

The advent of linear, high-quantum-efficiency, large-dynamic-range, and large-format charge-coupled devices (CCDs) in the 1980s made it possible for the first time to obtain reliable measurements of faint emission lines for bright nebulae. Meanwhile, the completion of the Opacity Project (Seaton, 1987; Cunto et al., 1993) has allowed the atomic data necessary to analyze those spectral features, specifically their effective recombination coefficients, to be calculated with high accuracy. Liu and Danziger (1993) obtained CCD measurements of the BJ for a sample of PNe and found that $T_e(BJ)$ is indeed systematically lower than $T_e([O\ III])$ obtained for the same object. They deduced that, on average, $t^2 = 0.035$, smaller than that found earlier by Peimbert (1971) but still significant, enough to cause the O^{++}/H^+ abundance ratio derived from the $\lambda\lambda 4959, 5007$ forbidden lines to be underestimated by a factor of 2.

Liu et al. (1995) presented high-quality Image Photon Counting System (IPCS) and CCD optical spectrophotometry for the legendary Saturn Nebula NGC 7009, as well as new effective recombination coefficients for O II ORLs. Nearly 100 O II ORLs were measured, yielding an O^{++}/H^+ ionic abundance that is consistently higher, by a factor of ~4.7, than the value deduced from the strong [O III] forbidden lines $\lambda\lambda 4959, 5007$. The close agreement of results deduced from a large number of O II ORLs from a variety of multiplets of different multiplicities, parities, and electron configurations vindicates the reliability of the recombination theory and rules out measurement uncertainties* or other effects, such as reddening corrections, line blending, or contamination of ORLs by other excitation mechanisms (fluorescence or charge-transfer reactions) as the cause of the large discr epancy between the ORL and CEL abundances. Further analysis of the carbon and nitrogen recombination spectra and of the neon recombination spectrum (Luo et al., 2001) show that in NGC 7009, abundances of these elements derived from ORLs are all higher than the corresponding CEL values by approximately a factor of 5. If one defines an abundance discrepancy factor (adf) as the ratio of ionic abundances X^{i+}/H^+ deduced from ORLs and from CELs, then in NGC 7009 the adf is ~5 for all four abundant second-row elements of the periodic table: C, N, O, and Ne. In NGC 7009, the Balmer discontinuity of the hydrogen recombination spectrum yields $T_e(BJ) = 8,100$ K, about 2,000 K lower than forbidden-line temperature $T_e([O\ III]) = 10,100$ K (Liu et al., 1995). The difference yields a Peimbert's t^2 value of 0.04, or temperature fluctuations of an amplitude of 20%.

DEEP SPECTROSCOPIC SURVEYS OF ORLS

Is the dichotomy observed in NGC 7009 between ORLs and CELs for nebular plasma diagnostics and abundance determinations ubiquitous among emission-line nebulae? What is the range and distribution of adf's for individual elements? Are there any correlations between the adf and other nebular properties (morphology, density, temperature, abundance, age, etc.) or properties of the ionizing source? What are the physical causes of the dichotomy?

To address those issues, several deep ORL spectroscopic surveys of faint heavy-element ORLs have been conducted. So far, more than 100 Galactic PNe have been surveyed, plus dozens of Galactic and extra-Galactic H II regions (e.g., Esteban et al., 2002; Tsamis et al., 2003a,b, 2004; Liu et al., 2004a,b; Wesson et al., 2005; Wang and Liu, 2007; for a complete list of references, please refer to a recent review by Liu, 2006a). Detailed comparisons contrasting ORL and CEL analyses show the following (Figure 7.3; see also Liu, 2006a):

1. Ionic abundances deduced from ORLs are *always* higher than CEL values, i.e., adf ≥ 1. Adf peaks at 0.35 dex, but with a tail extending to much higher values. About 20% and 10% of nebulae exhibit adf's higher than 5 and 10, respectively. For example, in the bright

---

* Mathis and Liu (1999) analyzed the observed relative intensities of the [O III], $\lambda\lambda 4959, 5007$ and the much fainter $\lambda 4931$ nebular lines, and demonstrated that accurate measurements have been achieved over a dynamic range of 10,000.

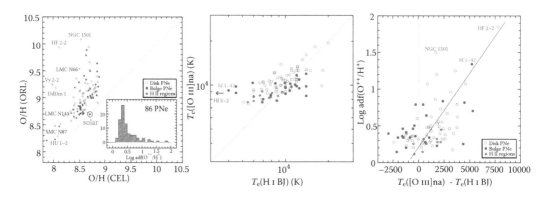

FIGURE 7.3 *Left*: O/H abundances deduced from ORLs plotted against those derived from CELs. The diagonal dotted line denotes $x = y$. The insert is a histogram of adf(O^{2+}/H^+). In the scenario of single-composition nebulae (see the next section for an alternative interpretation), uncertainties in O/H abundances, caused by observational and interpretive errors (e.g., atomic data), are expected to be typically less than 0.05 and 0.1 dex, respectively, for CEL and ORL results. *Middle*: T_e([O III]) versus T_e(BJ). Typical uncertainties of T_e([O III]) and T_e(BJ) are 5% and 10%, respectively. The diagonal dotted line denotes $x = y$. With T_e (BJ) = 900 K and T_e([O III]) = 8,820 K, Hf 2–2 falls off the left boundary of the plot. *Right*: log adf(O^{2+}/H^+) plotted against T_e([O III]) – T_e(BJ). The solid line denotes a linear fit obtained by Liu et al. (2001b) prior to the discovery of the very large adf in Hf 2–2. (Adapted from Liu, X.W., *Optical recombination lines as probes of conditions in planetary nebulae*. Cambridge University Press, Cambridge, 2006a.)

Galactic disk PN NGC 6153, adf = 9.2 (Liu et al., 2000), whereas in the bulge PN M 1–42, adf = 22 (Liu et al., 2001b). In the most extreme object discovered so far, Hf 2–2, adf (O^{2+}/H^+) reaches a record value of 71 (Figure 7.4).

2. While adf varies from object to object, for a given nebula, C, N, O, and Ne all exhibit comparable adf's (thus both CEL and ORL analyses yield compatible abundance ratios, such as C/O, N/O, and Ne/O, provided that lines of the same type, ORLs or CELs, are used for both elements involved in the ratio). Objects showing large adf's also tend to have high helium (ORL) abundances. However, magnesium, the only third-row element that has been analyzed using an ORL, shows no enhancement, even in high-adf objects (Barlow et al., 2003).

3. Excluding metal-poor nebulae in the Galactic halo and in the Large and Small Magellanic Clouds (LMC and SMC, respectively), oxygen abundances deduced from CELs for Galactic H II regions and PNe fall in a narrow range compatible with the solar value. In contrast, ORLs yield much higher abundances, more than 10 times the solar value in some cases.

4. Similarly, while the [O III] forbidden-line ratio yields values of T_e in a narrow range around 10,000 K, as one expects for a photoionized gaseous nebula of solar composition, the Balmer discontinuity yields some very low temperatures, below 1,000 K. In fact, the discrepancies in temperature and abundance determinations, using ORLs/continua on the one hand and CELs on the other, seem to be correlated—objects showing large adf's also exhibit very low T_e values.

5. Large, old PNe of low surface brightness tend to show higher adf's. In addition to this, spatially resolved analyses of a limited number of bright, extended nebulae with large adf's show that ORL abundances increase toward the nebular center, leading to higher adf near the center.

EVIDENCE OF COLD, H-DEFICIENT INCLUSIONS

What causes the ubiquitous, often alarmingly large discrepancies between the ORL and CEL plasma diagnostics and abundance determinations? Does the dichotomy imply that there are fundamental

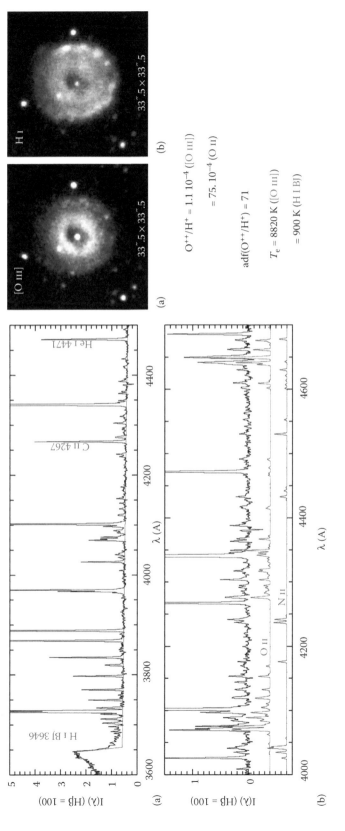

FIGURE 7.4 *Left:* The optical spectrum of Hf 2–2, which shows a record adf(O^{2+}/H^+) = 71 and an extremely low T_e(BJ) of 900 K. In the lower panel, also shown are two synthetic recombination line spectra of O II and N II, respectively. *Right:* Monochromatic images of Hf 2–2 in the light of [O III] λ5007 and Hα λ6563, respectively. (Adapted from Liu, X.W., Barlow, M.J., Zhang, Y., et al., *Monthly Notices of the Royal Astronomical Society*, 368, 1959–70, 2006.)

flaws in our understanding of the nebular thermal structure, or that we do not even understand basic processes such as the recombination of hydrogenic ions?

Can it be temperature fluctuations as originally postulated by Peimbert (1967)? This conjecture implicitly assumes that the higher abundances yielded by ORLs represent the true nebular composition as they are insensitive to temperature and temperature fluctuations. The discovery of nebulae exhibiting extreme values of adf and the abnormally high metal abundances implied by the observed strengths of ORLs, however, casts serious doubt on this paradigm. Given that PNe are descendants of low- and intermediate-mass stars, the very high ORL oxygen abundances recorded in objects of extreme adf's, if real and representative of the whole nebula, are extremely difficult to understand in the current theory of stellar evolution and nucleosynthesis.

Strong evidence against temperature fluctuations as the cause of the dichotomy between ORLs and CELs is provided by *Infrared Space Observatory* (*ISO*) measurements of mid- and far-IR fine-structure lines, such as the [Ne III] 15.5 μm and [O III] 52 and 88 μm (Liu et al., 2001a). Although collisionally excited, the IR fine-structure lines have, unlike their optical and UV counterparts, excitation energies less than ~1,000 K (Figure 7.2) and are therefore insensitive to temperature or temperature fluctuations. If temperature fluctuations are indeed at work, leading to an overestimated T_e([O III]) and consequently an underestimated λλ4959, 5007 O^{2+}/H^+ abundance, then one expects a higher abundance from the 52 and 88 μm fine-structure lines comparable to the value yielded by O II ORLs. *ISO* measurements and subsequent analyses, however, reveal otherwise. Figure 7.5 shows that in the case of NGC 6153, all CELs—UV, optical, and IR, regardless of their excitation energy and critical density—yield ionic abundances that are consistently a factor of ~10 lower than ORLs.

To account for the multiwaveband observations of NGC 6153, Liu et al. (2000) postulated that the nebula contains a previously unknown component of high-metallicity gas, presumably in the form of H-deficient inclusions embedded in the diffuse nebula of "normal" (i.e., about solar) composition. Because of the efficient cooling of abundant metals, this H-deficient gas has an electron temperature of only ~1,000 K, too low to excite any optical or UV CELs (thus invisible via the latter). Yet the high metallicity combined with a very low electron temperature makes those H-deficient inclusions powerful emitters of heavy-element ORLs. In this picture, ORLs and CELs yield discrepant electron temperatures and ionic abundances because they probe two different gas components that coexist in the same nebula, but have vastly different physical and chemical characteristics. Empirical analysis of Liu et al. (2000), as well as follow-up 1-dimensional photoionization modeling (Péquignot et al., 2002), shows that a small amount of H-deficient material, about 1 Jupiter mass, is sufficient to account for the strengths of heavy-element ORLs observed in NGC 6153.

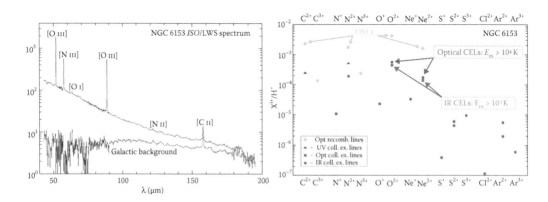

FIGURE 7.5 *Left*: The far-IR spectrum of NGC 6153. *Right*: Comparisons of ionic abundances of NGC 6153 deduced from ORLs, and from UV, optical, and IR CELs. (Adapted from Liu, X.W., Storey, P.J., Barlow, M.J., et al., *Monthly Notices of the Royal Astronomical Society*, 312, 585–628, 2000.)

The increasingly lower values of BJ temperature T_e(BJ) found for nebulae of larger adf's—6,000 K in NGC 6153 (adf = 9.2), 4,000 K in M 1-42 (adf = 22), and 900 K in Hf 2-2 (adf = 71)—provide the smoking-gun evidence that nebulae contain two regimes of vastly different physical properties. Further evidence is provided by careful analyses of the He I and heavy-element RL spectra, which show that the average emission temperatures of the He I and O II ORLs are even lower than indicated by the H I Balmer discontinuity (Liu, 2003). In general, it is found that T_e(O II) ≤ T_e(He I) ≤ T_e(BJ) ≤ T_e([O III]) (Table 7.1; c.f. Liu, 2006a, and references therein), as one expects in the dual-abundance scenario proposed by Liu et al. (2000).

Detailed 3-dimensional photoionization models of NGC 6153 with and without H-deficient inclusions have been constructed by Yuan et al. (2011) (see Figures 7.6a,b) using MOCASSIN, a Monte Carlo photoionization code capable of dealing with nebulae of arbitrary geometry and composition (Ercolano et al., 2003a). In their models, the main nebula was modeled with a chemically homogeneous ellipsoid of "normal composition" (i.e., about solar as yielded by the CEL analysis). To mimic the bipolar shape of the nebula, the density in the ellipsoid was allowed to decrease from the equator to the poles. In addition, to reproduce the strengths of low-ionization lines, such as [C I] λλ9824, 9850, [N I] λλ5198, 5200, [N II] λλ6548, 6584, [O I] λλ6300, 6363, and [O II] λλ3727, 3729, an equatorial ring of the same chemical composition but of a higher density was added. The presence of a high-density torus is supported by a high-resolution spectrum obtained with the Manchester Echelle Spectrograph mounted on the Anglo-Australian Telescope. The spectrum, centered on Hα and the [N II] λλ6548, 6584 lines and obtained with a long slit oriented in PA = 123 deg and through the central star, revealed two high-velocity emission spots at the positions of the bright shell, one on each side of the central star. The spots are particularly bright in [N II] and have, respectively, blue- and redshifted velocities relative

TABLE 7.1
Comparison of T_e's Deduced from CELs and from ORLs/Continua

Nebula	adf(O^{2+}/H$^+$)	Te([O III])(K)	Te(BJ)(K)	Te(He I)(K)	Te(O II)(K)
NGC 7009	4.7	9,980	7,200	5,040	420
H 1-41	5.1	9,800	4,500	2,930	<288
NGC 2440	5.4	16,150	14,000		<288
Vy 1-2	6.2	10,400	6,630	4,430	3,450
IC 4699	6.2	11,720	12,000	2,460	<288
NGC 6439	6.2	10,360	9,900	4,900	851
M 3-33	6.6	10,380	5,900	5,020	1,465
M 2-36	6.9	8,380	6,000	2,790	520
IC 2003	7.3	12,650	11,000	5,600	<288
NGC 6153	9.2	9,120	6,000	3,350	350
DdDm 1	11.8	12,300	11,400	3,500	
Vy 2-2	11.8	13,910	9,300	1,890	1,260
NGC 2022	16.0	15,000	13,200	15,900	<288
NGC 40	17.8	10,600	7,000	10,240	
M 1-42	22.0	9,220	4,000	2,260	<288
NGC 1501	31.7	11,100	9,400		
Hf 2-2	71.2	8,820	900	940	630

Source: Adapted from Liu, X.W., *Optical recombination lines as probes of conditions in planetary nebulae*. Cambridge University Press, Cambridge, 2006a.

Note: T_e(He I)'s were derived from the He I 5876/4472 and 6678/4472 line ratios and were typically accurate to 20% (c.f. Zhang et al., 2005). Values of Te(O II) were deduced from the O II 4089/4649 line ratio and had typical uncertainties of 30%.

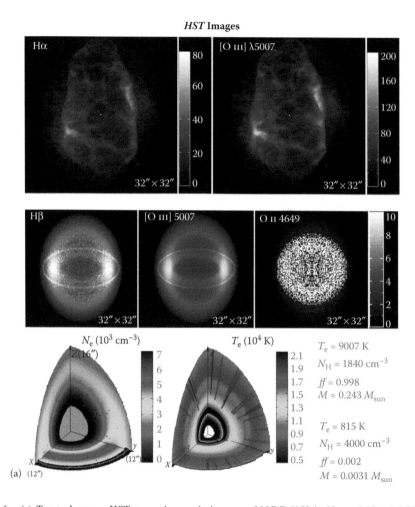

FIGURE 7.6 (a) *Top to bottom: HST* monochromatic images of NGC 6153 in Hα and [O III] λ5007 (north is up, east to the left); monochromatic images predicted by the best-fit model incorporating H-deficient inclusions; and the model distributions of electron density and temperature. (b) *Next page*: Comparison of model line intensities with observations for the best-fit chemically homogeneous model (*top*) and the model containing H-deficient inclusions (*bottom*). Transitions appended with plus (+) signs indicate intensities summed of several fine-structure components of the same multiplet. In the bottom panel, blue and red color bars denote, respectively, contributions of line fluxes from the cold, H-deficient inclusions and from the hot, diffuse gas of "normal composition." (Adapted from Yuan, H.-B., Liu, X.-W., Péquignot, D., et al., *Monthly Notices of the Royal Astronomical Society*, 411, 1035–52, 2011.)

to the Hα emission from the nearby nebular shell. The model fits the CEL measurements well, except for a few lines [see the top panel of Figure 7.6b]. The model still underestimates the [C I] λ9850 line by a factor of 3, even after the introduction of the high-density torus. The model also underestimates, by about a factor of 2, the fluxes of the [N III] λ1744 and [Ar III] 8.9 and 21.8 μm lines. The latter three lines were, however, only marginally detected, by the *International Ultraviolet Explorer* (*IUE*) and *ISO*, respectively, and the discrepancies may well be due to measurement errors. The largest discrepancy is found for the [Ne II] 12.8 μm line, where the model underestimates the observation by more than a factor of 5.

The chemically homogeneous model, however, fails to reproduce the strengths of all heavy-element ORLs by a factor of 10. To account for them, metal-rich inclusions were added in the form of clumps of H-deficient material. The clumps were distributed symmetrically in a spherical pattern

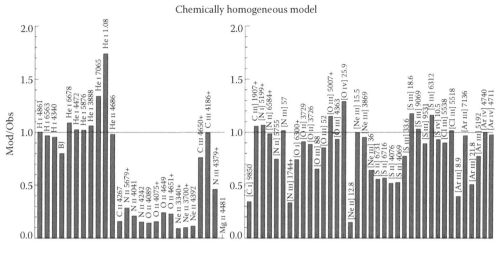

Chemically homogeneous model

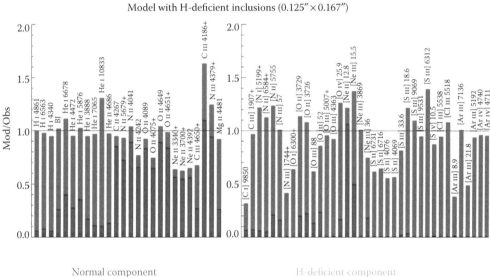

Model with H-deficient inclusions (0.125″ × 0.167″)

Normal component		
H: 10000	He: 1000	C: 3.20
N: 3.80	O: 5.53	Ne: 1.76

H-deficient component		
H: 10000	He: 5000	C: 177
N: 150	O: 440	Ne: 177

H-deficient knots are cooled by infrared fine-structure lines:

(b)   [O III] 52 μm, [Ne II] 12.8 μm and [Ne III] 15.5 μm

FIGURE 7.6   (Continued)

with a radial number-density profile that reproduced the O II and C II ORL surface brightness distributions deduced from the long-slit observations (Liu et al., 2000). Given that the H-deficient inclusions have not been resolved, even with the *Hubble Space Telescope* (*HST*)/Space Telescope Imaging Spectrograph (STIS) spectroscopy with a slit width of 0.2 arcsec (Yuan et al., 2011), they must have dimensions smaller than ~0.2 arcsec, or 300 AU at the assumed distance of 1.5 kpc of NGC 6153. The smallest size of inclusions that can be modeled with the MOCASSIN was limited by the memory available per processor of the computer employed. For NGC 6153, a grid of 48^3 cells, each of 1/4 arcsec by 1/3 arcsec, was used to model 1/8 of the nebula using an SGI Altix 330 cluster of eight dual-core processors and 8 GB memory per processor. Models that doubled the resolution were also constructed using the NASA/Ames Columbia Supercomputer. Results from one of these high-resolution models are shown in Figure 7.6a,b. The best-fit, dual-abundance model incorporating H-deficient inclusions matches all available observations reasonably well and successfully reproduces strengths of heavy-element ORLs. With a total mass of only 1 Jupiter mass, or just over 1% that of the whole nebula, and a helium abundance 5 times and CNONe abundances 40–100 times higher than the main diffuse nebula of "normal composition," the H-deficient inclusions account for approximately 5%, 35%, and 100% of the observed fluxes of H I, He I, and heavy-element ORLs, respectively, but produce essentially nil optical and UV CEL emission. Cooled to about 800 K by IR fine-structure lines of heavy-elemental ions, notably the [Ne II] 12.8 μm, [Ne III] 15.5 μm, and [O III] 52 μm lines, the inclusions are important contributors of those lines, as well as for the H I recombination continuum Balmer discontinuity (about 30%). The inclusions even dominate the [Ne II] 12.8 μm line flux. Note that gas densities in the H-deficient inclusions are not well constrained at the moment, partly because of the lack of suitable atomic data and diagnostic tools (see "The Need for New Atomic Data," below). This leads to some uncertainties in the deduced total mass of the H-deficient material. In the current treatment, the inclusions have densities and temperatures that are roughly in pressure equilibrium with the surrounding diffuse medium of higher temperatures and lower densities. Physically, one envisions individual inclusions to be optically thick such that they can survive long enough to have observational effects. Proper modeling of optically thick knots, however, will require a separate set of fine grids for each of them to resolve the ionization and thermal structures, and thus demands even more computing resources.

The average elemental abundances for the whole nebula of NGC 6153, including the diffuse nebula as well as H-deficient inclusions embedded in it, are 0.102, 3.9×10^{-4}, 4.4×10^{-4}, 7.3×10^{-4}, and 2.4×10^{-4} for He, C, N, O, and Ne, respectively. Except for helium, elemental abundances of the "normal" component are close to those deduced from the empirical method based on CELs (Liu et al., 2000) and are about 30% lower than the average abundances for the whole nebula. For helium, the abundance of the "normal" component is about 40% lower than derived from the empirical method based on the observed strengths of helium ORLs and actually comes closer to the average helium abundance of the whole nebula. That NGC 6153 has a lower helium abundance than implied by the empirical analysis is supported by a measurement of the near-IR He I λ10830 line, for which excitation from the 2s ^{3}S metastable level by electron impacts dominates the emission (Yuan et al., 2011).

Helium lines observable in the optical and UV wavelength ranges are all dominated by recombination excitation, and their strengths can be strongly enhanced by a small amount of H-deficient, ultracold plasma posited in the nebula. If H-deficient inclusions are, indeed, fully responsible for the ORL versus CEL dichotomy ubiquitously found in PNe and H II regions, then their presence may have a profound consequence on the helium abundances deduced for those objects, as well as on the primordial helium abundance determined by extrapolating the helium abundances of metal-poor H II galaxies to zero metallicity—the inferred primordial helium abundances from standard analyses would be too high (Zhang et al., 2004).

Corroborative evidence that ORLs arise from distinct regions of very low temperature plasma is provided by analyses of emission-line profiles. Spectroscopy at a resolution of 6 km/s of NGC 7009 shows that O II ORLs are significantly narrower than the [O III] forbidden lines (Liu, 2006b). Observations at an even higher resolution of 2 km/s of NGC 7009 and NGC 6153 yield complex multivelocity components and differing line profiles for the O II ORLs and for the [O III] forbidden lines. The profiles of the [O III] λ5007 nebular and λ4363 auroral lines also differ significantly. These results suggest that ionized regions of vastly different temperatures (and radial and thermal velocities) coexist in those nebulae (Barlow et al., 2006). Zhang (2008) shows that the differences in the observed line profiles of the O II ORLs and the [O III] CELs are hard to explain in a chemically homogeneous nebula.

SPECTROSCOPY OF PNE HARBORING H-DEFICIENT INCLUSIONS

H-deficient clumps were previously known to exist in a rare class of old PNe, including Abell 30, 58, and 78, and IRAS 15154-5258 and 18333-2357 (e.g., Harrington, 1996). They are identified as PNe

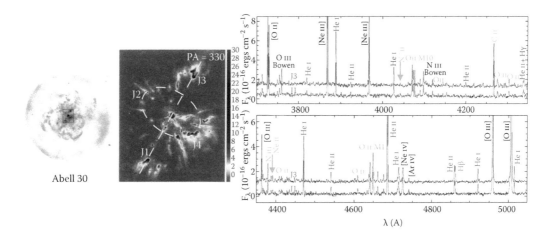

FIGURE 7.7 *Left*: The optical images of Abell 30, showing the H-deficient knots near the nebular center (Adapted from Borkowski, K.J., Harrington, J.P., Tsvetanov, Z., et al., *Astrophysical Journal*, 415, L47–50, 1993). *Right*: The optical spectrum of the polar knots J 1 and J 3 showing prominent ORLs of C, N, O, and Ne ions. (Adapted from Wesson, R., Liu, X.W., and Barlow, M.J., *Monthly Notices of the Royal Astronomical Society*, 340, 253–63, 2003.) Note the faintness of H β at 4861 Å.

that experience a last helium shell flash that brings them back to the AGB to repeat the PN evolution stage, the so-called born-again PNe (Iben et al., 1983). In Abell 30, *HST* imaging reveals a host of knots embedded near the center of a round, limb-brightened faint nebula of angular diameter 127 arcsec (Borkowski et al., 1993; see Figure 7.7). They include two point-symmetric polar knots along PA = 331 degrees at angular distances 6.66 and 7.44 arcsec from the central star, plus a number of knots loosely delineating an equatorial disk or ring. Deep optical spectra of knots J 1 and J 3 were obtained by Wesson et al. (2003). The spectra show prominent RLs from CNONe ions. These are remarkably similar to those seen in other PNe of large adf's, such as Hf 2-2 in Figure 7.4, except that in Abell 30 the strengths of those CNONe ORLs relative to H β are much higher. Detailed ORL analyses show that the ORLs are indeed emitted under very low temperatures of only a few hundred kelvin. The result is corroborated by detailed 3-dimensional photoionization modeling (Ercolano et al., 2003b), as well as by a similar analysis of the H-deficient knot found at the center of Abell 58 by Wesson et al. (2008).

ORIGINS OF H-DEFICIENT INCLUSIONS

The origin of H-deficient material is not well understood, and their existence in PNe, and possibly also in H II regions (see, for example, Esteban et al., 2002; Tsamis et al., 2003a,b), is not predicted by the current theory of stellar evolution. For PNe harboring H-deficient central stars, such as the Wolf-Rayet and PG 1159 stars, while the scenario of a single post-AGB star experiencing a last helium shell flash seems to be able to match the photospheric abundances of the star, not all PNe exhibiting large adf's have an H-deficient central star (e.g., NGC 7009). The process and mechanism by which H-deficient knots might be ejected in such systems are not understood. More importantly, detailed ORL abundance analyses by Wesson et al. (2008) show that the H-deficient knots found in Abell 30 and Abell 58 are oxygen-rich, not carbon-rich as one expects in the scenario of "born-again" PNe.

The precise colinearity of the two polar knots in Abell 30 with the central star, to within 5 arcmin, is hard to explain by single-star evolution and suggests instead the action of a bipolar jet from, for example, an accretion disk in a binary system (Harrington, 1996; De Marco, 2008). In this respect, it is interesting to note that Abell 58 is known to have experienced a nova-like outburst in 1917 (Clayton and De Marco, 1997). Hf 2-2 has also been found to be a close binary system with a period of 0.398571 days (Lutz et al., 1998). By comparing the known properties of Abell 58 with those of

Abell 30, Sakurai's Object, and several novae and nova remnants, Wesson et al. (2008) argue that the elemental abundances in the H-deficient knots of Abell 30 and Abell 58 have more in common with neon novae than with Sakurai's Object, which is believed to have undergone a final helium flash.

An alternative to the supposition of H-deficient inclusions being ejecta of nucleon-processed material is that they derive from metal-rich planetary material, such as icy planetesimals, that once orbited the PN progenitor star. As the star enters the PN phase by ejecting the envelope and evolves to become a luminous, hot white dwarf, the strong stellar winds and UV radiation fields begin to photoionize and strip gas off the planetesimals, now embedded in the PN (Liu, 2003, 2006a). As analyses show that metal-rich material of only a few Jupiter masses is required to explain the observed strengths of ORLs, the idea may not be so eccentric as it first looks. Evaporating protoplanetary disks around newly formed stars (such as the proplyds found in the Orion Nebula) may likewise provide a natural solution to the problem of ORL versus CEL dichotomy similarly found in H II regions.

THE NEED FOR NEW ATOMIC DATA

Atomic data relevant for the study of emission-line nebulae, including collision strengths and effective recombination coefficients, are generally calculated for a temperature range 5,000 K–20,000 K, typical for photoionized nebulae of solar composition. The finding that heavy-element ORLs emitted by the H-deficient knots in the "born-again" PNe Abell 30 and Abell 58 arise from ultracold plasma of only a few hundred kelvin, as well as the compelling evidence accumulated so far that points to the presence of similar cold H-deficient inclusions in other PNe, calls for new atomic data, in particular the effective recombination coefficients applicable in such low-temperature environments. New plasma diagnostics that use only ORLs, not CELs, need to be developed so that physical and chemical properties of the H-deficient knots/inclusions, such as temperature, density, elemental abundances, and mass, can be determined reliably, a prerequisite to unravel their origins.

The ratios of the intensities of ORLs from states of different orbital angular momenta show some temperature dependence and thus can be used to measure the average temperature under which the lines are emitted (Liu, 2003). In addition, as pointed out by Liu (2003), the dependence on density of the relative populations of the fine-structure levels of the ground spectral term of the recombining ions, such as O^{2+} $2p^2$ $^3P_{0,1,2}$, leads to variations of the relative intensities of RLs as a function of density. In the case of O II, for example, lines from high-l states, such as the $3p$ $^4D^o_{7/2}$–$3s$ $^4P_{5/2}$ $\lambda 4649$ transition (see Figure 7.2), weaken as density decreases as a result of the underpopulation of the O^{2+} $2p^2$ $^3P_2$ level under low densities. This effect opens up the possibility of determining electron density using ORLs.

New effective recombination coefficients, calculated down to temperatures of ~100 K and taking into account the dependence on density of level populations of the ground states of the recombining ion, have now been carried out for the O II (Bastin and Storey, 2006) and N II (Fang et al., 2011) recombination spectra. Figure 7.8 plots loci of the O II $I(\lambda 4076)/I(\lambda 4089)$ and $I(\lambda 4076)/I(\lambda 4070)$ intensity ratios for different temperatures and densities. The diagram also illustrates the importance of dielectronic recombination via the low-lying fine-structure auto-ionizing states under very low temperatures. Preliminary applications of these new atomic data and diagnostics to observations yield lower temperatures and higher densities than CELs, as one expects.

SUMMARY

To summarize, nearly 150 years after William Huggins's discovery of bright emission lines in the spectra of gaseous nebulae, enormous progress has been achieved in understanding them. In particular, we believe that we now understand the faint ORLs emitted by heavy-element ions, more than half a century after Wyse's pioneering work on O II. The theory of photoionized gaseous nebulae seems to be solid. Yet any progress in our understanding of their nature seems to be always

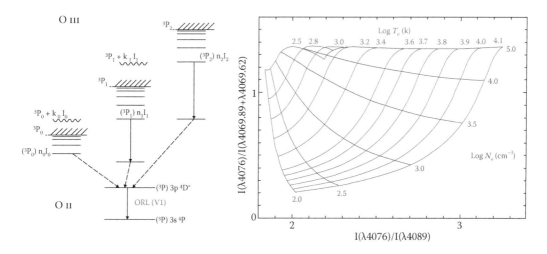

FIGURE 7.8 *Left*: Schematic diagram illustrating the low-temperature dielectronic recombination of O II via the low-lying fine-structure auto-ionizing states. Note that for $n_1 \geq 67$ and $n_2 \geq 37$, the doubly excited O II $2p^2(^3P_1)\,n_1l_1$ and $2p^2(^3P_2)\,n_2l_2$ states fall above the O III ground level $2p^2\,^3P_0$. Similarly, for $n_2 \geq 47$, O II $2p^2(^3P_2)$ n_2l_2 states have energies above the O III first excited fine-structure level $2p^2\,^3P_1$. For temperatures of a few hundred kelvin, dielectronic recombination via those low-lying fine-structure auto-ionizing states becomes an important process. *Right*: Loci of the O II recombination line ratios λ4076/λ4089 and λ4076/λ4070 for different T_e values and N_e values. (Based on unpublished data provided by Dr. P. J. Storey.)

accompanied by the emergence of new puzzles. Clearly, those beautiful heavenly objects, queens of the night, are more delicate than we think and are still withholding some fascinating secrets from us. With dedicated observations, more discoveries are sure to come.

REFERENCES

Aller, L.H. and Menzel, D.H. (1945). Physical processes in gaseous nebulae. XVIII. The chemical composition of the planetary nebulae. *Astrophysical Journal*, 102: 239–63.

Aller, L.H., Baker, J.G., and Menzel D.H. (1939). Physical processes in gaseous nebulae. VIII. The ultraviolet radiation field and electron temperature of an optically thick nebula. *Astrophysical Journal*, 90: 601–10.

Baker, J.G. and Menzel, D.H. (1938). Physical processes in gaseous nebulae. III. The Balmer decrement. *Astrophysical Journal*, 88: 52–64.

Baker, J.G., Menzel, D.H., and Aller, L.H. (1938). Physical processes in gaseous nebulae. V. Electron temperatures. *Astrophysical Journal*, 88: 422–28.

Barker, T. (1978). Spectrophotometry of planetary nebulae. I—Physical conditions. *Astrophysical Journal*, 219: 914–30.

Barker, T. (1991). The ionization structure of planetary nebulae. X—NGC 2392. *Astrophysical Journal*, 371: 217–325.

Barlow, M.J., Hales, A.S., Storey, P.J., et al. (2006). bHROS high spectral resolution observations of PN forbidden and recombination line profiles planetary nebulae. In: M.J. Barlow and R.H. Méndez (eds.), *Planetary Nebulae in our Galaxy and Beyond*, 367–68. Proceedings of the International Astronomical Union Symposium No. 234. Cambridge: Cambridge University Press.

Barlow, M.J., Liu, X.W., Péquignot, D., et al. (2003). PN recombination line abundances for magnesium, silicon and sulphur. In: S. Kwok, M. Dopita, and R. Sutherland (eds.), *Planetary Nebulae: Their Evolution and Role in the Universe*, 373–74. Proceedings of the International Astronomical Union Symposium No. 209. Astronomical Society of the Pacific.

Bastin, R.J. and Storey, P.J. (2006). Recombination line spectroscopy: The O II spectrum. In: M.J. Barlow and R.H. Méndez (eds.), *Planetary Nebulae in our Galaxy and Beyond*, 369–70. Proceedings of the International Astronomical Union Symposium No. 234. Cambridge: Cambridge University Press.

Borkowski, K.J., Harrington, J.P., Tsvetanov, Z., et al. (1993). *HST* imaging of hydrogen-poor ejecta in Abell 30 and Abell 78—wind-blown cometary structures. *Astrophysical Journal*, 415: L47–50.

Bowen, I.S. (1927a). Series spectra of boron, carbon, nitrogen, oxygen, and fluorine. *Physics Review*, 29: 231–47.

Bowen, I.S. (1927b). The origin of the nebulium spectrum. *Nature*, 120: 473.

Bowen, I.S. (1927c). The origin of the chief nebular lines. *Publications of the Astronomical Society of the Pacific*, 39: 295–97.

Bowen, I.S. (1934). The excitation of the permitted O III nebular lines. *Publications of the Astronomical Society of the Pacific*, 46: 146–48.

Bowen, I.S. (1935). The spectrum and composition of the gaseous nebulae. *Astrophysical Journal*, 81: 1–16.

Bowen, I.S. (1938). The image-slicer, a device for reducing loss of light at slit of stellar spectrograph. *Astrophysical Journal*, 88: 113–24.

Bowen, I.S. and Wyse, A. (1939). The spectra and chemical composition of the gaseous nebulae, NGC 6572, 7027, 7662. *Lick Observatory Bulletin*, 19: 1–16.

Burgess, A. (1958). The hydrogen recombination spectrum. *Monthly Notices of the Royal Astronomical Society*, 118: 477–95.

Clayton, G.C. and De Marco, O. (1997). The evolution of the final helium shell flash star V 605 Aquilae from 1917 to 1997. *Astronomical Journal*, 114: 2679–685.

Cunto, W., Mendoza, C., Ochsenbein, F., et al. (1993). Topbase at the CDS. *Astronomy & Astrophysics*, 275: L5–8.

De Marco, O. (2008). [WC] and PG 1159 central stars of planetary nebulae: The need for an alternative to the born-again scenario. In: K. Werner and T. Rauch (eds.), *Hydrogen-Deficient Stars,* 209–19. San Francisco: Astronomical Society of the Pacific.

Ercolano, B., Barlow, M.J., Storey, P.J. et al. (2003a). MOCASSIN: A fully three-dimensional Monte Carlo photoionization code. *Monthly Notices of the Royal Astronomical Society*, 340: 1136–152.

Ercolano, B., Barlow, M.J., Storey, P.J., et al. (2003b). Three-dimensional photoionization modelling of the hydrogen-deficient knots in the planetary nebula Abell 30. *Monthly Notices of the Royal Astronomical Society*, 344: 1145–54.

Esteban, C., Peimbert, M., Torres-Peimbert, S., et al. (2002). Optical recombination lines of heavy elements in giant extragalactic H II regions. *Astrophysical Journal*, 581: 241–57.

Fang, X., Storey, P.J., and Liu, X.-W. (2011). New effective recombination coefficients for nebular N II lines. *Astronomy and Astrophysics*, 530: A18 (12 pages).

Flower, D.R. (1969a). The ionization structure of planetary nebulae—VII. The heavy elements. *Monthly Notices of the Royal Astronomical Society*, 146: 171–85.

Flower, D.R. (1969b). The ionization structure of planetary nebulae—VIII. Models of the nebulae NGC 7662 and IC 418. *Monthly Notices of the Royal Astronomical Society*, 146: 243–63.

Goodson, W.L. (1967). Ionisationsstruktur planetarischer nebel. *Zeitschrift für Astrophysik*, 66: 118–55.

Harrington, J.P. (1968). Ionization stratification and thermal stability in model planetary nebulae. *Astrophysical Journal*, 152: 943–62.

Harrington, J.P. (1996). Observations and models of H-deficient planetary nebulae. In: C.S. Jeffery and U. Heber (eds.), *Hydrogen-Deficient Starsi*, 193–203. ASP Conferences Series Vol. 96. Astronomical Society of the Pacific.

Herschel, W. (1785). On the construction of the heavens. *Philosophical Transactions of the Royal Society of London,* 75: 213–66.

Hjellming, R. (1966). Physical processes in H II regions. *Astrophysical Journal*, 143: 420–51.

Huggins, W. (1881). Note on the photographic spectrum of the Great Nebula in Orion. *Proceedings of the Royal Society of London*, 33: 425–28.

Huggins, W. and Miller, W.A. (1864). On the spectra of some of the fixed stars. *Philosophical Transactions of the Royal Society of London*, 154: 437–44.

Iben, I., Jr., Kaler, J.B., Truran, J.W., et al. (1983). On the evolution of those nuclei of planetary nebulae that experience a final helium shell flash. *Astrophysical Journal*, 264: 605–12.

Kaler, J.B. (1966). Hydrogen and helium spectra of gaseous nebulae. *Astrophysical Journal*, 143: 722–42.

Kirchhoff, G. and Bunsen, R.W. (1860). Chemical analysis by observation of spectra. *Annalen der Physik und der Chemie (Poggendorff)*, 110: 161–89.

Liu, X.W. (2003). Probing the dark secrets of PNe with ORLs (invited review). In: S. Kwok, M. Dopita, and R. Sutherland (eds.), *Planetary Nebulae: Their Evolution and Role in the Universe*, 339–46. Proceedings of the International Astronomical Union Symposium No. 209. Astronomical Society of the Pacific.

Liu, X.W. (2006a). Optical recombination lines as probes of conditions in planetary nebulae. In: M.J. Barlow and R.H. Méndez (eds.), *Planetary Nebulae in Our Galaxy and Beyond*, 219–26. Proceedings of the International Astronomical Union Symposium No. 234. Cambridge: Cambridge University Press.

Liu, X.W. (2006b). Plasma diagnostics and elemental abundance determinations for PNe current status. In: L. Stanghellini, J.R. Walsh, and N.G. Douglas (eds.), *Planetary Nebulae beyond the Milky Way*, 169–82. Proceedings of the European Southern Observatory workshop held in Garching, Germany, 19–21 May 2004. *ESO Astrophysics Symposia*, B. Leibundgut (Series Editor). Berlin: Springer.

Liu, X.W. and Danziger, I.J. (1993). Electron temperature determination from nebular continuum emission in planetary nebulae and the importance of temperature fluctuations. *Monthly Notices of the Royal Astronomical Society*, 263: 256–66.

Liu, X.W., Barlow, M.J., Cohen, M., et al. (2001a). *ISO* LWS observations of planetary nebula fine-structure lines. *Monthly Notices of the Royal Astronomical Society*, 323: 343–61.

Liu, X.W., Barlow, M.J., Zhang, Y., et al. (2006). Chemical abundances for Hf 2-2, a planetary nebula with the strongest-known heavy-element recombination lines. *Monthly Notices of the Royal Astronomical Society*, 368:1959–970.

Liu, X.W., Luo, S.G., Barlow, M.J., et al. (2001b). Chemical abundances of planetary nebulae from optical recombination lines—III. The Galactic bulge PN M 1-42 and M 2-36. *Monthly Notices of the Royal Astronomical Society*, 327: 141–68.

Liu, X.W., Storey, P.J., Barlow, M.J., et al. (1995). The rich O II recombination spectrum of the planetary nebula NGC 7009: New observations and atomic data. *Monthly Notices of the Royal Astronomical Society*, 272: 369–88.

Liu, X.W., Storey, P.J., Barlow, M.J., et al. (2000). NGC 6153: A super-metal-rich planetary nebula? *Monthly Notices of the Royal Astronomical Society*, 312: 585–628.

Liu, Y., Liu, X.W., Barlow, M.J., et al. (2004a). Chemical abundances of planetary nebulae from optical recombination lines—II. Abundances derived from collisionally excited lines and optical recombination lines. *Monthly Notices of the Royal Astronomical Society*, 353: 1251–285.

Liu, Y., Liu, X.W., Luo, S.G., et al. (2004b). Chemical abundances of planetary nebulae from optical recombination lines—I. Observations and plasma diagnostics. *Monthly Notices of the Royal Astronomical Society*, 353: 1231–50.

Luo, S.G., Liu, X.W., and Barlow, M.J. (2001). Chemical abundances of planetary nebulae from optical recombination lines—II. The neon abundance of NGC 7009. *Monthly Notices of the Royal Astronomical Society*, 326: 1049–56.

Lutz, J., Alves, D., Becker, A., et al. (1998). V and R magnitudes for planetary nebula central stars in the MACHO project galactic bulge fields. *American Astronomical Society*, 192: 5309.

Mathis, J.S. and Liu, X.W. (1999). Observations of the [O III] $\lambda4931/\lambda4959$ line ratio and O^{+2} abundances in ionized nebulae. *Astrophysical Journal*, 521: 212–16.

Menzel, D.H. and Aller, L.H. (1941). Physical processes in gaseous nebulae. XVI. The abundance of O III. *Astrophysical Journal*, 94: 30–6.

Menzel, D.H., Aller, L.H., and Hebb, M.H. (1941). Physical processes in gaseous nebulae. XIII. *Astrophysical Journal*, 93: 230–5.

Miller, J.S. (1971). Photoelectric measurements of high-n Balmer lines in NGC 7027 and NGC 7662. *Astrophysical Journal*, 165: L101–5.

Miller, J.S. and Mathews, W.G. (1972). The recombination spectrum of the planetary nebula NGC 7027. *Astrophysical Journal*, 172: 593–607.

Osterbrock, D.E. (1974). *Astrophysics of Gaseous Nebulae*. Sausalito: University Science Books.

Osterbrock, D.E. and Ferland, G.J. (2005). *Astrophysics of Gaseous Nebulae and Active Galactic Nuclei*. Sausalito: University Science Books.

Pagel, B.E.J. (1997). *Nucleosynthesis and Chemical Evolution of Galaxies*. Cambridge: Cambridge University Press.

Peimbert, M. (1967). Temperature determinations of H II regions. *Astrophysical Journal*, 150: 825–34.

Peimbert, M. (1971). Planetary nebulae II. Electron temperatures and electron densities. *Boletín de los Observatorios de Tonantzintla y Tacubaya*, 6: 29–37.

Péquignot, D., Amara, M., Liu, X.W., et al. (2002). Photoionization models for planetary nebulae with inhomogeneous chemical composition. *Revista Mexicana de Astronomía y Astrofísica (Serie de Conferencias)*, 12: 142–3.

Rola, C. and Stasińska, G. (1994). The carbon abundance problem in planetary nebulae. *Astronomy & Astrophysics*, 282: 199–212.

Rubin, R.H. (1968). The structure and properties of H II regions. *Astrophysical Journal*, 153: 761–82.

Russell, H.N., Dugan, R.S., and Stewart, J.Q. (1927). *Astronomy*. Boston: Ginn and Co.

Schmidt, B. (1938). Ein lichtstarkes komafreies Spiegelsystem. *Mitteilungen der Hamburger Sternwarte in Bergedorf*, 7: 15–17 (#36).

Seaton, M.J. (1954). Electron temperatures and electron densities in planetary nebulae. *Monthly Notices of the Royal Astronomical Society*, 114: 154–71.

Seaton, M.J. (1959a). Radiative recombination of hydrogenic ions. *Monthly Notices of the Royal Astronomical Society*, 119: 81–9.

Seaton, M.J. (1959b). The solution of capture-cascade equations for hydrogen. *Monthly Notices of the Royal Astronomical Society*, 119: 90–7.

Seaton, M.J. (1960). H i, He i, and He ii intensities in planetary nebulae. *Monthly Notices of the Royal Astronomical Society*, 120: 326–37.

Seaton, M.J. (1987). Atomic data for opacity calculations: I. General description. *Journal of Physics, B*, 20: 6363–378.

Seaton, M.J. and Osterbrock, D.E. (1957). Relative [O ii] intensities in gaseous nebulae. *Astrophysical Journal*, 125: 66–83.

Shklovskii, I.S. (1957). Once more on the distances to planetary nebulae and the evolution of their nuclei. *Soviet Astronomy*, 1: 397–403.

Spitzer, L. (1948). The temperature of interstellar matter. I. *Astrophysical Journal*, 107: 6–33.

Strömgren, B. (1939). The physical state of interstellar hydrogen. *Astrophysical Journal*, 89: 526–47.

Tsamis, Y.G., Barlow, M.J., Liu, X.W., et al. (2003a). Heavy elements in Galactic and Magellanic Cloud H ii regions: Recombination-line versus forbidden-line abundances. *Monthly Notices of the Royal Astronomical Society*, 338: 687–710.

Tsamis, Y.G., Barlow, M.J., Liu, X.W., et al. (2003b). A deep survey of heavy element lines in planetary nebulae—I. Observations and forbidden-line densities, temperatures and abundances. *Monthly Notices of the Royal Astronomical Society*, 345: 186–220.

Tsamis, Y.G., Barlow, M.J., Liu, X.W., et al. (2004). A deep survey of heavy element lines in planetary nebulae—II. Recombination-line abundances and evidence for cold plasma. *Monthly Notices of the Royal Astronomical Society*, 353: 953.

Wang, W. and Liu, X.W. (2007). Elemental abundances of Galactic bulge planetary nebulae from optical recombination lines. *Monthly Notices of the Royal Astronomical Society*, 381: 669–701.

Wesson, R., Barlow, M.J., Liu, X.W., et al. (2008). The hydrogen-deficient knot of the 'born-again' planetary nebula Abell 58 (V605 Aql). *Monthly Notices of the Royal Astronomical Society*, 383: 1639–648.

Wesson, R., Liu, X.W., and Barlow, M.J. (2003). Physical conditions in the planetary nebula Abell 30. *Monthly Notices of the Royal Astronomical Society*, 340: 253–63.

Wesson, R., Liu, X.W., and Barlow, M.J. (2005). The abundance discrepancy—recombination line versus forbidden line abundances for a northern sample of galactic planetary nebulae. *Monthly Notices of the Royal Astronomical Society*, 362: 424–54.

Wood, R. (1910). The echellette grating for the infra-red. *Philosophical Magazine*, 20 (series 6): 770–78.

Wright, W.H. (1918). The wave lengths of the nebular lines and general observations of the spectra of the gaseous nebulae. *Publications of the Lick Observatory*, 13: 191–266.

Wyse, A.B. (1942). The spectra of ten gaseous nebulae. *Astrophysical Journal*, 95: 356–87.

Yuan, H.-B., Liu, X.-W., Péquignot, D., et al. (2011). Three-dimensional chemically homogeneous and bi-abundance photoionization models of the "super-metal-rich" planetary nebula NGC 6153. *Monthly Notices of the Royal Astronomical Society*, 411: 1035–52.

Zanstra, H. (1927). An application of the quantum theory to the luminosity of diffuse nebulae. *Astrophysical Journal*, 65: 50–70.

Zanstra, H. (1929). Luminosity of planetary nebulae and stellar temperatures. *Publications of the Dominion Astrophysical Observatory, Victoria*, 4: 209–60.

Zhang, Y. (2008). Emission line profiles as a probe of physical conditions in planetary nebulae. *Astronomy & Astrophysics*, 486: 221–28.

Zhang, Y., Liu, X.W., Liu, Y., et al. (2005). Helium recombination spectrum as temperature diagnostics for planetary nebulae *Monthly Notices of the Royal Astronomical Society*, 358: 457–67.

Zhang, Y., Liu, X.W., Wesson, R., et al. (2004). Electron temperatures and densities of planetary nebulae determined from the nebular hydrogen recombination spectrum and temperature and density variations. *Monthly Notices of the Royal Astronomical Society*, 351: 935–55.

Part III

Some Near-Term Challenges
in Astronomy

8 Can We Detect Dark Matter?

Elliott D. Bloom

CONTENTS

WHY DO WE NEED DARK MATTER?

The story begins in the 1930s with a Caltech astronomy professor. Fritz Zwicky, while analyzing his observations of the Coma Cluster in 1933, became the first to apply the virial theorem to infer the existence of unseen matter in this cluster. This unseen matter is now called dark matter (Zwicky, 1933; see also Zwicky, 1937). Using the virial theorem, he was able to infer the average mass of galaxies within the cluster, obtained a value 500 times greater than expected from their optical luminosity, and proposed that most of the matter was "dark (cold) matter" (DM; Rubin, 2001). Much more work of this type has been done since Zwicky, including the discovery by x-ray space telescopes of large amounts of hot gas in clusters through their x-ray emissions (not seen by Zwicky), and the result is that the missing mass, i.e., dark matter in galaxy clusters, is about 2.5–5 times the total directly observed luminosity mass (clearly baryonic mass). Thus, when these observations are extrapolated to the total mass of the Universe, this gives a fraction of mass for a flat universe $\Omega_M \sim$ 10%–30% (Reiprich and Böhringer, 2002). On the other hand, Big Bang nucleosynthesis (BBNS) sets limits on the baryonic matter fractions at $\Omega_b \sim$ 4%–5% (Spergel et al., 2007; for an entertaining overview of why we need dark matter, see Siegel, 2008).

Work has also been done on observing the rotation of individual spiral galaxies. The rotation curves of individual stars in spiral galaxies need a large extra dark mass to explain the radial dependence of their velocity around the center of the galaxy to very large distances from the center of the galaxy (Rubin, 1983). These measurements imply a large spherical dark matter halo that encloses the visible Galaxy with a radius up to about 10 times the visible disk radius. In typical spiral galaxies, the ratio of dark matter mass to luminosity mass is about 10, again indicating a strong need for dark matter.

The power density of galaxies over the observable Universe can be used to calculate how much total matter (Ω_M) and how much normal matter (Ω_b) there is normalized to the critical mass of the Universe. Using galaxy surveys such as the Sloan Digital Sky Survey (SDSS), one finds that Ω_M is about 0.3, and Ω_b is about 0.05 (the rest of the mass-energy density is thought to be in dark energy, Ω_Λ, i.e., $\Omega_M + \Omega_\Lambda = 1$). This measurement implies that dark matter constitutes about 25% of the mass-energy budget of the Universe (Tegmark et al., 2004).

The concordance model of Big Bang cosmology, or Λ cold dark matter (ΛCDM), uses measurements of the cosmic microwave background (CMB) (Spergel et al., 2007), as well as observations of the power density of galaxies (Tegmark et al., 2004), supernova observations of the accelerating expansion of the Universe (Perlmutter and Schmidt, 2003), and x-ray and gravitational lens measurements of galaxy cluster mass yielding direct cluster mass measurements (Allen, 1998; Allen et al., 2008). It is the simplest known and generally accepted model that is in excellent agreement with observed phenomena. Fits to ΛCDM of all of these data indicate $\Omega_M \sim 27\%$ and $\Omega_b \sim 5\%$.

However, a couple of loopholes need to be closed to fully believe that a nonbaryonic dark matter exists that dominates the mass in the Universe. There is strong evidence that these loopholes have been closed, but work is ongoing:

1. Maybe BBNS is wrong? Can we check for indications of hard-to-see colder baryonic matter that might fill in the matter deficit? Direct searches for cold baryonic matter, including black holes, via strong lensing have been extensive (e.g., massive compact halo object [MACHO] searches) and have come to the conclusion that the limits on numbers and mass of "cool bodies" of normal matter fall far short of accounting for the dark matter (MACHO Collaboration, 2000).

2. There are theories of alternative gravity that offer a counter-explanation to dark matter. These theories are phenomenological, such as modified Newtonian dynamics (MOND) (Sanders, 2003), or offer more complex theories of general relativity (Bekenstein, 2004). Strong and weak gravitational measurements combined with x-ray measurements of colliding clusters of galaxies make these alternate theories unlikely substitutes for the dark matter paradigm. The first to make it very difficult for a MOND explanation was the "Bullet cluster," showing a "recent" collision of two clusters, where one observes matter exerting gravity where no normal matter is seen to exist (Clowe et al., 2006). This type of research using colliding clusters is ongoing, with a number of additional examples seen since (Jee et al., 2007; Mahdavi et al., 2007), which makes it even more impossible for any currently proposed theory of alternative gravity to explain all of the observations without using some form of dark matter.

3. There are future prospects for using gravitational lensing to better constrain dark matter versus alternative gravitational theories. One method directly searches for substructure in dark matter halos, a strong prediction of the ΛCDM paradigm (see the next section) and not currently predicted by alternative gravities (Moustakas and Metcalf, 2003). Another is to constrain the ellipticity of galaxy-scale dark matter halos with weak lensing (Schrabback, 2008; work in progress and private communication). Dark matter halos can show nonspherically symmetric lensing effects (ellipticity). This is not possible with current alternative gravity theories with no dark matter.

WHAT IS THE DARK MATTER?

Currently, we have very little idea of what actually constitutes the dark matter from which we clearly see the gravitational effects. Figure 8.1 shows the broad range of possible particle constituents arising from particle physics theories. There is a huge range of well-motivated candidates.

One of the best motivated candidates is the axion, which is needed to prevent charge-parity (CP) violation in the strong interaction and is very light. Then there is the neutralino that arises in supersymmetry and is a perennial favorite of a large number of particle theorists. It is relatively heavy, more interactive, and easily and naturally gives the $\Omega_M \sim 1$ for the normalized dark matter density of the Universe. This is particularly true in the case of the "WIMP miracle" for the so-called generic weakly interacting massive particle (WIMP)* that has

---

* The prediction from particle physics theory that DM is a $\sim$100 GeV WIMP with a weak scale annihilation cross section and that this natural and untuned result in particle physics gives $\Omega_M \sim 1$ is called the "WIMP miracle."

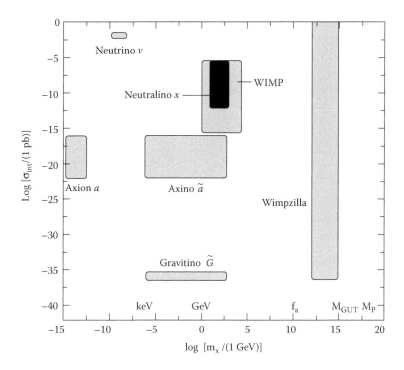

FIGURE 8.1 What constitutes dark matter? This figure shows a range of options for elementary particle constituents of dark matter based on particle physics ideas. It may be that more than one of these (or none) are what actually makes up dark matter. The abscissa shows the log of the normalized mass of the particle candidate, and the ordinate shows the log of the scattering cross section for DM on a nucleus normalized to 1 picobarn (1×10^{-36} cm^2). This sort of a plot is useful to gauge the progress of "direct detection" experiments (see "Brief Review of Direct Detection Experiments" in the text). NB: For WIMPS, the DM scattering cross section on a nucleus can be related to the WIMP annihilation cross section under general theoretical assumptions for many theories.

$$M_W \sim 100 \text{ GeV}, \text{ and} \langle \sigma_{annihilation} \times v \rangle \sim \text{few} \times 10^{-26} \text{ cm}^3/\text{s}. \tag{8.1}$$

Actually, there is no end of "natural" particle candidates for nonbaryonic dark matter from particle physics (e.g., little Higgs, sterile neutrinos, excited WIMPs). For a recent review on constraints on dark matter particle candidates see Boyanovsky et al. (2008).

Besides the very large uncertainty about the micro nature of dark matter, there is uncertainty about its macro distribution in the Universe. The latter problem has been addressed by a number of ΛCDM computer simulations. For example, the Via Lactea II dark matter simulation (Diemand et al., 2008; Kuhlen et al., 2008) has over 1 billion particles with a mass of only 4.1 thousand solar masses each and uses an improved, physical time-stepping method (Zemp et al., 2007). Via Lactea II took about 1 million CPU hours to finish and was run in November 2007 at the Oak Ridge National Laboratory on the Jaguar supercomputer. These simulations start just after recombination of the Universe with the primordial power spectrum imprinted as an initial condition on the dark matter distribution at that time. Initial conditions were generated with a modified, parallel version of the Gaussian random field community code (GRAFIC2; Bertschinger, 2001). The program steps through time using only gravitational interactions among the billion dark matter elements to simulate ultimately how the dark matter halo of a Milky Way-like galaxy would appear today (if one could actually see and resolve the dark matter structure). Figure 8.2 shows a picture of the result of

FIGURE 8.2 Projected dark matter density map of Via Lactea II. An approximately 800 kpc cube is shown. The Via Lactea II simulation has a mass resolution of 4,100 $M_\odot$ and a force resolution of 40 pc. It used over 1 million processor hours on the Jaguar Cray XT3 supercomputer at the Oak Ridge National Laboratory. A new method was employed to assign physical, adaptive time steps equal to one-sixteenth of the local dynamical timescale (but not shorter than 268,000 yr), which allows one to resolve very-high-density regions (From Diemand, J., Kuhlen, M., Madau, P., et al., *Nature*, 454, 735–8, 2008.).

these calculations. The dark matter structure formation is hierarchical with small dark matter halos merging to larger ones as the simulation proceeds.

Thus, in Figure 8.2 one sees the Milky Way halo composed of thousands of smaller halos varying in size from very small to very large. "The simulation reveals the fractal nature of dark matter clustering: Isolated halos and sub-halos contain the same relative amount of substructure and both have cuspy inner density profiles" (Diemand et al., 2008).

HOW CAN THE DARK MATTER PROBLEM BE SOLVED?

Solving the dark matter problem will require continuing the broad interdisciplinary approach that has been ongoing for the past decade. This strategy has encouraged trying to detect dark particles as particles in the Galaxy via direct and/or indirect detection. In addition, astronomers continue to probe the dark matter distribution in the Universe using optical telescopes via surveys and to study stellar velocity dispersion measurements that can establish new dark matter substructures in the Milky Way. Also, rapid progress is being made in gravitational strong and weak lensing with great promise for future work with new telescopes. Last, but not least, particle physicists are working hard to detect dark matter candidate particles in controlled environments at the Large Hadron Collider (LHC). To accomplish this daunting task, clearly one needs to combine data from astronomy, astrophysics, and high-energy physics.

The next decade promises dramatic improvements in implementing this strategy and thus greatly increasing our understanding as many new facilities have recently come online or will soon be online. These new facilities include the LHC; a new generation of indirect detection experiments such as the *Fermi Gamma-ray Space Telescope* (*FGST*, formerly called the *Gamma-Ray Large Area Space Telescope* [*GLAST*]), which was launched on June 11, 2008 (*Fermi*-LAT, 2009); and the Payload for Antimatter Matter Exploration and Light-nuclei Astrophysics (PAMELA) telescope, launched June 15, 2006.* They also include a relatively new generation of direct detection experiments such as the Cryogenic Dark Matter Search (CDMS) II;* Dark Matter/Large sodium Iodide Bulk for RAre processes (DAMA/LIBRA);* and XENON.* Also, dramatic improvements in optical survey experiments are beginning operation very soon, such as the Panoramic Survey Telescope and Rapid Response System (Pan-STARRS)* and the Dark Energy Survey (DES),* with the Large Synoptic Survey Telescope (LSST)* on the horizon for first light in 2016.

BRIEF REVIEW OF DIRECT DETECTION EXPERIMENTS

Direct detection experiments are a tour de force of low-noise experiments. The signal is WIMPs scattering elastically ("bumping") with a nucleus that then recoils with a velocity of $\sim 10^{-3}c$. The nucleus then excites the nearby atoms by transferring its kinetic energy of roughly tens of keV with some efficiency to make observable energy (phonons, light, ionized charge), which is measured in a manner depending on the detector medium (silicon/germanium, NaI, xenon). Unfortunately for direct detection experiments, one estimates that for $M_W \sim 100$ GeV there is about 1 WIMP per 300 cm^3 at the Earth (~ 0.4 GeV/cm^3), and the WIMP-nucleus elastic scattering cross section is a weak interaction scale, i.e., very tiny. Thus, these experiments are a sophisticated and large effort in designing and building for measuring very low signal-to-noise ratio, and in a very-low-noise environment. WIMPs are neutral, and so the main backgrounds are (low-energy) gamma rays and neutrons. Also, for some detector types, surface electrons from β-decay can mimic nuclear recoils. Gamma rays knock out atomic electrons that have recoil velocity $\sim 0.3c$. Neutron recoils only have a mean free path of a few centimeters. For a recent review of direct detection of dark matter, see Gaitskell (2004).

Figure 8.3 shows the recent status of direct detection searches (B. Cabrera, 2008, private communication; Trotta et al., 2007). The most sensitive of the experiments, CDMS II and XENON 10, are beginning to challenge the minimal supersymmetric standard model (MSSM) phase space (Trotta et al., 2007). However, there are orders of magnitude left in the MSSM phase space, which extends to 3 to 4 orders of magnitude smaller cross sections, to explore with future experiments. (There are also different models than MSSM to explore.) Another feature to note in this graph is the apparent disagreement of DAMA with ZEPLIN II, XENON 10, and CDMS II. Although MSSM fails to do so, some theories are able to accommodate all of the experimental results (see, e.g., Finkbeiner and Wiener, 2007; Feng et al., 2008).

DAMA/LIBRA (Bernabei et al., 2008) is an experiment that has been taking data in two configurations for more than a decade (DAMA for 7 years, DAMA/LIBRA for 4 years). This experiment is designed to make use of the annual modulation in a direct dark matter signal due to the motion of the Earth through the dark matter field as we go around the Sun (Drukier et al., 1986; Freese et al., 1988). With current detector technology, the annual modulation is the main model-independent signature for the DM signal. Although the modulation effect is expected to be relatively small, a suitably large mass detector embedded in a low-radioactive environment, with careful control of the running conditions and over a long enough time, could detect this modulation if dark matter particles exist with a sufficiently large interaction cross section. Optimizing to observe the modulation effect can dramatically improve the sensitivity of a direct detection experiment, and this has been in the design strategy of DAMA/LIBRA from its inception. If the modulation is observed, it must modulate according to a cosine through the year. The modulation is observed in a definite (low) energy range in the detector for single-hit events in a multielement detector. The phase maximum should be at about June 2 of each year when the Earth's velocity to the dark matter field is maximum

---

* For more information, see: http://pamela.roma2.infn.it/index.php?option=com_mjfrontpage&Itemid=159; http://cdms. berkeley.edu/; http://people.roma2.infn.it/~dama/web/home.html; http://xenon.astro.columbia.edu/; http://pan-starrs.ifa. hawaii.edu/public/home.html; https://www.darkenergysurvey.org/; and http://www.lsst.org/lsst.

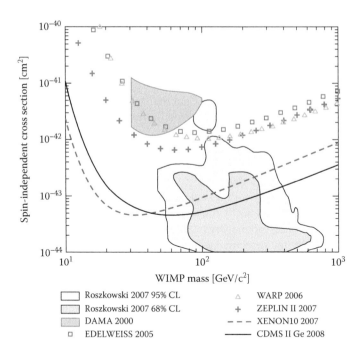

FIGURE 8.3   Status of direct detection searches from a number of competitive experiments. There is an apparent disagreement between DAMA, ZEPLIN II, XENON 10, and CDMS II, which is currently being explored by the theorists as described in the text.

for the year by adding to the Sun's velocity through the Galaxy ($t_0 = 152.5$ days as fitted to the data), as well as with signal modulation amplitude in the region of maximal sensitivity of <7% for usually adopted dark matter halo distributions (but it can be larger in the case of some possible scenarios). DAMA/LIBRA has claimed such a signal that has continued in two experimental setups maintaining the same phase for the past 11 years. This result does not appear to be a statistical fluctuation as it is currently at the ~8σ level statistically. As their many public presentations and recent publication show, Bernabei et al. (2008) have carefully examined many systematic effects and have found no obvious problems with their measurement. No other independent experiment has reproduced the DAMA/LIBRA result. It is currently left to future experiments to resolve this puzzle.

BRIEF REVIEW OF INDIRECT DETECTION EXPERIMENTS

Figure 8.4 shows the diversity of experiments that can contribute to the indirect detection of dark matter. The particle physics view is that dark matter is a particle, which is stable or at most has been decaying very slowly during the past 14 billion years. The most popular particle physics models posit annihilation of WIMPs that have been thermally produced in the early Universe and are essentially stable to decay. These annihilations produce final-state photons, protons, electrons/positrons, and neutrinos that can be observed as illustrated in the figure. However, there are models in which WIMPs of a different kind can decay, with long decay constants, and yield the same final-state particles, but with different relative probabilities.

To again illustrate the diversity of theoretical opinions and the resulting challenges to experimentalists, there are theories of dark matter that claim that axions are its main component. Axions that could make up a significant fraction of the dark matter are a product of nonthermal production during the quantum chromodynamics (QCD) phase transition in the early Universe. At this transition, free quarks were bound into hadrons and a very cold Bose condensate of axions formed; therefore, very light axions can also be CDM and are indistinguishable to cosmologists studying

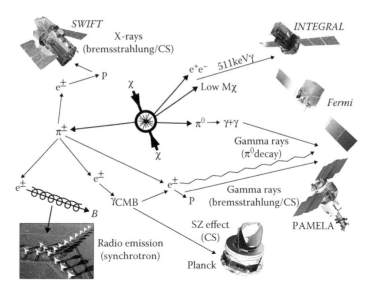

FIGURE 8.4 Representation of indirect detection. Dark matter annihilation in situ in the Universe via particle physics generic models feeds a large number of indirect detection methods. In these particle physics models, $<\sigma v>$ and the mass of dark matter particles are highly uncertain. The indirect detection methods shown span the electromagnetic spectrum from radio to very-high-energy gamma rays (air Cherenkov telescopes, such as Collaboration of Australia and Nippon (Japan) for a GAmma Ray Observatory (CANGAROO) III, Hess, and Very Energetic Radiation Imaging Telescope Array System (VERITAS), also contribute observations of $> \sim 100$ GeV photons and ~TeV [e^- + e^+], but are not shown explicitly in the figure). The presence of magnetic fields and/or stellar radiation fields is needed to generate the lower-energy signals. In addition, there are charged cosmic-ray signals in protons and electrons at high energy. PAMELA is focused more on the charged particles, while *Fermi* has been focused more on gamma rays; however, each can observe both, and, in fact, *Fermi* has excellent electron capability from the GeV range up to and greater than 1 TeV.

galaxy formation and the origin of large-scale structure (Kolb and Turner, 1990). Axions can also be indirectly observed in astrophysics experiments (see, e.g., Simet et al., 2008), as well as in ground-based experiments (Asztalos et al., 2006).*

I will focus on indirect searches for WIMP annihilation here, which is the idea that has generated a very broad range of searches in astrophysical settings (not to mention direct searches and searches at the LHC for WIMPs). Estimating the sensitivity of an experiment to WIMP annihilation somewhere in the Universe (typically in the Milky Way) involves four elements, besides calculating one's detector intrinsic sensitivity to the particles being measured. The first is the calculation of the integral of the energy spectrum of the particle over the energy range of interest,

$$\int \left(\sum_i \left(\frac{dN}{dE} \right)_i B_i \right) dE. \tag{8.2}$$

In equation 8.2, the index i denotes the particle species appearing in the decay that is being observed, and B_i is the branching ratio into that species. In the case of WIMP annihilation (or decay), this quantity can be estimated using computer programs such as DarkSUSY, which computes "SUperSYmmetric dark matter" properties numerically (Gondolo et al., 2004) combined with PYTHIA (Sjöstrand et al., 2009). As the first step, DarkSUSY provides (among other quantities) the

---

* For an informative and easily accessible review of axions, see http://en.wikipedia.org/wiki/Axion.

branching ratios and momentum distributions from the annihilation into the "fundamental" constituents of the standard model of particle physics, i.e., the various quarks, tau leptons (lighter leptons are neglected at this stage), and intermediate bosons (gluons, W, and Z) that can contribute. These branching ratios depend on the nature of the theory and its particular parameters. PYTHIA is then used by DarkSUSY to follow these "fundamental" constituents to their final observable particles by calculating hadronization of the quarks, gluons, and decays of the resulting unstable mesons, baryons, and leptons to the finally observed protons/antiprotons, electrons/positrons, photons, and neutrinos.

As an example, in the case of final-state photons, a topic I will review in some detail below for my favorite detector the *Fermi*-LAT, Table 8.1 shows the high-energy photon yield per final state $b\bar{b}$ pair. This example is for the annihilation of two WIMPs of 10, 100, and 1000 GeV. Note that in this calculation, it does not matter much which intermediate-state fundamental particles contribute; they all give very similar dN/dE distributions for the resulting photons from 100 MeV to the endpoint energy defined by the WIMP mass (for $M_{WIMP} > \sim 100$ GeV). The exception is decays dominated by tau lepton pairs, in which case the spectrum is noticeably harder (Cesarini et al., 2004). The potential detection of WIMP annihilation is enhanced as the final photon spectra calculated for this process tend to be harder than most astrophysical spectra and are not power laws. Also note that monoenergetic photon lines can be produced at the mass of the WIMP in annihilations and that for $M_{WIMP} > M_Z/2$ they can also have a monoenergetic photon from the γZ final state. This monoenergetic line is considered a "smoking gun" for the discovery of dark matter. The branching fractions to these lines in annihilations and decays theoretically range from 0.1 to 10^{-4}, with the most popular SUSY theories giving numbers in the smaller range. This is because in these theories higher loop diagrams are needed (Bergstrom and Ullio, 1997).

The second element in the calculation is $\langle\sigma v\rangle$, the annihilation cross section times the relative velocity of the WIMPs. For the "generic" WIMP case $\langle\sigma v\rangle \sim 3 \times 10^{-26}$ cm^3/sec; however, this cross section varies over orders of magnitude in various parts of the possible theoretical phase space. Recent theoretical speculation has suggested a considerably larger cross section (Arkani-Hamed et al., 2009), while standard MSSM admits much smaller cross sections (Gondolo et al., 2004).

The third element of the calculation is the contribution of the dark matter density distribution. This number density of the dark matter particles is given by

$$\frac{4\pi}{M_{WIMP}^2} \int \rho^2(r) r^2 dr \tag{8.3}$$

and depends on the dark matter clustering. Here $\rho(r)$ is the dark matter density distribution. A popular analytic $\rho(r)$ that was derived from computer simulations is the Navarro–Frenk–White (NFW) distribution (Navarro et al., 1996). This form is cuspy near $r = 0$, where $\rho(r) \sim 1/r$; it is also supported

---

TABLE 8.1

The Gamma-Ray Yield Per Final State $b\bar{b}$ Pair as a Function of the Mass of the WIMP

M_{WIMP}	Total# γ	>100 MeV	>1 GeV	>10 GeV
10 GeV	17.3	**12.6**	1.0	0
100 GeV	24.5	22.5	**12.4**	1.0
1 TeV	31.0	29.3	22.4	**12.3**

Note: The table shows the total number of gamma rays produced in the decay, and with energy >100 MeV, >1 GeV, and >10 GeV. The numbers in bold approximately show the constant multiplicity contour.

FIGURE 8.5 A far view of the Galaxy shining from dark matter annihilations only. The Galaxy shines in high-energy gamma rays from the annihilation of dark matter assuming a semianalytic computer simulation of ΛCDM (Taylor and Babul, 2005a,b) and a generic WIMP dark matter particle. The picture is for an "average" Milky Way Galaxy as determined by these computer simulations. The scale of the central part of the picture, i.e., the brightest part, yellow-red-light blue at the center, roughly corresponds to the visible Milky Way diameter in optical wavelengths. The rest of the picture is the dark matter halo dominated at larger distances by the dark matter clumps. The radius of the visible part of the Milky Way is about 20 kpc; the entire dark matter halo is ~100 kpc. The WIMP annihilation is proportional to density squared of the dark matter density, and that is what is visualized in this figure. The map is shown as a Hammer–Aitoff projection in Galactic coordinates. No normal matter is shown in this picture, and if shown it would dominate the gamma-ray intensity of the map for the Milky Way proper. The picture is a colorized version of that produced by E. Baltz (2005, private communication).

by the modern computer simulations that I previously discussed (Diemand et al., 2008). The last element of the puzzle is the distance to the object one is viewing as this gives the inverse square law flux factor, $1/4\pi d^2$. In combining all of these factors, there is clearly a great deal of uncertainty in the estimated flux of gamma rays at one's detector. Part of the progress in this field over the next decade will be to better understand and bracket each of the uncertain contributing elements.

Figure 8.5 shows the Galaxy shining in high-energy gamma rays from the annihilation of dark matter, assuming a semianalytic computer simulation of ΛCDM (Taylor and Babul, 2005a,b) and a generic WIMP dark matter particle. The map is shown as a Hammer–Aitoff projection in Galactic coordinates. No normal matter is shown in this picture, and if shown it would dominate the gamma-ray intensity of the map. The picture is a colorized version of that produced by E. Baltz (2005, private communication). In this figure, the more intense the radiation, the whiter it will be, while dark is the absence of radiation. The intensity is proportional to the square of the dark matter density. The center of the Galaxy is the brightest, and a number of Galactic dark matter satellites are prominent. Of course, as I have previously stressed, normal astrophysical sources of radiation at all wavelengths from radio to the highest-energy gamma rays dominate what we observe with current instruments. Dark matter radiations, if they exist, are but small fractions of the total and will take considerable time to untangle from the bulk. Setting progressively better limits is the expected outcome for some time. However, the tools we now have in hand and that are close on the horizon are dramatic improvements of past tools. In the next section I will discuss the results from two of these tools, *Fermi*-LAT and PAMELA, in more detail.

GLAST → *FERMI*: LAUNCH, FIRST RESULTS, AND SOME DARK MATTER PROSPECTS

GLAST was launched by NASA on June 11, 2008, from Cape Canaveral, Florida. The satellite went to low Earth orbit flawlessly and currently is in a circular orbit, 565 km altitude (96-min period) and 25.6 degree inclination. The satellite scans the entire sky every 192 min (two orbits). Standard data collection is the all-sky scanning mode, where *Fermi*-LAT is pointed 35 degrees toward the Earth's North Pole relative to Earth zenith at the satellite position on one orbit, and then –35 degrees toward

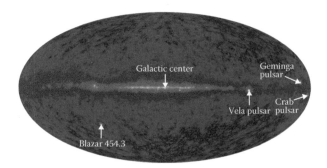

FIGURE 8.6 *Fermi*-LAT 3-month all-sky map collected in nominal all-sky scanning mode from August 4 to November 4, 2008. Some bright sources are indicated in the figure. The data shown have gamma-ray energy >200 MeV. This is a count map (1131 × 617) with 0.3-degree pixels, with Log scaling over the entire range, and is shown as a Hammer–Aitoff projection in Galactic coordinates. The map is corrected for exposure at 1 GeV (J. McEnery [The *Fermi*-LAT Collaboration], Invited Talk at the 2nd *Fermi* Symposium, Washington, DC, November 2–5, 2009. See http://fermi.gsfc.nasa.gov/science/symposium/2009/slides/day1/JMcEnery. pdf.).

the Earth's South Pole on the next. *GLAST* was renamed *Fermi* by NASA on August 26, 2008, after on-orbit commissioning was complete and nominal science operations had begun. The LAT was constructed and is being operated by the *Fermi*-LAT Collaboration; it is described in Atwood et al. (2009).* The telescope has been optimized to measure gamma rays from 20 MeV to 300 GeV, has unprecedented angular resolution in this energy range compared with previous gamma-ray missions, and views 20% of the entire sky at any instant. *Fermi*-LAT achieves about 30 times the sensitivity of Energetic Gamma Ray Experiment Telescope (EGRET) in the EGRET energy range, 100 MeV–10 GeV, and extends measurements well beyond the EGRET energy range. The *Fermi* mission requirement (NASA) is 5 years, with a 10-year goal. Ten years seems quite feasible as the instrument uses no consumables. LAT's potential for making systematics-limited measurements of cosmic ray (CR) electrons was recognized during the initial phases of the LAT design (Moiseev et al., 2007), and we have indeed found that *Fermi*-LAT is an excellent CR (electron + positron) detector for energies in the range 20 GeV–1 TeV.

The *Fermi*-LAT is not really a telescope in the standard sense. However, it is an astounding machine—a massive particle physics detector in orbit. It is 1.8 × 1.8 m², 3 metric tons, and moves at 17,000 mph. The detector's position is known to a few meters in orbit, its attitude to ~10 arcsec, and time to <10 μs. Yet, as the LAT instrument architect, Bill Atwood, says, "It uses less power than a toaster and we talk to it over a telephone line." Note that the average downlinked data rate is considerably higher at an average event rate to the ground of about 500 Hz ~ megabit/s. The CR background is intense and requires multilevel onboard filtering of events to achieve the data rate to the ground; the average trigger rate before onboard filtering is ~2.5 kHz.

Figure 8.6 shows a *Fermi*-LAT 3-month all-sky map collected in nominal all-sky scanning mode from August 4 to November 4, 2008. The data shown have gamma-ray energy >200 MeV. Some bright sources are indicated in the figure, and many other sources are evident. The Galactic disk dominates the picture, with many sources seen in the central bulge region (shown in numerous talks by *Fermi*-LAT Collaboration members; see, e.g., Abdo et al., 2010). This is the all-sky image that will get clearer with time, from which one will need to dig out any dark matter signal that might be there.

Tables 8.2 and 8.3 show active *Fermi*-LAT Collaboration efforts for dark matter searches using the LAT, as well as multiwavelength connections for other telescopes. The *Fermi*-LAT prelaunch

---

* The collaboration membership and information about *Fermi*-LAT science can be found on our Website: http://www-glast.stanford.edu/.

TABLE 8.2
Ongoing Indirect Dark Matter Searches Using Photons

Focus of Search	Advantages	Challenges	Experiments
Galactic Center region—WIMP[a]	Good statistics	Source confusion, astrophysical background	ACTs,[b] *Fermi*, *WMAP*[c] (Haze), Integral, X-ray, radio
DM Galactic satellites/ Dwarfs/Black hole mini spikes—WIMP	Low background	Low statistics, follow-up multiwavelength observations, astrophysical uncertainties	ACTs (guided by *Fermi*), *Fermi*, Optical telescopes
Milky Way halo—WIMP	High statistics	Galactic diffuse modeling	*Fermi*
Spectral lines—WIMP	No astrophysical backgrounds	Low statistics in many models	*Fermi*, ACTs (GC)[d]
Extra-Galactic background —WIMP	High statistics	Galactic diffuse modeling, instrumental backgrounds	*Fermi*

Source: Details of the potential sensitivity for these searches for *Fermi*-LAT have been published. (From Baltz, E.A., Berenji, B., Bertone, G., et al., Pre-launch estimates for *GLAST* sensitivity to dark matter annihilation signals. *Journal of Cosmology and Astroparticle Physics*, 07, 013, 2008.)

Note: The searches, a brief description of the pros and cons for each search, and the multiwavelength contributions are indicated.

[a] Weakly Interacting Massive Particle
[b] Air Cherenkov Telescopes (ACTs)
[c] *Wilkinson Microwave Anisotropy Probe (WMAP)*
[d] ACTs observing the Galactic Center

sensitivity estimates for most of the searches listed in the tables have been published (Baltz et al., 2008). The various telescopes listed in the last column of the tables contribute complementary multiwavelength information for the dark matter searches. In Table 8.2, the Milky Way satellite search and the WIMP line search stand out as potential "smoking guns" for dark matter if a signal is discovered. In the case of dark matter satellites, optical telescope surveys can find them by the peculiar motions of their stars also giving mass/light ratios and accurate locations (see, e.g., Simon and Geha, 2007). With bigger telescopes, e.g., Keck, one learns more details about the putative dark matter distribution of the satellite from much better spectrographic observations of the associated stars. *Fermi* can examine the locations of these known satellites and set limits on dark matter models improved by the more detailed knowledge of the dark matter distribution. Also, *Fermi* can search the sky for unknown dark matter satellites (Baltz et al., 2008). If found, optical follow-up would be important in understanding the structure of these *Fermi*-found dwarf galaxies.

Most of the *Fermi* dark matter searches using photons will take deep exposures over 5 years or more and considerable work in other wavelengths to produce significant limits on current theories of dark matter. If the LHC were to discover a particle candidate in this time frame with a well-specified mass, this could dramatically improve *Fermi*'s, as well as other telescopes', chances for establishing this potential candidate as the dark matter particle of the Universe. On the other hand, the LHC alone cannot do this. LHC experiments cannot measure the lifetime of a putative dark matter candidate particle and can set lifetime limits that are only on the order of perhaps seconds, less than the age of the Universe by many orders of magnitude.

COSMIC-RAY RESULTS FROM ATIC, *FERMI*, AND PAMELA

The indirect search for dark matter has been the subject of recent excitement with the release of new results from the PAMELA experiment on the antiproton/(proton + antiproton) ratio (Adriani et al., 2009b) and positron/(electron + positron) ratio from 1 to 100 GeV (Adriani et al., 2009a).

TABLE 8.3

Ongoing Searches for Dark Matter using Different Sorts of Astrophysical Photon Sources and Cosmic Rays (CRs)

Focus of Search	Advantages	Challenges	Experiments
High-latitude neutron stars—KK[a] graviton	Low background	Astrophysical uncertainties, instrument response ~100 MeV	*Fermi*
AGN[b] jet spectra—axions	Many point sources, good statistics	Understanding details of AGN[b] jet physics and spectra	ACTs,[c] *Fermi*, x-ray, radio (multiwavelength)
e[+] + e[-], or e[+]/e[-]	Very high statistics	Charged particle propagation in galaxy, astrophysical uncertainties	*Fermi*, PAMELA,[d] AMS[e]
Antiprotons/Protons	″	″	PAMELA, AMS

Source: Hannestad, S., and Raffelt, G.G. Supernova and neutron-star limits on large extra dimensions reexamined. *Physical Review* D, 67, 125008, 2003; Sánchez-Conde, M.A., Paneque, D., Bloom, E.D., et al. Hints of the existence of axionlike particles from the gamma-ray spectra of cosmological sources. *Physical Review* D, 79, 123511, 2009.

Note: This table considers two searches for dark matter with photons that are a bit unusual compared with those in Table 8.2. The searches, a brief description of the pros and cons for each search, and the multiwavelength contributions are indicated. The 1st is a search for large extra dimensions using older neutron pulsars (Hannestad and Raffelt, 2003), and the second uses AGN jet spectra to search for axions (Sánchez-Conde et al., 2009). The last two searches use positrons, electrons, antiprotons, and protons from CRs.

[a] Kaluza-Klein
[b] Active galactic nucleus
[c] Air Cherenkov Telescopes (ACTs)
[d] Payload for Antimatter Matter Exploration and Light-nuclei Astrophysics (PAMELA) telescope
[e] Alpha Magnetic Spectrometer

The antiproton/(proton + antiproton) ratio measurements fit the expectations of CR models assuming pure secondary production of antiprotons during the propagation of CRs in the Galaxy (Ptuskin et al., 2006). However, the positron/(electron + positron) ratio does not fit the currently favored CR model (Moskalenko and Strong, 1998). The PAMELA positron ratio increases from 0.055 at 10.2 GeV to 0.14 at 82.6 GeV, or an increase of a factor of ~2.5, while in this energy range the Moskalenko and Strong model shows a rapid decrease in the ratio. In their *Nature* paper, the PAMELA Collaboration concludes that, to explain these data, "a primary source, be it an astrophysical object or dark matter annihilation, is necessary." The experiment is continuously taking data, and the increased statistics will allow the measurement of the positron fraction to be extended up to about 300 GeV in the future.

Since the *New Vision 400* conference took place in October 2008,* and before the time of this writing, two new developments have considerably heated up interest in the indirect search for dark matter ignited by the PAMELA results. Although these results were published after the conference, first reports were made in the summer conferences of 2008 and so were known at the time of *NV400*. In the first development, the Advanced Thin Ionization Calorimeter (ATIC) balloon experiment reported observing a peak in the (electron + positron) CR spectrum at an energy of about 600 GeV (Chang et al., 2008). In their *Nature* paper, the ATIC Collaboration reported "an excess of Galactic CR electrons at energies of about 300–800 GeV, which indicates a nearby source of energetic electrons [plus positrons]. Such a source could be an unseen astrophysical object (such as a pulsar or micro-quasar) that accelerates electrons to those energies, or the electrons could arise from the annihilation of dark matter particles (such as a Kaluza–Klein particle with a mass of about 620 GeV)."

The second development came from *Fermi*-LAT. The *Fermi*-LAT Collaboration has measured the (electron + positron) spectrum from 20 GeV to 1 TeV with very high statistical precision (Abdo et al., 2009). Figure 8.7 shows the results of the CR (e⁻ + e⁺) measurement from the *Fermi*-LAT Collaboration, along with measurements from other experiments, from 20 GeV to 1 TeV. This spectrum contains more than 4 million (e⁻ + e⁺) events, and so the statistical errors are very small compared with the systematic errors indicated in the figure. The details of the analysis and how the systematic errors were estimated are discussed in some detail in *Fermi*-LAT (2009) and Ackermann et al. (2010). The main conclusion to be drawn from these data is two-fold: (1) The *Fermi*-LAT does not confirm the ATIC peak at about 600 MeV. If this peak were present at the strength reported by ATIC, the *Fermi*-LAT analysis would have reproduced it, but with ~7000 (e⁻ + e⁺) in a peak above the spectrum shown in Figure 8.7. (2) The *Fermi*-LAT spectrum is much harder than expected in conventional Galactic diffusive models (Strong, Moskalenko, and Reimer, 2004). A simple power-law fit to the data in the *Fermi*-LAT energy band gives a spectral index of –3.04 with small errors and a $\chi^2 = 9.7$ for 24 degrees of freedom; this is a very good fit. The reason the fit is seemingly too good, $\chi^2/\text{d.o.f} = 0.4$, is that the *Fermi* team has taken the systematic errors, represented by the gray band in the figure, and added them in quadrature with the statistical errors. The team considers this to be the conservative thing to do at this time. A future long paper using more data will explore this issue again.

Combined with the PAMELA positron fraction discussed above, the *Femi* result still poses a serious problem to the conventional Galactic diffusive models and strongly reinforces the need for relatively local galactic sources of electrons and positrons. Two such sources of electrons and positrons have so far been considered: pulsars and dark matter annihilation or decays. An example of comparisons of these very different models with the *Fermi* and PAMELA data can be found in a recent *Fermi*-LAT Collaboration publication (Grasso et al., 2009). Good fits are obtained in both models, but much more needs to be learned from the experiments before a choice of mechanism is finally made.

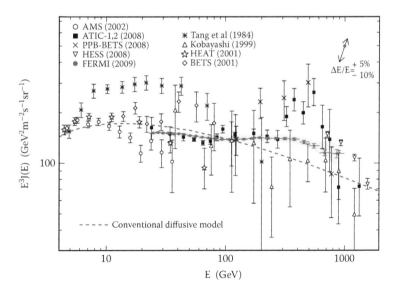

FIGURE 8.7   The *Fermi*-LAT CR electron spectrum (*red filled circles*). Systematic errors are shown by the gray band. The two-headed arrow in the top right corner of the figure gives size and direction of the rigid shift of the spectrum implied by a shift of +5%–10% of the absolute energy, corresponding to the present estimate of the uncertainty of the LAT energy scale. Other high-energy measurements and a conventional diffusive model are shown. (From Strong, A.W., Moskalenko, I.V., and Reimer, O., *Astrophysical Journal*, 613, 962–76, 2004.)

SUMMARY AND CONCLUSIONS

A number of new experiments have recently come online, or are coming online over the next few years, that will greatly enhance the discovery space for direct and indirect detection of dark matter. These experiments include new strong and weak gravitational lensing techniques that will soon be making an impact on understanding DM structure in galaxies and, in particular, Milky Way dwarf galaxies. The LHC will also start beam collisions for doing science soon, and the potential discovery of new high-mass particles would give strong impetus to targeted astrophysical dark matter searches. Thus, the next 5–10 years should be a "golden age" for expanding our knowledge of the nature of dark matter. We hope that we will actually discover what the stuff of this mysterious dark matter is! Maybe DAMA, *Fermi*, PAMELA, ATIC, and other dark matter search experiments are close? Tune in as the adventure unfolds.

ACKNOWLEDGMENTS

I would like to thank the organizers of *NV400* for arranging and executing an excellent and very informative conference that took place in a convenient, warm, and supportive environment. I thank the John Templeton Foundation for the financial support provided to me to attend this conference. I thank the *Fermi*-LAT Collaboration, of which I am a member, for the unpublished materials from *Fermi* that I used in writing this chapter.

REFERENCES

Abdo, A.A., Ackermann, M., Ajello, M., et al. (2009). Measurement of the cosmic ray $e^+ + e^-$ spectrum from 20 GeV to 1 TeV with the Fermi Large Area Telescope. *Physical Review Letters*, 102: 181101 (6 pages).

Abdo, A.A., Ackermann, M., Ajello, M., et al. (The *Fermi*-LAT Collaboration) (2010). Fermi Large Area Telescope First Source Catalog. *Astrophysical Journal Supplement Series*, 188: 405–36.

Ackermann, M., Ajello, M., Atwood, W.B., et al. (The *Fermi*-LAT Collaboration) (2010). *Fermi*-LAT observations of cosmic-ray electrons from 7 GeV to 1 TeV. *Physical Review D*, 82: 092004 (20 pages).

Adriani, O., Barbarino, G.C., Bazilevskaya, G.A., et al. (2009a). Observation of an anomalous positron abundance in the cosmic radiation. *Nature,* 458: 607–9.

Adriani, O., Barbarino, G.C., Bazilevskaya, G.A., et al. (2009b). A new measurement of the antiproton-to-proton flux ratio up to 100 GeV in the cosmic radiation. *Physical Review Letters*, 102: 051101.

Allen, S.W. (1998). Resolving the discrepancy between X-ray and gravitational lensing mass measurements for clusters of galaxies. *Monthly Notices of the Royal Astronomical Society*, 296: 392–406.

Allen, S.W., Rapetti, D.A., Schmidt, R.W., et al. (2008). Improved constraints on dark energy from Chandra X-ray observations of the largest relaxed galaxy clusters. *Monthly Notices of the Royal Astronomical Society*, 383: 879–96.

Arkani-Hamed, N., Finkbeiner, D.P., Slatyer, T.R., et al. (2009). A theory of dark matter. *Physical Review D,* 79: 015014 (22 pages).

Asztalos, S.J., Rosenberg, L.J., van Bibber, K., et al. (2006). Searches for astrophysical and cosmological axions. *Annual Review of Nuclear and Particle Science,* 56: 293–326.

Atwood, W.B., Abdo, A.A., Ackermann, M., et al. (2009). The Large Area Telescope on the *Fermi Gamma-Ray Space Telescope* mission. *Astrophysical Journal*, 697: 1071–102.

Baltz, E.A., Berenji, B., Bertone, G., et al. (2008). Pre-launch estimates for *GLAST* sensitivity to dark matter annihilation signals. *Journal of Cosmology and Astroparticle Physics,* 07: 013 (49 pages) [arXiv:0806.2911v2 [astro-ph]].

Bekenstein, J.D. (2004). Relativistic gravitation theory for the modified Newtonian dynamics paradigm. *Physical Review D,* 70: 083509 (28 pages). (Erratum 2005, *Physical Review D,* 71: 069901 [1 page].)

Bergström, L. and Ullio, P. (1997). Full one loop calculation of neutralino annihilation into two photons. *Nuclear Physics B,* 504: 27–44.

Bernabei, R., Belli, P., Cappella, F., et al. (2008). First results from DAMA/LIBRA and the combined results with DAMA/NaI. *The European Physical Journal C,* 56: 333–55.

Bertschinger, E. (2001). Multiscale Gaussian random fields for cosmological simulations. *Astrophysical Journal Supplement Series,* 137: 1 (38 pages).

Boyanovsky, D., de Vega, H.J., and Sanchez, N. (2008). Constraints on dark matter particles from theory, galaxy observations, and *N*-body simulations. *Physical Review D,* 77: 043518 (20 pages).

Cesarini, A., Fucito, F., Lionetto, A., et al. (2004). The Galactic center as a dark matter gamma-ray source. *Astroparticle Physics,* 21: 267–85.

Chang, J., Adams, J.H. Jr, Ahn, H.S., et al. (2008). An excess of cosmic ray electrons at energies of 300–800 GeV. *Nature,* 456: 362–65.

Clowe, D., Bradač, M., Gonzalez, A.H., et al. (2006). A direct empirical proof of the existence of dark matter. *Astrophysical Journal,* 648: L109.

Diemand, J., Kuhlen, M., Madau, P., et al. (2008). Clumps and streams in the local dark matter distribution. *Nature,* 454: 735–38.

Drukier, A.K., Freese, K., and Spergel, D.N. (1986). Detecting cold dark-matter candidates. *Physical Review D,* 33: 3495–508.

Feng, J.L., Kumar, J., and Strigari, L.E. (2008). Explaining the DAMA signal with WIMPless dark matter. *Physics Letters B,* 670: 37–40.

Finkbeiner, D.P. and Wiener, N. (2007). Exciting dark matter and the INTEGRAL/SPI 511 keV signal. *Physical Review D,* 76: 083519 (9 pages).

Freese, K., Friedman, J., and Gould, J. (1988). Signal modulation in cold-dark-matter detection. *Physical Review D,* 37: 3388–405.

Gaitskell, R.J. (2004). Direct detection of dark matter. *Annual Review of Nuclear Particle Science,* 54: 315–59.

Gondolo, P., Edsjö, J., and Ullio, P. (2004). DarkSUSY: Computing supersymmetric dark matter properties numerically. *Journal of Cosmology and Astroparticle Physics,* 0407: 008 (35 pages). Available at: http://www.darksusy.org.

Grasso, D., Profumo, S., Strong, A.W., et al. (2009). On possible interpretations of the high energy electron–positron spectrum measured by the *Fermi* Large Area Telescope. *Astroparticle Physics,* 32: 140–51.

Hannestad, S., and Raffelt, G.G. (2003). Supernova and neutron-star limits on large extra dimensions reexamined. *Physical Review D,* 67: 125008 (10 pages). (Erratum 2004, *Physical Review D,* 69: 029901 [1 page].)

Jee, M.J., Ford, H.C., Illingworth, G.D., et al. (2007). Discovery of a ringlike dark matter structure in the core of the Galaxy Cluster Cl 0024+17. *Astrophysical Journal,* 661: 728–49.

Kolb, E.W. and Turner, M.S. (1990). The early Universe. In: *Frontiers in Physics,* Vol. 69. Boulder, CO: Westview Press.

Kuhlen, M., Diemand, J., and Madau, P. (2008). The dark matter annihilation signal from Galactic substructure: predictions for *GLAST. Astrophysical Journal,* 686: 262–78.

MACHO Collaboration. (2000). The MACHO Project: Microlensing results from 5.7 years of LMC observations. *Astrophysical Journal,* 542: 281–307.

McEnery, J. (The *Fermi*-LAT Collaboration) (2009). Invited talk at the 2nd *Fermi* Symposium, Washington, DC, November 2–5, 2009. Available at: http://fermi.gsfc.nasa.gov/science/symposium/2009/slides/day1/JMcEnergy.pdf.

Mahdavi, A., Hoekstra, H., Babul, A., et al. (2007). A dark core in Abell 520. *Astrophysical Journal,* 668: 806–14.

Moiseev, A.A., Ormes, J.F., and Moskalenko, I.V. (2007). Measuring 10–1000 GeV cosmic ray electrons with GLAST/LAT. *Proceedings of the International Cosmic Ray Conference* (Merida), 2: 449–52 and references therein [arXiv:0706.0882v1 [astro-ph]].

Moskalenko, I.V. and Strong, A.W. (1998). Production and propagation of cosmic-ray positrons and electrons. *Astrophysical Journal,* 493: 694–707.

Moustakas, L.A. and Metcalf, R.B. (2003). Detecting dark matter substructure spectroscopically in strong gravitational lenses. *Monthly Notices of the Royal Astronomical Society,* 339: 607–15.

Navarro, J., Frenk, C., and White, S. (1996). The structure of cold dark matter halos. *Astrophysical Journal,* 462: 563–75.

Perlmutter, S. and Schmidt, B.P. (2003). Measuring cosmology with supernovae. In: K. Weiler (ed.), *Supernovae and Gamma Ray Bursters,* 195–218. Berlin: Springer.

Ptuskin, V.S., Moskalenko, I.V., Jones, F.C., et al. (2006). Dissipation of magnetohydrodynamic waves on energetic particles: Impact on interstellar turbulence and cosmic-ray transport. *Astrophysical Journal,* 642: 902–16.

Reiprich, T.H. and Böhringer, H. (2002). Constraining cosmological models with the brightest galaxy clusters in x-ray sky. In: M. Gilfanov, R. Sunyaev, and E. Churazov (eds.), *Lighthouses of the Universe: The Most Luminous Celestial Objects and Their Use for Cosmology,* 84–93. Berlin: Springer.

Rubin, V. (1983). The rotation of spiral galaxies. *Science,* 220: 1339–44.

Rubin, V. (2001). A brief history of dark matter. In: Mario Livio (ed.), *The Dark Universe: Matter, Energy and Gravity. Proceedings of the Space Telescope Science Institute Symposium*, 1–13. Cambridge: Cambridge University Press.

Sánchez-Conde, M.A., Paneque, D., Bloom, E.D., et al. (2009). Hints of the existence of axionlike particles from the gamma-ray spectra of cosmological sources. *Physical Review D*, 79: 123511 (16 pages).

Sanders, R.H. (2003). Modified Newtonian dynamics and its implications. In: Mario Livio (ed.), *The Dark Universe: Matter, Energy and Gravity. Proceedings of the Space Telescope Science Institute Symposium*, 62–76. Cambridge: Cambridge University Press.

Siegel, E. (2008). http://startswithabang.com/?p=109.

Simet, M., Hooper, D., and Serpico, P.D. (2008). The Milky Way as a kiloparsec-scale axionscope. *Physical Review D*, 77: 063001 (7 pages).

Simon, J.D. and Geha, M. (2007). The kinematics of the ultra-faint Milky Way satellites: Solving the missing satellite problem. *Astrophysical Journal*, 670: 313–31.

Sjöstrand, T., Mrenna, S., and Skands, P. (2009). Available at: http://home.thep.lu.se/~torbjorn/Pythia.html.

Spergel, D.N., Bean, R., Doré, O., et al. (2007). Three-Year *Wilkinson Microwave Anisotropy Probe (WMAP)* observations: Implications for cosmology. *Astrophysical Journal Supplement Series*, 170: 377–408.

Strong, A.W., Moskalenko, I.V., and Reimer, O. (2004). Diffuse Galactic continuum gamma rays: A model compatible with EGRET data and cosmic-ray measurements. *Astrophysical Journal*, 613: 962–76.

Taylor, J.E. and Babul, A. (2005a). The evolution of substructure in galaxy, group and cluster haloes – II. Global properties. *Monthly Notices of the Royal Astronomical Society*, 364: 515–34.

Taylor, J.E. and Babul, A. (2005b). The evolution of substructure in galaxy, group and cluster haloes – III. Comparison with simulations. *Monthly Notices of the Royal Astronomical Society*, 364: 535–51.

Tegmark, M., Blanton, M.R., Strauss, M.A., et al. (2004). The three-dimensional power spectrum of galaxies from the Sloan Digital Sky Survey. *Astrophysical Journal*, 606: 702–40.

Trotta, R., Ruiz de Austri, R., and Roszkowski, L. (2007). Prospects for direct dark matter detection in the Constrained MSSM. *New Astronomy Reviews*, 51: 316–20.

Zemp, M., Stadel, J., Moore, B., et al. (2007). An optimum time-stepping scheme for N-body simulations. *Monthly Notices of the Royal Astronomical Society*, 376: 273–86.

Zwicky, F. (1933). Die Rotverschiebung von extragalaktischen Nebeln. *Helvetica Physica Acta*, 6: 110–27.

Zwicky, F. (1937). On the masses of nebulae and of clusters of nebulae. *Astrophysical Journal*, 86: 217.

9 Can We Understand Dark Energy?

Mark Sullivan

CONTENTS

INTRODUCTION

Determining the nature of our Universe and its constituents is one of the grandest and most fundamental questions in modern science and drives the research underlying many of the chapters in this book. The newest puzzle is the observed acceleration in the rate at which the Universe is expanding. The knowledge that the Universe is not a static and unchanging place, but is growing with time, was attained in the beginning of the 20th century. The early work of Slipher, Hubble, and Humason (Slipher, 1917; Hubble, 1929; Hubble and Humason, 1931) showed that nearby "spiral nebulae" are receding from the Earth in every direction on the sky at velocities proportional to their inferred distance. The only viable interpretation of these observations is that the Universe is getting larger over time, expanding in every direction.

For a universe filled with ordinary matter and radiation, the theory of general relativity (GR) predicts that the gravitational attraction of the material in that universe should lead to a reduction, or *deceleration*, in its expansion rate as it ages—matter should pull back or slow down the speed of the expansion. However, work in the last decade has shown the exact opposite: the rate of the expansion is increasing with time; the expansion of the Universe is *accelerating*. This "cosmic acceleration" has been confirmed with a wide variety of different astrophysical observations, and the data indicating this acceleration are now not seriously in question. *However, the underlying physical reason for the observed cosmic acceleration remains a complete mystery.*

There are two broad possibilities generally considered. The first is that around 70% of the matter-energy density of the Universe exists in an as yet unknown form, what we call: "dark energy," the key characteristic of which is a strong negative pressure that pushes the Universe apart. However, there exists no compelling or elegant explanation for the presence or nature of this dark energy, or the magnitude of its observed influence, although various theoretical possibilities have been postulated (Copeland et al., 2006; Peebles and Ratra, 2003). The Universe may be filled with a vacuum energy, constant in space and time—a "cosmological constant." Alternatively, dark energy may be dynamical, a rolling scalar energy field that varies with both time and location ("quintessence" theories).

The second possibility is that the observed cosmic acceleration is an artifact of our incomplete knowledge of physical laws of gravity in the Universe, in particular that the laws of GR, a foundation of modern physics, simply break down on the largest scales. The implication of this is that the cosmological framework in which we interpret astronomical observations is simply incorrect, and this is manifested, and interpreted, in observational data as acceleration or dark energy. These ideas are collectively known as "modified gravity" theories. Such theories are constrained in that they must be essentially equivalent to GR on scales of the Solar System, where GR is stunningly successful, and also in the early Universe where the predictions of standard cosmology match observational effects such as the properties of the cosmic microwave background (CMB) and the growth of large-scale structure (Lue et al., 2004). Evidently, a confirmation of this alternative explanation for the observed acceleration would be as profound as the existence of dark energy itself.

Either of these possibilities would revolutionize our understanding of the laws governing the physical evolution of the Universe. Understanding the cosmic acceleration has therefore rapidly developed over the last decade into a key goal of modern science (Albrecht et al., 2009; Frieman et al., 2008a; Peacock and Schneider, 2006; Trotta and Bower, 2006). This chapter is aimed at the observational part of this effort. I will discuss the latest astrophysical research and observational results from the current generation of experiments and review the possibilities for future progress in its understanding. Deliberately, this chapter does not tackle the various theoretical possibilities for explaining the cosmic acceleration in any great detail; excellent reviews of these can be found elsewhere (e.g., see Copeland et al., 2006). For simplicity, much of this chapter is written in the context of GR. In other words, cosmic acceleration is cast in terms of the unknown nature of dark energy, rather than in terms of modified gravity. However, from an observational perspective, many of the concepts discussed are generic to both, and data from the techniques can be analyzed in either context.

DISCOVERY

Despite the excitement over the last decade, cosmic acceleration is neither a new nor a novel concept. Its history can be traced back to the development of the theory of GR, and the idea has reemerged several times in the intervening century (for a "pre-1998" review, see Carroll et al., 1992). At the time of the publication of GR, contemporary thinking indicated that the Universe was a static place. Einstein perceived that solutions to the field equations of GR did not allow for these static solutions where space is neither expanding nor contracting, but rather is dynamically stable. The effects of gravity in any universe containing matter would cause that universe to eventually collapse. Hence, Einstein famously added a repulsive "cosmological constant" term to his equations—Λ.

A cosmological constant (Λ) has the same effect mathematically as an intrinsic energy density of the vacuum with an associated pressure. A positive vacuum energy density implies a negative pressure (i.e., in effect it acts repulsively) and vice versa. If the vacuum energy density is positive, this negative pressure will drive an accelerated expansion of empty space, acting against the slowing force of gravity. Hence, static universe solutions in GR could now be permitted, at least in principle. Following observations a few years later that the Universe was not a static place, but instead expands with time, the perceived need for a Λ term in GR was removed. Einstein famously remarked in his later life that modifying his original equations of GR to include Λ was his "biggest blunder."

Despite Einstein's retraction of Λ, in the early 1990s it was realized that the existence of Λ could potentially explain many puzzling observational effects in astronomical data. Many cosmologists were disturbed by the low matter density implied by observations of the large-scale structure of the Universe—if $\Omega_M < 1$, where was the rest of the matter-energy? Was the Universe non-flat, or was the interpretation of the observations at fault? The apparent ages of globular clusters were another puzzle, seemingly older than the accepted age of the Universe in the then-standard cosmological models. This generated a renewed interest in Λ, which could explain many of these inconsistencies (e.g., see Efstathiou et al., 1990; Krauss and Turner, 1995; Ostriker and Steinhardt, 1995). However, the first direct evidence did not come until a few years later following observations of a particular kind of exploding star known as a Type Ia supernova (SN Ia).

SNe Ia are a violent endpoint of stellar evolution, the result of the thermonuclear destruction of an accreting carbon-oxygen (C/O) white dwarf star approaching the Chandrasekhar mass limit, the maximum theoretical mass that a white dwarf star can attain before the electron degeneracy pressure supporting it against gravitational collapse is no longer of sufficient strength. As the white dwarf star gains material from a binary companion and approaches this mass limit, the core temperature of the star increases, leading to a runaway fusion of the nuclei in the white dwarf's interior. The kinetic energy release from this nuclear burning—some 10^{44} J—is sufficient to dramatically unbind the star. The resulting violent explosion, nucleosynthesis, and radioactive decay contribute to a luminosity that appears billions of times brighter than our Sun, comfortably outshining even many galaxies.

These cosmic explosions have a remarkable and useful property in cosmology—they usually explode with nearly the same brightness (to within a factor of 2, although extreme cases can result in differences upward of a factor of 5) everywhere in the Universe. This amazing property is presumably due to the similarity of the triggering white dwarf mass (i.e., the Chandrasekhar mass) and, consequently, the amount of nuclear fuel available to burn. This makes them the best, or at least the most practical, examples of "standard candles" in the distant Universe. Standard candles are objects to which a distance can be inferred from only a measurement of the apparent brightness on the sky. The luminosity distance d_L to an object of known intrinsic bolometric luminosity L and observed bolometric flux density f can be derived from the well-known inverse-square law:

$$d_L = \sqrt{\frac{L}{4\pi f}}. \tag{9.1}$$

Hence, by observing the apparent brightness of many SNe Ia at different redshifts, their luminosity distances can be accurately measured in a manner that is completely independent of any particular cosmological world model.

In fact, a factor of 2 dispersion in the observed peak brightness is not particularly useful for standard candles. The key development in the use of SNe Ia was the realization that their luminosities could be further homogenized, or standardized, using simple empirical techniques and correlations. Raw SN Ia luminosities are strongly correlated with the width of the SN light curve (the time it takes the observed flux to rise and fall) and the SN color—intrinsically brighter Type Ia supernovae (SNe) typically have wider (slower) light curves and a bluer optical color than their fainter counterparts (e.g., see Phillips, 1993). Applying the various calibrating relationships to SN Ia measurements provides distance estimates precise to ~6%–7% (or 0.12–0.14 mag). Such a precision is easily capable of discriminating different cosmological models with only a few 10s of events.

For SNe Ia, a combination of their extreme brightness, uniformity, and a convenient month-long duration made them extremely observationally attractive and practical as calibratable standard candles. Yet, for many years following the realization of this potential, finding distant events in the numbers required for meaningful constraints was a considerable logistical and technological challenge. Years of searching were required to discover only a handful of distant SNe Ia (e.g., see Norgaard-Nielsen et al., 1989; Perlmutter et al., 1997). The field came of age only through improving technology: the advent of large-format charge-coupled device (CCD) cameras on 4m-class (13 ft) telescopes, capable of efficiently scanning large areas of sky, and the simultaneous development of sophisticated image-processing routines and powerful computers capable of rapidly analyzing the volume of data produced.

The substantial search effort culminated in the late 1990s, when two independent surveys for distant SNe Ia (Perlmutter et al., 1997; Schmidt et al., 1998) made the same remarkable discovery: the high-redshift SNe Ia appeared about 40% fainter—or equivalently more distant—than expected in a flat, matter-dominated universe (Riess et al., 1998; Perlmutter et al., 1999; see Figure 9.1). This indicated that the expansion of the Universe had been speeding up over the last ~4–5 Gyr, providing compelling direct evidence for an accelerating universe. When these observations were combined with analyses of the CMB, a consistent picture emerged of the Universe as spatially flat and dominated by a "dark energy" responsible for ~70%–75% of its energy, opposing the slowing effect of gravity and accelerating the Universe's rate of expansion.

This evidence for the accelerating Universe sparked an intense observational effort. The first attempts to confirm or refute the unpredicted SN-based result were rapidly replaced by concerted observational programs to place the tightest possible constraints on the cosmic acceleration in the hope that a theoretical understanding might follow. The next section describes the concepts underlying the observational techniques for studying dark energy.

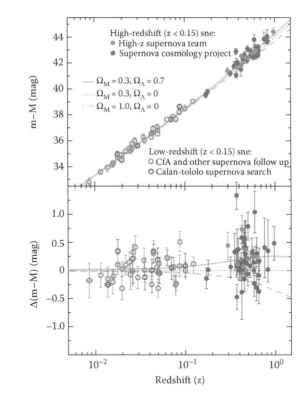

FIGURE 9.1 The original "discovery data" that directly indicated the accelerating Universe. This is the original Type Ia SN Hubble Diagram compiled from data taken by the Supernova Cosmology Project (Perlmutter, S., Aldering, G., Goldhaber, G., et al., *Astrophysical Journal*, 517, 565–86, 1999.) and the High-z Supernova Search Team (Riess, A.G., Filippenko, A.V., Challis, P., et al., *Astronomical Journal*, 116, 1009–38, 1998.). The bottom panel shows the residuals in the distance modulus relative to an open universe. The SNe Ia lie above and are inconsistent with (fainter than) the nonaccelerating Universe lines. The color coding indicates SNe from different surveys, two at high redshift and two at low redshift. (From Frieman et al., 2008a. Reprinted from Annual Review of Astronomy and Astrophysics, Volume 46, © 2008 by Annual Reviews (available at: www.annualreviews.org). With permission.)

HOW TO MEASURE DARK ENERGY?

The key problem with understanding dark energy is that it has a very low density—less than 10^{-29} g/cm^3—which makes detecting it in a laboratory on Earth, let alone studying it in detail, extremely challenging (in fact, currently impossible). The only reason that dark energy has such an important measurable effect on the physical evolution of the Universe is that it is thought to uniformly fill the vast vacuum of space. When the low density of dark energy is integrated over the sheer volume of the Universe, its effect dominates over that of matter, which tends to be extremely clustered in stars and galaxies. The influence of dark energy can therefore be observed only over cosmological scales, making astronomy the only field currently capable of effectively studying it.

Although SNe Ia were the original tools by which dark energy was discovered, many other independent techniques have been developed over the last decade. The breadth of these new ideas is a testament to both the recognition of the importance of dark energy and the ingenuity of astronomers in developing methodology to tackle the problem of understanding it. The SN Ia technique itself has been modified and improved, particularly with regard to assessing systematic errors in the measurement. Other observational probes have been introduced that both complement and reinforce the original observations. This section contains a brief overview of the underlying concepts behind the different techniques.

THE EQUATION OF STATE

The key observational measurement is the determination of the "equation-of-state parameter" of the dark energy, w, the ratio of its pressure to energy density ($w = p/\rho$), a concept closely related to the equation of state common in thermodynamics. In the solutions to Einstein's unmodified field equations of GR (i.e., without the additional Λ term) known as the Friedmann–Lemaître–Robertson–Walker (FLRW) metric, the Universe is described as homogeneous and isotropic, possibly expanding or contracting, and filled with a perfect fluid, one that can be completely characterized by an energy density ρ and an isotropic pressure p. In these solutions, the growth of the Universe over time is parameterized by a dimensionless scale factor parameter $a(t)$, essentially describing how the Universe "stretches" over time, and defined so that at the present day $a = 1$. The equation that governs the rate of expansion, or rate of change of a, $\dot{a} = da/dt$, is usually known as the Friedmann equation

$$\left(\frac{\dot{a}}{a}\right)^2 \equiv H^2(a) = \frac{8\pi G \rho(a)}{3} - \frac{k}{a^2}. \tag{9.2}$$

The left-hand side of Equation 9.2 is the Hubble parameter $H(a)$, which measures the relative expansion rate of the Universe as a function of time (or a). Despite a contentious history, the present-day value of H, H_0, is now generally agreed to be close to 70 (km/s)/Mpc, resulting from the use of a wide variety of different techniques (e.g., see Freedman et al., 2001). The right-hand side of Equation 9.2 determines the expansion rate from the matter-energy contents of the Universe. The expression $\rho(a)$ describes the mean density of each of the different components making up the Universe—ordinary baryonic matter, dark matter, radiation, neutrinos, dark energy, and so forth (G is Newton's gravitational constant). The effect of spatial curvature is parameterized by k. A flat universe is indicated by $k = 0$.

The density of each of the different components of ρ evolves with the scale factor a as

$$\rho(a) \propto a^{-3(1+w)}, \tag{9.3}$$

with w the equation-of-state parameter of each given component. (In this case, the equations of state of each component are assumed constant with a, but more general descriptions for varying equations of state are easy to derive.)

More conveniently, each of the different components of ρ can be written in terms of energy density parameters Ω, defined as a fraction of the "critical energy density" ρ_c, the current energy density of a flat $k = 0$ universe:

$$\Omega \equiv \frac{\rho}{\rho_c} = \frac{8\pi G \rho}{3H^2}. \tag{9.4}$$

Ordinary, non-relativistic matter, such as the atoms making up planets and stars, as well as dark matter, has an equation state of $w = 0$. From Equation 9.3, its energy density Ω_M will therefore be diluted as the Universe expands as a^{-3}, or by the volume. Ultra-relativistic matter, such as radiation and neutrinos, has $w = 1/3$. Its energy density Ω_R is diluted more quickly by the expansion than matter as a^{-4}, decreasing faster than a simple volume expansion as radiation has momentum and therefore a wavelength, which is stretched by a factor of a. The final component Ω_{DE}, dark energy, must have a strong negative pressure to explain the observed cosmic acceleration, and hence a negative w. Equation 9.2 is then written as

$$H^2(a) = H_0^2 \left[\Omega_M a^{-3} + \Omega_R a^{-4} + \Omega_k a^{-2} + \Omega_{DE} a^{-3(1+w)} \right], \tag{9.5}$$

where w is the (unknown) equation-of-state parameter of the dark energy component. Here $\Omega_k = -K/H^2$ describes the large-scale curvature of the Universe. Current evidence points to Ω_k being very close to 0, which constitutes a flat universe.

The expansion history of the Universe can therefore be thought of as a straight "competition" between these different components (Figure 9.2). At early times, from around 3 sec after the Big Bang until an age of 50,000 years (a cosmological redshift of ~3,500), the Universe was dominated by radiation. As the Universe expanded and the radiation energy density dropped off as a^{-4}, the Universe entered a matter-dominated era, where gravitational attraction due to matter caused a period of deceleration. The energy density due to ordinary matter falls as a^{-3}, and at an age of about 9 billion years (a redshift of ~0.45) the effect of dark energy became dominant over that of gravity (although the effects of dark energy can be observed well before this redshift). This dark-energy-dominated era is the one in which we still live and that will presumably continue into the distant future—and is a period marked by cosmic acceleration. However, the cosmic acceleration is a relatively recent phenomenon in the expansion history of the Universe, as dark energy was simply not important in terms of the expansion history at early times.

For the dark energy density term, the simplest solution is Λ, mathematically identical to a vacuum energy with a negative pressure exactly equal to its energy density unchanging with time: $w = -1$. This is equivalent to the Λ term introduced into GR by Einstein, and for that reason dark energy is often denoted by that term. In this case the expansion properties of a universe containing dark energy can be described by three parameters, w, Ω_{DE} and Ω_M, and one parameter fewer if the Universe is assumed flat, $\Omega_{DE} + \Omega_M = 1$. However, attempts to calculate the vacuum energy density from the zero-point energies of quantum fields result in estimates that are many orders of magnitude too large—a challenge to theories of fundamental physics.

Alternatively to vacuum energy, dark energy may be a scalar energy field of unknown physical origin that varies over both time and space, either decreasing or increasing in energy density, the latter leading to a "big rip" eventually tearing apart all structure. In these cases there is no *a priori* reason to assume that w is not changing with redshift, and many reasons to think that it is.

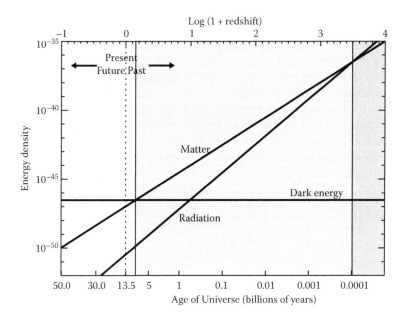

FIGURE 9.2 The importance of the different components of the energy density of the Universe as a function of the age of the Universe and cosmological redshift for a flat universe. The three epochs discussed in the text are in different shades of gray. The early Universe, at $z > 3,500$, is radiation dominated. Between $0.45 < z < 3,500$ is a matter-dominated era, and at $z < 0.45$ the Universe is dark energy-dominated.

Finally, it should be noted that the framework discussed above is relevant only in the context of the FLRW metric and solutions to GR. If instead cosmic acceleration is an artifact or indication of problems with GR, then of course this concept of w is meaningless. Typically, approaches along these lines involve changes to the Friedmann equation (Equation 9.2) and the evolution of $a(t)$.

Parameterizing w

The variety of possibilities capable of explaining cosmic acceleration make comparisons between observations and theory challenging. Ideally, the energy density of dark energy would be measured smoothly as a function of time, but in practical terms this is not yet possible. Instead, measuring $w(a)$ requires a parameterization of its form with a. The simplest method is to assume that w is constant: experiments then measure some average value w. This is particularly valuable for assessing whether cosmic acceleration is consistent with vacuum energy (Λ): is w consistent with -1? However, it is not particularly well motivated for other models of dark energy where w may change with time.

For these varying w models, more complicated parameterizations must be used. Many simple, but useful, "two-parameter" parameterizations have been suggested with a linear dependence on either a or redshift z. The form $w(a) = w_0 + w_a(1 - a)$ is often used (Albrecht et al., 2006; Linder, 2003). Other more general and complicated parameterizations are clearly possible (e.g., see Corasaniti and Copeland, 2003), including a principal component approach where $w(a)$, for example, can be measured over discrete intervals (Huterer and Starkman, 2003). Other model-independent approaches such as direct reconstruction have also been examined (e.g., see Sahni and Starobinsky, 2006) and tested against real data (e.g., see Daly et al., 2008). Each approach has advantages and drawbacks. Simpler parameterizations are easier to measure observationally, but harder to compare with models other than Λ. More complicated parameterizations and consequently more free parameters result in more poorly constrained measurements.

DISTANCE-REDSHIFT RELATIONS

Three main tools are used when constraining the cosmological parameters through the observational effects of dark energy. The first is to measure the expansion history and compare it with Equation 9.5. The scale factor a is easy to measure. When distant astronomical objects are observed, the photon wavelengths of the radiation that they emit are stretched ("redshifted") by the expansion by a factor $1/a = 1 + z$, where z is the cosmological redshift.

The rate of change of a, $\dot{a}$, is trickier, as time is not directly observable. Instead, distances to objects as a function of redshift are used, which are themselves intimately related to the expansion history. The comoving distance d (the distance between two points measured along a path defined at present) to an object at redshift z is

$$d = \int_0^z \frac{c}{H(z')} dz' = \frac{c}{H_0} \int_0^z \frac{dz'}{\sqrt{\Omega_M\left(1+z'\right)^3 + \Omega_k\left(1+z'\right)^2 + \Omega_{DE}\left(1+z'\right)^{3(1+w)}}}, \tag{9.6}$$

where $H(z)$ is the Hubble parameter from Equation 9.5, and a has been recast in terms of z. Related to this comoving distance are a variety of other distance definitions depending on the manner in which the distance measurement is made. In particular, the luminosity distance d_L of Equation 9.5 is

$$d(1+z) \equiv d_L = \sqrt{\frac{L}{4\pi f}}. \tag{9.7}$$

The power of distance measures is now clear: when L, f, and z are all known from measurements of a set of astrophysical objects, the only remaining unknowns are the cosmological parameters—including w. Thus, measuring a large set of astrophysical objects distributed in redshift that are known to be standard candles (such as SNe Ia) can directly measure parameters of interest and trace out the expansion history. In practice, even knowledge of the absolute luminosity L is not required. Instead, *relative* distances between local and distant standard candles can be measured, which has the advantage of removing any dependence on H_0.

The size of the variation in the apparent magnitude of a standard candle versus redshift for different cosmological models is shown in Figure 9.3. For a simple measurement of w, the "sweet spot" region is around $z = 0.6$, where the differences between different models are the largest, and the redshift is still small enough that high-quality data can be obtained. Above $z = 1$, the relative effect of a change in w in terms of apparent magnitude difference from that at $z = 1$ is very small: at these epochs, the Universe was still decelerating, and dark energy had only a minor influence on its evolution. Clearly, when trying to measure $w(a)$, samples of standard candles are required across the entire redshift range: the problem is quite degenerate if only a limited range in redshift can be observed. Figure 9.3 shows the variation assuming a simple linear function in $w(a)$.

Note that several of the cosmological parameters enter the distance calculation: the matter density Ω_M, the energy density of dark energy Ω_{DE} and its equation of state w, and the amount of curvature in the Universe Ω_k. Other complementary observations are therefore useful in conjunction with standard candles (see Figure 9.5 below), which can place constraints, or priors, on the matter density Ω_M (e.g., observations of large-scale structure) or spatial flatness Ω_k (e.g., observations of the CMB discussed below).

A closely related technique to standard candles uses a different distance measure and the concept of "standard rulers," objects of known dimensions, rather than known luminosity. Such sizes can be compared with the angular diameter distance d_A, the ratio of an object's (transverse) physical size to its angular size. It is related to the luminosity distance d_L as $d_A = d_L/(1 + z)^2 = d/(1 + z)$ and can probe the expansion history in a very similar way as standard candles. The method of measuring

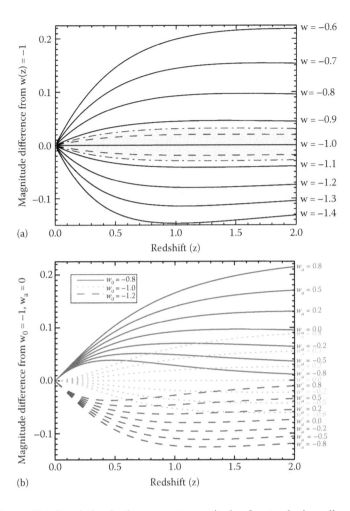

FIGURE 9.3 The predicted variation in the apparent magnitude of a standard candle versus redshift for various cosmological models. On the top are different models assuming a constant equation of state, from $w = -0.6$ to $w = -1.4$. The current best constraints in $\langle w \rangle$ are shown in the gray shaded area. The upper dot-dashed line shows the constraints including systematic errors, and the dashed line shows just the statistical error. On the bottom is the same plot, but assuming a variable w according to $w(a) = w_0 + w_a(1-a)$.

baryon acoustic oscillations (BAOs) in the Galaxy power spectrum, for example, exploits this idea. Generically, distance-redshift relations $d(z)$ provide very strong constraints on dark energy as they directly track the expansion history.

The Cosmic Microwave Background

The second diagnostic is the CMB. The CMB is a nearly isotropic background radiation discovered in the 1960s (Penzias and Wilson, 1965), which also has a near-perfect blackbody spectrum, peaking in the radio with a temperature of $\simeq 2.7$ K. The radiation originates from the early Universe and an epoch when the Universe was much hotter and denser and almost entirely ionized—photons and baryons (i.e., protons and neutrons) were tightly coupled to one another, essentially opaque to radiation. Some 380,000 years after the Big Bang at a cosmological redshift near $z \sim 1,100$, the Universe had expanded sufficiently and adiabatically cooled to a temperature near 3,000 K, where electrons and protons are able to (re)combine to form neutral hydrogen ("the epoch of recombination"), decoupling the photons and baryons. The photons, free from the baryons, then propagate through the Universe and appear to us now as the CMB. As the Universe has expanded by a factor of about 1,100 since the epoch of recombination when the CMB was emitted, the CMB photons appear considerably less energetic, redshifted into the microwave spectral region.

Although the CMB is extremely isotropic, it has small temperature fluctuations of the order of one-thousandth of 1%. Before recombination, any initial density fluctuations, or perturbations, excited gravity-driven sound wave or acoustic oscillations in the relativistic ionized plasma of the early Universe. The matter and radiation were attracted, by gravity, into these regions of high density. A gravitational collapse then followed until photon pressure support became sufficient to halt the collapse, causing the overdensity to rebound because of the nonzero pressure of the gas, generating acoustic waves. These two effects competed to create oscillating density perturbations, driven by gravity and countered by photon pressure. At recombination, as the photons are decoupled, those photons originating in overdense regions will appear hotter than average, while those from less dense regions will appear colder. These small density fluctuations in the Universe at that time are therefore imprinted directly onto the photons of the CMB, appearing to us as small temperature fluctuations, or a temperature anisotropy.

These temperature differences can be "routinely" measured from the CMB power spectrum, the fluctuation in the CMB temperature (anisotropy) as a function of angular scale on the sky. This angular power spectrum of the CMB temperature anisotropy (Dunkley et al., 2009; Nolta et al., 2009; Figure 9.4) contains a series of peaks and troughs arising from the gravity-driven acoustic oscillations of the coupled photon-baryon fluid in the early Universe. In particular, a strong peak is seen in the power spectrum on an angular scale corresponding to the sound horizon (r_s, the maximum distance sound waves can travel before recombination), where a perturbation crossed this horizon at exactly the time of recombination—the scale that was first feeling the causal effects of gravity at that epoch. Smaller scales had been oscillating for longer and are manifest as weaker peaks in the angular power spectrum.

A wealth of cosmological information is contained in positions and heights of the series of peaks and troughs (e.g., see Bond and Efstathiou, 1987; Peebles and Yu, 1970). For example, the first peak, corresponding to the physical length of the sound horizon at recombination, depends on the curvature of space. If space is positively curved, then this sound horizon scale r_s will appear larger on the sky than in a flat universe (the first peak will move to the left in Figure 9.4); the opposite is true if space is negatively curved. The third peak can be used to help constrain the total matter

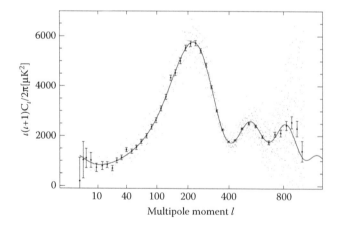

FIGURE 9.4 The temperature anisotropy angular power spectrum from the WMAP-5 data from the *Wilkinson Microwave Anisotropy Probe (WMAP)* (From Dunkley, J., Komatsu, E., Nolta, M.R., et al., *Astrophysical Journal Supplement Series*, 180, 306–29, 2009. With permission.). The gray dots represent the unbinned data, and the black points represent the binned data with 1σ error bars. The red line is the best-fit ΛCDM cosmological model. (Reproduced from American Astronomical Society (AAS). With permission.)

density. However, the CMB by itself provides little direct constraint on dark energy—it is, after all, a snapshot of the Universe at an epoch when dark energy was only a small contributor to the total energy density (Figure 9.2). Crucially, the CMB does provide a significant contribution in constraining curvature Ω_k and the total matter density Ω_M (as well as the size of the sound horizon) for use in conjunction with other dark energy probes that more directly measure dark energy.

THE GROWTH OF STRUCTURE

The final tool is the growth of large-scale structure in the Universe. The large-scale structure that we observe is a product of the random density fluctuations in the early Universe at the time of recombination. At this epoch, when the photon pressure that was supporting the overdense regions from gravitational collapse disappears, the matter can then fall into the overdense regions. This mass grows over time to form the structures we see today—a process known as the "growth of structure."

Dark energy has an effect on how the initial density fluctuations at the time of recombination subsequently grow and evolve. In a non-expanding universe, overdense regions would continue to increase in density, but in an expanding universe, the gravitational collapse is countered by the expansion. A more rapid expansion, caused by dark energy, reduces this increase in density, or growth of structure, more strongly than a slower expansion. The more dark energy, the earlier in the Universe dark energy dominates, and the earlier the growth of the linear perturbations is ended. The strength of the density fluctuations at recombination can be accurately measured from the CMB, so measuring the amplitude of the matter fluctuations as a function of redshift or scale factor, the growth factor $g(z)$, provides additionally observable constraints on dark energy.

CONSTRAINING MODIFIED GRAVITY?

The two different approaches to measuring dark energy described above—$d(z)$ and $g(z)$—can together combine to provide a method for testing for changes in the laws of gravity. The evolution of the matter density fluctuations can be described with GR and linear perturbation theory as a second-order differential equation:

$$\ddot{\delta} + 2H(a)\dot{\delta} = 4\pi G\rho_m\delta = \frac{3\Omega_M H_0^2}{2a^3}\delta. \tag{9.8}$$

Here δ is a fractional density excess or perturbation relative to the mean density ρ_m. The right-hand side depends directly on the theory of gravity and, as written in Equation 9.8, is only correct assuming the veracity of GR. The left-hand side is dependent on the expansion history $H(a)$. The solution to this equation describes the growth of density fluctuations. Given a measurement of $H(a)$ (from, for example, a standard candle experiment), the growth factor $g(a)$ is uniquely predicted. Any discrepancies between an observed growth function and the one predicted by GR would indicate potential problems with the theory of GR.

OBSERVATIONAL TECHNIQUES

Modern experiments are now capable of using the concepts of the last section to provide sensitive constraints on the nature of dark energy through measurement of its equation of state, w, using either the distance-redshift relations $d(z)$ or the growth of structure $g(z)$ combined with observations of the CMB. This section discusses the four main techniques in current use and the limitations and prospects for each. This is not intended as an exhaustive review or a full discussion of the ultimate potential of each method, for which complex numerical forecasting should be used (Albrecht et al., 2009). Multiple reviews are available that discuss these techniques in more detail (Albrecht et al., 2006; Peacock and Schneider, 2006; Trotta and Bower, 2006).

TYPE IA SUPERNOVAE

The quantity and quality of SN Ia data have dramatically improved since the original surveys. Dedicated allocations of observing time on 4m-class (13 ft) telescopes, such as the Canada-France-Hawaii Telescope (CFHT) and the Cerro Tololo Inter-American Observatory (CTIO) Blanco Telescope, have provided homogeneous light curves of more than 500 distant SN Ia events over $z = 0.3$ to -1.0. The principal advances in this redshift range have come from the Supernova Legacy Survey (SNLS; Astier et al., 2006) and the Equation of State SupErNovae trace Cosmic Expansion (ESSENCE) supernova survey (Wood-Vasey et al., 2007). At higher redshifts above $z = 1$, the *Hubble Space Telescope* has been used to locate ~25 SN events probing the expected epoch of deceleration (Riess et al., 2004, 2007). The latter observations also rule out the invocation of "gray dust" to explain the SN data in place of acceleration.

Lower-redshift SN Ia samples are also important and are often neglected. The absolute luminosity of a SN Ia is not known precisely and cannot be used *a priori*. The SN Ia method therefore relies on sets of local SNe at $0.015 < z < 0.10$, where the effect of varying the cosmological parameters is small, and which essentially anchor the analysis and allow relative distances to the more distant events to be measured. (At redshifts lower than $\simeq 0.015$, the peculiar velocities of the SN Ia host galaxies, or bulk flows, can make the measurement both noisier and biased if not corrected for.) The Sloan Digital Sky Survey (SDSS; York et al., 2000) of SNe fills in the region from ($0.1 < z < 0.3$), and many hundreds of lower-redshift SNe Ia in the nearby Hubble flow ($0.03 < z < 0.1$) are either available or upcoming (e.g., see Hamuy et al., 2006; Hamuy et al., 1996; Hicken et al., 2009; Jha et al., 2006). The result of these new SN data will be a comprehensive set of well-calibrated events uniformly distributed from the local Universe out to $z > 1$. The first results from some of these new samples can be found in Figure 9.5.

Although SNe Ia provided the first direct evidence for dark energy and still provide the most mature and constraining measurements, the technique has a number of potential drawbacks—the apparently simple standard candle concept has several unapparent difficulties. These difficulties are fundamentally related to the precision required (Figure 9.3): detecting departures in dark energy from $w = -1$ requires an extremely sensitive experiment. A 10% difference in w from -1 is

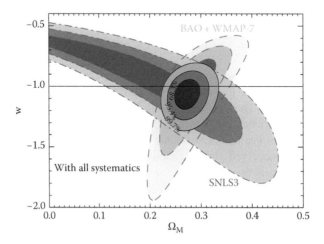

FIGURE 9.5 Latest constraints on the nature of dark energy from SNe Ia and other techniques (From Sullivan, M., Guy, J., Conley, A., et al., arXiv:1104.1444, 2011). The contours show the joint 1 and 2σ constraints in w and Ω_M from SN Ia, baryon acoustic oscillations (From Percival, W.J., Reid, B.A., Eisenstein, D.J., et al., *Monthly Notices of the Royal Astronomical Society*, 401, 2148–68, 2010), and the CMB from WMAP-7 (From Komatsu, E., Dunkley, J., Nolta, M.R., et al., *Astrophysical Journal Supplement Series*, 180, 330–76, 2009.). A flat universe is assumed. The contours include all errors, both statistical and systematic. (Reproduced from the American Astronomical Society. With permission.)

equivalent to a change in SN Ia (or standard candle) brightness at $z = 0.6$ of only 0.04 mag. The absolute calibration of the SN Ia fluxes measured in different filters also demands at least this level of precision—a 1%–2% level of absolute precision is perhaps not routinely achieved in astronomy. The challenge is even more complex when fitting variable w models. Therefore, the SN experiments are not simply about obtaining more data, but about obtaining higher-quality data on which these experimental systematics can be controlled.

While the challenge of photometrically calibrating the physical SN Ia fluxes is considerable, this is at least a well-defined and tractable problem on which substantial progress can be made. Of more concern is the possibility of intrinsic variability in the SN Ia population. The most significant is the unknown astrophysical nature of the SN Ia events (e.g., see Hillebrandt and Niemeyer, 2000). Although the consensus of an exploding near-Chandrasekhar mass C/O white dwarf star residing in a binary system is commonly accepted, the configuration of the progenitor system is hotly debated. The companion star to the progenitor white dwarf could be a second white dwarf star ("double degenerate") or a main-sequence or giant star ("single degenerate"). Evidence from observations or theory for and against these two possibilities is ambiguous. The physics of the accretion process and explosion propagation are also uncertain. These unknowns are some of the biggest drawbacks of the SN Ia technique—in the absence of any theoretical guidance, it must be assumed that any variation can be empirically controlled using the various light-curve width and color relations.

There are also open questions as to how the metallicity or age of the progenitor star may influence the observed properties and luminosities of the SN Ia explosion, leading to possible biases as the demographics of the SN Ia population shifts slightly with look-back time (Howell et al., 2007; Sarkar et al., 2008). Recent evidence has shown the first indications of a variation in SN Ia luminosity with host galaxy properties, even after corrections for light curve shape have been made (Kelly et al., 2010; Sullivan et al., 2010). However, new empirical techniques have been developed that allow these effects to be calibrated in a cosmological analysis, even if the physical cause remains a mystery (Sullivan et al., 2011).

The local environment in which SNe Ia explode can also cloud their interpretation in a cosmological context. Their host galaxies span the full range of age, from dwarf irregular galaxies through giant ellipticals, and contain vastly different amounts of dust. This dust has the effect of dimming the light from objects as it passes through, preferentially in the ultraviolet and blue spectral regions. This makes SNe appear both fainter and redder than they are intrinsically (e.g., see Tripp, 1998) and must be carefully corrected for in cosmological studies. As with most of the SN Ia field, this can be attempted only empirically by correlating SN color with SN luminosity. Surprisingly, these analyses give quite different results than expected based on the known properties of Milky Way dust (e.g., see Conley et al., 2007). Either dust in external galaxies is different from that observed in the Milky Way, or SNe Ia possess some intrinsic relationship between color and luminosity that cannot yet be separated from the effects of dust. Probably, the question of dust represents the most serious challenge to SN Ia cosmology. Observing SNe at redder wavelengths where the effect of dust is smaller is an obvious potential solution (Wood-Vasey et al., 2008); theory also suggests that any intrinsic variability in the population is smaller at these wavelengths (Kasen and Woosley, 2007).

While these potential systematics may appear serious, in part this is because the SN Ia technique is the most mature and tested probe of dark energy. Despite many decades of intensive testing, no fatal flaw has yet been identified—and SNe Ia have, so far, passed many detailed examinations of systematic effects with flying colors.

BARYON ACOUSTIC OSCILLATIONS

BAOs are closely related to the oscillations seen in the CMB angular power spectrum (Figure 9.4). Following the epoch of recombination at $z \sim 1,100$, the immediate loss of photon pressure led to a consequent reduction in the effective sound speed of the baryons. The acoustic waves excited by the gravitationally unstable density fluctuations became "frozen" into the matter distribution with a characteristic size equal to their total propagation distance—the sound horizon scale r_s. As discussed above, this r_s can be seen in the power spectrum of the CMB temperature anisotropy, but additionally these sound waves remain "imprinted" in the baryon distribution and, through gravitational interactions, in the dark matter distribution as well. As galaxies (roughly) trace the dark matter distribution, observations of galaxy clustering can uncover this characteristic scale. Making this observation at different redshifts therefore allows this scale r_s to be used as a standard ruler—just as SNe Ia trace $d(z)$ using $d_L(z)$, BAOs measure $d_A(z)$ (e.g., see Blake and Glazebrook, 2003; Seo and Eisenstein, 2003).

Power spectra analyses of galaxy redshift surveys contain the acoustic oscillations and are used to measure the cosmological parameters: the conversion of redshift data into real space requires a cosmology to be assumed, and an incorrect choice will distort the power spectrum, with the acoustic peaks appearing in incorrect places. Observations of the CMB play a critical role here, as this same characteristic scale can be calibrated accurately by observations of anisotropy in the CMB imprinted at the same epoch. This scale can be precisely measured from the angular scale of the first acoustic peak in the CMB power spectrum (Figure 9.4) and is determined to be $r_s = 146.8 \pm 1.8$ Mpc (e.g., see Page et al., 2003).* This observed, calibrated scale can therefore be used as a geometric probe of the expansion history—a measurement at low redshift provides an accurate measurement of the distance ratio between that redshift and $z \simeq 1,100$. Spectroscopic redshift BAO surveys can also measure the change of this characteristic scale radially along the line of sight as well as in the transverse direction, in effect a direct measurement of $H(z)$.

However, measurements of the power spectra or correlation function of galaxies are challenging. The oscillations appear as a series of bumps with an amplitude of only about 10%. This is substantially more subtle than the acoustic oscillations observed in the power spectrum of the CMB anisotropies because the impact of baryons on the far larger dark matter component is relatively small. Hence, enormous galaxy spectroscopic redshift surveys covering substantial volumes are required to make a constraining measurement. For example, the first detections of peaks in the galaxy power spectrum required nearly 50,000 SDSS luminous galaxy redshifts at z ~ 0.35 (Eisenstein et al., 2005) and ~200,000 galaxies at lower redshifts from the 2-degree Field Galaxy Redshift Survey (2dFGRS; Cole et al., 2005). Because such large numbers of galaxies are needed, BAO measurements provide distance estimates that are coarsely grained in redshift. Photometric redshift surveys could in principle also be used and cheaply add hundreds of thousands of galaxies; this comes at the expense of a measurement of $H(z)$ and reduces the ability to measure $d_A(z)$ as a result of systematic errors and the higher noise of photometric redshifts over spectroscopic measures.

Although using BAOs to measure dark energy with precision requires enormous survey volumes and millions of galaxies, numerical simulations suggest that systematic uncertainties associated with BAO measurements are small—this method is currently believed to be relatively unaffected by systematic errors. The physics underlying the standard ruler can be understood from first principles. The main systematic uncertainties that are present in any interpretation of BAO measurements are the effects of nonlinear gravitational evolution and scale-dependent differences between the clustering of galaxies and of dark matter (known as bias). For spectroscopic redshift surveys, redshift distortions of the clustering can also shift the BAO features. However, studies suggest that the resulting shift of the scale of the BAO peak in the galaxy power spectrum is 1% or less (e.g., Seo and Eisenstein, 2007).

WEAK GRAVITATIONAL LENSING

Gravitational lensing by massive clusters of galaxies is responsible for some of the most stunning images in astronomy. Lensing occurs when the passage of photons from distant objects is deflected by mass (primarily the dark matter in large-scale structure) concentrations that they pass

---

* See http://lambda.gsfc.nasa.gov/product/map/current/parameters.cfm for the *WMAP* Cosmological Parameters Model/Dataset Matrix.

in proximity to, causing the apparent position of the distant background object as seen on the Earth to be moved from its true position. The magnitude of this deflection depends on the amount of deflecting mass and the ratios of the various distances among observer, lens, and source. The most common case is "weak lensing," where the deflection becomes observable as a "shear" in the shape of the distant lensed galaxy, making it appear slightly more elliptical than its intrinsic shape. The typical size of the effect is about 2% and thus requires very large numbers of galaxies for a signal to be robustly detected, given the intrinsic variety in galaxy shapes and sizes. But, by measuring and averaging these shears across many objects, structures in the dark matter distribution can be effectively mapped out.

The key observational diagnostic is the shear angular power spectrum. It is the sensitivity of weak lensing to the ratios of the various distances involved (as well as the projected mass density along the line of sight) that allows dark energy to be measured. The distribution of the dark matter and its evolution with redshift probes the effect of dark energy on the growth of structure, and the distances can provide estimates of $d(z)$. Further, weak-lensing data allow internal tests for many (but not all) potential systematic errors. Thus, weak lensing is, in principle, an extremely powerful probe of dark energy, and the direct connection to gravity via the dark matter means that it can also be used to probe modified gravity theories.

However, extracting the lensing signal is a challenging task. In contrast to SN Ia cosmology and galaxy cluster counting (see below), where astrophysical systematics are becoming the dominant uncertainties, weak-lensing studies are still completely dominated by measurement systematics (Huterer et al., 2006). The problem is that the shapes and distortions of millions of galaxies need to be accurately measured without bias. This requires an accurate knowledge of the image distortions introduced by the camera and telescope optics, any telescope tracking errors, and the effects of atmospheric blurring or "seeing," particularly at large angular scales where the correlations to be measured are quite weak. These "point-spread functions" (PSFs) tend to be non-Gaussian, varying both temporally during an integration and spatially across the field of view of a camera, and controlling for this PSF variation is a daunting technical issue. Most of these systematic errors can be identified as they introduce certain types of shear patterns into the data that cannot have an astrophysical origin, although accurately correcting for these effects is still problematic. Many techniques have been proposed, and concerted programs are in place to assess the potential of each on large samples of fake data (Bridle et al., 2009; Heymans et al., 2006), but none have yet been demonstrated to reach the final goal of a <1% calibration error.

Astrophysical systematics are also present. Any tendency for galaxies to align on small scales with their neighbors, known as intrinsic alignment, can be confused into a dark energy signal and must be corrected for. Source or lens clustering can also impact small-scale lensing measures. The most powerful and useful technique of "3D lensing" requires accurate photometric redshifts of individual galaxies, and any systematics in these can introduce biases into the dark energy constraints and prevent an accurate control of, for example, intrinsic alignments. As with SN Ia cosmology, a photometric calibration at around the percent level is probably required. The result is that while the future potential of weak-lensing measurements in constraining dark energy is considerable, at the time of this writing the best constraints on dark energy are still weak (Hoekstra et al., 2006), really capable of only showing consistency with Λ cold dark matter (ΛCDM) or other probes, rather than making constraining measurements. Instead, lensing surveys have concentrated on measuring Ω_M and σ_8, the amplitude of the density fluctuations. The weak-lensing field is one that may not truly come of age until the effects of the atmosphere can be removed, by going either into space or, in the shorter term, into the upper atmosphere on a balloon mission.

GALAXY CLUSTER COUNTING

Galaxy clusters are the largest structures in the Universe to have gravitationally collapsed and are now the largest virialized objects in the Universe. They therefore originated from the largest density fluctuations in the early Universe at recombination. The space density of the dark matter halos around which they form depends sensitively on the cosmological parameters, and in particular on dark energy. Computer N-body simulations of the growth of structure from the density perturbations can predict the number density of cluster-sized dark matter halos that should be observed as a function of redshift: $dN/(dzdMd\Omega)$. This is then dependent on dark energy in two different ways: the mass function itself, $dN(z)/dM$, is extremely sensitive to the growth of structure, and the counts depend on the comoving volume element.

The challenge then is to measure or count the number of clusters in some area of sky as a function of redshift and, critically, as a function of cluster mass. Clusters of galaxies can be detected in several ways. The most obvious is to look for their constituent member galaxies, overdensities on the sky of galaxies of a particular color or at similar redshifts. The hot gas in clusters can also be detected from the x-ray emission. Another technique is to search for their effect on the CMB: hot intracluster gas Compton scatters any CMB photons as they pass through the cluster, generating a distortion on the CMB spectrum (the Sunyaev-Zel'dovich effect, or SZE; Sunyaev and Zeldovich, 1970). Finally, the effect of gravitational lensing by the cluster on background galaxies can also be used to detect clusters.

Mass estimation is the most problematic element of using clusters to measure dark energy and the most susceptible to systematic errors—the cluster mass is not measured directly by most of these techniques. Instead, some other quantity is directly measured, such as temperature or x-ray luminosity, galaxy number counts, SZE decrement, or weak-lensing shear, which must then be calibrated with scaling relations or associated with a cluster mass. This is sensitive to the uncertainties in the baryonic physics at work in the clusters and hence can introduce systematic errors into any measurement of dark energy. A calibration in these relations of about 1% is probably required from future surveys. Nonetheless, the latest techniques are capable of providing interesting constraints on dark energy and w (Mantz et al., 2008).

OTHER TECHNIQUES

As well as the four techniques described above, many other probes can also be used to trace dark energy, although most are not yet as well developed. One of these probes is long-duration gamma-ray bursts (GRBs), exceptionally luminous explosions that have, for many years, been postulated to be used as standard candles at the highest redshifts, beyond the grasp of SNe Ia. Several empirical calibrations have been suggested (Ghirlanda et al., 2006; Schaefer, 2007), but the lack of local counterparts makes calibration difficult, and they are not yet competitive with SNe Ia (Friedman and Bloom, 2005). Techniques using SNe Ia to calibrate GRBs may assist with these issues (Liang et al., 2008). In the future, with the advent of more advanced gravitational wave detectors, extreme-mass-ratio inspiral events should be readily detected to $z = 1$ (Gair et al., 2004) and may be usable as standard sirens with, or even without, a direct redshift estimation (Dalal et al., 2006; MacLeod and Hogan, 2008). Finally, the integrated Sachs–Wolfe (ISW) effect (Sachs and Wolfe, 1967), which introduces CMB anisotropies from gravitational redshifting between the surface of last scattering and an observer on the Earth, can be used to provide independent confirmation of dark energy (e.g., see Giannantonio et al., 2008).

CURRENT STATUS

Combinations of measurements of $d(a)$, $g(a)$, and the CMB have given rise to the now standard model of cosmology, known as the ΛCDM cosmological model. This model, a universe with ~70% of its mass-energy density as dark energy and ~25% as dark matter, seems capable of accounting for all current cosmological observations, and as such it is an outstandingly successful predictive model (albeit with 95% of the mass-energy of the Universe in an unknown form!).

The current status of cosmological measurements of dark energy can be seen in Figure 9.5 (Sullivan et al., 2011). This uses around 500 SNe Ia distributed in redshift from different SN Ia surveys, observations of the CMB from the WMAP-7 data release (Dunkley et al., 2009), and BAO measurements from the SDSS (Eisenstein et al., 2005). Using this combination of techniques, the latest results show that w is consistent with -1 with a sub-5% statistical precision. Systematics probably increase this total error to about 6%–7%, although only systematics from the SN Ia analysis are included in this error estimate (Conley et al., 2011). Of particular note is that, at present, the BAO measurements provide an almost orthogonal constraint to SNe Ia in Ω_M/w space (Figure 9.5). Without either the SN Ia constraints or the BAO, our measurement of the equation of state of dark energy would be considerably weakened. This is currently generically true—no single technique can yet place tight constraints on dark energy in isolation.

Nearly all current observational results are consistent with the ΛCDM Universe and a $w = -1$ cosmological model (Figure 9.5; see also Komatsu et al., 2009; Kowalski et al., 2008). One exception to this is the recent work of Percival et al. (2007), who used the BAO technique on the two different redshift surveys in which the BAO signature was originally detected (the SDSS and the 2dFGRS). The brightest, luminous red galaxies on the SDSS measure galaxy clustering and the distance-redshift relation (sound horizon scale) at $z = 0.35$. The main galaxy samples of both surveys perform the same measurement at $z = 0.20$. Thus, if the BAOs are matched to have the same measured scale at both redshifts, a distance ratio can be measured and compared with predictions in different cosmological models.

In a flat universe and assuming a constant w, this differential distance measurement currently favors models with $w < -1$. SN Ia data over the same redshift range do not show this trend (Astier et al., 2006), indicating a tension at 2.4σ between the SNLS SN Ia data and BAO measurements. More recent analysis with a larger SDSS galaxy sample shows a smaller discrepancy (Percival et al., 2010). Only further data will resolve this, either from improved SN Ia data in this redshift region (e.g., the SDSS data set; Frieman et al., 2008b) or from new and larger-volume galaxy redshift information.

FUTURE PERSPECTIVES

A 5%–6% measurement of the equation of state of dark energy represents major progress since its discovery a decade ago: similar advances over the coming decade require substantial investment to assemble the large astrophysical data sets required to make the next-generation measurement. Many new surveys are either just commencing or are actively being planned, which should provide these opportunities.

For SNe Ia, new low-redshift surveys such as the Palomar Transient Factory (Law et al., 2009; Rau et al., 2009) and SkyMapper will provide exquisite data sets in which astrophysical systematics can be examined and that can anchor the next generation of high-redshift samples. The Dark Energy Survey (DES) will provide higher-redshift data, leading into more long-term future projects such as a proposed space mission (at the time of this writing, the *Joint Dark Energy Mission* [*JDEM*] and/ or *Euclid*) or 8 m (26 ft) ground-based survey (using the Large Synoptic Survey Telescope [LSST]). BAOs will benefit from several new surveys. WiggleZ (Drinkwater et al., 2010) uses the Anglo-Australian Telescope to collect ~400,000 galaxy redshifts over $0.5 < z < 1$. The Baryon Oscillation Sky Survey (BOSS; Schlegel et al., 2007) is using the SDSS telescope ~1.5 million galaxy spectra to $z = 0.7$ and, using the Lyman-α forest, to $z = 2.5$. DES will attempt a photometric-redshift measurement, and, again, a space-based measurement will probably extend to beyond $z = 2$. Weak-lensing ground-based surveys include DES and the Panoramic Survey Telescope and Rapid Response System (Pan-STARRS). But the real breakthrough in weak-lensing constraints may have to wait for a space-based mission, free from the complicated distorting effects of the atmosphere. New surveys of distant galaxy clusters will be assembled using the SZE, combined with optical imaging data, in the Atacama Cosmology Telescope and the South Pole Telescope. In terms of the CMB, the *Planck* space mission (Tauber, 2004) will observe the power spectrum to unprecedented accuracy, as well as measure CMB polarization. It too will also provide large samples of clusters via the SZE effect. The prognosis and likelihood for an even more accurate measurement than shown in Figure 9.5 are therefore excellent—if not inevitable.

At present, the simplest explanation for dark energy, that of vacuum energy, is consistent with the vast majority of data. Virtually all measurements show that the average value of w is consistent with -1, or Λ. In the absence of any theoretical guidance, a natural question from an observational perspective is, at what point should the measurements be considered "close enough" to -1 to imply diminishing returns from further study? Clearly, that point has not yet been reached—although a 6% measure of w is impressive, this is still only an average measure based almost entirely on data at $z < 1$. Only weak constraints have yet been made on the variation of w with redshift (those that do exist are consistent with an unchanging w).

Therefore, the generation of new experiments described above over the next decade or so should lead to substantial progress in measuring dark energy in terms of much tighter constraints on the parameters describing its form and evolution. Whether this leads to an improved physical understanding probably depends on the results from these surveys and whether $w(a)$ is shown to be consistent with -1. If departures from -1 are identified, or if GR is shown to break down on large scales, this would be an extraordinary payback in terms of the proposed investment of resources and could lead to rapid theoretical advances. If, on the other hand, all the results continue to point to $w(a) = -1$, then although the new measurements will still be a critical test of the standard cosmological model, theoretical progress may be slower. However, for either outcome the additional resources for astronomy that will be attained—vast imaging and spectroscopic surveys of galaxies and SNe in the distant Universe of exquisite quality—will ensure dramatic progress across many diverse areas of astrophysics.

REFERENCES

Albrecht, A., Bernstein, G., Cahn, R., et al. (2006). Report of the Dark Energy Task Force. arXiv:astro-ph/0609591v1.

Albrecht, A., Amendola, L., Bernstein, G., et al. (2009). Findings of the *Joint Dark Energy Mission* Figure of Merit Science Working Group. arXiv: 0901.0721v1 [astro-ph.IM].

Astier, P., Guy, J., Pain, J., et al. (2006). The Supernova Legacy Survey: Measurement of Ω_M, Ω_Λ and w from the first year data set. *Astronomy & Astrophysics*, 447: 31–48.

Blake, C. and Glazebrook, K. (2003). Probing dark energy using baryonic oscillations in the galaxy power spectrum as a cosmological ruler. *Astrophysical Journal*, 594: 665–73.

Bond, J.R. and Efstathiou, G. (1987). The statistics of cosmic background radiation fluctuations. *Monthly Notices of the Royal Astronomical Society*, 226: 655–87.

Bridle, S., Shawe-Taylor, J., Amara, A., et al. (2009). Handbook for the Great08 Challenge: An image analysis competition for cosmological lensing. *Annals of Applied Statistics*, 3: 6–37.

Carroll, S.M., Press, W.H., and Turner, E.L. (1992). The cosmological constant. *Annual Review of Astronomy & Astrophysics*, 30: 499–542.

Cole, S., Percival, W.J., Peacock, J.A., et al. (2005). The 2dF Galaxy Redshift Survey: Power–spectrum analysis of the final data set and cosmological implications. *Monthly Notices of the Royal Astronomical Society*, 362: 505–34.

Conley, A., Carlberg, R.G., Guy, J., et al. (2007). Is there evidence for a Hubble Bubble? The nature of Type Ia supernova colors and dust in external galaxies. *Astrophysical Journal*, 664: L13–16.

Conley, A., Guy, J., Sullivan, M., et al. (2011). Supernova constraints and systematic uncertainties from the first three years of the Supernova Legacy Survey. *Astrophysical Journal Supplement*, 192: 1–29.

Copeland, E.J., Sami, M., and Tsujikawa, S. (2006). Dynamics of dark energy. *International Journal of Modern Physics D*, 15: 1753–1935.

Corasaniti, P.S. and Copeland, E.J. (2003). Model independent approach to the dark energy equation of state. *Physical Review D*, 67: 063521, 1–5.

Dalal, N., Holz, D.E., Hughes, S.A., et al. (2006). Short GRB and binary black hole standard sirens as a probe of dark energy. *Physical Review D*, 74: 063006, 1–9.

Daly, R.A., Djorgovski, S.G., Freeman, K.A., et al. (2008). Improved constraints on the acceleration history of the universe and the properties of dark energy. *Astrophysical Journal*, 677: 1–11.

Drinkwater, M.J., Jurek, R.J., Blake, C., et al. (2010). The WiggleZ Dark Energy Survey: Survey design and first data release. *Monthly Notices of the Royal Astronomical Society*, 401: 1429.

Dunkley, J., Komatsu, E., Nolta, M.R., et al. (2009). Five-year *Wilkinson Microwave Anisotropy Probe* (*WMAP*) observations: Likelihoods and parameters from the WMAP data. *Astrophysical Journal Supplement Series*, 180:306–29.

Efstathiou, G., Sutherland, W.J., and Maddox, S.J. (1990). The cosmological constant and cold dark matter. *Nature*, 348: 705–7.

Eisenstein, D.J., Zehavi, I., Hogg, D.W., et al. (2005). Detection of the baryon acoustic peak in the large-scale correlation function of SDSS luminous red galaxies. *Astrophysical Journal*, 633: 560–74.

Freedman, W.L., Madore, G.F., Gibson, B., et al. (2001). Final results from the *Hubble Space Telescope* Key Project to measure the Hubble Constant. *Astrophysical Journal*, 553: 47–72.

Friedman, A.S. and Bloom, J.S. (2005). Toward a more standardized candle using gamma-ray burst energetics and spectra. *Astrophysical Journal*, 627: 1–25.

Frieman, J.A., Turner, M.S., and Huterer, D. (2008a). Dark energy and the accelerating universe. *Annual Review of Astronomy & Astrophysics*, 46: 385–432.

Frieman, J.A., Bassett, B., Becker, A., et al. (2008b). The Sloan Digital Sky Survey – II Supernova Survey: technical summary. *Astronomical Journal*, 135: 338–47.

Gair, J.R., Barack, L., Creighton, T., et al. (2004). Event rate estimates for *LISA* extreme mass ration capture sources. *Classical and Quantum Gravity*, 21: S1595–1606.

Ghirlanda, G., Ghisellini, G., and Firmani, C. (2006). Gamma-ray bursts as standard candles to constrain the cosmological parameters. *New Journal of Physics*, 8: 123 (34 pages).

Giannantonio, T., Scranton, R., Crittenden, R.G., et al. (2008). Combined analysis of the Integrated Achs – Wolfe effect and cosmological implications. *Physical Review D*, 77: 123520 (22 pages).

Hamuy, M., Folatelli, G., Morrell, N.I., et al. (2006). The Carnegie Supernova Project: The Low – Redshift Survey. *Publications of the Astronomical Society of the Pacific*, 118: 2–20.

Hamuy, M., Phillips, M.M., Suntzeff, N.B., et al. (1996). BVRI light curves for 29 Type IA supernovae. *Astronomical Journal*, 112: 2408–437.

Heymans, C., Van Waerbeke, L., Bacon, D., et al. (2006). The Shear Testing Programme – I. Weak lensing analysis of simulated ground-based observations. *Monthly Notices of the Royal Astronomical Society*, 368: 1323–339.

Hicken, M., Challis, P., Jha, S. et al. (2009). CfA3: 185 Type Ia Supernova light curves from the CfA. *Astrophysical Journal*, 700: 331–57.

Hillebrandt, W. and Niemeyer, J.C. (2000). Type IA supernova explosion models. *Annual Review of Astronomy & Astrophysics*, 38: 191–230.

Hoekstra, H., Mellier, Y., van Waerbeke, L., et al. (2006). First cosmic shear results from the Canada-France-Hawaii Telescope Wide Synoptic Legacy Survey. *Astrophysical Journal*, 647: 116–27.

Howell, D.A., Sullivan, M., Conley, A., et al. (2007). Predicted and observed evolution in the mean properties of Type Ia supernovae with redshift. *Astrophysical Journal*, 667: L37–40.

Hubble, E. (1929). A relation between distance and radial velocity among extra-galactic nebulae. *Proceedings of the National Academy of Sciences*, 15: 168–173.

Hubble, E. and Humason, M.L. (1931). The velocity-distance relation among extra-galactic nebulae. *Astrophysical Journal*, 74: 43–80.

Huterer, D. and Starkman, G. (2003). Parametrization of dark-energy properties: A principal-component approach. *Physics Review Letters*, 90: 031301 (4 pages).

Huterer, D., Takada, M., Bernstein, G., et al. (2006). Systematic errors in future weak-lensing surveys: Requirements and prospects for self-calibration. *Monthly Notices of the Royal Astronomical Society*, 366: 101–14.

Jha, S., Kirshner, R.P., Challis, P., et al. (2006). *UBVRI* light curves of 44 Type Ia supernovae. *Astronomical Journal*, 131: 527–54.

Kasen, D. and Woosley, S.E. (2007). On the origin of the Type Ia supernova width-luminosity relation. *Astrophysical Journal*, 656: 661–65.

Kelly, P.L., Hicken, M., Burke, D.L., et al. (2010). Hubble residuals of nearby Type Ia supernovae are correlated with host galaxy masses. *Astrophysical Journal*, 715: 743–56.

Komatsu, E., Dunkley, J., Nolta, M.R., et al. (2009). Five-year *Wilkinson Microwave Anisotropy Probe* (*WMAP*) observations: Cosmological interpretation. *Astrophysical Journal Supplement Series,* 180: 330–76.

Kowalski, M., Rubin, D., Aldering, G., et al. (2008). Improved cosmological constraints from new, old, and combined supernova data sets. *Astrophysical Journal*, 686: 749–78.

Krauss, L.M. and Turner, M.S. (1995). The cosmological constant is back. *General Relativity and Gravitation*, 27: 1137–144.

Law, N., Kulkarni, S.R., Dekany, R.G., et al. (2009). The Palomar Transient Factory: System overview, performance, and first results. *Publications of the Astronomical Society of the Pacific*, 121: 1395–1408.

Liang, N., Xiao, W.K., Liu, Y., et al. (2008). A cosmology-independent calibration of gamma-ray burst luminosity relations and the Hubble diagram. *Astrophysical Journal*, 685: 354–60.

Linder, E.V. (2003). Exploring the expansion history of the universe. *Physics Review Letters*, 90: 091301 (4 pages).

Lue, A., Scoccimarro, R., and Starkman, G. (2004). Differentiating between modified gravity and dark energy. *Physical Review D*, 69: 044005 (9 pages).

MacLeod, C.L. and Hogan, C.J. (2008). Precision of Hubble constant derived sing black hole binary absolute distances and statistical redshift information. *Physical Review D*, 77: 043512 (8 pages).

Mantz, A., Allen, S.W., Ebeling, H., et al. (2008). New constraints on dark energy from the observed growth of the most X-ray luminous galaxy clusters. *Monthly Notices of the Royal Astronomical Society*, 387: 1179–192.

Nolta, M.R., Dunkley, J., Hill, R.S., et al. (2009). Five-year *Wilkinson Microwave Anisotropy Probe* (*WMAP*) observations: Angular power spectra. *Astrophysical Journal Supplement Series*, 180: 296–305.

Norgaard-Nielsen, H.U., Hansen, L., Jorgensen, H.E., et al. (1989). The discovery of a type IA supernova at a redshift of 0.31. *Nature*, 339: 523–25.

Ostriker, J.P. and Steinhardt, P.J. (1995). The observational case for a low-density Universe with a non-zero cosmological constant. *Nature*, 377: 600–2.

Page, L., Nolta, M.R., Barnes, C., et al. (2003). First-Year *Wilkinson Anisotropy Probe* (*WMAP*) observations: Interpretation of the TT and RE angular power spectrum peaks. *Astrophysical Journal Supplement Series*, 148: 233–41.

Peacock, J. and Schneider, P. (2006). The ESO-ESA Working Group on Fundamental Cosmology. *The Messenger*, 125: 48–50.

Peebles, P.J. and Ratra, B. (2003). The cosmological constant and dark energy. *Reviews of Modern Physics*, 75: 559–606.

Peebles, P.J.E. and Yu, J.T. (1970). Primeval adiabatic perturbation in an expanding universe. *Astrophysical Journal*, 162: 815–36.

Penzias, A.A. and Wilson, R.W. (1965). A measurement of excess antenna temperature at 4080 Mc/s. *Astrophysical Journal*, 142: 419–21.

Percival, W.J., Cole, S., Eisenstein, D.J., et al. (2007). Measuring the baryon acoustic oscillation scale using the Sloan Digital Sky Survey and 2DF Galaxy Redshift Survey. *Monthly Notices of the Royal Astronomical Society*, 381: 1053–66.

Percival, W.J., Reid, B.A., Eisenstein, D.J., et al. (2010). Baryon acoustic oscillations in the Sloan Digital Sky Survey Data Release 7. *Monthly Notices of the Royal Astronomical Society*, 401: 2148–168.

Perlmutter, S., Gabi, S., Goldhaber, G., et al. (1997). Measurements of the cosmological parameters omega and lambda from the first seven supernovae at $z > = 0.35$. *Astrophysical Journal*, 483: 565–81.

Perlmutter, S., Aldering, G., Goldhaber, G., et al. (1999). Measurements of Ω and Λ from 42 high-redshift supernovae. *Astrophysical Journal*, 517: 565–86.

Phillips, M.M. (1993). The absolute magnitudes of Type Ia supernovae. *Astrophysical Journal*, 413: L105–8.

Rau, A., Kulkarni, S.R., Law, N.M., et al. (2009). Exploring the optical transient sky with the Palomar Transient Factory. *Publications of the Astronomical Society of the Pacific*, 121: 1334–351.

Riess, A.G., Filippenko, A.V., Challis, P., et al. (1998). Observational evidence from supernovae for an accelerating universe and a cosmological constant. *Astronomical Journal*, 116: 1009–38.

Riess, A.G., Strolger, L.-G., Tonry, J., et al. (2004). Type Ia supenova discoveries at $z > 1$ from the *Hubble Space Telescope*: Evidence for past deceleration and constraints on dark energy evolution. *Astrophysical Journal*, 607: 665–87.

Riess, A.G., Strolger, L.-G., Casertano, S., et al. (2007). New *Hubble Space Telescope* discoveries of Type Ia supernovae at $z >= 1$: Narrowing constraints on the early behavior of dark energy. *Astrophysical Journal* 659: 98–121.

Sachs, R.K. and Wolfe, A.M. (1967). Perturbations of a cosmological model and angular variations of the microwave background. *Astrophysical Journal* 147: 73–90.

Sahni, V. and Starobinsky, A. (2006). Reconstructing dark energy. *International Journal of Modern Physics D*, 15: 2105–132.

Sarkar, D., Amblard, A., Cooray, A., et al. (2008). Implications of two Type Ia supernova populations for cosmological measurements. *Astrophysical Journal*, 684: L13–16.

Schaefer, B.E. (2007). The Hubble diagram to redshift > 6 from 69 gamma-ray bursts. *Astrophysical Journal,* 660: 16–46.

Schlegel, D.J., Blanton, M., Eisenstein, D. et al. (2007). SDSS-III: The Baryon Oscillation Spectroscopic Survey (BOSS). *Bulletin of the American Astronomical Society,* 38: 966.

Schmidt, B.P., Suntzeff, N.B., Phillips, M.M., et al. (1998). The high-Z supernova search: Measuring cosmic deceleration and global curvature of the universe using Type Ia supernovae. *Astrophysical Journal,* 507: 46–63.

Seo, H-J. and Eisenstein, D.J. (2003). Probing dark energy with baryon acoustic oscillations from future large galaxy redshift surveys. *Astrophysical Journal,* 598: 720–40.

Seo, H-J. and Eisenstein, D.J. (2007). Improved forecasts for the baryon acoustic oscillations and cosmological distance scale. *Astrophysical Journal,* 665: 14–24.

Slipher, V.M. (1917). Nebulae. *Proceedings of the American Philosophical Society,* 56: 403–9.

Sullivan, M., Conley, A., Howell, D.A., et al. (2010). The dependence of Type Ia supernovae luminosities on their host galaxies. *Monthly notices of the Royal Astronomical Society,* 406: 782–802.

Sullivan, M., Guy, J., Conley, A., et al. (2011). SNLS3: Constraints on dark energy combining the Supernova Legacy Survey three year data with other probes. arXiv: 1104–1444.

Sunyaev, R.A. and Zeldovich, Y.B. (1970). The spectrum of primordial radiation, its distortions and their significance. *Comments on Astrophysics and Space Physics,* 2: 66–73.

Tauber, J.A. (2004). The Planck mission. *Advances in Space Research,* 34: 491–96.

Tripp, R. (1998). A two-parameter luminosity correction for Type Ia supernovae. *Astronomy & Astrophysics,* 331: 815–20.

Trotta, R. and Bower, R. (2006). Surveying the dark side. *Astronomy and Geophysics,* 47: 20–7.

Wood-Vasey, W.M., Miknaitis, G., Stubbs, C.W., et al. (2007). Observational constraints on the nature of dark energy: First cosmological results from the ESSENCE Supernova Survey. *Astrophysical Journal,* 666: 694–715.

Wood-Vasey, W.M., Friedman, A.S., Bloom, J.S., et al. (2008). Type Ia supernovae are good standard candles in the near infrared: Evidence from PAIRITEL. *Astrophysical Journal,* 689: 377–90.

York, D.G., Adelman, J., Anderson, J.E., et al. (2000). The Sloan Digital Sky Survey: Technical summary. *Astronomical Journal,* 120: 1579–587.

10 Astrophysical Black Holes in the Physical Universe

Shuang-Nan Zhang

CONTENTS

INTRODUCTION

In modern astronomy, the mystery of black holes (BHs) attracts extraordinary interest for both researchers and the general public. Through the 1930s, the applications of general relativity and quantum mechanics to the studies of the late evolution of stars predicted that stars with different initial masses, after exhausting their thermal nuclear energy sources, may eventually collapse to become exotic compact objects, such as white dwarfs, neutron stars, and BHs. A low-mass star, such as our Sun, will end up as a white dwarf, in which the degeneracy pressure of the electron gas balances the gravity of the object. For a more massive star, the formed compact object can be more massive than around 1.4 solar masses ($M_\odot$), the so-called Chandrasekhar limit, in which the degeneracy pressure of the electron gas cannot resist the gravity, as pointed out by Chandrasekhar. In this case, the compact object has to further contract to become a neutron star, in which most of the free electrons are pushed into protons to form neutrons and the degeneracy pressure of neutrons balances the gravity of the object, as suggested by Zwicky and Landau. Then as Oppenheimer and others noted, if the neutron star is too massive, for example, more than around 3 $M_\odot$, the internal pressure in the object also cannot resist the gravity and the object must undergo catastrophic collapse and form a BH.

Up to now, about 20 BHs with masses around 10 $M_\odot$, called stellar-mass BHs, have been identified observationally. On the other hand, the concept of a BH has been extended to galactic scales. Since the discovery of quasars in the 1960s, these BHs with masses between 10^5 and 10^{10} $M_\odot$, which are called supermassive BHs, are believed to be located in the centers of almost all galaxies. Therefore, tremendous observational evidence supporting the existence of BHs in the Universe is gradually permitting the uncovering of the mysteries of BHs. BH astrophysics has become a fruitful, active, and also challenging frontier research field in modern astrophysics.

Despite tremendous progress in BH research, many fundamental characteristics of astrophysical BHs in the physical Universe remain not fully understood or clarified. In this chapter, I will try to address the following questions: (1) What is a BH? (2) Can astrophysical BHs be formed in the physical Universe? (3) How can we prove that what we call astrophysical BHs are really BHs? (4) Do we have sufficient evidence to claim the existence of astrophysical BHs in the physical Universe? (5) Will all matter in the Universe eventually fall into BHs?

Disclaimer: I will not discuss quantum or primordial BHs. Reviews on theoretical models and observations are intended to be very brief, and thus I will miss many references. Some of the discussions, especially on the question: Will all matter in the Universe eventually fall into BHs?, are quite speculative.

WHAT IS A BLACK HOLE?

I classify BHs into three categories: *mathematical BHs*, *physical BHs*, and *astrophysical BHs*.

A *mathematical BH* is the vacuum solution of Einstein's field equations of a point-like object, whose mass is completely concentrated at the center of the object, i.e., the singularity point. It has been proven that such an object may possess only mass, angular momentum (spin), and charge, the so-called three hairs. Because of the relatively large strength of the electromagnetic force, BHs formed from gravitational collapse are expected to remain nearly neutral. I therefore discuss only electrically neutral BHs in this chapter. Figure 10.1 is an illustration of the structure of a mathematical BH. The event horizon surrounding the object ensures that no communications can be carried out across the event horizon; therefore, a person outside the event horizon cannot observe the singularity point.

Birkhoff's theorem further ensures that the person outside the event horizon cannot distinguish whether the mass and charge of the object are concentrated at the singularity point or distributed within the event horizon. Therefore, I define a *physical BH* as an object whose mass and charge are all within its event horizon, regardless of the distribution of matter within.

Consequently, a physical BH is not necessarily a mathematical BH. This means that a physical BH may not have a singularity at its center. I further define an *astrophysical BH* as a physical BH that can be formed through astrophysical processes in the physical Universe and within a time much shorter than or at most equal to the age of the Universe. Figure 10.2 is an illustration of a possible process of forming an astrophysical BH through gravitational collapse of matter. So far, all observational studies of BHs have been made on astrophysical BHs. Therefore, the rest of this chapter is focused on them.

CAN ASTROPHYSICAL BLACK HOLES BE FORMED IN THE PHYSICAL UNIVERSE?

About 70 years ago, Oppenheimer and Snyder studied this problem in their seminal paper "On Continued Gravitational Contraction" (Oppenheimer and Snyder, 1939). Because of the historical and astrophysical importance of this paper, I include a facsimile of the abstract of this paper as Figure 10.3. In the beginning of the abstract, Oppenheimer and Snyder wrote, "When all thermonuclear sources of energy are exhausted a sufficiently heavy star will collapse. Unless . . . [see abstract] this contraction will continue indefinitely." This statement assures that the contraction

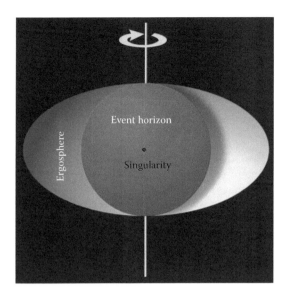

FIGURE 10.1 Illustration of the structure of a mathematical black hole (BH), which is rotating and has its mass concentrated at its singularity point. The existence of an ergosphere is due to the spin of the BH; a test particle in the ergosphere, although still outside the event horizon, cannot remain stationary. This figure is adapted from artwork in the Wikimedia Commons (available at: http://en.wikipedia.org/wiki/File:Ergosphere.svg).

process illustrated in Figure 10.2 can indeed take place in the physical Universe. In the end of the abstract, Oppenheimer and Snyder arrived at two conclusions that have deeply influenced our understanding of astrophysical BH formation ever since. (1) "The total time of collapse for an observer comoving [called comoving observer in the rest of this chapter] with the stellar matter is finite." This process is depicted in the last frame of Figure 10.2. This is the origin of the widespread and common belief that astrophysical BHs can be formed through gravitational collapse of matter. However, it should be realized that the observer is also within the event horizon with the collapsing matter, once a BH is formed. (2): "An external observer sees the star asymptotically shrinking to its gravitational radius." This means that the external observer will never witness the formation of an astrophysical BH. Given the finite age of the Universe and the fact

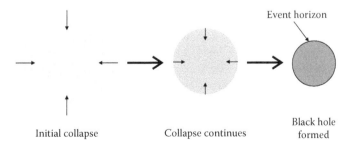

Initial collapse   Collapse continues   Black hole formed

FIGURE 10.2 Illustration of a possible formation process of an astrophysical black hole (BH). A spherically symmetric cloud of gas collapses under its self-gravity, assuming no internal pressure of any kind. The gas gradually contracts, the size getting smaller and smaller and density getting higher and higher, and eventually falls within the event horizon; it is at this point that a BH is formed. Apparently, not all mass has necessarily arrived at its center at the moment when all matter has just crossed the event horizon; therefore, at least at this moment, this astrophysical BH is just a physical BH and not a mathematical one.

SEPTEMBER 1, 1939   · PHYSICAL REVIEW   VOLUME 56

On Continued Gravitational Contraction

J. R. OPPENHEIMER AND H. SNYDER
University of California, Berkeley, California
(Received July 10, 1939)

When all thermonuclear sources of energy are exhausted a sufficiently heavy star will collapse. Unless fission due to rotation, the radiation of mass, or the blowing off of mass by radiation, reduce the star's mass to the order of that of the sun, this contraction will continue indefinitely. In the present paper we study the solutions of the gravitational field equations which describe this process. In I, general and qualitative arguments are given on the behavior of the metrical tensor as the contraction progresses: the radius of the star approaches asymptotically its gravitational radius; light from the surface of the star is progressively reddened, and can escape over a progressively narrower range of angles. In II, an analytic solution of the field equations confirming these general arguments is obtained for the case that the pressure within the star can be neglected. The total time of collapse for an observer comoving with the stellar matter is finite, and for this idealized case and typical stellar masses, of the order of a day; an external observer sees the star asymptotically shrinking to its gravitational radius.

FIGURE 10.3   Abstract of the seminal work on astrophysical black hole (BH) formation by Oppenheimer and Snyder (1939). (Reprinted with permission from Oppenheimer, J.R. and Snyder, H., *Physical Review*, 56(5), 455–9, 1939. Copyright 1939 by the American Physical Society.)

that all observers are necessarily external, the second, and last, conclusion of Oppenheimer and Snyder (1939) seems to indicate that astrophysical BHs cannot be formed in the physical Universe through gravitational collapse.

If, according to Oppenheimer and Snyder, an external observer sees matter asymptotically approach, but never quite cross, the event horizon, then matter must be continually accumulated just outside the event horizon and appear frozen there. Therefore, a gravitationally collapsing object has also been called a "frozen star" (Ruffini and Wheeler, 1971). In fact, the "frozen star" is a well-known novel phenomenon predicted by general relativity, i.e., a distant observer (O) sees a test particle falling toward a BH moving slower and slower, becoming darker and darker, and it is eventually frozen near the event horizon of the BH. This situation is shown in Figure 10.4, in which the velocity of a test particle, as observed from an external observer, approaches zero as it falls toward the event horizon of a BH. This process was also vividly described and presented in many popular science writings (Ruffini and Wheeler, 1971; Luminet, 1992; Thorne, 1994; Begelman and Rees, 1998) and textbooks (Misner et al., 1973; Weinberg, 1977; Shapiro and Teukolsky, 1983; Schutz, 1990; Townsend, 1997; Raine and Thomas, 2005). A fundamental question can be asked: Does a gravitational collapse form a frozen star or a physical BH?

In a recent paper, my student (Yuan Liu) and I summarized the situation as follows (Liu and Zhang, 2009):

> Two possible answers [to the above question] have been proposed so far. *The first one* is that since [the comoving observer] O' indeed has observed the test particle falling through the event horizon, then in reality (for O') matter indeed has fallen into the BH ... However, since [the external observer] O has no way to communicate with O' once O' crosses the event horizon, O has no way to "know" if the test particle has fallen into the BH ... *The second answer* is to invoke quantum effects. It has been argued that quantum effects may eventually bring the matter into the BH, as

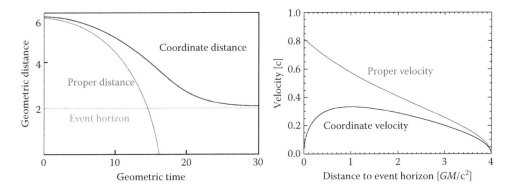

FIGURE 10.4   Calculation of the motion of a test particle free-falling toward a black hole (BH) starting at rest from $r = 6$ GM/c^2, where M is the mass of the BH and c is the speed of light in vacuum. Here "proper" and "coordinate" refer to the comoving and external observers, respectively. A set of rigid rulers or milestones are placed everywhere in the system; both the comoving and external observers get the coordinate of the infalling test particle this way. However, the comoving and external observers use their own wristwatches, which are no longer synchronized once the freefall starts. The left panel shows that a test particle takes finite or infinite time to cross the event horizon of the BH, for the comoving and external observers, respectively. The right panel shows that the comoving observer measures the test particle (in fact the observer himself) crossing the event horizon with a high velocity; however, the external observer measures that the test particle stops just outside the event horizon, i.e., is "frozen" to the event horizon. (Left panel adapted from Figure 3 in Ruffini, R. and Wheeler, J.A., *Physics Today*, 30–41, 1971. Copyright 1971 by the American Physical Society. With permission.)

seen by O (Frolov and Novikov, 1998). However, as pointed out recently (Vachaspati et al., 2007), even in that case the BH will still take an infinite time to form and the pre-Hawking radiation* will be generated by the accumulated matter just outside the event horizon. Thus this does not answer the question in the real world. Apparently O cannot be satisfied with either answer. In desperation, O may take the attitude of "who cares?" When the test particle is sufficiently close to the event horizon, the redshift is so large that practically no signals from the test particle can be seen by O and apparently the test particle has no way of turning back, therefore the "frozen star" does appear "black" and is an infinitely deep "hole." For practical purposes O may still call it a "BH," whose total mass is also increased by the infalling matter. Apparently this is the view taken by most people in the astrophysical community and general public, as demonstrated in many well-known textbooks (Misner et al., 1973; Hawking and Ellis, 1973; Weinberg, 1977; Shapiro and Teukolsky, 1983; Schutz, 1990; Townsend, 1997; Raine and Thomas, 2005) and popular science writings (Ruffini and Wheeler, 1971; Luminet, 1992; Thorne, 1994; Begelman and Rees, 1998). However when two such "frozen stars" merge together, strong electromagnetic radiations will be released, in sharp contrast to the merging of two genuine BHs (i.e. all their masses are within their event horizons); the latter can only produce gravitational wave radiation (Vachaspati, 2007). Thus this also does not answer the question in the real world.

The fundamental reason for the above "frozen star" paradox is that the "test particle" calculations have neglected the influence of the mass of the test particle. In reality, the infalling matter has finite mass, which certainly influences the global spacetime of the whole gravitating system, including the infalling matter and the BH. Because the event horizon is a global property of a gravitating system,

---

* Hawking radiation is a quantum mechanical effect of black holes (BHs) due to vacuum fluctuations near the event horizon of a BH. The radiation is thermal and blackbody-like, with a temperature inversely proportional to the mass of the BH. Therefore, Hawking radiation is not important at all for the astrophysical BHs we have discussed in this chapter. Pre-Hawking radiation of a BH is in fact not the radiation from the BH, but is hypothesized to come from the matter accumulated just outside the event horizon of the BH. For a remote observer, it may not be possible to distinguish between Hawking radiation and pre-Hawking radiation (even if it does exist) unless we know precisely the properties of the BH and the matter accumulated just outside its event horizon.

the infalling matter can cause non-negligible influence to the event horizon. In Figure 10.5, the infalling process of a spherically symmetric and massive shell toward a BH is calculated (Liu and Zhang, 2009) within the framework of Einstein's general relativity. In this calculation, all gravitating mass of the whole system, including both the BH and the massive shell, is taken into account consistently by solving Einstein's field equations. For the comoving observer, the shell can cross the event horizon and arrive at the singularity point within a finite time. *For the external observer, the body of the shell can also cross the event horizon within a finite time but can never arrive at the singularity point*, and its outer surface can only asymptotically approach the event horizon. Compared with the case of the infalling process of a test particle as shown in the left panel of Figure 10.4, the qualitative difference is the expansion of the event horizon as the shell falls in, which does not take place for the test particle case. It is actually the expansion of the event horizon that swallows the infalling shell. Therefore, matter cannot accumulate outside the event horizon of the BH if the influence of the gravitation of the infalling massive shell is also considered.

The calculations shown in Figure 10.5 still neglected one important fact for real astrophysical collapse. There is always some additional matter between the observer and the infalling shell being observed (we call it the inner shell), and the additional matter is also attracted to fall inward by the inner shell and the BH. We thus modeled the additional matter as a second shell (we call it the outer shell) and calculated the motion of the double-shell system. Our calculations show that in this case the inner shell can cross the event horizon completely even for the external observer, but it can still never arrive at the central singularity point (Liu and Zhang, 2009). Based on these calculations, we can conclude that *real astrophysical collapses can indeed form physical BHs*, i.e., all mass can cross the event horizon within a finite time for an external observer, and thus no "frozen stars" are formed in the physical Universe. A rather surprising result is that matter can never arrive at the singularity point, according to the clock of an external observer. This means that *astrophysical BHs in the physical Universe are not mathematical BHs because, given the finite age of the Universe, matter cannot arrive at the singularity point* (Liu and Zhang, 2009). This justifies my classifications of BHs into three categories.

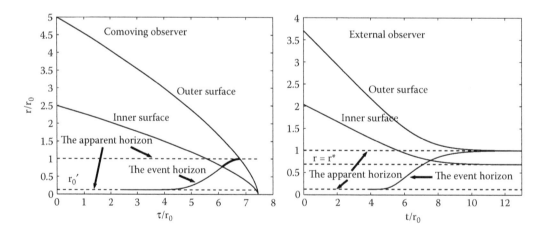

FIGURE 10.5   The infalling process of a spherically symmetric and massive shell toward a black hole (BH), calculated within the framework of Einstein's theory of general relativity. The left and right panels show the observations made by a comoving observer and an external observer, respectively; the two solid lines mark the inner and outer surfaces of the shell, respectively. The expansion of the event horizon as the shell falls in is also shown. For the comoving observer, the shell can cross the event horizon and arrive at the singularity point within a finite time. For the external observer, the body of the shell can also cross the event horizon within a finite time, but it can never arrive at the singularity point, and its outer surface can only asymptotically approach the event horizon. (This figure is adapted from panels (a) and (b) of Figure 2 in Liu, Y. and Zhang, S.N., *Physics Letters B*, 679, 88–94, 2009. Copyright 2009 by Elsevier. With permission.)

HOW CAN WE PROVE THAT WHAT WE CALL ASTROPHYSICAL BLACK HOLES ARE REALLY BLACK HOLES?

The defining characteristic of an astrophysical BH is that all its gravitating mass is enclosed by its event horizon, and consequently all infalling matter will fall into its event horizon within a finite time of an external observer. Therefore, it has been commonly believed that the final and unambiguous confirmation of the detection of BHs requires direct evidence for the existence of the event horizon of a BH. However, by virtue of the very definition of the event horizon that no light can escape from it to infinity, direct evidence for the existence of the event horizon of a BH can never be obtained by a distant observer. However, in science direct evidence is not always what leads to the discovery of something. For example, we never "see" directly many particles created in accelerator experiments, whose existence is usually inferred by their decay products. Actually, quarks do not even exist in free forms, and very few scientists today question that quarks exist. Searching for dark matter* particles, which may be created in CERN's Large Hadron Collider (LHC) experiments, is currently under way.

However, even if dark matter particles are being produced there, these particles have no chance of annihilating or even interacting in these detectors. Therefore, only indirect evidence, such as "missing mass," can be used to demonstrate detection of dark matter particles in accelerator experiments. In astronomy, similar situations exist. For example, no "direct" evidence exists for dark matter and dark energy in the Universe. However, dark matter and dark energy are widely believed to exist, from a collection of many pieces of indirect evidence. Do we have a collection of indirect evidence to prove that what we call astrophysical BHs are really BHs? Because in astronomy we are dealing with astrophysical BHs with masses over a range of at least eight orders of magnitude and located in very different astrophysical environments, here I suggest five criteria, or parameters, in determining whether astronomers have found astrophysical BHs:

1. The concept and theoretical model based on astrophysical BHs can be used to explain a series of common observational phenomena known previously.
2. The same concept and theoretical model based on astrophysical BHs can be used to explain the ever-increasing volume of new observational phenomena.
3. No counterevidence comes forward against the model based on astrophysical BHs.
4. The BH formation and evolution scenario inferred from those observational phenomena are self-consistent and physically and astrophysically reasonable.
5. There is no alternative theoretical model that can also explain the same or even more phenomena with the same or even better success than the astrophysical BH model.

Although general, the above five criteria meet the highest standard for recognizing new discoveries in experimental physics and observational astronomy. As a matter of fact, these criteria also meet Carl Sagan's principle that "extraordinary claims require extraordinary evidence" because of the importance and impacts of discovering BHs in the Universe. Indeed, it is debatable that the discoveries of very few, if any, astrophysical objects meet such stringent and extensive requirements.

DO WE HAVE SUFFICIENT EVIDENCE TO CLAIM THE EXISTENCE OF ASTROPHYSICAL BLACK HOLES IN THE PHYSICAL UNIVERSE?

Having given up the hope of finding "direct" evidence for the existence of the event horizon of a BH, we must search for other supporting evidence for the existence of BHs, following the five criteria I proposed in the previous section. The next hope is to study what happens when matter or light

---

* This is a kind of matter believed to dominate the total mass of the Universe, but it does not produce any electromagnetic radiation. For details on dark matter, please refer to Bloom's chapter in this volume.

gets sufficiently close to or even falls into BHs and then explains in this way as many observational phenomena as possible. Around a BH, several important effects might be used to provide indirect evidence for the existence of the BH:

1. The surface of a BH or matter hitting it does not produce any radiation detectable by a distant observer; this is a manifestation of the event horizon of a BH.
2. There exists an innermost stable circular orbit for a BH, beyond which matter will free-fall into the BH; this orbital radius is a monotonic function of the angular momentum of a BH, as shown in Figure 10.6. In some cases this general relativistic effect can be used to measure the spin of a BH, for example, by fitting the continuum spectrum or relativistically blurred lines produced from the inner region of an accretion disk around a BH (Loar, 1991; Zhang et al., 1997).
3. The very deep gravitational potential around a BH can produce strong gravitational lensing effects; an isolated BH may be detected this way.
4. The very deep gravitational potential around a BH can cause matter accreted toward a BH to convert some of its rest mass energy into radiation; an accreting BH may be detected this way. In Figure 10.7, I show the conversion efficiency of different kinds of BH accretion systems, in comparison with the conversion efficiencies of other astrophysical systems.
5. For a spinning BH, its ergosphere (as shown in Figure 10.1) will force anything (including magnetic field lines) within the ergosphere to rotate with it; the Penrose or magnetic Penrose mechanism may allow the spin energy of a BH to be extracted to power strong outflows (Blandford and Znajek, 1977). Sometimes outflows can also be produced from accretion disks around non-spinning BHs (Blandford and Payne, 1982).

LUMINOUS ACCRETING BLACK HOLES

If there is a sufficient amount of matter around a BH, matter under the gravitational attraction of the BH will be accreted toward it, and in this process an accretion disk can be formed surrounding the BH. Under certain conditions a geometrically thin and optically thick accretion

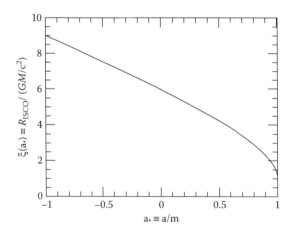

FIGURE 10.6 The radius of the innermost stable circular orbit (R_{ISCO}) of a black hole (BH) as a function of the spin parameter (a_*) of the BH, i.e., the dimensionless angular momentum; a negative value of a_* represents the case that the angular momentum of the disk is opposite to that of the BH. The spin angular momentum of a BH, the seond parameter for a BH, can be measured by determining the inner accretion disk radius if the inner boundary of the disk is the innermost stable circular orbit of the BH.

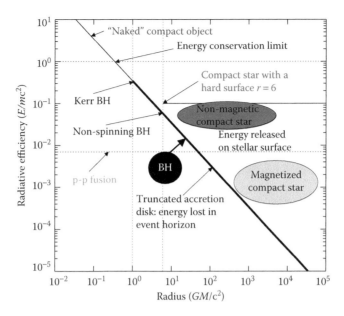

FIGURE 10.7 For an accretion disk around a black hole (BH), the radiative efficiency (ratio between radiated energy and the rest mass energy of accreted matter) is approximately inversely proportional to the inner boundary radius of the accretion disk. Here it is assumed that the radiation produced by the accreted matter (in almost free fall) between the disk boundary and the event horizon of the BH is negligible; however, the radiative efficiency is slightly higher if the very weak emission from the matter between the disk boundary and the event horizon of the BH is also considered (Mahadevan, 1997; also see the caption for Figure 10.9). The diagonal line shows a $1/r$ scaling, calibrated to take the value of 0.057 when $r = 6$. The thick black line is for strongly suspected BH accreting systems. The range of $r = 1$–9 corresponds to the innermost stable circular orbit of a BH with different spin, assuming that the disk extends all the way there; the radiative efficiency ranges from a few to several tens of percent, far exceeding the p-p fusion radiative efficiency taking place in the Sun. The case for $r > 9$ corresponds to a truncated accretion disk, whose radiative efficiency can be extremely low, because energy is lost into the event horizon of the BH. The thin solid black horizontal line is for the 10% efficiency when matter hits the surface of a neutron star where all gravitational energy is released as radiation. The thin solid black diagonal line above the point marked for "Kerr BH" (Kerr black hole) is for a speculated "naked" compact object whose surface radius is extremely small, and thus the radiative efficiency can be extremely high.

disk can be formed (Shakura and Sunyaev, 1973), which is very efficient in converting the gravitational potential energy into thermal radiation. The radiative efficiency (ratio between radiated energy and the rest mass energy of accreted matter) is approximately inversely proportional to the inner boundary radius of the accretion disk, as shown in Figure 10.7, because the matter between the inner disk boundary and the event horizon of the BH is free-falling, and almost all the kinetic energy is carried into the BH. Please refer to the caption of Figure 10.7 for detailed explanations.

Figure 10.8 describes accreting disks surrounding a Kerr (spinning) BH (*left*) and a Schwarzschild (non-spinning) BH (*right*); the inner boundary of the disk stops at the innermost stable circular orbit of the BH when the accretion rate is around 10% of the Eddington rate. Such high radiation efficiency is commonly observed in the luminous state of a binary system suspected to contain a BH of several solar masses as the accretor (Remillard and McClintock, 2006), or in a quasi-stellar object (QSO) (also called a quasar or active galactic nucleus [AGN]) suspected to harbor at the center of a galaxy a supermassive BH of millions to billions of solar masses as the accretor (Yu and Tremaine, 2002). The BH accretion model, with essentially only three parameters (two for the mass and spin of a BH, and one for the accretion rate of the disk),

FIGURE 10.8   Accretion disks around non-spinning (*left*) and spinning (*right*) black holes (BHs). For the spinning BH, both its inner disk and event horizon radii are smaller, thus providing a deeper gravitational potential well for a more efficient energy conversion, reaching a maximum efficiency of about 42% (Page, D.N., and Thorne, K.S., *Astrophysical Journal*, 191, 499–506, 1974). (Courtesy of NASA/CXC/M. Weisskoff http://chandra.harvard.edu/photo/2003/bhspin/.)

can explain the many observed properties of dozens of BH binary systems in the Milky Way and countless AGNs in the Universe (Zhang, 2007b). Currently, no single alternative model can be used in a systematic and consistent way to explain these same observations in those binary systems and AGNs.

FAINT ACCRETING BLACK HOLES

When the radiation of the disk is substantially below 10% Eddington luminosity, the optically thin and geometrically thick disk tends to retreat away from the BH, and the central region is replaced by some sort of radiatively inefficient accretion flow, for example, the advection-dominated accretion flow (Narayan and Yi, 1994). Generically, this corresponds to the case for $r > 9$ in Figure 10.7, i.e., a truncated accretion disk, whose radiative efficiency can be extremely low because almost all gravitational potential energy is converted into the kinetic energy of the accreted matter that free-falls into the BH and thus is lost into the event horizon of the BH. This model has been used to explain the extremely low luminosity of the quiescent state of BH binaries (Shahbaz et al., 2010), the inferred supermassive BHs in the center of the Milky Way, and many nearby very-low-luminosity AGNs (Ho, 2008); normally, $r > 100$ for these extremely underluminous systems. Recently, evidence has been found for the truncation radius in the range of $r = 10$–100 for binary systems in their normal, but slightly less luminous, states, for example, around 0.01 to 0.1 Eddington luminosity (Gierliński et al., 2008). The top panel of Figure 10.9 shows a theoretical calculation of the expected truncation radius as a function of accretion rate $\dot{M}$ (Liu and Meyer-Hofmeister, 2001), i.e., roughly $r \propto \dot{M}^{-1/2}$. The bottom panel of Figure 10.9 shows the observed accretion disk luminosity L as a function of observationally inferred disk truncation radius (Shahbaz et al., 2010), i.e., roughly $L \propto r^{-3}$. Therefore, the radiative efficiency $\eta = L/\dot{M} \propto r^{-3}r^2 \propto 1/r$, as shown in Figure 10.7. Once again, the BH accretion disk model is so far the only one that can explain all these observations across huge dynamic ranges of mass, time, space, environment, and luminosity.

THE SUPERMASSIVE BLACK HOLE AT THE CENTER OF THE MILKY WAY

A single strong case for a BH lies at the center of the Milky Way. As shown in the top panel of Figure 10.10, the mass of the central object is measured to be around 4 million solar masses by observing the stellar motions very close to it; the closest distance between the S2 star (the

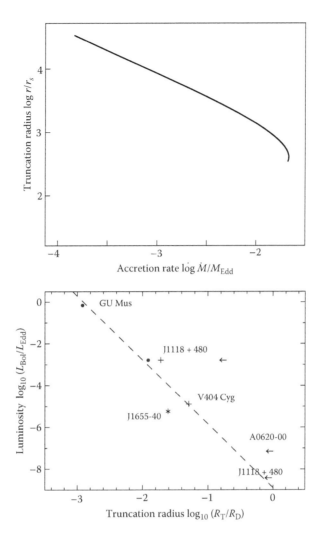

FIGURE 10.9 Accretion disk truncation radius and luminosity for faint (low-luminosity) accreting black holes (BHs). The top panel presents a theoretical calculation of the expected truncation radius (normalized to the radius of the event horizon of a BH) as a function of accretion rate (normalized to the Eddington rate). The bottom panel presents accretion disk luminosity (normalized to the Eddington luminosity) as a function of the observationally inferred disk truncation radius (normalized to an arbitrary unit). (The top and bottom panels give roughly $r \propto \dot{M}^{-1/2}$ and $L \propto r^{-3}$, respectively.* Therefore, the radiative efficiency is $\eta = L/\dot{M} \propto r^{-3}r^2 \propto 1/r$, as shown in Figure 10.7. The data points on the bottom panel are for suspected BH accretion systems. (top panel adapted from Figure 1 in Liu, B.F. and Meyer-Hofmeister, E., *Astronomy and Astrophysics*, 372, 386–90, 2001. Copyright (2001) by *Astronomy and Astrophysics*. With permission. Bottom panel adapted from Figure 8 in Shahbaz, T., Dhillon, V.S., Marsh, T.R., et al., *Monthly Notices of the Royal Astronomical Society*, 403, 2167–75, 2010. Copyright (2010) by *Wiley*. With permission.) (*More precisely, the top panel gives $r \propto \dot{M}^{-2/3}$, thus $\eta = L/\dot{M} \propto r^{-3}r^{3/2} \propto r^{-3/2}$, consistent with the prediction of the advection-dominated accretion flow model if the emission between the disk boundary and the event horizon of the BH is not negligible [Mahadevan, 1997].)

currently known nearest star to the center) and the center is around 2,100 times the radius of the event horizon of the BH, thus excluding essentially all known types of single astrophysical objects as the compact object there. The bottom panel of Figure 10.10 shows the extremely compact size of the radio signal-emitting region, which is merely several times the radius of the event horizon of the BH, ruling out a fermion star model and also disfavoring a boson star

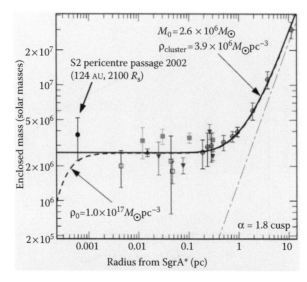

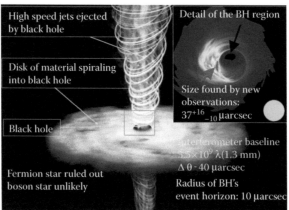

FIGURE 10.10 Mass and size of the supermassive black hole (BH) at the center of the Milky Way. The top panel shows the enclosed mass as a function of radius from the dynamic center of the Milky Way. The bottom panel illustrates our current understanding of what is going on around the suspected supermassive BH at the center of the Milky Way. The inset at the upper-right corner of the bottom panel shows that the angular resolution of the observation is about 40 μarcsec (marked as the green circular area), obtained with the $\lambda = 1.3$ mm wavelength interferometer with a baseline of 3.5×10^9. The inferred size of the radio signal-emitting region (red arrow) is about 37 μarcsec, comparable to the size of the event horizon of this supermassive BH, which is about 10 μarcsec (black arrow). This suggests that the compact object must be at least smaller than several times the size of the event horizon of the suspected supermassive BH, thus ruling out a fermion star model and disfavoring a boson star model. (Top panel reprinted from Schödel, R., Ott, T., Genzel, R., et al., *Nature*, 419, 694–6, 2002. Copyright (2002) by Macmillan Publishers Ltd. With permission. Bottom panel adapted from the online supplementary material of Doeleman, S.S., Weintroub, J., Rogers, A.E.E., et al., *Nature*, 455, 78–80, 2008. Copyright (2008) by Macmillan Publishers Ltd. With permission.)

model. In fact, the extremely low radiation efficiency of this object requires that the central object cannot have a surface, i.e., the majority of the gravitational energy is converted to the kinetic energy of the accreted matter and subsequently lost into the BH (Broderick et al., 2009). Putting all these pieces of supporting evidence together does not leave much room for a non-BH object as the central compact object of the Milky Way. The properties of this system can be well explained with the same BH accretion model used to explain the quiescent-state properties of other low-luminosity AGNs and galactic BH binaries (Yuan et al., 2003).

COMPARISON WITH ACCRETING NEUTRON STARS

The thin solid black horizontal line in Figure 10.7 is for the 10% efficiency when matter hits the surface of a neutron star where all gravitational energy is released as radiation. Essentially, all accreted matter can reach the surface of a neutron star if the surface magnetic field of the neutron star is so low that the magnetic field pressure does not play a significant role in blocking the accreted matter from reaching the surface of the neutron star; in this case, the radiation efficiency is not much below 10%. However, the radiation efficiency can be substantially below 10%, when the accretion rate is very low such that the surface magnetic field of the neutron star can block the accreted matter through the so-called propeller effect (Zhang et al., 1998). If the accretion disk around the neutron star at very low accretion rate is in the advection-dominated flow state, some of the accreted matter can still reach the surface of the neutron star and produce a non-negligible amount of radiation from the surface of the neutron star (Zhang et al., 1998; Menou et al., 1999). Therefore, for two binary systems with a BH and a neutron star as the accretors, respectively, of material from a normal star, the neutron star binary will appear brighter, even if their accretion disks are exactly the same, as shown in Figure 10.7. This expectation has been observationally confirmed for all known BH and neutron star binaries at their quiescent states as shown in Figure 10.11 (Narayan and McClintock, 2008). Therefore, the simple accreting BH (and neutron star) model can explain nicely a large collection of observations.

ISOLATED BLACK HOLES

Clearly, for an isolated astrophysical BH, which is not surrounded by dense medium and thus is not actively accreting matter, the only way to detect it is through a gravitational lensing effect (Paczynski, 1986, 1996). So far, several candidate BHs have been found this way (Bennett et al., 2002; Mao et al., 2002). However, practically speaking, lensing observations can find only

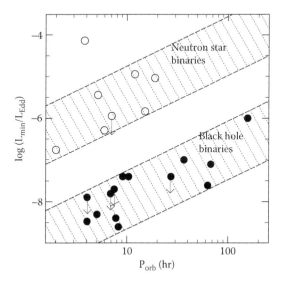

FIGURE 10.11 Comparison of the quiescent bolometric luminosity between neutron star and black hole (BH) binaries. In an accreting binary, its quiescent-state luminosity (lowest luminosity state) is scaled positively with its compact object mass and orbital period, regardless of whether the accretor is a neutron star or a BH. The main difference between neutron star and BH accretors is that the surface radiation of the neutron star makes the neutron star system brighter (in units of the Eddington luminosity) for the same orbital period. (Adapted from Narayan, R. and McClintock, J.E., *New Astronomy Reviews*, 51, 733–51, 2008. Copyright 2008 by Elsevier. With permission.)

candidate BHs because it is extremely difficult to exclude all other possibilities responsible for the detected lensing events. Additional evidence supporting the BH nature of the candidate object must be sought, e.g., x-ray emission from accreted interstellar medium onto the putative BH (Agol and Kamionkowski, 2002). Currently, only an upper limit on the anticipated x-ray emission from one candidate has been observed, indicating that the radiative efficiency is as low as around 10^{-10}—10^{-9}, assuming that the putative BH is located in the normal interstellar medium (ISM) (Nucita et al., 2006); this efficiency is far below the range shown in Figure 10.7. However, as recently found, all microquasars are located in parsec-scale cavities with density lower by at least three orders of magnitude than the normal ISM (Hao and Zhang, 2009). Then the estimated radiative efficiency upper limit might be increased by at least three orders of magnitude, if this putative BH is also located in a very-low-density cavity. Even in this case, the radiative efficiency would still be in the lowest end in Figure 10.7, thus indicating that the majority of the kinetic energy of the accreted matter is lost into the event horizon of the BH.

LUMINOUS "NAKED" COMPACT OBJECTS?

In Figure 10.7, the thin solid black diagonal line above the point marked for "Kerr BH" is for a speculated "naked" compact object, whose surface radius is extremely small but not enclosed by an event horizon. The concept for a "naked" compact object is related to "naked" singularity, which is not enclosed by an event horizon; a "naked" singularity can be formed in a variety of gravitational collapse scenarios (Pankaj, 2009), thus breaking Penrose's cosmic censorship.* A key characteristic for an accreting "naked" singularity is that radiation can escape from it, in sharp contrast to an accreting BH, as illustrated in Figure 10.12. Following the arguments I made when answering the question, Can astrophysical BHs be formed in the physical Universe?, "naked" compact objects, rather than "naked" singularities, might be formed in the physical Universe. In this case, the radiative efficiency can be very high, depending on the radius of the "naked" compact object. For extremely small radii, the efficiency may exceed 100%, implying that the energy of the "naked" compact object is extracted. Unfortunately, so far there has been no observational evidence supporting this conjecture. However, this possibility, if true, may have fundamental impacts regarding the evolution and fate of the Universe, as I will discuss at the end of this chapter.

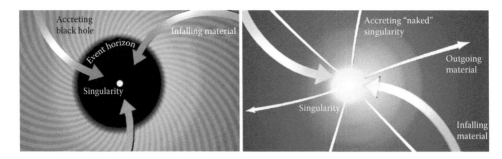

FIGURE 10.12 Comparison between an accreting black hole (*left*) and an accreting "naked singularity" (*right*), which can be luminous for a distant observer. (Adapted from the online slides of *Scientific American* (available at http://www.scientificamerican.com/slideshow.cfm?id=naked-singularities&photo_id=DC1F7444-DCC7-F2E4-2EF03074D470B687 and http://www.scientificamerican.com/slideshow.cfm?id=naked-singularities&photo_id=DC1F8C9A-0E60-3C59-5CD90FA1B4505784). Copyright (2009) by Alfred T. Kamajian. With permission.)

---

* Penrose's cosmic censorship conjectures that each and every singularity in the Universe is enclosed by an event horizon, i.e., there is no "naked" singularity in the Universe.

Relativistic Jets

A spinning BH can also power relativistic jets, as observed commonly from AGNs (or quasars) and galactic BH binaries (or microquasars), as shown in Figure 10.13. This can happen when large-scale magnetic fields are dragged and wound up by the ergosphere (see Figure 10.1) of a spinning BH, as shown in Figure 10.14. The twisted and rotating magnetic field lines can then accelerate the infalling plasmas outward along the spin axis of the BH to relativistic speeds (Blandford and Znajek, 1977), producing powerful relativistic jets that can carry a substantial amount of the accretion power and travel to distances far beyond these binary systems or their host galaxies. Recent studies have shown that the BHs in microquasars are indeed spinning rapidly (Zhang et al., 1997; Mirabel, 2010; McClintock et al., 2009). Once again, a conceptually simple BH accretion model can explain the observed relativistic jets from accreting BH systems with very different scales.

Gamma-Ray Bursts

Gamma-ray bursts (GRBs) (Klebesadel et al., 1973; Fishman and Meegan, 1995; Gehrels et al., 2009) are strong gamma-ray flashes with an isotropic energy between 10^{50} and 10^{54} ergs released in seconds

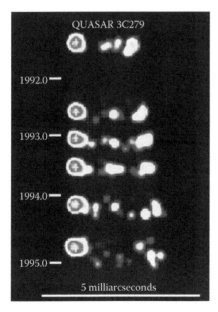
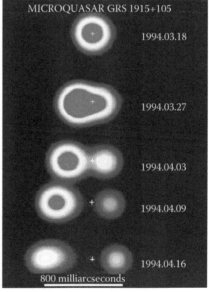

FIGURE 10.13 Relativistic jets from the quasar 3C 279 (*left panel*: an active galactic nucleus with a redshift of $z = 0.536$) and the microquasar GRS 1915+105 (*right panel*: a Galactic black hole [BH] binary). The radio images from top to bottom are observed sequentially at different times; in the left panel, the starting time of each year is marked as a short bar, and in the right panel the date when each observation was made is shown. Radio signals are synchrotron radiation from entrained particles in higher-density portions of the jets illustrated elsewhere in this chapter. The crosses mark the locations of the BHs, providing reference points for measuring the proper motions of jets. The lengths of the long horizontal bars (5 and 800 mas in the left and right panels, respectively) near the bottom of each panel show the angular size scales of the jets on them. The Galactic object (right panel) shows a two-sided jet; the color scale uses redder colors for higher intensity. The jet coming toward us is relativistically Doppler boosted and thus is brighter than the counter jet. The quasar is at cosmological distance; counter jets are not normally observed in such cases because they are very faint. The inferred intrinsic velocities of the jets for both systems are more than 98% of the speed of the light. (Adapted from Mirabel, I.F. and Rodriguez, L.F., *Nature*, 371, 46–8, 1994. Copyright 1994 by Macmillan Publishers Ltd. With permission.)

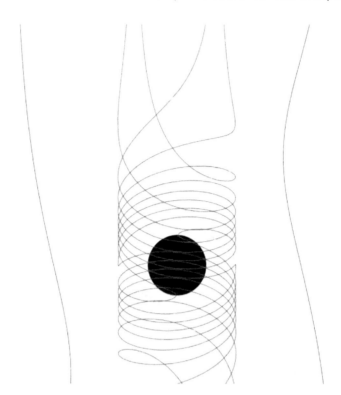

FIGURE 10.14 Illustration of the production process of a relativistic jet, similar to that shown in Figure 10.10, by an accreting spinning black hole (BH). The magnetic field lines are wound up by the ergosphere (see Figure 10.1) of the spinning BH, because nothing can stay stationary there and must rotate with the spinning BH. Accreted matter into this region is spun out with relativistic speeds along the spin axis of the BH, because the accreted matter is fully ionized and must move along these wound up magnetic field lines. (Reprinted from Figure 4d in Meier et al. [2001]. With permission from the American Association for the Advancement of Science.)

or shorter for each event. They originate at redshifts as high as 8.3 (Salvaterra et al., 2009; Tanvir et al., 2009) or even beyond 10 (thus seen as they were at only a few percent of the age of the Universe [Lin et al., 2004]). GRBs are the biggest explosions in the Universe since the Big Bang and can be used to probe the evolution of the Universe. At least some of the "long" GRBs, with duration approximately more than 2 sec, are believed to be produced from spinning BHs accreting at extremely high rates (Gehrels et al., 2009; Mézáros, 2009; Zhang, 2007a). In this picture, a spinning BH is formed as a massive star ends its life in a gravitational collapse; the fallback matter after the accompanying supernova (SN) explosion forms an accretion disk around the BH. In an extremely violent process similar to that shown in Figure 10.14, super-relativistic jets, with Lorenz factors of hundreds to thousands, are produced, which produce luminous and also highly beamed gamma-ray emissions.*

Putting It All Together: Astrophysical Black Holes Have been Detected

Therefore, the BH accretion (and outflow) model can be used to explain a vast array of astrophysical phenomena across huge dynamical ranges of time, space, mass, luminosity, and astrophysical environments.

The first collection of "indirect" evidence for the existence of BHs is with the radiative efficiency when matter falls toward a central compact object. As we have proven (Liu and Zhang,

---

* For more details on supernovae and gamma-ray bursts, please refer to the chapter by Filippenko in this volume.

2009), matter in a gravitational potential well must continue to fall inward (but cannot be "frozen" somewhere), either through the event horizon of a BH or hitting the surface of a compact object not enclosed by an event horizon but with a radius either larger or smaller than the event horizon of the given mass (called a compact star or "naked" compact object, respectively). No further radiation is produced after the matter falls through the event horizon of the BH; thus, the majority of the kinetic energy of the infalling matter is carried into the BH. On the other hand, surface emission will be produced when matter hits the surface of the compact star or "naked" compact object, because it is not a BH. Therefore, the radiative efficiencies for these different scenarios are significantly different, as shown in Figure 10.7. Currently, all observations of the strongly suspected accreting BHs in binary systems or at the centers of many galaxies agree with the BH accretion model, over a huge range of accretion rates.

The second collection of "indirect" evidence for the existence of BHs is with the relativistic jets from microquasars (accreting BH binaries), quasars (accreting supermassive BHs), and GRBs (also called collapsars, i.e., accreting BHs just formed in a special kind of SN event). In Figure 10.15, a unified picture of BH accretion and outflow is presented for these three seemingly very different kinds of systems. The key ingredient of the model is that the combination of the deep gravitational potential well and the ergosphere of a spinning BH extracts both the potential energy and the spinning energy of the BH, producing strong electromagnetic radiation and powerful relativistic outflows. This model explains current observations satisfactorily.

Among all competing models (many of them can only be used to explain some of these phenomena), the BH accretion (and outflow) model is the simplest, and the astrophysical BHs are also the simplest objects, with only two physical properties (mass and spin). The BH masses and spin parameters, found by applying the BH accretion model to many different kinds of

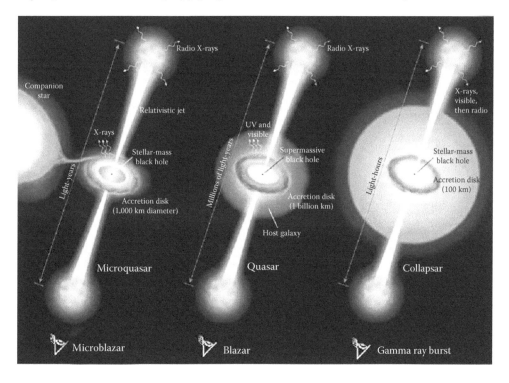

FIGURE 10.15 Unified picture of black hole accretion and outflow model for three kinds of astrophysical systems with very different observational characteristics and extremely different scales of mass, time, size, luminosity, and astrophysical environments. (Reprinted from Mirabel and Rodriguez, *Sky & Telescope*, p. 32, 2002. With permission.)

data, are physically and astrophysically reasonable and also well understood so far. The mass of a stellar-mass BH comes from the gravitational collapse of the core of a massive star and the subsequent matter in-falling process; some of the core-collapse supernovae and GRBs are manifestations of this process. A supermassive BH grows up by accreting matter in its host galaxy; the active accretion process makes the galaxy show up as a QSO. The BH accretion process can efficiently increase the spin of a BH, by transferring the angular momentum of the accreted matter to the BH.

Is the model falsifiable? If surface emission is detected from the putative BH in any of the above systems, one can then confidently reject the validity of the BH accretion model, at least for that specific system. For the only other two kinds of compact objects known, i.e., white dwarfs and neutron stars, surface emissions have been commonly detected. Yet so far this has not happened to any of the putative accreting BH systems we discussed above. Therefore, there is no counterevidence against the BH accretion model used to explain all phenomena discussed in this chapter.

Positive identification of astrophysical BHs in those objects also satisfies the principle of Occam's razor, i.e., that "entities should not be multiplied unnecessarily," commonly interpreted as "take the simplest theory or model among all competitors." However, the history of science tells us that Occam's razor should be used only as a heuristic to guide scientists in the development of theoretical models rather than as an arbiter between published models; we eventually accept only models that are developed based on existing data but can also make falsifiable predictions, are confirmed with additional data, and can explain new data or new phenomena. This is indeed what has happened to the BH accretion model. In this sense, the BH accretion (and outflow) model has survived all possible scrutiny.

I therefore conclude that *we now have sufficient evidence to claim that we have found astrophysical BHs, at least in some galactic binary systems, at the center of almost every galaxy, and as the central engines of at least some long GRBs.*

WILL ALL MATTER IN THE UNIVERSE EVENTUALLY FALL INTO BLACK HOLES?

In the previous sections, I have emphasized the importance of BH accretion and actually relied on the BH accretion model to argue in favor of the existence of astrophysical BHs in the physical Universe. It is then not accidental to ask the following question: Will all matter in the Universe eventually fall into BHs? As a matter of fact, I have indeed been asked this question numerous times by nonprofessional researchers when I gave public talks on BHs; somehow only the professional researchers hesitate to ask this question. Each time I have almost randomly used one of three answers: "yes," "no," or "I don't know." Here I attempt to provide some rather speculative discussions on this question.

Ignoring the Hawking radiation of a BH and assuming that no "naked" singularities (compact stars) exist in the physical Universe (i.e., that Penrose's cosmic censorship holds), indeed it is inevitable that all matter (including dark matter and perhaps all forms of energy) will eventually fall into BHs if the Universe is not expanding (i.e., is stationary) and does not have a boundary. This is because regardless of how small the probability is for a particle or a photon to fall into a BH, it eventually has to fall into a BH after a sufficiently large number of trials. A universe made of only BHs is of course an eternally dead universe. An eternally expanding universe will save some matter from falling into BHs because eventually particles or even light escaping from a galaxy or those (such as dark matter and hot baryons and electrons) that are already in intergalactic regions may never reach another galaxy and thus not fall into any BH. However, whatever is left in a given galaxy will still eventually fall into one of the BHs in the galaxy. Therefore, each galaxy will be made of only BHs, and these BHs may collide with one another to become a huge BH. It is inevitable that in the end each galaxy will be just a huge

BH. Then eventually the expanding universe will be made of numerous huge BHs moving apart from one another, with some photons and particles floating between them and never quite catching them. This universe is still a dead one. If at some point the universe begins to contract, then particles (including dark matter) and photons outside BHs will begin to be sucked into BHs, and BHs will also begin to merge with each other. Eventually, the whole universe may become just a single huge BH.

Can the Hawking radiation intervene to rescue our Universe from an eternal death? It is easy to calculate that for a 10 $M_\odot$ BH, its Hawking temperature is below 10^{-7} K, far below the current temperature of cosmic microwave background radiation (CMBR). Therefore, the Hawking radiation of BHs will not be effective before most CMB photons are absorbed into BHs or the Universe has expanded to decrease the CMB temperature below that of the Hawking radiation of the BHs. Eventually (after almost an eternal time), the Universe will be in equilibrium between the photons trapped by the BHs and the Hawking radiation at a temperature below 10^{-7} K. Such a universe is not much better than a dead universe made of essentially only BHs.

Mathematically, wormholes and white holes may be able to dig out the energy and matter lost in BHs. However, with our current knowledge of physics and astrophysics we do not yet know how wormholes and white holes can be produced in the physical Universe. Although I cannot reject this possibility, this is not favored by me, because I do not want to rescue the Universe from eternal death by relying on unknown physics and astrophysics.

As I discussed briefly in the last section, if Penrose's cosmic censorship is broken, "naked" compact objects may quite possibly exist in the physical Universe (similarly, astrophysical BHs can also be turned into "naked" compact objects), although they have not been identified so far. As shown in Figure 10.7, for "naked" compact objects with extremely small radii, radiative efficiency exceeding 100% is possible. For an external observer, this is equivalent to extracting energy from the "naked" compact object, because globally and on the average energy conservation is required. This situation is similar to the Hawking radiation: the vacuum fluctuations around a BH lead to the escape of particles from just outside the event horizon of a BH, but globally this is equivalent to consuming the energy (mass) of the BH as a result of global energy conservation. Likewise, the energy extracted from the "naked" compact object can be turned into matter through various known physical processes. This scenario is just the re-cycling of the previously accreted matter in BHs. Therefore, with "naked" compact objects, if they do exist, the Universe can indeed be rescued from an eternal death caused by all matter being sucked into BHs. I call this the *"naked" compact object re-cycle conjecture*.

Therefore, my final answer to this question is mixed: *Almost all matter indeed will fall into astrophysical BHs; however, "naked" compact objects can re-cycle matter out, if astrophysical BHs can somehow be turned into "naked" compact objects.*

SUMMARY, CONCLUDING REMARKS, AND FUTURE OUTLOOKS

In this chapter, I have focused on asking and answering the following questions:

- What is a BH? Answer: There are three types of BHs, namely, *mathematical BHs, physical BHs,* and *astrophysical BHs*. An astrophysical BH, with mass distributed within its event horizon but not concentrated at the singularity point, is not a mathematical BH.
- Can astrophysical BHs be formed in the physical Universe? Answer: Yes, at least this can be done with gravitational collapse.
- How can we prove that what we call astrophysical BHs are really BHs? Answer: Finding direct evidence of the event horizon is not the way to go. Instead, I proposed five criteria that meet the highest standard for recognizing new discoveries in experimental physics and observational astronomy.

- Do we have sufficient evidence to claim the existence of astrophysical BHs in the physical Universe? Answer: Yes, astrophysical BHs have been found at least in some galactic binary systems, at the center of almost every galaxy, and as the central engines of at least some long GRBs.
- Will all matter in the Universe eventually fall into BHs? Answer: Probably "no," because "naked" compact objects, if they do exist with radii smaller than the radii of event horizons for their mass but are not enclosed by event horizons, can rescue the Universe from an eternal death by re-cycling out the matter previously accreted into astrophysical BHs. I call this the *"naked" compact object re-cycle conjecture.*

The main conclusion of this chapter is thus that we have confidence to claim discoveries of astrophysical BHs in the physical Universe with the developments of theoretical calculations and modeling of astrophysical BH formation, accretion, and outflows and the applications of these theories to the ever-increasing amount of astronomical observations of many different types of objects and phenomena. This should be considered as a major verification of Einstein's general relativity, given that the Schwarzschild BH is the very first analytic solution of Einstein's field equations. With this, general relativity has prevailed at the gravity (or curvature) level from the Solar System, where the general relativity correction over the Newtonian gravity is small but still non-negligible, to the vicinity of a BH, where the general relativity effects dominate.

It is then interesting to ask this question: Do we need a quantum theory of gravity in order to further understand astrophysical BHs? My answer is: *Probably no.* There are three reasons for giving this perhaps surprising (and perhaps not welcome) answer:

1. Quantum effects outside astrophysical BHs are unlikely to be important because of their macro scales.
2. No information from matter fallen into an astrophysical BH can be obtained by an external observer.
3. For an external observer, matter inside an astrophysical BH is distributed, but not concentrated at its very center, and thus no physical singularity exists even inside it.

However, a quantum theory of gravity is probably needed to understand the behavior of stellar-mass "naked" compact objects, if Penrose's cosmic censorship is broken, because their densities can be extremely high such that quantum effects will be very important. Therefore, a quantum theory of gravity is needed to understand the "naked" compact object re-cycle conjecture I proposed here.

Finally, I ask one more question: What additional astronomical observations and telescopes are needed to make further progress on our understanding of astrophysical BHs and perhaps also "naked" compact objects? The answer to this question can be extremely long, but I try to be very brief here. Personally, I would like to see two types of major observational breakthroughs:

1. X-ray timing and spectroscopic observations of astrophysical BHs with throughputs at least an order of magnitude higher than the existing *Chandra* and *X-ray Multi-Mirror Mission* (*XMM*)-*Newton* x-ray observatories. This would allow detailed examinations of the structure around astrophysical BHs; detailed mapping; and an understanding of the rich physics of accretion, radiation, and outflows under the extreme physical conditions there, as well as exact measurements of BH masses and spin parameters in many systems. For stellar-mass BHs in binaries, these measurements will help us understand their formation mechanism and evolution of massive stars. For actively accreting supermassive BHs in AGNs, these measurements will be very important for understanding the active interactions between astrophysical BHs and their surrounding

environments, as well as the formation, evolution, and growth of their host galaxies. This is a major goal of the *International X-ray Observatory* (*IXO*) (http://ixo.gsfc.nasa. gov/) being proposed in the US, Europe, and Japan; this is also the main scientific objective of the proposed *X-ray Timing and Polarization* (*XTP*) space mission within the Diagnostics of Astro-Oscillation (DAO) Program on China's Space Science Road Map (Guo and Wu, 2009).

2. Imaging astrophysical BHs with telescopes of extremely high angular resolving power. Seeing a hole or a shadow of the size of the event horizon of a BH in any accreting BH system would remove any doubt of the existence of the BH for even the most conservative people. Practically, perhaps the supermassive BH at the center of the Milky Way is the first accreting astrophysical BH to be imaged at an angular resolution capable of resolving its event horizon scale. Sub-millimeter interferometers with very long baselines on the Earth or even in space may be able to do just this in the next decade or so. Theoretically, the best and also technically feasible angular resolution can be achieved with space x-ray interferometer telescope arrays, which can obtain direct images of the smallest x-ray-emitting region just outside the event horizon of a BH, the goal of NASA's proposed BH imager mission *MicroArcsecond X-ray Imaging Mission* (*MAXIM*) (http://maxim.gsfc.nasa.gov/). Imaging astrophysical BHs is also a goal of the Portraits of Astro-Objects (PAO) Program on China's Space Science Road Map (Guo and Wu, 2009).

These two types of observational breakthroughs, to be made with future extremely powerful telescopes in space and on the ground, would revolutionize our understanding of astrophysical BHs. With astrophysical BHs as probes of stellar, galactic, and cosmic evolution, observational and theoretical studies of astrophysical BHs in the physical Universe will play increasingly important roles in astronomy, astrophysics, and fundamental physics.

ACKNOWLEDGMENTS

I am indebted to Don York for his push, patience, and encouragement on writing this chapter; his many insightful comments and suggestions on the manuscript have clarified several points and improved readability. My student Yuan Liu made a substantial contribution to some of the research work used here (mainly on the question, Can astrophysical BHs be formed in the physical Universe?). I appreciate the discussions (mainly on the question, Will all matter in the Universe eventually fall into BHs?) made with my former student Sumin Tang. My colleague Bifang Liu provided me a literature reference and also made some interesting comments on the radiative efficiency of the advection-dominated accretion flow model. Some of our research results included in this chapter are partially supported by the National Natural Science Foundation of China under Grant Nos. 10821061, 10733010, and 0725313 and by 973 Program of China under Grant No. 2009CB824800.

REFERENCES

Agol, E. and Kamionkowski, M. (2002). X-rays from isolated black holes in the Milky Way. *Monthly Notices of the Royal Astronomical Society,* 334: 553–62.

Begelman, M.C. and Rees, M.J. (1998). *Gravity's Fatal Attraction—Black Holes in the Universe.* New York: Scientific American Library.

Bennett, D.P., Becker, A.C., Quinn, J., et al. (2002). Gravitational microlensing events due to stellar-mass black holes. *Astrophysical Journal,* 79: 639–59.

Blandford, R.D. and Payne, D.G. (1982). Hydromagnetic flows from accretion discs and the production of radio jets. *Monthly Notices of the Royal Astronomical Society,* 199: 883–903.

Blandford, R.D. and Znajek, R.L. (1977). Electromagnetic extraction of energy from Kerr black holes. *Monthly Notices of the Royal Astronomical Society,* 179: 433–56.

Broderick, A.E., Loeb, A., and Narayan, R. (2009). The event horizon of Sagittarius A*. *Astrophysical Journal*, 701: 1357–366.

Doeleman, S.S., Weintroub, J., Rogers, A.E.E., et al. (2008). Event-horizon-scale structure in the supermassive black hole candidate at the Galactic Centre. *Nature*, 455: 78–80.

Fishman, C.J. and Meegan, C.A. (1995). Gamma-ray bursts. *Annual Review of Astronomy and Astrophysics*, 33: 415–58.

Frolov, V.P. and Novikov, I.D. (1998). *Black Hole Physics: Basic Concepts and New Developments*. Dordrecht and Boston: Kluwer.

Gehrels, N., Ramirez-Ruiz, E., and Fox, D.B. (2009). Gamma-ray bursts in the *Swift* era. *Annual Review of Astronomy & Astrophysics*, 47: 567–617.

Gierliński, M., Done, C., and Page, K. (2008.) X-ray irradiation in XTE J1817-330 and the inner radius of the truncated disc in the hard state. *Monthly Notices of the Royal Astronomical Society*, 388: 753–60.

Guo, H. and Wu, J. (eds.) (2009). *Space Science & Technology in China: A Roadmap to 2050*. Beijing: Science Press, Springer.

Hao, J.F. and Zhang, S.N. (2009). Large-scale cavities surrounding microquasars inferred from evolution of their relativistic jets. *Astrophysical Journal*, 702: 1648–661.

Hawking, S.W. and Ellis, G.F.R. (1973). *The Large Scale Structure of Space-Time*. Cambridge: Cambridge University Press.

Ho, L.C. (2008). Nuclear activity in nearby galaxies. *Annual Review of Astronomy & Astrophysics*, 46: 475–539.

Klebesadel, R.W., Strong, I.B., and Olson, R.A. (1973). Observations of gamma-ray bursts of cosmic origin. *Astrophysical Journal*, 182: L85–8.

Lin, J.R., Zhang, S.N., and Li, T.P. (2004). Gamma-ray bursts are produced predominately in the early universe. *Astrophysical Journal*, 605(2): 819–22.

Liu, B.F. and Meyer-Hofmeister, E. (2001). Truncation of geometrically thin disks around massive black holes in galactic nuclei. *Astronomy and Astrophysics*, 372: 386–90.

Liu, Y. and Zhang, S.N. (2009). The exact solution for shells collapsing towards a pre-existing black hole. *Physics Letters B*, 679: 88–94.

Loar, A. (1991). Line profiles from a disk around a rotating black hole. *Astrophysical Journal*, 376: 90–4.

Luminet, J.P. (1992). *Black Holes*. Cambridge: Cambridge University Press.

Mahadevan, R. (1997). Scaling laws for advection-dominated flows: Applications to low-luminosity galactic nuclei. *Astrophysical Journal*, 477: 585.

Mao, S., Smith, M.C., Woźniak, P., et al. (2002). Optical Gravitational Lensing Experiment OGLE-1999-BUL-32: the longest ever microlensing event—evidence for a stellar mass black hole? *Monthly Notices of the Royal Astronomical Society*, 329: 349–54.

McClintock, J.E., Narayan, R., Gou, L., et al. (2009). Measuring the spins of stellar black holes: A progress report. arXiv: 0911.5408.

Meier, D.L., Koide, S., and Uchida, Y. (2001). Magnetohydrodynamic production of relativistic jets. *Science*, 291: 84–92.

Menou, K., Esin, A.A., Narayan, R., et al. (1999). Black hole and neutron star transients in quiescence. *Astrophysical Journal*, 520: 276–91.

Mézáros, P. (2009). Gamma-ray bursts: Accumulating afterglow implications, progenitor clues, and prospects. *Science*, 291: 79–84.

Mirabel, I.F. (2010). Microquasars: Summary and outlook. *Lecture Notes in Physics*, 794: 1–15.

Mirabel, I.F. and Rodriguez, L.F. (1994). A superluminal source in the Galaxy. *Nature*, 371: 46–8.

Mirabel, I.F. and Rodriguez, L.F. (2002). *Sky and Telescope*, May 2002, p. 32.

Misner, C.W., Thorne, K.S., and Wheeler, J.A. (1973). *Gravitation*. New York: W. H. Freeman.

Narayan, R. and McClintock, J.E. (2008). Advection-dominated accretion and the black hole event horizon. *New Astronomy Reviews*, 51: 733–51.

Narayan, R. and Yi, I. (1994). Advection-dominated accretion: A self-similar solution. *Astrophysical Journal*, 428: L13–16.

Nucita, A.A., De Paolis, F., Ingrosso, G., et al. (2006). An *XMM-Newton* search for x-ray emission from the microlensing event MACHO-96-BLG-5. *Astrophysical Journal*, 651: 1092–7.

Oppenheimer, J.R. and Snyder, H. (1939). On continued gravitational contraction. *Physical Review*, 56(5): 455–59.

Paczynski, B. (1986). Gravitational microlensing by the galactic halo. *Astrophysical Journal*, 304: 1–5.

Paczynski, B. (1996). Gravitational microlensing in the Local Group. *Annual Review of Astronomy and Astrophysics*, 34: 419–60.

Page, D.N. and Thorne, K.S. (1974). Disk-accretion onto a black hole: Time-averaged structure of accretion disk. *Astrophysical Journal,* 191: 499–506.

Pankaj, S.J. (2009). Do naked singularities break the rules of physics? *Scientific American Magazine*, February 2009.

Raine, D. and Thomas, E. (2005). *Black Holes—An Introduction*. London: Imperial College Press.

Remillard, R.A. and McClintock, J.E. (2006). X-ray properties of black-hole binaries. *Annual Review of Astronomy & Astrophysics,* 44(1): 49–92.

Ruffini, R. and Wheeler, J.A. (1971). Introducing the black hole. *Physics Today*, January 1971, 30–41.

Salvaterra, R., Della Valle, M., Campana, S., et al. (2009). GRB090423 at a redshift of z~8.1. *Nature,* 461: 1258–260.

Schödel, R., Ott, T., Genzel, R., et al. (2002). A star in a 15.2-year orbit around the supermassive black hole at the centre of the Milky Way. *Nature,* 419: 694–96.

Schutz, B.F. (1990). *A First Course in General Relativity*. Cambridge: Cambridge University Press.

Shahbaz, T., Dhillon, V.S., Marsh, T.R., et al. (2010). Observations of the quiescent x-ray transients GRS 1124–684 (=GU Mus) and Cen X-4 (=V822 Cen) taken with ULTRACAM on the VLT. *Monthly Notices of the Royal Astronomical Society*, 403: 2167–175.

Shakura, N.I. and Sunyaev, R.A. (1973). Black holes in binary systems: Observational appearance. *Astronomy & Astrophysics,* 24: 337–55.

Shapiro, S.L. and Teukolsky, S.A. (1983). *Black Holes, White Dwarfs and Neutron Stars*. New York: John Wiley & Sons.

Tanvir, N.R., Fox, D.B., Levan, A.J., et al. (2009). A γ-ray burst at a redshift of z ~ 8.2. *Nature,* 461: 1254–257.

Thorne, K.S. (1994). *Black Holes & Time Warps—Einstein's Outrageous Legacy*. New York: W. W. Norton & Company.

Townsend, P. (1997). Lecture notes for a "Part III" course "Black Holes" given in Cambridge. gr-qc/9707012.

Vachaspati, T. (2007). Black stars and gamma ray bursts. arXiv: 0706.1203v1.

Vachaspati, T., Stojkovic, D., and Krauss, L.M. (2007). *Physical Review D,* 76: 024005.

Weinberg, S. (1977). *Gravitation and Cosmology: Principles and Applications of the General Theory of Relativity*. New York: Basic Books.

Yu, Q. and Tremaine, S. (2002). Observational constraints on growth of massive black holes. *Monthly Notice of the Royal Astronomical Society,* 335(4): 965–76.

Yuan, F., Quataert, E., and Narayan, R. (2003). Nonthermal electrons in radiatively inefficient accretion flow models of Sagittarius A*. *Astrophysical Journal,* 598: 301–12.

Zhang, B. (2007a). Gamma-ray bursts in the Swift era. *Chinese Journal of Astronomy and Astrophysics,* 7: 1–50.

Zhang, S.N. (2007b). Similar phenomena at different scales: Black holes, the Sun, γ-ray bursts, supernovae, galaxies and galaxy clusters. *Highlights of Astronomy,* 14: 41–62.

Zhang, S.N., Cui, W., and Chen, W. (1997). Black hole spin in x-ray binaries: Observational consequences. *Astrophysical Journal,* 482(2): L155–58.

Zhang, S.N., Yu, W., and Zhang, W. (1998). Spectral transitions in Aquila X-1: Evidence for "propeller" effects. *Astrophysical Journal Letters,* 494: L71–4.

11 Ultrahigh Energy Cosmic Rays

Glennys R. Farrar

CONTENTS

INTRODUCTION

In 1911, Victor Hess discovered cosmic rays (CRs) quite unexpectedly. This was in the years after the discovery of radioactivity when all aspects of radiation were being studied. Becquerel had invented a method of detecting radiation by observing the rate at which an electroscope would discharge. Using this method, a general background radiation was discovered.

Hess was following up on the discovery of a diffuse background radiation not associated with any specific laboratory source. He understood that knowing how the level of the background radiation decreases as a function of altitude would help identify the nature of the source. To do this, Hess took an electroscope with him as he flew in a hot-air balloon—to the remarkable altitude of 4880 m (16,000 ft)! To Hess's amazement, he found that although the flux of radiation decreased up to about 1 km, above that it started to *increase* with altitude. Figures 11.1 and 11.2 show Hess setting out, evidently causing great interest among the locals. The fact that the flux of radiation increases with altitude was an indication that a component is "cosmic" in origin—hence the term "cosmic rays."

CRs are now known to consist primarily of electrons, protons, and atomic nuclei, along with a sprinkling of their antiparticles. Presumably, they also have a large component of neutrinos, but we can detect only a very few of these because of their extremely weak interactions, so less is known about them. When CRs hit the Earth, the low-energy charged rays are deflected or captured by the Earth's magnetic field. The higher-energy CRs interact with nuclei in the atmosphere, losing energy in such collisions and producing secondary particles. Most CRs are absorbed in the atmosphere and never reach ground. Those that do reach ground are primarily muons. Muons are produced when charged pions—the most common secondaries of the collisions—decay ($\pi^\pm \to \mu^\pm \nu$). Another common type of secondary of the collisions is the very short-lived π^0, which decays almost instantaneously into photons ($\pi^0 \to \gamma\gamma$). These photons then interact with nuclei and produce electrons and positrons, some of which eventually reach ground level. At sea level, the total flux of CRs is of the order of 100/m^2/sec (not counting neutrinos), roughly half of which are muons and half electrons.

When the primary particle in a collision has a very high energy, many secondaries are produced, and these, in turn, initiate collisions. Depending on the energy of the collision, other "hadrons"* besides pions are also produced, which can also be seen in high-energy particle experiments at

---

* "Hadron" is the generic term for particles made of quarks and/or antiquarks—for instance, a π^+ is a bound state of "up" (charge = +2/3) and "anti-down" (charge = +1/3). Hadrons are the main products of particle collisions.

FIGURE 11.1   Hess and his balloon (1912). (From *Early History of Cosmic Ray Studies* by Y. Sekido and H. Elliot, 1985, Dordrecht: Reidel. With permission.)

FIGURE 11.2   Hess balloon launch (1911–12). (Courtesy of the Museum of Military History, Vienna.) (From *Early History of Cosmic Ray Studies* by Y. Sekido and H. Elliot, 1985, Dordrecht: Reidel. With permission.)

accelerators. The secondaries mostly interact again or may decay. This process results in a cascade or "extensive air shower." For the highest-energy CRs, known as ultrahigh energy cosmic rays (UHECRs), the number of particles in the cascade reaches 10^{10} (10 billion) at its maximum size. As the shower continues to develop, the energies of the particles become low enough that they begin to be absorbed, and the shower attenuates at greater depth in the atmosphere.

The rarity of CRs increases very rapidly with their energy, primarily because of the difficulty of accelerating particles to high energies, but also—at the highest energies—because of energy losses that they experience as they propagate. The decrease is described quantitatively via the *spectrum* of CRs at the surface of the Earth, as shown in Figure 11.3. This is a "log-log" plot showing the log-base-10 of the flux vs. the log-base-10 of the energy.* It covers a remarkable range. The observed energies vary by a

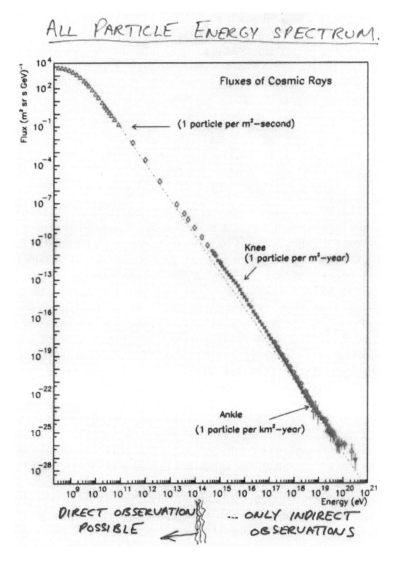

FIGURE 11.3   The spectrum of cosmic rays (CRs), courtesy of S. Swordy, a pioneer in CR physics who died prematurely in 2010. The handwritten notes are from a slide he used in presenting the figure in talks.

---

* Energies are given in electron volts (eV); this is the energy an electron gets when accelerated by a 1 V potential difference. The beam energy in the LHC is expected to reach 7 TeV or 7×10^{12} eV.

factor of a trillion—10^{12}—and the flux varies by the unimaginably large factor of 10^{32}. At the highest energies, the flux is one particle per square kilometer *per century*! The highest-energy CRs have an energy 10^8 (100 million) times higher than in the highest energy accelerator created by humankind, corresponding to a "center of mass" collision energy up to hundreds of times higher than at the Large Hadron Collider (LHC).

Here are some "UHECR trivia":

1. The highest-energy CR ever recorded had an energy of 320 EeV (1 EeV is 10^{18} eV)—more than the energy carried by a golf ball hit by a pro in a single elementary particle!
2. The speed of such a UHECR is 99.99999999999999999999% of the speed of light.
3. Relativistic time dilation, predicted by Einstein's theory of special relativity, says that a neutron traveling at that speed—whose lifetime is about 15 min in the laboratory—would take about a million years to decay.

Low-energy CRs, $E <$ 10 GeV (10^{10} eV), are produced mainly by the Sun, and medium-energy CRs are thought to be produced mainly in supernovae in our Galaxy. It is not known how the highest-energy CRs are produced—more on that later—but it is quite certain that they are not produced in our Galaxy. Charged CRs are deflected by magnetic fields and travel approximately along spiral paths with a "Larmor radius" of

$$R_L = 1\,\text{kpc}\,\frac{E_{\text{EeV}}}{ZB_{\mu\text{G}}}. \tag{11.1}$$

In this formula, E_{EeV} is the energy of the CR in exa electron volts (EeV), Z is its charge in units of the proton's charge, and $B_{\mu\text{G}}$ is the field transverse to the direction of propagation in microgauss, a typical value for our Galaxy. One kiloparsec is 3,000 lt-yr—several times the thickness of the Galactic disk of stars forming the Milky Way that we see in the sky. At lower energies, CRs produced in the Galaxy are trapped by the magnetic fields, but by 1 EeV they stream out. At the highest energies, the Galactic magnetic field (GMF) deflects UHECRs only slightly, thus even if there were a sufficiently powerful accelerator to produce them within our Galaxy, they would, with very high probability, escape without hitting our detector on Earth. Presumably, UHECRs are produced by some exotic and very powerful process in rare and, therefore, generally distant galaxies.

Figure 11.4 shows the spectrum of CRs (Hörandel, 2008) as measured at different facilities. In contrast to Figure 11.3, it has been multiplied by E^3 so it is easier to see the "fine structure" in the steeply falling spectrum. For energies up to about 100 TeV (10 times the LHC energy), the flux is big enough that it can be measured with particle physics detectors small enough to be sent up on satellites, allowing a direct measurement of the composition. At higher energies, up to about 0.1 EeV, the flux is such that CRs can be detected using a compact, well-instrumented array of particle physics detectors on the ground, which allows the composition to be constrained with the help of models of the air shower. KASCADE (KArlsruhe Shower Core and Array DEtector) is the premier instrument of this type. Curves representing the contributions of different nuclei to the total spectrum are indicated. For a given energy, protons have the largest Larmor radius—26 times bigger than iron, for which $Z = 26$. Thus, the protons are the first species of CRs to escape from the Galactic magnetic confinement as the energy increases; at higher energies, successively heavier nuclei are lost. It remains something of a mystery why the CR spectrum continues without a break, even above the energy where Fe drops off, and exactly where the transition to extra-Galactic sources occurs.

Many details of Galactic CR production remain to be understood, but the most exciting topics today concern the highest-energy CRs. What type of particles are they? Where do they come from? How are they accelerated?

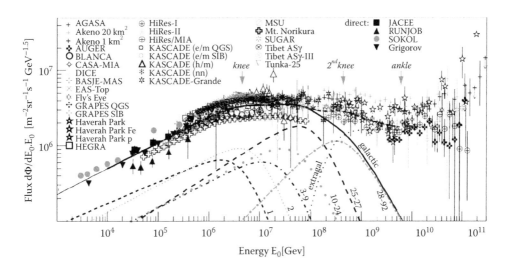

FIGURE 11.4 The spectrum of cosmic rays, multiplied by E^3, with the inferred contributions of different nuclear species indicated. (From Hörandel, J.R., *Advances in Space Research*, 41, 442–63, 2008. With permission.)

COSMIC-RAY TELESCOPES

Figure 11.5 shows a schematic representation of the Pierre Auger Observatory in the Argentinian pampas, overlaid on a map of the region (Abreu et al., [The Pierre Auger Collaboration] 2010).* "Auger South" consists of 1,600 instrumented water tanks spread out over 3,000 km^2 in a triangular pattern with 1.5 km spacing. In addition, four air fluorescence detector (FD) telescopes that operate on clear moonless nights, surveying the horizon over the surface detector (SD) array are shown. The extensive air shower that is produced when a UHECR hits the upper atmosphere can be broken down into two parts: a core and a halo. Roughly speaking, the FD measures the core of the shower and the SD measures the halo. Figure 11.6 shows one of the water tanks of the Auger SD in the foreground, with the sheds housing one of the fluorescence telescopes in the distance. Each FD actually consists of seven separate Schmidt telescopes, each with an 11 m^2 mirror imaged onto an array of 440 photomultiplier tubes (PMTs). Figure 11.7 shows an Auger fluorescence telescope. To illustrate other CR observatories, Figure 11.8 is a historic photograph of the construction of the very first air fluorescence telescope at Cornell, and Figure 11.9 shows the ARGO detector (ARGO-YBJ, Astroparticle-physics Research at Ground-based Observatory Yangbajing) in Tibet, which provides detailed particle identification in an immense indoor detector array, complemented with a more extensive outdoor array, shown in Figure 11.10.

Of the events observed by Auger, 90% are seen with the SD, because only about 10% of the time is it dark and clear enough for the FD to be used. Even this small fraction of "hybrid" events—those seen by both FD and SD—provides an extremely powerful tool for understanding the atmospheric showers produced by UHECRs. Very occasionally, all four FD telescopes see the same shower; the first example of such an event is shown in Figure 11.11. In this case, 15 of the SD tanks saw a signal. The color coding represents the arrival time of the signal at each detector, which is used to determine the precise direction of arrival of the event. Angular resolution with the SD alone is ≈ 1 degree, but for hybrid events it is much better. Figure 11.12 shows the signal readouts for a typical hybrid event, with SD above and FD below.

---

* The Pierre Auger Collaboration consists of several hundred scientists from 18 countries, including faculty, postdoctoral and research scientists, graduate students, and technical staff.

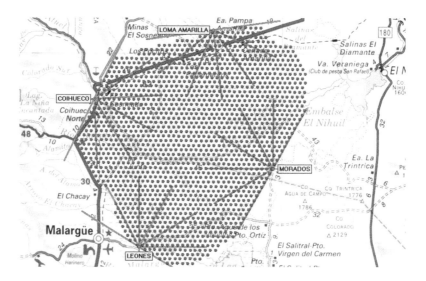

FIGURE 11.5   The layout of Auger South; completed water tanks colored red, with fluorescence telescopes shown as green rays. (Courtesy of the Pierre Auger Observatory.)

FIGURE 11.6   View of Auger South, with a surface detector tank in the foreground and a fluorescence telescope in the background. (Courtesy of the Pierre Auger Observatory.)

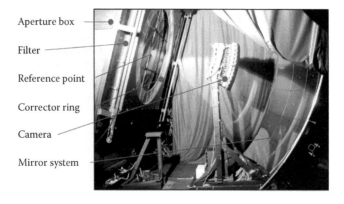

FIGURE 11.7   One of the Auger fluorescence telescopes. (Courtesy of the Pierre Auger Observatory.)

FIGURE 11.8   The first air fluorescence detector was built in 1964 by S. Ozaki, G. Tanahashi, A. Bunner, and E. Jenkins under the direction of K. Greisen at Cornell; this is a photograph of an identical one that they built in 1966. (Courtesy of Ed Jenkins.)

IDENTIFYING THE SOURCES OF UHECRS

It is intuitively clear that acceleration of particles to the energies of UHECRs must take a very special environment. The following candidates have been proposed:

- A cataclysmic event, such as a massive star collapsing into a black hole (BH), responsible for gamma-ray bursts (GRBs)
- Jets in quasars, BL Lacertae objects (BL Lacs), or powerful radio galaxies

FIGURE 11.9   The ARGO detector at Yangbajing Observatory, Tibet.

FIGURE 11.10   The Yangbajing Observatory, Tibet.

- Gradual acceleration in large-scale magnetic shocks taking hundreds of millions of years
- New physics, e.g., the decay of some invisible, super-heavy particle created in the Big Bang

Despite the fact that the sources must be remarkable, discovering what they are may prove very difficult. In traditional astronomy, the photons observed point to their source, and the redshift of spectral lines combined with the known cosmological expansion history gives distance information on a given source candidate. The resolution of telescopes is excellent and improving, with

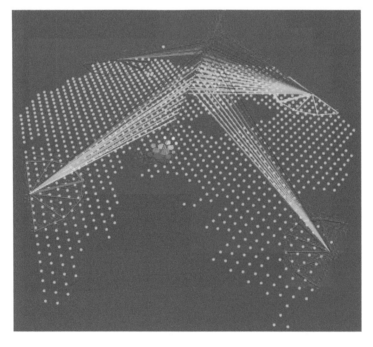

FIGURE 11.11   The first Auger event seen in all four fluorescence telescopes (May 20, 2007), $E \sim 10^{19}$ eV. Fifteen surface detector tanks also fired. (Courtesy of the Pierre Auger Observatory.)

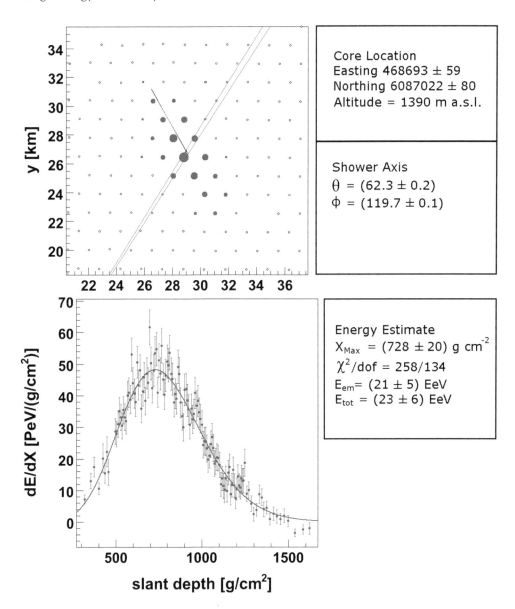

FIGURE 11.12 Readout of a hybrid event. The upper panel shows the signal in the surface detectors that fired, with the area of the dots proportional to the signal strength, and the coordinates indicating the position of the tanks in the array; the arrow indicates the arrival direction, determined by the time the different tanks fired. The lower panel shows the energy deposited per unit length (roughly proportional to the number of electrons and positrons in the shower) along the shower axis, inferred from the signal in the fluorescence detector. The reconstructed properties of the primary CR are shown in the boxes to the right. (Courtesy of the Pierre Auger Observatory.)

arcsecond or better accuracy routine at many wavelengths. Thus, source candidates can be located in three dimensions, at least roughly.

UHECR astronomy is more difficult. The first problem is that the arrival direction of the observed CR is more difficult to translate into the direction of the source. The resolution with which the UHECR's arrival direction is measured is only ≈ 0.5 to 1 degree. Furthermore, despite their extremely high energy, charged CRs are subject to deflection in the GMF and extra-Galactic magnetic fields (EGMFs) when traveling over large distances. Equation 11.1 gives the Larmor radius.

The local magnitude of the GMF is several microgauss, but the extent of the GMF and its degree of coherence are poorly constrained. In general, when the field is turbulent, the deflections tend to average out such that if a UHECR travels a distance $D = 100\,D_{100}$ Mpc through a turbulent magnetic field of typical strength $B = B_{nG}\,10^{-9}$ G, and correlation length λ_{Mpc} Mpc in the case of the EGMF, its typical angular deflection is

$$\delta\theta \approx 2.5°ZE_{20}^{-1}\sqrt{D_{100}B_{nG}^2\lambda_{Mpc}}, \tag{11.2}$$

or typically a few degrees for a proton.

As we will see later, in at least one direction the net magnetic deflection and dispersion seems to be $\lesssim 1$ degree, but more data are needed to tell how representative this is. Simulations of structure formation have been used to estimate the spatial distribution of the EGMF, formed by shocks when matter falls onto overdense regions. Figure 11.13 shows the map of predicted deflections of 40 EeV CR protons, arriving from a sphere of radius 100 Mpc, in one such simulation (Dolag et al., 2005); another simulation finds a roughly similar structure, but larger deflections (Sigl et al., 2003).

The other complication in dealing with UHECRs is estimating the distance to the source. Fortunately, for CRs above about 50 or 60 EeV, some distance information is embedded in its energy, at least in a probabilistic sense, thanks to the "GZK effect," named after Greisen, Zatsepin, and Kuzmin, who, shortly after the discovery of cosmic microwave background radiation (CMBR), realized that CMBR places an effective upper limit on the distance that UHECRs can propagate and retain their energy. Figure 11.14 shows the energy attenuation length for a proton as a function of its energy in two different models for cosmic background radiation (CBR). The CBR is dominated by CMBR, which is very well measured, but CBR is less well known at other wavelengths. Below about $10^{19.6}$ eV, the dominant energy loss for protons is due to reactions in which e^+e^- pairs are created, such as $p + \gamma \rightarrow p + e^+ + e^-$, while at higher energies the losses are predominantly due to pion production: $p + \gamma \rightarrow p + \pi^0$ or $\rightarrow n + \pi^+$.

The way to interpret Figure 11.14 can be illustrated by considering a 200 EeV proton ($\log E = 20.3$). It has an energy attenuation length of about 30 Mpc, which means it loses energy approximately as $E(x) = E(0)e^{(-x/30\,Mpc)}$.

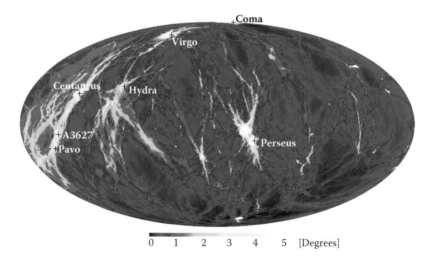

FIGURE 11.13 Model extra-Galactic magnetic deflections. The names of prominent clusters of galaxies appear on the figure. (From Dolag, K., Grasso, D., Sprigel, V. et al., *Journal of Cosmology and Astroparticle Physics*, 0501, 009, 1–37, 2005. With permission.)

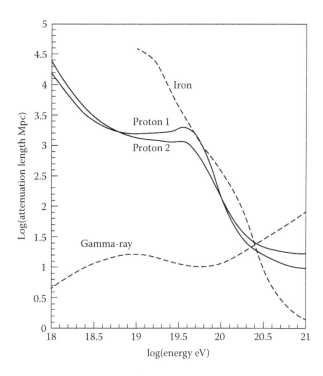

FIGURE 11.14 Energy attenuation length for protons, iron, and photons as a function of energy. (From Nagano, M. and Watson, A., *Reviews of Modern Physics*, 72, 689–732, 2000. With permission.)

Thus, in a distance of 3 Mpc, it would lose about 10% of its energy. The fractional energy loss in each pion production interaction is about 10%, but the process is stochastic, so a given CR may lose more or less than average. At lower energies, below the threshold for pion production, the main energy loss mechanism is e^+e^- production, for which the fractional energy loss in each interaction is much smaller, resulting in a smoother process. The energy at which losses become rapidly large is called the "GZK energy," and this is often used to define the meaning of ultrahigh energy in the CR context. It begins to turn on at about 50 EeV and then rapidly becomes a very strong effect at higher energies.

Nuclei are also subject to energy losses through photodissociation, e.g., $He + \gamma \rightarrow dd$ and $O + \gamma \rightarrow \alpha + C$. For our purposes, it is enough to know that the intermediate-mass nuclei are much more fragile and thus break down relatively rapidly to protons, while Fe is rather stable and its "horizon" is similar to that for protons. For simplicity, sometimes these energy-loss processes are collectively referred to as "GZK energy losses" below.

The steepening of the spectrum above the GZK energy has been observed by both the High-Resolution Fly's Eye (HiRes) experiment (Abbasi et al., 2004) and Auger (Abraham et al., 2008a). Figure 11.15 shows the spectrum measured by Auger, which is consistent with the steepening expected for protons and also for a mixed composition (Allard et al., 2008). However, this does not prove that the GZK effect has been observed; we cannot be sure that the spectrum is not simply cutting off because of reaching the limit of accelerating power of the sources. Another result from the Auger collaboration, if confirmed with more data, will establish that the steepening in the spectrum is due to the GZK effect, namely, discovery of the "GZK horizon." Because of the GZK effect, it is very unlikely that the source of a CR of 100 EeV is farther than a few tens of megaparsecs away. Thus, for any UHECR data set with a given threshold energy, one can define the GZK horizon, commonly taken to be the distance within which, on average, 90% of the events originate. The GZK horizon for protons with a threshold energy of 60 EeV is about 200 Mpc (Harari et al., 2006).

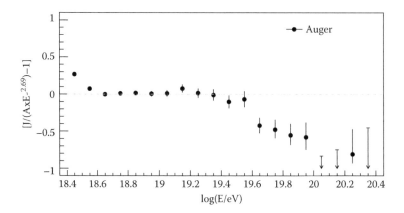

FIGURE 11.15 Auger spectrum relative to an $E^{-2.69}$ power law, showing a cutoff above $10^{19.5}$ eV. (From Abraham, J. et al. [The Pierre Auger Collaboration], *Physical Review Letters*, 101, 061101, 1–7, 2008.)

To establish the GZK effect, the Auger collaboration examined (Abraham et al., 2007) the correlations between the arrival directions of UHECRs and a catalog of nearby active and emission-line galaxies compiled by Véron-Cetty and Véron (VCV) (Véron-Cetty and Véron, 2006). Because the GMF and EGMF are poorly known, the degree of magnetic deflection could not be predicted. Moreover, the absolute energy calibration of the detector is still quite rough, so the mapping between the energy attributed to the CR and its GZK horizon is only approximate. To deal with the uncertainties surrounding the typical angular separation between source and UHECR, and at what nominal energy the correlation would be maximized, the Auger team devised a "scan" analysis in which the threshold energy E_{min}, maximum redshift of the VCV galaxy z_{max}, and maximum angular separation ψ were varied. For each choice, the number of correlations were counted and a "probability measure" P was computed. (P is just the cumulative binomial probability of finding the observed number of "hits" in the exposure-weighted total area covered by the VCV galaxies for the given angular separation radius.) Using UHECR data above 40 EeV through May 27, 2006 and VCV galaxies out to $z = 0.024$, the most significant correlation was found for $\psi = 3.2$ degrees, $z_{max} = 0.018$ (about 75 Mpc), and $E_{min} = 57$ EeV, for which 11 of 14 events correlated with VCV galaxies. (A correlation is simply defined as a UHECR falling within the specified angular separation of a VCV galaxy.) Then, independent data taken from June 1, 2006 to August 30, 2007 were used to do an *a priori* correlation study with the same energy threshold, maximum redshift, and angular separation. Out of 13 new events above 57 EeV, nine were found to correlate, with approximately two expected by chance. Taking into consideration the way the experiment was designed, this corresponds to a chance probability of <1% (Abraham et al., 2007). In the full data set with 27 events, approximately five to six chance correlations would be expected if the arrival direction of the UHECRs were isotropic because VCV covers ~21% of the sky at a 3.2-degree radius, while 20 were observed. Restricting to $|b| > 10$ degrees; where the VCV catalog is more complete, there were 22 UHECRs of which 19 were correlated. Figure 11.16 shows the arrival directions of the 27 Auger events above 56 EeV (open circles with a 3.1 degree radius) and the VCV galaxies nearer than 75 Mpc. The map is in Galactic coordinates; the shading shows the Auger exposure with each value of grayscale covering an area with the same total exposure.

A correlation between highest-energy CRs and VCV galaxies shows that:

1. UHECRs come from *outside* our Galaxy (otherwise, they would not reflect the distribution of extra-Galactic matter).
2. There is a GZK horizon (otherwise, UHECRs would come from such a large volume that the projected Galaxy distribution is uniform and there would be no correlation with nearby galaxies).
3. Some UHECRs come from active galaxies in the VCV catalog.

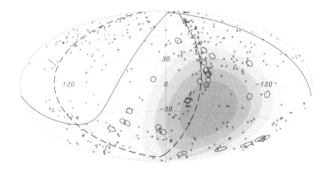

FIGURE 11.16 Sky map of UHECR arrival directions (open circles of 3.1 degree radius) and Véron-Cetty and Véron galaxies within 75 Mpc. (From Abraham, J. et al. [The Pierre Auger Collaboration], *Science*, 318, 938–43, 2007.)

The latter statement goes beyond what Auger claimed, given that the Auger analysis could not exclude that VCV galaxies are merely "tracers" of the sources of UHECRs. Although the VCV catalog is not pure, complete, or homogeneously defined (Zaw et al., 2009), the galaxies it contains are mostly active galactic nuclei (AGNs)—point-like sources of emission due to accretion onto a supermassive BH. Powerful AGNs are among the prime candidates for producing UHECRs, so it is very interesting to know whether the UHECR sources are galaxies in the VCV catalog. However, galaxies of all types are clustered rather than being isotropically distributed because of the large-scale structure of the Universe. Thus, UHECRs could be produced by other types of galaxies and correlated with VCV galaxies only indirectly, given their clustering with other galaxies, including those that host UHECR sources.

By comparing the degree of correlation between UHECRs and VCV galaxies, and between UHECRs and randomly chosen galaxies, it was possible to establish that the correlation with VCV galaxies was much stronger than with randomly chosen galaxies (Zaw et al., 2011). Specifically, 1,000 mock source catalogs were created by randomly drawing the same number of galaxies as in VCV from the 2MASS (Two Micron All Sky Survey; Skrutskie et al., 2006) Redshift Survey (2MRS) out to $z = 0.024$. Then, the scan analysis was performed using these as the source catalog instead of the VCV catalog. For each scan, the probability measure P was calculated. Figure 11.17 is Figure 4 from Zaw et al. (2011). It shows a histogram of the values of $\log_{10}(P)$ obtained using the mock source catalogs from 2MRS (open) and using 1,000 isotropic catalogs (hatched), as well as the P value for the full published data set (Abreu et al., 2010) using VCV. (To avoid a bias from the data prior to the experiment, only the "postprescription" UHECRs are used.) Using the original Auger data, released in Abraham et al. (2008a), gave the red line and black histogram. One sees that it is indeed more likely to get the observed P value by chance from random galaxies than from isotropically distributed sources, but that a correlation as strong as that seen with VCV galaxies occurs only 0.7% of the time. This shows that sources of UHECRs are concentrated in the VCV catalog compared with random galaxies and that at least some UHECRs come from AGNs.

UHECR ACCELERATION AND THEORETICAL CONSTRAINTS

The leading candidates for UHECR accelerators are the powerful jets that are produced by intensely accreting systems or the shocks created when such jets hit the surrounding intergalactic material. Examples of such jets are found in AGNs or GRBs. Figure 11.18 shows the jet and radio lobes of a powerful quasar. (The terms "AGN" and "quasar" are sometimes used interchangeably, but here "quasar" will be reserved for the most powerful examples that have luminosities in excess of 10^{45} erg/sec. The correlated AGNs in the VCV catalog have luminosities as low as 10^{42} erg/sec and as high as 10^{45} erg/sec.)

Fermi originated the idea that CRs could be accelerated by bouncing them back and forth across a shock, like a Ping-Pong ball caught between converging walls. Figure 11.19 shows a schematic

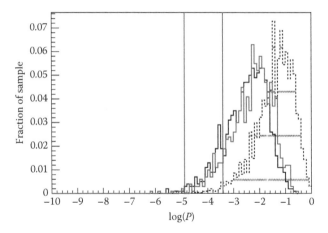

FIGURE 11.17 Vertical lines are the P values obtained from correlating the original postprescription events with $|b| > 10$ degrees (red) and the entire published postprescription data set (blue), with galaxies in the Véron-Cetty and Véron catalog as detailed in Zaw et al. (2011). The black (blue) histograms are the corresponding distribution of P values correlating the same sets of events with subsamples of 2MRS galaxies identical in number to VCV. The hatched histogram displays the P values corresponding to the black histogram, but using isotropic sources. The P-value distributions for the two UHECR data sets cannot be compared directly because of the different number of UHECRs in the two cases. The isotropic distribution is qualitatively similar for both cases, thus it is not shown for the enlarged data set for visual clarity.

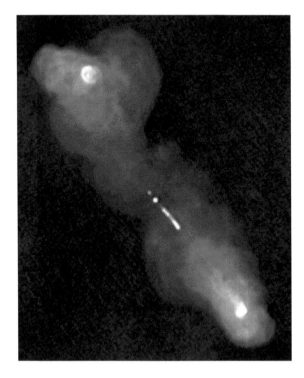

FIGURE 11.18 Jets and lobes of a powerful radio galaxy, 3C219; the lobes extend hundreds of kiloparsecs (nearly a million light-years). (Courtesy of the National Radio Astronomy Observatory.)

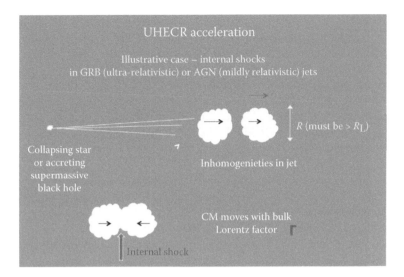

FIGURE 11.19 Schematic diagram of UHECR acceleration in internal shocks in jets.

drawing of UHECR acceleration in shocks in the relativistic jet of a quasar (mildly relativistic—Lorentz factors of a few) or a GRB (ultra-relativistic—Lorentz factors in the hundreds). In order for this mechanism to work, the magnetic fields in the material on either side of the shock must be strong enough to turn the CR around. That means that the size of the accelerating region must be larger than the Larmor radius of the CR. Using Equation (11.1) for the Larmor radius and denoting the characteristic size of the accelerating region by R, this condition implies that

$$RB \gtrsim 3 \times 10^{17} (E_{20}/Z) \text{ G cm}, \tag{11.3}$$

where E_{20} is the energy in units of 10^{20} eV. But the product RB determines the (isotropic equivalent) Poynting luminosity of the magnetic field:

$$L \sim \frac{1}{6} c \Gamma^4 B^2 R^2 \gtrsim 10^{45} \Gamma^2 (E_{20}/Z)^2 \text{ erg/sec}. \tag{11.4}$$

Evidence from the Haverah Park and HiRes experiments indicates that extra-Galactic CRs become increasingly proton dominated as the energy increases. If that extends to the highest energies, then Equation 11.4 requires (Farrar and Gruzinov, 2009) that the sources of UHECRs have luminosities in excess of 10^{45} erg/sec—luminosities achieved only in GRBs or super-Eddington accretion onto a supermassive BH. Only one or two of the AGNs correlated with UHECRs have such high luminosities, so the requirement of Equation 11.4 presents a conundrum.

In addition to the minimum luminosity constraint on UHECR accelerators, three other important constraints on the sources of UHECRs must be noted:

1. The source density cannot be too low because, if it were, the high-energy UHECRs would be clustered, with multiple events coming from a few individual sources. The minimum number of sources to account for the Auger events above 57 EeV is about 70 (Abraham et al., 2008b), which translates into a minimum source density of 3×10^{-5} Mpc^{-3} given the z_{max} found in the scan correlation analysis (Farrar and Gruzinov, 2009).
2. The energy-injection rate—the energy-weighted production rate of UHECRs per unit volume—can be inferred from the observed flux of UHECRs and the GZK energy losses.

It is rather poorly constrained because GZK losses are very sensitive to the energy of the CR, even though the energy calibration of the data has uncertainties of order $\pm30\%$, and also because it depends on the shape of the spectrum at the source, although estimates give 10^{44-45} erg/Mpc^3/year (Waxman, 1995; Berezinsky, 2008).

3. The observed correlation between UHECRs and the VCV catalog indicates that typical deflections of UHECRs above 57 EeV are a few degrees. This generates an estimate for the time delay between the actual CRs and a neutral particle that could travel in a straight line: $\Delta T \approx \frac{1}{2}\delta\theta^2 T$, where T is the straight-line travel time. From (2) and using the z_{max} to estimate $T \approx 200$MYR, we infer that arrival times of UHECRs from a burst are spread out over $\Delta T \approx 10^5$ year.

But these constraints make it unlikely that most UHECRs are produced by the two most popular candidates for UHECR acceleration: powerful quasars and GRBs. They are just too rare! Only a few quasars have luminosity above 10^{45} erg/sec within 100 Mpc, nowhere near the >70 sources implied by the Auger observations (Farrar and Gruzinov, 2009). Moreover, of the 14 VCV galaxies that are confirmed AGNs that correlate with the highest-energy Auger events, only two have luminosities near 10^{45} erg/sec—rather, the luminosities of most of the correlated AGNs are in the 10^{42-43} erg/sec range. Although individual GRBs meet the luminosity requirement of Equation 11.4, GRBs are very rare in the nearby universe, so to explain the observed flux of UHECRs, most of the energy output of individual GRBs would have to be in UHECRs rather than gamma rays, which is difficult to imagine theoretically.

An exciting prospect is that a new type of phenomenon is responsible for UHECRs—giant flares of AGNs emanating from stellar tidal disruption or disk instabilities (Farrar and Gruzinov, 2009). About once every 10,000–100,000 years, on average, a star is predicted to get too near the supermassive BH at the center of its galaxy, whereupon it gets torn up by tidal forces and then swallowed up (Magorrian and Tremaine, 1999). Flares attributable to such events have been observed in galaxies without AGNs (Gezari et al., 2008), consistent with the predicted properties. Whether such events would produce a sufficiently high accretion rate to produce jets and thereby accelerate UHECRs is not clear. However, if the supermassive BH has even a thin preexisting accretion disk, the rate at which material would be accreted should easily exceed the 10^{45} erg/sec required for UHECR acceleration. Putting in numbers for the rate and power of such giant AGN flares, assuming that 1% of the available energy goes into producing the jet and the required Poynting flux and that the energy in CRs is comparable to that in photons, predicts an energy-injection rate and effective-source density in excellent agreement with observations (Farrar and Gruzinov, 2009)! This naturally accounts for why there is a correlation between UHECR arrival directions and AGNs, yet most of the correlated AGNs have much lower luminosities than required for UHECR acceleration: we are seeing the AGN in a quiescent state, $\approx10^5$ years after the giant flare that accelerated UHECRs.

During the flare, the galaxy becomes a powerful quasar for a period of 10 days to several months (Farrar and Gruzinov, 2009). Remarkably, these flares would have not yet been observed, although an amount of order 10 is predicted to have been recorded among the multiple-imaged Sloan Digital Sky Survey (SDSS) galaxies. Sjoert van Velzen, the author's graduate student at New York University, has recently searched the archives for such events and found two unambiguous examples (van Velzen et al., 2010). Another promising vehicle for searching for such flares is the *Fermi Gamma-ray Space Telescope*, formerly referred to as the *Gamma-ray Large Area Space Telescope (GLAST)*. The predicted detection rate is of the order of 10 per year, assuming that the luminosity in gamma rays is the same as the luminosity in UHECRs. Thus, the giant flare scenario should be tested within a few years, unless unexpectedly the luminosity of AGN bursts in photons at optical wavelengths or *Fermi-GLAST* energies are low compared with the UHECR luminosity.

BURSTING SOURCES

How can one identify the source of a UHECR if it is not steady—e.g., if the source is a GRB or giant AGN flare? In this case, the source in its normal state may be too dim to be seen, and the CRs arrive

10,000 or 100,000 years after the photon burst, so it is not possible to observe both the photon burst and the associated UHECR burst. Statistical analyses showing that UHECRs are more strongly correlated with special classes of galaxies that, e.g., are particularly likely to host GRBs or giant flares, would certainly be a smoking gun pointing to the source. And the correlation between UHECRs and low-power AGNs, which may have hosted a giant flare, is circumstantial evidence in favor of the giant AGN flare scenario. However, there is a direct way to determine whether the source of a UHECR is bursting or steady. To do this, enough UHECRs from a single source are required to measure the spectrum of events from that source. With even six events from each of several individual sources, it will be possible to decide whether the events were produced in a burst or by a steady source. This is because of the magnetic deflections that the UHECRs experience, which are inversely proportional to their energies. The highest-energy events generally take a shorter path and arrive early, whereas lower-energy events experience more deflection and arrive later. Thus, at any moment, the spectrum of a bursting source is peaked, while that of a continuous source replicates the spectrum at the source (apart from GZK distortions) (Waxman and Miralda-Escude, 1996). Figure 11.20 shows the spectrum of a bursting source, without taking into account GZK distortions (Farrar, 2008).

So far, nature has presented us with just one cluster of events from a single source with enough events to give a meaningful spectrum, the Ursa Major UHECR cluster (Farrar, 2008). In that case, four or possibly five events come from a single source (with the chance probability of seeing such strong clustering being on the order of 0.3%). The events were seen by the Akeno Giant Air Shower Array (AGASA) (three events, with energies 52, 55, and 77 EeV) and HiRes (two events, with energies $\approx$15 and 38 EeV). The angular separation is nearly consistent with there being no magnetic smearing. Luckily, the region was surveyed by SDSS. An analysis of the galaxy distribution in that direction reveals a major void in the foreground followed by a filament just within the GZK distance for the given energies. The spectrum is consistent with a bursting source, whose power is consistent with that expected in the GAF scenario. A *Chandra* observation reveals an AGN with luminosity of about 10^{42} erg/sec, at a distance of 200 Mpc, and within the expected angular domain. Figure 11.21 shows the UHECRs of the Ursa Major cluster with gray disks containing 68% of the arrival direction probability given the event's angular resolution. The size of the asterisks is proportional to the event energy. Dots represent galaxies in SDSS, with galaxy clusters colored according to their redshifts, as explained in the caption.

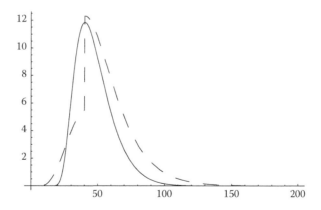

FIGURE 11.20 Theoretical spectrum of events from the Ursa Major Cluster, if it is a bursting source (solid), weighted with the AGASA plus HiRes acceptance (dashed); note that GZK losses are not included here. The horizontal axis is the energy in exa electron volts; the vertical axis is arbitrary. (From Farrar, G.R. [2008]. Evidence that a cluster of UHECRs was produced by a burst or flare. Proceedings of the International Cosmic Ray Conference 2007, Mérida, Mexico; see arXiv:0708.1617v1 [astro-ph], 2007 [Aug. 12]: 1–4.)

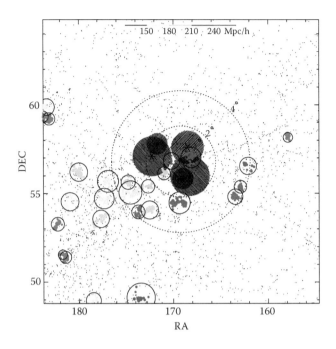

FIGURE 11.21   Five events in the Ursa Major UHECR cluster shown as disks whose size reflects the angular resolution of the detector; the asterisk is in proportion to its energy, plotted with SDSS galaxies (black dots) with galaxy clusters identified (colored dots, color-coded according to redshift). (From Farrar, G.R., Berlind, A.A., and Hogg. D.W., *Astrophysical Journal*, 642, L89–94, 2006. With permission.)

A prime objective of the proposed Auger North Observatory in Colorado is to measure the spectra of individual sources. It is designed to have 3,000 tanks covering 10,000 km^2 and about a seven-fold larger event rate than Auger South. If approved, it could be operational in 2013.

OPEN QUESTIONS AND CONCLUSIONS

This is an extremely exciting time for UHECR physics and astrophysics. The key issues are the following:

1. What are UHECRs? At present, the composition is uncertain and the evidence is conflicting, with some indications of a proton composition and other data suggesting that, at the highest energies, heavier nuclei become more important.
2. What are the sources of UHECRs? Are they bursting or continuous? (The spectra will tell us!) Can UHECRs be correlated with well-defined astrophysical catalogs? (Homogeneous, all-sky catalogs of relevant candidate types are urgently needed!)
3. We must understand UHE particle physics better! Discrepancies of ~50% exist between simulations of atmospheric showers and observations—i.e., fixing the energy of a hybrid event using the fluorescence profile leads to a predicted SD signal about 50% lower than observed. Is this an indication of new physics that plays an essential role in UHE showers? (Recall that the center-of-mass energy of the primary collisions is more than a factor of 100 larger than at the LHC.)
4. We can use UHECRs to measure GMFs and EGMFs. This avoids reliance on poorly known electron and relativistic electron densities needed to interpret Faraday rotation measures and polarized synchrotron emission data. The results will be important for the *Wilkinson Microwave Anisotropy Probe* (*WMAP*) and *Planck's* subtraction of the polarized foreground to access fundamental information about the Big Bang.

In the next few years, Auger South can probably determine the composition of UHECRs and decide whether new UHE particle physics is needed to explain their showers. Considerable progress can be anticipated in identifying the sources of at least some of the UHECRs. Auger North should have about an order of magnitude larger aperture and should be able to access the spectra of individual sources, tremendously increasing the range of questions that can be addressed. In particular, with much larger UHECR data sets it should be possible to determine whether the sources are bursting or continuous and to measure the magnetic deflections. *Fermi-GLAST*, SDSS, Quasar Equatorial Survey Team (QUEST), and other photon telescopes will complement the search for the sources of UHECRs by looking for photon counterparts of UHECR bursts and providing well-characterized and uniform all-sky catalogs of AGNs and other source candidates needed to do quantitative studies of sources. Yet another exciting frontier may be around the corner—the age of neutrino astrophysics—but that's a story for another occasion.

REFERENCES

Abbasi, R.U., Abu-Zayyad, T., Amann, J. F. et al. (The Pierre Auger Collaboration) (2004). Measurement of the flux of ultrahigh energy cosmic rays from monocular observations by the High Resolution Fly's Eye experiment. *Physical Review Letters, 92*, 151101: 1–4.

Abraham, J., Abreu, P., Aglietta, M. et al. (The Pierre Auger Collaboration) (2007). Correlation of the highest energy cosmic rays with nearby extragalactic objects. *Science, 318*: 938–43.

Abraham, J., Abreu, P., Aglietta, M. et al. (The Pierre Auger Collaboration) (2008a). Observation of the suppression of the flux of cosmic rays above 4×10^{19} eV. *Physical Review Letters, 101*, 061101: 1–7.

Abraham, J., Abreu, P., Aglietta, M. et al. (The Pierre Auger Collaboration) (2008b). Correlation of the highest-energy cosmic rays with the positions of nearby active galactic nuclei. *Astroparticle Physics, 29* (12): 188–204.

Abreu, P., Aglietta, M., Ahn, E.J. et al. (The Pierre Auger Collaboration) (2010). Update on the correlation of the highest energy cosmic rays with nearby extragalactic matter. *Astroparticle Physics, 34* (5): 314–26.

Allard, D., Busca, N., Decarprit, G. et al. (2008). Implications of the cosmic ray spectrum for the mass composition at the highest energies. *Journal of Cosmology and Astroparticle Physics, 0810*, 033: 1–17.

Berezinsky. V. (2008). Propagation and origin of ultra high-energy cosmic rays. *Advances in Space Research, 41*: 2071–78.

Dolag, K., Grasso, D., Sprigel, V. et al. (2005). Constrained simulations of the magnetic field in the local Universe and the propagation of ultrahigh energy cosmic rays. *Journal of Cosmology and Astroparticle Physics, 0501*, 009: 1–37.

Farrar, G.R. (2008). Evidence that a cluster of UHECRs was produced by a burst or flare. Proceedings of the International Cosmic Ray Conference 2007, Mérida, Mexico; see arXiv: 0708.1617v1 [astro-ph], 2007 (Aug. 12): 1–4.

Farrar, G.R. and Gruzinov, A. (2009). Giant AGN flares and cosmic ray bursts. *Astrophysical Journal, 693*: 329–32.

Farrar, G.R., Berlind, A.A., and Hogg. D.W. (2006). Foreground and source of a cluster of ultra-high energy cosmic rays. *Astrophysical Journal, 642*: L89–94.

Gezari, S., Basa, S., Martin, D.C. et al. (2008). UV/optical detections of candidate tidal disruption events by *GALEX* and CFHTLS. *Astrophysical Journal, 676*: 944–69.

Harari, D., Mollerach, S., and Roulet, E. (2006). On the ultra-high energy cosmic ray horizon. *Journal of Cosmology and Astroparticle Physics, 0611*, 012: 1–10.

Hörandel, J.R. (2008). Cosmic-ray composition and its relation to shock acceleration by supernova remnants. *Advances in Space Research, 41*: 442–63.

Magorrian, J. and Tremaine, S. (1999). Rates of tidal disruption of stars by massive central black holes. *Monthly Notices of the Royal Astronomical Society, 309*: 447–60.

Nagano, M. and Watson, A. (2000). Observations and implications of the ultrahigh-energy cosmic rays. *Reviews of Modern Physics, 72*: 689–732.

Sekido, Y. and Elliot, H. (1985). *Early History of Cosmic Ray Studies: Personal Reminiscences with Old Photographs.* Dordrecht: Reidel.

Sigl, G., Miniati, F., and Ensslin, T.A. (2003). Ultrahigh energy cosmic rays in a structured and magnetized universe. *Physical Review D, 68* (4): 043002–11.

Skrutskie, M.F., Cutri, R.M., Stiening, R. et al. (2006). The two micron all sky survey (2MASS). *Astronomy Journal,* 131: 1163–83.

van Velzen, S., Farrar, G.R., Gezari, S., et al. (2010). Optical discovery of stellar tidal disruption flares. eprint: arXiv: 1009.1627: 1–21.

Véron-Cetty, M.-P. and Véron, P. (2006). A catalogue of quasars and active nuclei. 12th edition. *Astronomy & Astrophysics,* 455: 773–77.

Waxman, E. (1995). Cosmological gamma-ray bursts and the highest energy cosmic rays. *Physical Review Letters,* 75: 386–89.

Waxman, E. and Miralda-Escude, J. (1996). Images of bursting sources of high-energy cosmic rays: Effects of magnetic fields. *Astrophysical Journal,* 472: L89–92.

Zaw, I., Farrar, G.R., and Berlind, A.A. (2011). Testing the correlations between ultrahigh energy cosmic rays and AGNs and the Véron-Cetty and Véron catalogue of quasars and AGNs. *Monthly Notices of the Royal Astronomical Society,* 410: 263–72.

Zaw, I., Farrar, G.R., and Greene, J. (2009). Galaxies correlating with ultrahigh energy cosmic rays. *Astrophysical Journal,* 696: 1218–229.

Part IV

Technologies for Future Questions

12 New Technologies for Radio Astronomy

K. Y. Lo and Alan H. Bridle

CONTENTS

INTRODUCTION

In this 400th year since the invention of the optical telescope, and in celebration of the *New Vision 400* conference held in Beijing in 2008,* we discuss how new radio astronomical techniques have led to important discoveries in astronomy. We then describe how further technical developments are leading to new facilities such as the Atacama Large Millimeter/submillimeter Array (ALMA), the largest ground-based astronomical facility now under construction, and to the next-generation radio astronomy facilities, collectively named the Square Kilometre Array (SKA), or the SKA Program.

The advent of new technologies in radio astronomy, ironically, has little to do with the invention of the optical telescope 400 years ago. Developed from techniques of radio broadcasting and communication, and especially propelled by the rapid advancement of radar techniques during World War II, radio astronomy opened up the first electromagnetic window into the Universe beyond the visible wavelengths and transformed a view of the Universe that had previously been based entirely on optical studies of stars and nebulae. The most fundamental discoveries made by radio astronomers involve phenomena not observable in visible light, and four Nobel Prizes in Physics have been awarded for discoveries enabled by radio astronomy techniques.

DISCOVERIES ENABLED BY RADIO ASTRONOMY TECHNIQUES

The advent of new techniques leading to new discoveries in the Universe is a pervading theme in astronomy. The transformation of our understanding of the Universe due to radio astronomy is rather

---

* See: http://nv400.uchicago.edu/.

profound. Table 12.1 gives a list of major new astronomical phenomena discovered as a result of new radio astronomy techniques, as well as the resulting new understanding of the Universe.

Radio astronomy from the ground covers a very wide wavelength range: $\lambda \sim 30$ m to ~ 300 μm, corresponding to a frequency range of $\nu \sim 10$ MHz to ~ 1 THz. In radio astronomy, the techniques of reception and detection of the electromagnetic wave are different from those at optical wavelengths. Whereas geometric optics governs the design of optical telescopes, electromagnetic wave theory governs the design of radio telescopes, taking account of the wavelength explicitly because it is no longer negligible compared with the dimensions of the telescope. The detection of radio waves is based on heterodyne detection (or coherent detection) of the electric field E, whereas the detection of visible light is via incoherent detection of E^2, the intensity of the electric field—i.e., photon counting. Heterodyne detection of the radio wave involves the use of a nonlinear detector that produces a beat signal between the radio signal and a local oscillator (reference) signal. In a domestic radio receiver, the beat signal is the sound waves we hear. Because the wavelength of radio waves is some millions of times greater than that of light, the diffraction limit of a single-dish radio telescope, λ/D, is typically in the range of degrees to arcminutes, whereas the angular resolution of ground-based optical telescopes without adaptive optics is limited by atmospheric fluctuations to about 0.5 arcsec.

By the 1920s, radio communication techniques were developed to the extent that there were commercial radio broadcasting, a worldwide network of commercial and government radiotelegraphic stations, and extensive use of radiotelegraphy by ships for both commercial purposes and passenger messages. In 1932, Karl Jansky at Bell Laboratories was assigned to investigate the sources of radio

TABLE 12.1
New Astronomical Phenomena Discovered via Radio Astronomy Techniques

Discovery	Impact	Year	Ref.
Milky Way radio noise	Cosmic radio emission exists	1933	1
Solar radio noise	Radio emissions of normal star	1945	2
21 cm line of atomic hydrogen	Interstellar medium important component of galaxies	1951	3
Double nature of radio galaxies	Need for large-scale energy transport from AGNs	1953	4
Cosmic microwave background (CMB)	Remnant heat of Big Bang	1965	5
Pulsars	Neutron stars exist	1968	6
Polyatomic interstellar molecules	Astrochemistry, precursors of life in space	1968	7
Molecular clouds	Birthplaces of stars and planets	1971	8
Superluminal motion in AGNs	Relativistic potential wells in Galactic nuclei	1971	9
Galactic center source Sgr A*	Subparsec-scale structure at center of Milky Way	1974	10
Flat H I rotation curve of M31	Dark matter halos of galaxies	1975	11
Binary pulsar	Gravitational radiation	1975	12
Anisotropy of CMB	Origins of cosmological structure	1989	13
Pulsar companions	Exoplanets	1992	14

References: (1) Jansky (1933); (2) Southworth (1945); Appleton (1945); (3) Ewen and Purcell (1951); (4) Jennison and Das Gupta (1953); (5) Penzias and Wilson (1965); (6) Hewish et al. (1968); (7) Cheung et al. (1968); (8) Buhl (1971); (9) Cohen et al. (1971); (10) Balick and Brown (1974); (11) Roberts and Whitehurst (1975); (12) Hulse and Taylor (1975); (13) Smoot et al. (1992); (14) Wolszczan and Frail (1992).

FIGURE 12.1   Karl Jansky (*left*) and the movable dipole array antenna (*right*) with which he discovered cosmic radio noise at Bell Laboratories in 1932. (Courtesy of the NRAO Archives.)

noise that might interfere with transatlantic radio–telephone transmission. He built a directional radio antenna (Figure 12.1) made up of a phased array of vertical dipoles producing a fan beam near the horizon in the direction perpendicular to the length of the antenna so he could locate the sources of interfering radio emissions. His serendipitous discovery of radio emission from the center of the Milky Way (Jansky, 1933) was clearly a case of a new technology unexpectedly matching a natural phenomenon. At that time, radio emission from an astronomical object was unexpected because astronomers had been familiar only with thermal radiation that is very weak in radio wavelengths at stellar temperatures. As a result, the mechanism of astronomical radio emission from beyond the Solar System remained a puzzle for many years.

It was later realized (Ginzburg, 1951; Shklovsky, 1953) that cosmic radio waves could be produced nonthermally by synchrotron radiation (Alfvén and Herlofson, 1950; Kiepenheuer, 1950) from relativistic electrons spiraling in magnetic fields. Astronomers then became aware of the enormous energy reservoirs that must underlie the nonthermal radio emission that occupies volumes hundreds of kiloparsecs in extent (Jennison and Das Gupta, 1953) around radio galaxies (Figure 12.2) and, later, radio-loud quasars. As the most luminous of these extra-Galactic radio sources can be detected at look-back times that are significant fractions of the age of the Universe, their discovery immediately extended the "reach" of observational cosmology. The need to *resolve* these radio structures in order to elucidate the physics of their prodigious energy supply motivated the development of the first high-resolution radio interferometers.

A key innovation at Cambridge University in the 1960s, for which a share of the 1974 Nobel Prize in Physics was awarded to Martin Ryle, was the use of *Earth-rotation aperture synthesis* to make images of the radio sky. Ryle (1962) married techniques from Fourier optics and phase-stable interferometry to then-emerging methods in "fast" digital computing, to realize "synthetic apertures" a few kilometers in diameter D. To obtain still better resolution λ/D, it was necessary to exploit the fact that measurements in radio astronomy are far from being photon-limited. This allows *multi-element* interferometers many kilometers in extent to be used with the principle of *phase closure** (Jennison, 1958) to adaptively correct images for the effects of atmospheric and instrumental

---

* The "closure" phase is a quantity derivable from the phases measured for an incoherent source by a *triplet* of Michelson interferometers independently of instrumental errors. Use of closure-phase information underpins many algorithms for removing instrumental and atmospheric effects from sky images made using multi-element interferometers.

FIGURE 12.2   The radio galaxy Fornax A (1.4 GHz continuum intensity observed with the VLA shown in orange) superposed on an optical (STScI/POSS II) image of a 0.97-degree region around the giant elliptical galaxy NGC 1316. The active nucleus of NGC 1316 has energized two radio "lobes" each ~180 kpc in extent. (From radio data of E.B. Fomalont, R.D. Ekers, W. van Breugel, and K. Ebneter. Image © NRAO/Associated Universities, Inc.)

fluctuations. In the early 1980s, new "self-calibration" algorithms (Pearson and Readhead, 1984) and faster digital computers allowed the 27-element Very Large Array (VLA; Figure 12.3) to surpass the resolving power of the great optical telescopes while making high-quality radio images. Using high-bandwidth data recorders, atomic clocks as local oscillators, and custom-built digital correlators, very long baseline interferometry (VLBI) (Bare et al., 1967; Broten et al., 1967; Moran et al., 1967) extended these methods to synthesize Earth-sized apertures (and larger, using orbiting antennas). VLBI enabled the highest-resolution imaging and the most precise astrometry that has ever been realized for astronomy. These technical advances were used to show that the enormous extra-Galactic radio sources are powered by relativistically moving jets that originate on subparsec scales deep within active galactic nuclei (AGNs). This led to acceptance of the idea that the AGN power source is the extraction of gravitational and rotational energy via accretion of matter (and magnetic fields) into the relativistic potential wells of supermassive black holes (BHs) at the centers of galaxies, as originally proposed by Salpeter (1964) and Lynden-Bell (1969).

Also at Cambridge University, a dipole antenna array designed to study time-variable signals due to interplanetary scintillation of small-diameter radio sources at 81.5 MHz (3.7 m wavelength) came into operation in 1967. Its large (~4.5 acre, i.e., ~18,600 m^2) area collected sufficient signal to detect variability with time constants as short as 0.5 sec. Although aimed at investigating the structure of small-diameter radio sources by studying signal fluctuations caused by irregularities in the solar wind, this array and its instrumentation were ideally suited to detecting short pulses of radio emission spaced ~1.3 sec apart from a previously unknown source, which was fleetingly suspected to be due to extraterrestrial intelligence. Such pulsing sources (Hewish et al., 1968) were subsequently identified with neutron stars (Gold, 1968; Pacini and Salpeter, 1968), whose existence had been theoretically predicted based on supernovae (Baade and Zwicky, 1934). Hewish shared the 1974 Nobel Prize in Physics for the discovery of the "pulsars."

Another exceptionally large antenna, the 305 m (1,000 ft)-diameter spherical reflector at the Arecibo Observatory in Puerto Rico (Figure 12.4), was later used at 430 MHz to detect the first millisecond pulsar in a binary system (Hulse and Taylor, 1975). This discovery opened the way to precise measurements of the merging of the close binary system that are consistent with energy loss by

FIGURE 12.3   The NRAO's VLA on the plains of San Agustin near Socorro, New Mexico. The VLA uses 27 25 m (82 ft) antennas to make images of the radio sky at centimeter wavelengths by Earth-rotation aperture synthesis. (From © NRAO/Associated Universities, Inc.)

gravitational radiation, for which Hulse and Taylor were awarded the 1993 Nobel Prize in Physics. Pulsars, especially millisecond pulsars, which are very accurate clocks, and particularly the 22 ms double pulsar (Burgay et al., 2003) discovered with the 210 ft (64 m) antenna at Parkes, Australia, have become very important for exploring a wide range of fundamental physics issues. These issues include nuclear equations of state, general relativity in the strong-field limit, and the indirect and direct detection of gravitational waves (GWs), primarily based on precise measurements of the arrival time of pulsar pulses with accuracy now achievable near the 100 ns level.

Centimeter-wave and meter-wave astronomy of nonthermal sources at high angular and time resolution thus made relativistic astrophysics an *observational* science by revealing new classes of object whose measurable properties are dominated by effects of special relativity (aberration, beaming) or of general relativity (lensing, gravitational radiation). Equally fundamental progress was made through new techniques applied to observations of thermal emission and to radio spectroscopy.

The detection of the 2.7 K cosmic microwave background radiation (CMBR; Penzias and Wilson, 1965), now accepted as the remnant heat of the Big Bang, was technically achievable owing to the combined use of a highly sensitive maser amplifier receiver and a "horn" antenna designed to minimize stray radiation from the ground (Figure 12.5). Because of the resultant high sensitivity of the antenna-receiver system, the unexplained excess signal was impossible to dismiss and was ultimately identified as the CMBR, resulting in the award of a share of the 1978 Nobel Prize in Physics to Penzias and Wilson. In 1989, the *Cosmic Background Explorer* (*COBE*) satellite documented the precise blackbody spectrum of the CMBR, confirming its interpretation as the residual heat from the primordial explosion of the Big Bang. *COBE* also measured the level of anisotropies in the CMBR that indicate the seeds of structure that evolved into galaxies and clusters of galaxies (Smoot et al.,

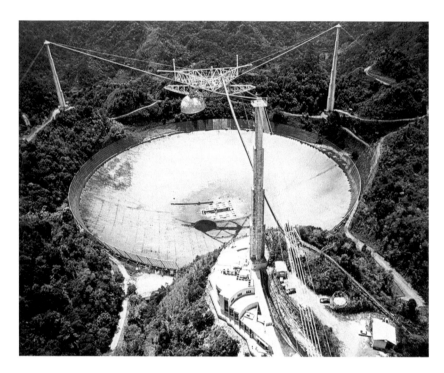

FIGURE 12.4   The Arecibo 305 m (1,000 ft) telescope at the National Astronomy and Ionosphere Center in Puerto Rico. (Courtesy of the NAIC–Arecibo Observatory, a facility of the NSF.)

1992). More than 40 years after the initial discovery by Penzias and Wilson, detailed ground-based (Halverson et al., 2002; Mason et al., 2003), balloon-borne (de Bernardis et al., 2000; Hanany et al., 2000), and space-based (Bennett et al., 2003) observations of the anisotropies in the CMBR, for which John Mather and George Smoot shared the 2006 Nobel Prize in Physics, led to a precise determination of all the cosmological parameters, including the age, the curvature, and the energy densities of the various components of the Universe (e.g., Spergel et al., 2007). Perhaps most importantly, when combined with the discovery of the acceleration of the cosmic expansion via the Type

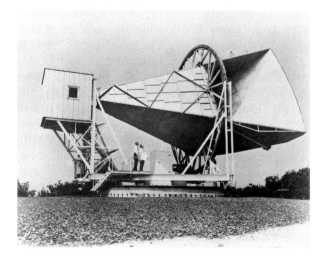

FIGURE 12.5   The horn antenna used by Penzias and Wilson at Bell Laboratories in 1963 to discover the cosmic microwave background. (From NASA photograph.)

Ia supernova Hubble diagram (Knop et al., 2003; Astier et al., 2006; Riess et al., 2007), the CMBR measurements helped to identify dark energy, a term denoting a hitherto completely unknown component of the Universe, making up almost three-quarters of the cosmological energy content.

Another major astronomical discovery made possible by radio techniques was the extensive presence of neutral hydrogen atoms in interstellar space. To begin with, the realization that detecting a radio frequency spectral line would allow its Doppler shift to be used to map motions in the Milky Way led Jan Oort in Leiden to assign his student Hendrik van de Hulst the task of identifying candidate lines for study. This led to the suggestion (van de Hulst, 1945) that the 21 cm hyperfine transition in the ground state of the hydrogen atom might be observable if the lifetime of the upper state is not too great. The detection of this H I line from the Milky Way was accomplished in 1951 (Ewen and Purcell, 1951; Muller and Oort, 1951). Specifically, the novel technique developed by Ewen and Purcell that enabled the detection of the line was "frequency switching," by which the reference local oscillator signal was periodically switched between two frequencies. Frequency switching turned the detection of the line into a differential measurement instead of a much more challenging detection of a weak spectral signal amidst a high background continuum signal.

Subsequent observations demonstrated that atomic hydrogen gas is extensively distributed in the Milky Way and in external galaxies, showing that the interstellar medium (ISM), while dark, is by no means empty, and also providing evidence from Galactic rotation curves for the presence of dark matter halos in galaxies (e.g., Roberts and Whitehurst, 1975).

In the late 1960s and early 1970s, the application of laboratory microwave spectroscopic techniques to astronomy enabled the discovery of inorganic and organic molecules in interstellar space, ushering in the new field of astrochemistry in the ISM, which may yet reveal the pervasive existence of building blocks of life, such as amino acids, in interstellar space. These studies also revealed the extensive existence of molecular clouds in the ISM within the Milky Way and in external galaxies. The technical innovations at that time involved the availability of large high-frequency telescopes (e.g., the National Radio Astronomy Observatory's [NRAO]* 43 m [140 ft]) and high-precision millimeter-wave telescopes (e.g., the NRAO 36 ft [later 12 m] and the University of Texas Millimeter Wave Observatory 5 m [16 ft]); of digital correlator spectrometers (Weinreb, 1963); and of low-noise centimeter-wave amplifiers, such as electron-beam parametric amplifiers (Adler et al., 1959) and maser amplifiers (e.g., Alsop et al., 1959). The invention of sensitive superconductor-insulator-superconductor (SIS) mixer millimeter-wave receivers (Tucker, 1979; Phillips and Dolan, 1982) was also crucial to this field.

ALMA†

The most important astronomical consequence of discovering the significant molecular component of the ISM was the subsequent demonstration that the birth of stars takes place within molecular clouds. Understanding the physical mechanisms of star formation is basic to understanding the formation of stars, planets, and galaxies. The recognition of the importance of studying such processes within molecular clouds in the Milky Way, and in galaxies going back to the epoch of reionization when the first luminous objects were formed, has led to the construction in Chile of what is currently the largest telescope facility for ground-based astronomy—ALMA.

---

* The National Radio Astronomy Observatory (NRAO) is a facility of the National Science Foundation (NSF) operated under cooperative agreement by Associated Universities, Inc. (AUI).

† An international astronomy facility, ALMA is a partnership of Europe, North America, and East Asia in cooperation with the Republic of Chile. It is funded in Europe by the European Organization for Astronomical Research in the Southern Hemisphere (ESO); in North America by the US NSF in cooperation with the National Research Council of Canada (NRC) and the National Science Council of Taiwan (NSC); and in East Asia by the National Institutes of Natural Sciences (NINS) of Japan in cooperation with the Academia Sinica (AS) in Taiwan. ALMA construction and operations are led on behalf of Europe by ESO; on behalf of North America by NRAO, which is managed by AUI; and on behalf of East Asia by the National Astronomical Observatory of Japan (NAOJ). The Joint ALMA Observatory (JAO) provides the unified leadership and management of the construction, commissioning, and operation of ALMA. More information can be found at http://www.nrao.edu/ and http://www.almaobservatory.org/.

When completed by the end of 2012, ALMA will be among the most powerful telescopes ever built. With unprecedented sensitivity, resolution, and imaging capability, it will explore the Universe via millimeter- and submillimeter-wavelength light, one of astronomy's last frontiers. ALMA will open a new window on celestial origins, capturing new information about the very first stars and galaxies in the Universe, and directly imaging the formation of planets. Located at Llano de Chajnantor in the Andes in northern Chile (Figure 12.6), one of the world's best sites for astronomy, ALMA will reside at an elevation of 16,500 ft (5,000 m) above sea level and include at least 66 high-precision submillimeter-wave telescopes. ALMA, an international astronomy facility, is a partnership of Europe, North America, and East Asia in cooperation with the Republic of Chile.

Scientific Case

The primary science goals that have been used to develop the technical specifications of ALMA are (1) the ability to detect spectral line emission from CO or C+ in a normal galaxy, such as the Milky Way at a redshift of $z = 3$, in less than 24 hr of observation; (2) the ability to image the gas kinematics in a solar-mass protoplanetary disk at a distance of 150 pc (roughly the distance of the star-forming clouds in Ophiuchus or Corona Australis), enabling one to study the physical, chemical, and magnetic field structure of the disk and to detect the tidal gaps created by planets undergoing formation; and (3) the ability to provide precise images at an angular resolution of 0.1 arcsec. Here the term "precise image" means an accurate representation of the sky brightness at all points where the brightness is greater than 0.1% of the peak image brightness. This last requirement applies to all sources visible to ALMA that transit at an elevation greater than 20 degrees.

To meet these scientific goals, ALMA will have the following superior capabilities:

- at least fifty 12 m (39 ft) submillimeter-wave telescopes for sensitive, high-resolution imaging;
- four additional 12 m (39 ft) telescopes, providing total-power data, and twelve 7 m (23 ft) telescopes making up the ALMA Compact Array (ACA), enhancing the fidelity of wide-field imaging;
- imaging ability in all atmospheric windows from 3.6 to 0.3 mm (84–950 GHz), with coverage down to 10 mm (30 GHz) possible through future receiver development;

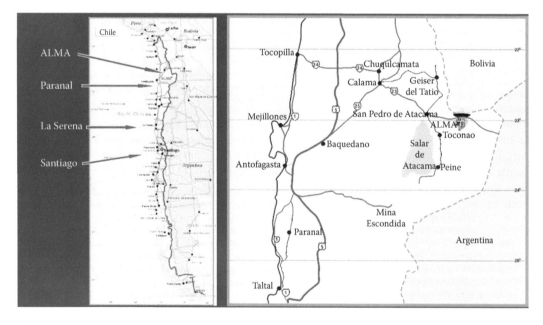

FIGURE 12.6   Geographic location of ALMA in Northern Chile. (From © ALMA (ESO/NAOJ/NRAO).)

- array configurations with maximum baselines from approximately 150 m to 15 km;
- ability to image sources many arcminutes across at arcsecond resolution;
- top spatial resolution of 5 mas (better than the VLA and *Hubble Space Telescope* [*HST*]);
- top velocity resolution better than 0.05 km/sec.

ALMA will be a complete astronomical imaging and spectroscopic instrument for the millimeter/submillimeter wavelength range. It will provide scientists with capabilities and wavelength coverage that complement those of other research facilities of its era; under construction, such as the Expanded Very Large Array (EVLA) and the *James Webb Space Telescope* (*JWST*); or being planned, such as the Thirty Meter Telescope (TMT), the Giant Magellan Telescope (GMT), the European Extremely Large Telescope (E-ELT), and SKA. Specifically, ALMA will fill in a crucial scientific gap by providing a sensitive, high-resolution probe of the properties of cold gas and dust in star-forming regions in our Galaxy and other galaxies, as well as in protoplanetary disks. Given that these regions are obscured at visible wavelengths, ALMA will complement shorter-wavelength observations by providing a complete picture of these cold regions in which stars and planets are formed.

TECHNICAL CHALLENGES

ALMA presents many technical challenges, notably the high-precision submillimeter telescopes, the quantum SIS mixers, phase-stable fiberoptic transmission of signals over 15 km, and the pairwise digital correlation of the signals from all the telescopes. The ALMA telescopes are the highest-precision radio telescopes ever built, and they must maintain their accurate shape under the strains of open-air operation on the high-altitude Llano de Chajnantor site near the oasis town of San Pedro de Atacama in northern Chile. This site offers the exceptionally dry and clear sky required to operate at millimeter/submillimeter wavelengths, but it also experiences large diurnal temperature variations and strong midday winds. Other major performance requirements of each antenna are 2 arcsec absolute pointing over the whole sky, 0.6 arcsec (~10^{-5} radian) offset pointing, a 25 μm RMS overall surface accuracy, and the ability to change its pointing over a 2-degree range in less than 1.5 sec and operable under winds up to 30 km/hr. In addition, these telescopes have to preserve their specifications after repeated moves that are needed to reconfigure ALMA and to survive earthquakes up to magnitude ~8. In the early planning stages of ALMA, such requirements posed serious concerns about whether they can be met in practice, and three different prototype telescopes were built to demonstrate the feasibility of constructing the ALMA telescopes.

Receiving systems on the ALMA telescopes will cover the entirety of the electromagnetic spectrum observable from the Earth's surface from 31.3 to 950 GHz (9.6–0.32 mm in wavelength). At the heart of the receiving system are SIS quantum tunnel junction mixers, operating at 4 K (–269°C) with sensitivities below a few times the quantum limit: single sideband $T \leq 6$–10 $h\nu$/k for $\nu \leq 1$ THz. Such detectors are not commercially available and can be fabricated only in a handful of laboratories throughout the world, requiring the mastery of the techniques of planar circuit design, thin-film deposition, lithography, and cryogenics. The production of hundreds of such detectors for ALMA at different locations in Europe, North America, and East Asia is technically and logistically very challenging and unprecedented.

All the telescopes of ALMA must operate with constant phase relationship relative to one another. As a result of the unprecedented combination of high observing frequencies (up to 950 GHz) and long baselines (up to 15 km), this poses a particularly difficult challenge. ALMA can be thought of as 66 radio receivers, with the main "tuner" for each of the radios (the source of the reference signal) located in a central technical building. The reference signal to each telescope is transmitted over an optical fiber with lengths up to 15 km. This central "tuner" must tune from 27 to 122 GHz by differencing the frequencies of two very-high-frequency oscillators (lasers!), and the low jitter is achieved by phase-locking the lasers to very low noise microwave references.

However, even if this central "tuner" were a perfect clock, the distribution of the ALMA main tuner (reference) signals must be distributed to the telescopes using fiberoptic transmission with an electrical length maintained to an accuracy of <3.6 μm out of 15 km (corresponding to the RMS phase variation, $\delta\varphi$, to be ≤0.55 degree at 119 GHz or a timing error of ≤13 fs) for a fractional stability ratio of 2.4×10^{-10}! The approach adopted by ALMA to maintain the phase stability uses a very accurate yardstick—the master laser—to probe the small changes in the fiberoptic delay to each antenna and to use an optoelectronic line-length compensator to continuously correct the small changes in the delay. The master laser accuracy must be better than the required fiber path-length accuracy, and the delivered unit had accuracy better than 10^{-12} fractional stability ratio over timescales of typical ALMA observing periods (1–1,000 sec).

ALMA continuously correlates the signals from all the pairs of telescopes in the array (there are 1,225 antenna pairs in the main 50-telescope array and 66 pairs in the 12-telescope compact array). From each antenna, signals across a bandwidth of 16 GHz will be received from the astronomical object being observed. The electronics will digitize and correlate these signals at a rate of over 1.6×10^{16} operations per second. Astronomical images will then be numerically constructed from the correlation of the signals of all the antenna pairs via Fourier transform after suitable calibration and corrections to the "raw" correlation of the signals.

Current Status

JAO in Chile, which operates ALMA on behalf of the three regions (see footnote 4), consists of three components: the Array Operations Site (AOS), the Operations Support Facilities (OSF), and the Central Offices in Santiago. The AOS (Figure 12.7), where the telescopes and technical building (which houses the correlator) are located, is at 5,000 m elevation (16,400 ft). The array will be operated, and maintenance of the telescopes and electronics will be carried out, 33 km from the AOS at the OSF (Figure 12.7), just below 3,000 m (9,800 ft) elevation.

The construction of ALMA is well on its way, having achieved first light and first closure phase with three telescopes (Figure 12.8) by December 2009. Production of the antennas and electronics is proceeding steadily, with more than two dozen telescopes in various stages of assembly in Chile in early 2010 (Figure 12.9). It is planned that ALMA, with at least 50 telescopes, will be inaugurated by the end of 2012.

SKA*

While ALMA is a transformative astronomy facility covering the millimeter- and submillimeter-wavelength range, many astronomical issues require a revolutionary facility covering the meter- and centimeter-wavelength range. Current forefront centimeter- and meter-wave radio telescopes typically have a collecting area A ~10,000 m^2 and receiver system temperature T on the order of 50 K. As the sensitivity of a radio telescope is characterized by A/T and T is nearing the theoretical limit to within a factor of a few, any significant advance of sensitivity must henceforth be obtained by increasing A. In order to carry out scientific explorations far beyond current sensitivities, the SKA (Figure 12.10) aims to achieve a 100-fold sensitivity increase over the current centimeter- and meter-wave telescope facilities by building an array with a collecting area of 1 km^2 or 10^6 m^2, as indicated by the name. Furthermore, the SKA will also be optimized for surveys and therefore designed to achieve a very wide field of view. To achieve the requisite high resolution, the largest dimension of the SKA will reach ~3,000 km. The frequency range will span from ~70 MHz to ~22 GHz (corresponding to a wavelength range of ~4 m to ~1.4 cm).

---

* The SKA Program is an international effort; current information about the program can be found at http://www.skatele-scope.org/.

FIGURE 12.7 *Top and middle*: ALMA Operations Support Facilities at the 2,900 m (9,500 ft) level near San Pedro de Atacama. *Bottom*: The Array Operations Site Technical Building at the 5,000 m (16,400 ft) level. (From © ALMA (ESO/NAOJ/NRAO).)

SCIENCE CASE

The SKA will be one of a suite of new, large telescopes for the 21st century that will be used to probe fundamental physics, the origin and evolution of the Universe, the structure of the Milky Way, the formation and distribution of planets, and astrobiology. The science case for the SKA has been documented in the book *Science with the Square Kilometre Array* (Carilli and Rawlings, 2004) and can be grouped into five themes:

1. *Cradle of life*: Are there Earth-like planets around other stars? Do they host intelligent life? By observing the process of planet building through detection of structure in the dusty disks that form around nearby young stars, the SKA at short-centimeter wavelengths can play a unique role in telling us how Earth-like planets are formed. Its great sensitivity will also open up the possibility of detecting extraterrestrial intelligence via unintentional "leakage" radio transmissions from planets around the closest stars.

2. *Probing the Dark Ages*: The combination of the absorption spectra of quasars at redshift $z > 6$ and the *WMAP* measurement of the surprisingly large electron scattering optical depth to the CMBR implies that the Dark Ages, before the formation of the first luminous

FIGURE 12.8 Phase closure was achieved with these three telescopes at the ALMA Array Operations Site in Chile in December 2009. (From © ALMA (ESO/NAOJ/NRAO).)

objects in the Universe, ends at $z \sim 20$ when the epoch of reionization began. The SKA will provide detailed pictures of structure formation and reionization during this period of Dark Ages and epoch of reionization that ends at $z \sim 6$, through observations of the redshifted 21 cm line of neutral hydrogen. Such observations will allow us to separate the contributions from different redshifts to make fully 3-dimensional maps of the neutral gas in the Universe that will be crucial for studying the time dependence of reionization.

3. *The origin and evolution of cosmic magnetism*: In spite of their importance to the evolution of stars, galaxies, and galaxy clusters, the origin of cosmic magnetic fields is still an open problem in both fundamental physics and astrophysics. Did significant primordial fields exist before the first stars and galaxies were formed? If not, when and how were the magnetic fields of galaxies, stars, and planets subsequently generated, and what now maintains them? The great sensitivity of the SKA will allow it to survey the Faraday rotation of the plane of polarization of radiation from distant polarized extra-Galactic sources. An

FIGURE 12.9 ALMA Vertex telescopes at various stages of assembly at the Operations Support Facilities. (From © ALMA (ESO/NAOJ/NRAO).)

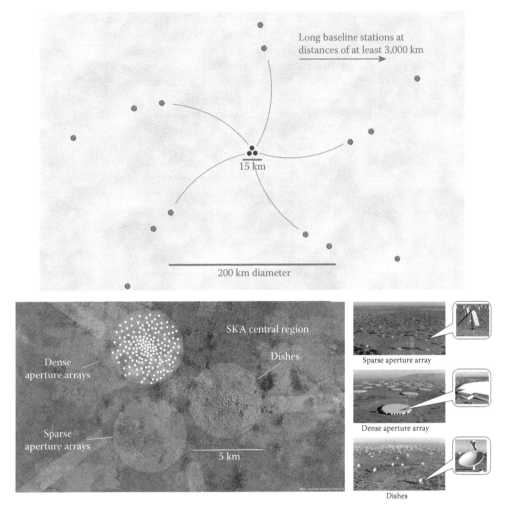

FIGURE 12.10 Artist's renditions of SKA design concepts. *Bottom left*: The central region will be densely packed with stations containing ~50% of the collecting area. *Top*: Other stations will be laid out in a logarithmic spiral pattern extending to the maximum baseline of ~3,000 km (98,400 ft). *Bottom right*: At low frequencies, the array will consist of many phased-aperture arrays, while at high frequencies it will consist of ~3,000 10–15 m-class (33–49 ft) parabolic dishes with new technology feed systems. (From © Swinburne Astronomy Productions and SKA Project Development Office. With permission.)

all-sky Faraday rotation measure survey by the SKA will be a powerful probe for studies of the geometry and evolution of Galactic and intergalactic magnetic fields, to investigate connections between the formation of magnetic fields and the formation of structure in the early Universe, and thus to answer questions about when and how the first magnetic fields in the Universe were generated.

4. *Strong field tests of gravity using pulsars and BHs*: Through its sensitivity, sky coverage, and frequency coverage, the SKA will discover—besides extra-Galactic pulsars—a very large fraction of the pulsars in galaxies. It will also have the sensitivity to time pulsars at the 100 ns precision needed to probe the strong-field realm of gravitational physics and to detect the distortion of spacetime as a result of a stochastic background of *GWs* due to mergers of supermassive BHs or to GWs generated in the inflation era after the Big Bang.

5. *Galaxy evolution, cosmology, and dark energy*: The SKA will enable revolutionary progress in studies of the evolution of galaxies and of large-scale structure in the Universe by
tracing the kinematics, merger history, and environments of galaxies at great distances
using the 21 cm transition of neutral hydrogen. Once a galaxy has been detected in the
21 cm line, the observed wavelength of the line provides an accurate redshift and locates
the object in the 3-dimensional cosmic web. With the SKA, it will be possible to detect the
21 cm line emission from typical galaxies at redshifts $z \sim 3$ in a reasonable integration time
and thus to pursue such studies at distances that are almost entirely inaccessible to current
instrumentation. The SKA will become the premier machine for surveying the large-scale
structure of the Universe and for testing theoretical models for the growth of that structure.
Together with CMBR anisotropy data, SKA measurements of the matter power spectrum
of galaxies will provide powerful constraints on the equation of state of the dark energy
that causes the cosmic expansion to accelerate.

SKA SPECIFICATIONS

The current concept of the SKA Program involves three components:

1. *SKA-low*: Covering a frequency range of roughly ~70 MHz to 0.3 GHz (wavelengths of
~4 m to 1 m) the low-frequency component of the SKA Program will investigate the early
Universe and Galactic transient sources such as BH binaries and flare stars.
2. *SKA-mid*: Covering a frequency range of roughly 0.3–3 GHz or higher (~1 m to ~10 cm
wavelength), the mid-frequency component will be primarily a survey instrument, exploring the evolution of galaxies, dark energy, transient sources, pulsars, and the realm of
strong gravity.
3. *SKA-high*: Covering a frequency range from 3 GHz or higher to 25–50 GHz (~10 to ~1 cm
wavelength), the high-frequency component will explore the formation of stars and planets,
test strong gravity, and search for extraterrestrial intelligence.

Approximately 50% of the collecting area of the SKA is to be contained within a centrally condensed inner array of 5 km diameter to provide ultrahigh brightness sensitivity at arcsecond-scale
resolution for studies of the faint spectral line signatures of structures in the early Universe. Another
25% of the collecting area will be located within a diameter of 150 km, and the remainder out to
3,000 km or more. The nature of the SKA antenna elements will depend on the frequency range. For
the lower frequencies, the SKA is currently conceived to be made up of planar aperture arrays that
have no moving parts and are pointed electronically. Individual array stations would have a sparse
layout of individual elements at the lowest frequencies and denser packing for the mid-frequency
range (see Figure 12.10). For high frequencies, the array is envisaged as being made up of ~3,000
10–15 m-class (33–49 ft) dishes with solid or meshed surface, equipped with "smart feeds" (focal-
plane arrays or wideband feeds). The final design, including the optimal frequency ranges for using
each type of element, will be determined from the outcome of an extensive prototyping exercise
that is now under way. The candidate sites for SKA-low and SKA-mid are in western Australia and
South Africa, both of which have very low radio frequency interference from artificial sources.

TECHNICAL CHALLENGES AND CURRENT STATUS

Aside from beam-forming array receiver technology (Van Ardenne et al., 2005), the technical
requirements of the individual SKA antennas and associated electronics are not particularly difficult. The challenges are primarily due to the scale of the SKA: a large (~3,000) number of antenna
elements, broadband signal transmission over long distances, and the processing and storage of an
enormous volume of data. Given the expected sensitivity of the SKA, calibration and systematic

errors become the technical limits to achieving the theoretical imaging dynamic range (74 db: ~2.5 × 10⁷).

Last but not least, the cost-per-unit-collecting-area of the construction, operation, and maintenance of the SKA has to be significantly lower than that of current facilities in order for it to be affordable. As a reference, the power consumption is estimated to be ~100 MW. Very importantly, attention is now being devoted to the use of solar power as a practical solution to the power requirements, especially given that the potential sites of the SKA are in areas where weather conditions are such that sunshine is abundant. The current cost estimate of the construction of SKA-mid is ~2 billion Euros, and the estimated cost of operations is ~100 million Euros per year.

NOTABLE FACILITIES LEADING UP TO THE SKA

Before the realization of the SKA Program starting in the early 2020s, a number of notable meter- and centimeter-wave facilities that are pathfinders or demonstrators toward the SKA are undergoing construction, nearing completion, or have recently been completed. They include the following:

1. the Giant Meter-wave Radio Telescope (GMRT) in India, an aperture synthesis array with a maximum baseline of 25 km using 30 inexpensive, lightweight 45 m (148 ft) parabolic reflectors formed by stretch mesh attached to rope trusses (SMART), operable from 50 to 1,500 MHz and in use for astronomy since the late 1990s (Rao, 2002);
2. the Allen Telescope Array (ATA-42) in the US, a 42-element pathfinder for the use of large numbers of small-diameter (6 m [20 ft]) paraboloids in a centimeter-wave (0.5–11 GHz) aperture synthesis telescope as envisaged for SKA-high, in use for astronomical surveys and in the Search for Extraterrestrial Intelligence (SETI);
3. the Low-Frequency Array (LOFAR) in Europe, a project to greatly increase sensitivity for imaging at 10–250 MHz using a large number of inexpensive phased arrays of omnidirectional dipole antennas over 1,000 km baselines with digital beam forming as a pathfinder for SKA-low;
4. the Precision Array for Probing the Epoch of Reionization (PAPER), a 128-element interferometer operating from 100 to 200 MHz developed to detect the signal from redshifted H I at the epoch of reionization, to be deployed in South Africa;
5. the Murchison Wide-field Array (MWA), an international (US/Australia/India) SKA-low pathfinder project using a 512-tile array with a maximum baseline of 3 km designed to detect the redshifted H I from the epoch of reionization, to survey the dynamic radio sky at 80 to 300 MHz, and to make measurements of the Sun and heliospheric plasma;
6. the Long Wavelength Array (LWA) in the US, another SKA-low pathfinder, using inexpensive phased arrays of crossed dipoles to achieve Galactic-noise-limited sensitivity from 20 to 80 MHz with baselines up to 400 km for wide-field imaging of both compact and complex sources;
7. the EVLA in the US, greatly improving the sensitivity and expanding the spectroscopic capabilities of the existing 27-element VLA (Figure 12.3) at 1–50 GHz using modern digital wide-bandwidth correlator technology as a pathfinder for SKA-high;
8. the Australian SKA Pathfinder (ASKAP) including field-of-view enhancement by focal-plane phased arrays on new-technology 12 m-class (39 ft) parabolic reflectors to greatly increase survey speed and sensitivity in an SKA-mid-pathfinder at 0.7–1.8 GHz;
9. the Five-hundred meter Astronomical Spherical Telescope (FAST; Figure 12.11) in China, extending the spherical reflector concept used at Arecibo to a larger aperture, increasing sensitivity for single-dish spectroscopy, pulsar, and VLBI observations at 70 MHz to 3 GHz, and as a pathfinder for possible use of extensive karst landforms in an SKA-mid design;

FIGURE 12.11 *Left*: The FAST concept. *Right*: A 50 m (164 ft)-diameter demonstration model built at the Miyun Station of the National Astronomical Observatories, Chinese Academy of Sciences. (Courtesy of the FAST Project Team.)

10. the Karoo Array Telescope (MeerKAT; "meer" means "more" in Afrikaans) in South Africa, a Southern Hemisphere complement to the EVLA using composite, one-piece 12 m (39 ft) reflectors; single-pixel wideband receivers; and low-cost, high-reliability cryogenic systems in an SKA-high precursor that will be optimized for high-fidelity imaging of extended low-brightness emission.

Table 12.2 summarizes the specifications and the status of these facilities, which collectively constitute significant advances in the capabilities of meter/centimeter-wave telescopes and will address many outstanding scientific issues, as well as exploring new antenna, receiver, correlator, and software technologies that will be needed to realize the goals of the SKA Program.

TABLE 12.2
Notable Facilities Leading Up to the SKA

Facility	Date	Country	Frequency Range	Type, Largest Dimension	Ref.
GMRT	1999	India	50 MHz–1.4 GHz	Thirty 45 m antennas, 25 km	1
ATA-42	2007	US	0.5–11.2 GHz	Forty-two 6 m antennas, 300 m	2
LOFAR	2010	Netherlands	10–250 MHz	7,000-element array, 1,500 km	3
PAPER	2010	South Africa	125–200 MHz	128 dipoles	4
MWA	2010	Australia	80–300 MHz	8,192 elements in 512 tiles, 3 km	5
EVLA	2012	US	1–50 GHz	Twenty-seven 25 m antennas, 25 km	6
LWA	2012	US	10–88 MHz	Fifty-three stations, 400 km	7
ASKAP	2013	Australia	700 MHz–1.8 GHz	Thirty-six 12 m antennas, 6 km	8
FAST	2014	China	300 MHz–2 GHz	500 m spherical reflector	9
MeerKAT	2015	South Africa	0.6–14.5 GHz	Eighty-seven 12 m antennas, 60 km	10

Website references: (1) http://www.gmrt.ncra.tifr.res.in/; (2) http://ral.berkeley.edu/ata/; (3) http://www.lofar.org/; (4) http://astro.berkeley.edu/~dbacker/eor/; (5) http://mwatelescope.org/; (6) https://science.nrao.edu/facilities/evla; (7) http://lwa.phys.unm.edu/; (8) http://www.atnf.csiro.au/projects/askap/; (9) http://www.skatelescope.org/publications/; (10) http://www.ska.ac.za/meerkat/.

SUMMARY

Radio astronomy is entering a very exciting era, with many new facilities that embody the latest technologies and make possible the exploration of the latest frontiers of astronomy. At the same time, there are continuing efforts all over the world dedicated to developing novel techniques and technologies needed for the next-generation radio astronomy facilities. Because of the scale of the SKA, collaboration with industry will be essential in the future, changing the way technical development and construction are carried out in the radio astronomy community. The many ambitious projects of radio astronomers, and of astronomers generally, perhaps constitute the best illustration of the essential interplay between the development of new technologies and the unbounded curiosity and imagination of man in the quest to unravel the workings of the Universe.

REFERENCES

Adler, R., Hrbek, G., and Wade, G. (1959). The quadrupole amplifier, a low-noise parametric device. *Proceedings of the Institute of Radio Engineers,* 47: 1713–723.

Alfvén, H. and Herlofson, N. (1950). Cosmic radiation and radio stars. *Physical Review,* 78: 616.

Alsop, L.E., Giordmaine, J.A., Mayer, C.H., et al. (1959). Observations of discrete sources at 3-cm wavelength using a maser. In: R.N. Bracewell (ed.), *Paris Symposium on Radio Astronomy,* IAU Symposium No. 9 and URSI Symposium No. 1, 69–74. Stanford: Stanford University Press.

Appleton, E.V. (1945). Departure of long-wave solar radiation from black-body intensity. *Nature,* 156: 534–35.

Astier, P., Guy, J., Regnault, N., et al. (2006). The Supernova Legacy Survey: Measurement of Ω_M, Ω_Λ and w from the first year data set. *Astronomy and Astrophysics,* 447: 31–48.

Baade, W. and Zwicky, F. (1934). Cosmic rays from super-novae. *Proceedings of the National Academy of Sciences of the USA,* 20: 259–63.

Balick, B. and Brown, R.L. (1974). Intense sub-arcsecond structure in the Galactic center. *Astrophysical Journal,* 194: 265–70.

Bare, C., Clark, B.G., Kellermann, K.I., et al. (1967). Interferometer experiment with independent local oscillators. *Science,* 157: 189–91.

Bennett, C.L., Halpern, M., Hinshaw, G., et al. (2003). First-year *Wilkinson Microwave Anisotropy Probe* (*WMAP*) observations: Preliminary maps and basic results. *Astrophysical Journal Supplement Series,* 148: 1–27.

Broten, N.W., Legg, T.H., Locke, J.L., et al. (1967). Long Base Line Interferometry: A new technique. *Science,* 156: 1592–593.

Buhl, D. (1971). Chemical constituents of interstellar clouds. *Nature,* 234: 332–24.

Burgay, M., D'Amico, N., Possenti, A., et al. (2003). An increased estimate of the merger rate of double neutron stars from observations of a highly relativistic system. *Nature,* 426: 531–33.

Carilli, C. and Rawlings, S. (eds.), (2004). *Science with the Square Kilometre Array. New Astronomy Reviews,* Vol. 48. Amsterdam: Elsevier.

Cheung, A.C., Rank, D.M., Townes, C.H., et al. (1968). Detection of NH_3 molecules in the interstellar medium by their microwave emission. *Physical Review Letters,* 21: 1701–705.

Cohen, M.H., Cannon, W., Purcell, G.H., et al. (1971). The small-scale structure of radio galaxies and quasi-stellar sources at 3.8 centimeters. *Astrophysical Journal,* 170: 207–17.

de Bernardis, P., Ade, P.A.R., Bock, J.J., et al. (2000). A flat Universe from high-resolution maps of the cosmic microwave background radiation. *Nature,* 404: 955–59.

Ewen, H.I. and Purcell, E.M. (1951). Observation of a line in the Galactic radio spectrum—radiation from Galactic hydrogen at 1,420 Mc./sec. *Nature,* 168: 356–57.

Ginzburg, V.L. (1951). Kosmicheskie luchi kak istochnik galakticheskogo radioielueniia (Cosmic rays as the source of galactic radio emission). *Doklady Akademii Nauk SSSR* 76: 377–80.

Gold, T. (1968). Rotating neutron stars as the origin of the pulsating radio sources. *Nature,* 218: 731–32.

Halverson, N.W., Leitch, E.M., Pryke, C., et al. (2002). Degree angular scale interferometer first results: A measurement of the cosmic microwave background angular power spectrum. *Astrophysical Journal,* 568: 38–45.

Hanany, S., Ade, P., Balbi, A., et al. (2000). MAXIMA-1: A measurement of the cosmic microwave background anisotropy on angular scales of 10'-5°. *Astrophysical Journal,* 545: L5–9.

Hewish, A., Bell, S.J., Pilkington, J.D., et al. (1968). Observation of a rapidly pulsating radio source. *Nature*, 217: 709–13.

Hulse, R.A. and Taylor, J.H. (1975). Discovery of a pulsar in a binary system. *Astrophysical Journal*, 195: L51–3.

Jansky, K.G. (1933). Radio waves from outside the solar system. *Nature*, 132: 66.

Jennison, R.C. (1958). A phase sensitive interferometer technique for the measurement of the Fourier transforms of spatial brightness distributions of small angular extent. *Monthly Notices of the Royal Astronomical Society*, 118: 276–84.

Jennison, R.C. and Das Gupta, M.K. (1953). Fine structure of the extra-terrestrial radio source Cygnus 1. *Nature*, 172: 996–97.

Kiepenheuer, K.O. (1950). Cosmic rays as the sources of general Galactic radio emission. *Physical Review*, 79: 738–39.

Knop, R.A., Aldering, G., Amanullah, R., et al. (2003). New constraints on Ω_M, Ω_Λ, and w from an independent set of 11 high-redshift supernovae observed with the *Hubble Space Telescope*. *Astrophysical Journal*, 598: 102–37.

Lynden-Bell, D. (1969). Galactic nuclei as collapsed old quasars. *Nature*, 223: 690–94.

Mason, B.S., Pearson, T.J., Readhead, A.C.S., et al. (2003). The anisotropy of the microwave background to $l = 3500$: Deep field observations with the Cosmic Background Imager. *Astrophysical Journal*, 591: 540–55.

Moran, J., Crowther, P.P., Burke, B.F., et al. (1967). Spectral line interferometry with independent time standards at stations separated by 845 kilometers. *Science*, 157: 676–77.

Muller, C.A. and Oort, J.H. (1951). The interstellar hydrogen line at 1,420 Mc./sec., and an estimate of galactic rotation. *Nature*, 168: 357–58.

Pacini, F. and Salpeter, E.E. (1968). Some models for pulsed radio sources. *Nature*, 218: 733–34.

Pearson, T.J. and Readhead, A.C.S. (1984). Image formation by self-calibration in radio astronomy. *Annual Review of Astronomy and Astrophysics*, 22: 97–130.

Penzias, A.A. and Wilson, R.W. (1965). A measurement of excess antenna temperature at 4080 Mc/s. *Astrophysical Journal*, 142: 419–21.

Phillips, T.G. and Dolan, G.J. (1982). SIS mixers. *Physica B+C*, 110B: 2010–119.

Rao, A.P. (2002). GMRT—current status. In: A.P. Rao, G. Swarup, and Gopal-Krishna (eds.), *The Universe at Low Frequencies*. IAU Symposium No. 199, 439–46.

Riess, A.G., Strolger, L.-G., Casertano, S., et al. (2007). New *Hubble Space Telescope* discoveries of Type Ia supernovae at $z >= 1$: Narrowing constraints on the early behavior of dark energy. *Astrophysical Journal*, 659: 98–121.

Roberts, M.S. and Whitehurst, R.N. (1975). The rotation curve and geometry of M31 at large galactocentric distances. *Astrophysical Journal*, 201: 327–46.

Ryle, M. (1962). The new Cambridge radio telescope. *Nature*, 194: 517–18.

Salpeter, E.E. (1964). Accretion of interstellar matter by massive objects. *Astrophysical Journal*, 140: 796–800.

Shklovsky, I.S. (1953). Problema kosmicheskogo radioizlucheniia (The problem of cosmic radio waves). *Astronomicheskii Zhurnal*, 30: 15–36.

Smoot, G.F., Bennett, C.L., Kogut, A., et al. (1992). Structure in the *COBE* differential microwave radiometer first-year maps. *Astrophysical Journal*, 396: L1–5.

Southworth, G.C. (1945). Microwave radiation from the sun. *Journal of the Franklin Institute*, 239: 285–97.

Spergel, D.N., Bean, R., Doré, O., et al. (2007). Three-year *Wilkinson Microwave Anisotropy Probe* (*WMAP*) observations: Implications for cosmology. *Astrophysical Journal Supplement Series*, 170: 377–408.

Tucker, J. (1979). Quantum limited detection in tunnel junction mixers. *IEEE Journal of Quantum Electronics*, 15: 1234–258.

Van Ardenne, A., Wilkinson, P.N., Patel, P.D., et al. (2005). Electronic Multi-beam Radio Astronomy Concept: EMBRACE a demonstrator for the European SKA Program. *Experimental Astronomy*, 17: 65–77.

van de Hulst, H.C. (1945). Radiogolven uit het wereldruim; herkomst der radiogolven. *Nederlands Tijdschrift voor Natuurkunde* 11: 210–21.

Weinreb, S. (1963). A digital spectral analysis technique and its application to radio astronomy. *M.I.T. Research Lab of Electronics Tech Report*, 412.

Wolszczan, A. and Frail, D. (1992). A planetary system around the millisecond pulsar PSR1257+12. *Nature*, 355: 145–47.

13 Advanced Optical Techniques in Astronomy

Michael Shao

CONTENTS

INTRODUCTION

With the current generation of large (8–10 m-class [26–32 ft]) telescopes, adaptive optics (AO) to provide diffraction-limited wavefronts became a standard tool. Astronomers are now working on, and in many cases have completed, technology development on the next generation of devices, from multi-conjugate adaptive optics (MCAO), extreme AO (very precise wavefront control), long-baseline interferometry, and ultra-precise astrometry. The most precise wavefront measurement and control technologies will be applied to space observatories, with sub-angstrom precision to enable microarcsec astrometry and very-high-dynamic-range (10^{10}) imaging of planets a fraction of an arcsec away from their parent star.

As technology has advanced, astronomers have built larger and larger telescopes. The angular resolution of a telescope is ultimately limited by diffraction, to $1.2 * \lambda/D$. But ground-based telescopes larger than about 20 cm (7–8 in) are not limited by diffraction, but by atmospheric turbulence. The idea of correcting the effects of turbulence dates back perhaps 60 years. But it was not until 1981 that military researchers were able to demonstrate active correction of atmospheric turbulence on a few-millisecond timescale (Hardy, 1998).

It took another 10–15 years before technology advanced to the point where the scientific community could afford to deploy AO on major telescopes. The first generation of AO systems used light from a bright star to "sense" the fluctuations in the atmosphere and commanded a deformable mirror (DM) to remove the phase fluctuations (Tyson, 2000). At present, virtually all large (8–10 m [26–32 ft]) telescopes and a large number of (4–5 m-class [26–32 ft]) telescopes have AO systems. With large telescopes, AO does not just provide higher angular resolution, but can dramatically reduce the cost of instruments. AO on a large telescope dramatically reduces the Area*Solid angle product (A*Omega) of the starlight, which means that an instrument such as a high-resolution spectrometer can be dramatically smaller, and hence cheaper.

A*Omega is a conserved quantity in all imaging optical systems. Because of atmospheric turbulence, the image of a star at a mountain-top observatory is typically ~1 arcsec or 5 microradians. In order for a back-end instrument, such as a camera or a spectrograph, to focus 100% of the starlight onto the detector, a larger A*Omega product means the camera/spectrograph has to be correspondingly larger. As a result, as telescopes got larger, the corresponding back-end instruments also got proportionally larger. AO produces a diffraction-limited image, which at 2.2 μm is ~25 times smaller than an image for a 10 m (32 ft) telescope. A much more compact camera or spectrograph is correspondingly less expensive.

As the astronomy community starts to think about the next generation of giant telescopes, standard AO is taken for granted. But standard AO has a number of limitations. It has a small field of view, it needs a bright guide star, and it achieves a low strehl image compared with a perfect diffraction-limited image. The strehl ratio, often abbreviated to strehl, is a commonly used metric for the quality of an AO-corrected image. When the wavefront errors are much less than a wavelength, the shape of the image of a star is governed by diffraction. The strehl ratio is the ratio of the peak of the image, divided by the peak of the image from a "perfect" wavefront going through the same telescope. For wavefront errors that have a Gaussian random distribution, the strehl is $\sim\exp(-\sigma^2)$ where σ is the rms wavefront error in radians.

For those who want even higher angular resolution, coherently combining the light from multiple telescopes is another option. The two largest telescope facilities, the Keck Telescope and the Very Large Telescope (VLT) of the European Southern Observatory (ESO), are both able to operate as interferometers with, in the case of the VLT, baselines of several hundred meters.

This chapter will give a brief description of standard AO and laser-guide star AO (LGSAO) as an introduction to the next generation of advanced optical techniques. Several advanced AO concepts will be mentioned, but the focus will be on the techniques and science that are enabled when astronomers have control of the wavefront with $\lambda/1,000-\lambda/10,000$ precision. One area is ultra-precise astrometry with a space interferometer, and the second is the area of extreme AO, both for ground-based and space-based instruments.

Astrometry with an interferometer implies the ability to measure optical paths with single-digit picometer precision. After many years of NASA support, this has been demonstrated in the laboratory in an environment that is similar to an orbiting observatory. Sub-microarcsec astrometry enables one to detect the astrometric wobble of a star from an orbiting Earth-mass planet in the habitable zone. (See the paper by Charles Beichman in this volume for more detail on the scientific applications of astrometry to planets and an explanation of the "habitable zone.")

For direct detection of exoplanets, one needs not only to measure the wavefront with high precision, but to control it. The next generation of ground-based extreme AO coronagraphic systems, to become operational sometime in late 2011 to early 2012, hopes to detect dim objects 10^{-7} as bright as the star ~1/4 arcsec away. Achieving this performance requires controlling optical imperfections in the telescope and instrument at the ~1 nm level. Laboratory experiments for space-based coronagraphs have demonstrated control of the wavefront at 0.1~0.2 nm level to achieve 10^{-9} contrast.

With 10^{-7} contrast, it should be possible to detect self-luminous Jovian planets less than ~0.1 Gyr old in the near-infrared (IR). A contrast of 10^{-9} would enable the direct detection of planets of Jupiter's mass in reflected light, and in the more distant future, 10^{-10} contrast would enable detection of planets of Earth's mass in reflected light.

BASICS OF ADAPTIVE OPTICS

Atmospheric turbulence causes the wavefront from a star to be corrupted. Wavefront errors significantly larger than an optical wavelength give rise to an image of a star whose diameter is much larger than the diffraction limit of large modern telescopes. AO was designed to correct the errors caused by atmospheric turbulence. Three technologies make AO possible: (1) the wavefront sensor (WFS) that measures the corruption of the wavefront caused by passage through 10 km of turbulent atmosphere; (2) the DM, which is used to correct the wavefront; (3) the computer that takes the data from the WFS and calculates the necessary commands to the DM. A conventional natural guide star AO system is shown in Figure 13.1, although without the computer (3, above).

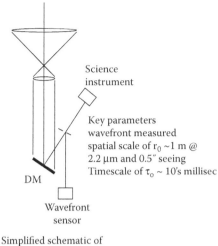

Science
instrument

Key parameters
wavefront measured
spatial scale of r_0 ~1 m @
2.2 µm and 0.5″ seeing
Timescale of τ_0 ~ 10's millisec

DM

Wavefront
sensor

Simplified schematic of
conventional AO system

FIGURE 13.1 Simplified schematic of AO system.

All AO systems are complex and expensive, and hence are deployed only on the largest telescopes. In the early days of astronomical AO, the components of the WFS and the DM had developed only to the point where diffraction-limited images were feasible in the near-IR. An 8–10 m (26–32 ft) telescope at visible wavelengths would need a WFS and DM with 2,500 elements to correct the atmosphere with 0.5 arcsec seeing at visible wavelengths. At 2.2 µm, a DM with <200 elements would be sufficient. AO made diffraction-limited imaging possible from ground-based telescopes, often with angular resolution exceeding that of smaller space-based telescopes such as the *Hubble Space Telescope* (*HST*). An example of the difference AO makes is shown in Figure 13.2 (Wizinowich, 2000).

LIMITATIONS OF NATURAL GUIDE STAR AO: LASER GUIDE STARS

Natural guide star AO uses light from a moderately bright star to "sense" the wavefront error caused by turbulence. A moderately bright star is needed because the wavefront has to be measured at 200–300 locations in the telescope pupil every ~10 msec. The idea of using lasers to provide artificial guide stars was again first pioneered by the military. The initial concept was to use high-power pulsed UV lasers and Rayleigh scattering off the atmosphere. As the pulses left the telescope, light would be scattered

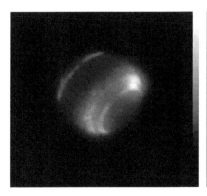

FIGURE 13.2 Neptune with and without AO.

FIGURE 13.3   Sodium laser at the Palomar 5 m (16 ft) telescope.

back. A fast WFS detected would be time-gated, so only the light scattered at a distance of ~10 km would be detected. Most astronomical laser guide stars (LGSs), however, are sodium guide stars, where a laser at 589 nm would excite sodium atoms at an altitude of ~90 km, producing a spot that the WFS can use to sense the atmospheric turbulence located in the lowest 10 km of the atmosphere.

LGSs, unlike actual stars, are not infinitely far away. The spot is not a point source, and the wavefront is spherical, not planar. The path traversed by the laser light on its way down is not exactly the same as the path the starlight traverses. In general, the quality of the wavefront correction with LGSs is not as good as with a bright natural guide star. However, with LGSs, one is not limited to looking within ~20 arcsec of a bright star. Figure 13.3, shows a sodium LGS at the Palomar 5 m (16 ft) telescope.

EVEN HIGHER ANGULAR RESOLUTION, COHERENT COMBINATION FROM MULTIPLE TELESCOPES

AO enables one to obtain diffraction-limited images from a large ground-based telescope. To get even higher angular resolution, one has to coherently combine the light from multiple telescopes. The largest stellar interferometers are the Keck Interferometer, which combines the light from the two 10 m (32 ft) Keck telescopes, and the VLT Interferometer (VLTI), which can combine the light from the four 8 m (26 ft) VLTs or an additional four moveable 1.8 m (6 ft) telescopes.

NEXT-GENERATION AO CONCEPTS

Conventional AO measures the wavefront distortion of a star. The atmospheric turbulence that produces that distortion is distributed over the lower 10 km of the atmosphere. Different stars at different locations will not have the same wavefront distortion. Conventional AO usually achieves, under the best circumstances, a strehl of ~70% in the near-IR. Several different types of next-generation AO systems are now under construction to overcome some of these limitations. One class of the new AO systems is aimed at obtaining a much larger high-resolution field of view than normal AO. The other is aimed at obtaining very precise correction of the wavefront, with a strehl typical for a space observatory.

WIDE-FIELD AO

Two types of AO systems are being developed to enlarge the field of view where higher-resolution images can be obtained. Both make use of atmospheric tomography. Figure 13.4 illustrates the basic principle behind MCAO.

Two or more—usually more—WFSs measure the wavefront from multiple stars (Rigaut et al., 2000). Each star probes a slightly different combination of lower- and upper-atmospheric turbulence. With a sufficient number of stars, one can reconstruct the 3-dimensional profile of turbulence.

This information is used differently in MCAO and ground-layer AO (GLAO). At most observatories, the turbulence is worst within the lowest ~100 m of air above the ground. GLAO uses a single DM to remove the turbulence in this "ground" layer. With the ground layer removed, it is possible to improve image quality by perhaps a factor of 2, but over a very wide field. Diffraction-limited imaging is not possible because the turbulence beyond ~100 m is not corrected.

MCAO, on the other hand, aims to correct the 3-dimensional nature of the turbulence. Figure 13.4 shows a two-layer system, where DM's reimaged to two different altitudes above the telescope to produce diffraction-limited images over a wider field of view than normal AO. MCAO is by far the most complex and ambitious of the next-generation AO systems. In June 2011, the Gemini AO team successfully demonstrated an MCAO system using a constellation of five LGSs in an engineering run at the Gemini South

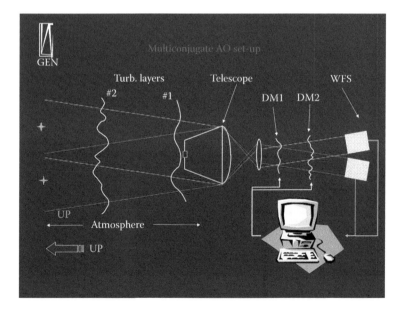

FIGURE 13.4   Schematic of a multi-conjugate AO system. (From Rigaut, F., Ellerbroek, B., and Flicker, R. *MCAO for Dummies: Principles, Performance and Limitations of Multi-conjugate Adaptive Optics, SPIE Conference* 2007, Munich. Available at: http://www.gemini.edu/sciops/instruments/adaptiveOptics/docs/ MCAO4DUMMIES.ppt, 2000.)

observatory (see http://www.gemini.edu/node/11647). Diffraction-limited images over an ~85 arcsec field was demonstrated, although optimization of the MCAO system won't start until early 2012.

VERY PRECISE WAVEFRONT CONTROL (EXTREME AO)

Extreme AO is basically a conventional AO system, but one that has a much larger number of correction elements, typically ~2500–3000 subapertures in the WFS and a matching number of actuators in the DM. In addition, the computer is faster. While a conventional AO system will measure and update the DM when turbulence changes the wavefront by ~$\lambda/6$, typically 10 msec, the goal of extreme AO is to obtain strehls of 90%–95%, updating a much denser DM at several kHz.

A conventional AO system with 60% strehl at 2.2 μm would have a wavefront error of ~240 nm. An extreme AO system with 95% strehl ~1.6 μm would have an rms wavefront error of ~50 nm. The main goal of extreme AO is to directly image exoplanets around nearby stars. Planets are much dimmer than the stars they orbit. Viewed from outside our Solar System at 10 pc, Jupiter would be 1 billion times dimmer than the Sun, just 0.5 arcsec away. The Earth would be 10 billion times fainter and just 0.1 arcsec away.

Ground-based extreme AO, at least in the near term, cannot image a Jupiter like the one existing today. However, a very young Jupiter, 100 million years old, would be self-luminously radiating the energy in the near-IR. Figure 13.5 is a simulation of an image from an extreme AO coronagraph. The AO system controls the wavefront at the pupil. The image is the Fourier transform of the pupil. A DM with 64*64 actuators creates a "dark hole" at the output of the coronagraph that is 64*64 $(\lambda/D)^2$ in area, where D is the diameter of the telescope. The bright light outside the dark (square) hole represents spatial frequencies that cannot be controlled by the DM because it only has 64*64 actuators. Ground-based extreme AO coronagraphs hope to achieve a contrast of 10^7, that is, to be able to detect a planet 10 million times fainter than the nearby star. The small circle in the center is the inner working angle (IWA) of the coronagraph. The coronagraph in blocking the light of the star will also block the light from the planet if it is closer to the star than the IWA. The dark hole in Figure 13.5 is not 10^7 dimmer than the star; the AO system would have to have a strehl better than 99.9% for that to be the case. The light in the dark hole is not uniform and will fluctuate on timescales of milliseconds. These fluctuations have to average down and be smooth to 10^{-7} after an hour of integration.

Extreme AO instruments, such as the Gemini Planet Imager (Gemini) and Sphere (VLT), will be operational by sometime in 2012 with a contrast of $10^6 \sim 2 \times 10^7$. Extreme AO coronagraphs on future giant telescopes such as Thirty Meter Telescope (TMT), Giant Magellan Telescope (GMT),

FIGURE 13.5 Simulated image from an extreme AO coronagraph. (From Macintosh, B., Graham, J., Palmer, D., et al., *The Gemini Planet Imager*, Vol. 6272, 62720L, 2006.)

and Extremely Large Telescope (ELT) might get to a 10^8 contrast, a significant improvement, but still far from what is possible in space.

ULTRA PRECISE WAVEFRONT CONTROL (IN SPACE)

The fundamental limitation of all AO systems is photon noise. The atmosphere varies on a timescale of milliseconds. The AO system must measure the wavefront and correct it on a timescale faster than it changes. Ultimately, this is limited by photon noise from the guide star. In space, there is no atmosphere. But the optics telescope is not perfect. In ground-based extreme AO, the wavefront is corrected to <50 nm rms, $\lambda/30$ at 1.6 μm. The optics for the *Hubble* telescope are precise to ~$\lambda/60$. Telescopes, such as *HST*, are passively precise. If we use active wavefront control techniques, such as AO, wavefront precision far better than $\lambda/60$ can be achieved. But why bother? The images are already diffraction-limited with *HST*.

There are two astronomical applications, both in the detection of exoplanets, that need wavefront measurement or control at the $\lambda/1{,}000$–$\lambda/10{,}000$ level. More than 1,200 planet candidates have been detected outside our Solar System, but no Earth-mass planet has been detected in the habitable zone of a star. Astrometric detection of an Earth–Sun system requires sub-microarcsec astrometry. The *Space Interferometry Mission* (*SIM*), a space-based astrometric interferometer with a 6 m (19 ft) baseline shown in Figure 13.6, needs to measure optical paths to a few picometers precision in order to do sub-microarcsec astrometry to detect the wobble of the stars because of an orbiting Earth in a ~1 year orbit. (Shao and Nemati, 2009). The technology to do so has been demonstrated in the

FIGURE 13.6   *Space Interferometry Mission* (*SIM*) astrometric observatory.

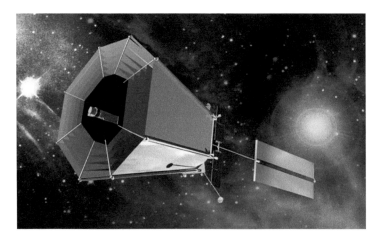

FIGURE 13.7 *Terrestrial Planet Finder (TPF)* coronagraph concept.

laboratory under environmental conditions typical for a space observatory in solar orbit. *SIM* could detect a planet of Earth's mass in the habitable zone of around ~60 of the nearest stars.

Even harder is the direct detection of the light from a planet of Earth's mass orbiting a star. As stated before, the contrast between a planet of Earth's mass in reflected starlight and the parent star is ~10 billion, and the separation at most 0.1 arcsec. While astrometry requires "measurement" of optical paths to a few picometers, a space coronagraph that can detect a planet of Earth's mass has to control the wavefront to ~30 picometer accuracy. Despite the difficulty, a tremendous amount of progress has been made in this area, and scientists are within an order of magnitude of demonstrating this technology in the laboratory (Trauger, 2007). Several different types of coronagraphs have been studied with NASA funding to detect planets orbiting nearby stars and measure their spectra (Shao et al., 2009). One example is the *Terrestrial Planet Finder (TPF)* coronagraph concept shown in Figure 13.7.

SUMMARY

Advanced optical techniques, specifically precise wavefront control, have enabled diffraction-limited imaging through atmospheric turbulence. Natural guide star AO has been operational at almost all of the major ground-based observatories. The next-generation AO systems, such as MCAO and extreme AO, are under construction and are expected to be operational in the near future. A 3,000-actuator AO system is being commissioned in the summer of 2011 at Palomar. A similar extreme AO system for Gemini is expected to be taken to the telescope in early 2012. For MCAO, the first demonstration of a five-star LGS constellation was demonstrated on Gemini South in January 2011. The entire MCAO system will be integrated and commissioned throughout 2011 and into 2012. In 2012, the system should begin providing remarkably sharp images for the study of a wide variety of topics ranging from the birth and evolution of stars to the dynamics of distant galaxies.

In the 400 years since the invention of the telescope, the recent past has seen dramatic improvements. For most of this 400-year history, larger telescopes have given us bigger collecting areas and, hence, the ability to see fainter objects. But for most of those 400 years, the angular resolution was limited by atmospheric turbulence to ~1 arcsec. With AO, a 10 m (32 ft) telescope will now have angular resolution of 0.01 arcsec in the visible. In the near future, ultra-precise control of the wavefront for both ground and space telescopes will dramatically improve the dynamic range of large telescopes, letting us image faint planets a fraction of an arcsec from their bright parent star. From the ground, we should be able to image young self-luminous planets of Jupiter's mass 0.5 arcsec from a star at a contrast of 10^7, and a bit farther into the future, in space we should be able to image Earth-like planets around Sun-like stars with a contrast of 10^{10}.

REFERENCES

Hardy, J. (1998). *Adaptive Optics for Astronomical Telescopes*. Oxford: Oxford University Press.

Macintosh, B., Graham, J., Palmer, D., et al. (2006). The Gemini Planet Imager. In: B.L. Ellerbroek and D.B. Calia (eds.), *Advances in Adaptive Optics II, Proceedings of Society of Photo-Optical Instrumentation Engineers (SPIE) Conference*, Vol. 6272, 62720L.

Rigaut, F., Ellerbroek, B., and Flicker, R. (2000). *MCAO for Dummies: Principles, Performance and Limitations of Multi-conjugate Adaptive Optics, SPIE Conference 2007*, Munich. Available at: http://www.gemini. edu/sciops/instruments/adaptiveOptics/docs/MCAO4DUMMIES.ppt.

Shao, M. and Nemati, B. (2009). Sub-microarcsecond astrometry with SIM-Lite: A testbed-based performance assessment. *Publications of the Astronomical Society of the Pacific*, 121: 41–44.

Shao, M., Deems, E., Fletcher, L., et al. (2009). *Astro2010: Response to Request for Information for Proposed Activities Dilute Aperture Visible Nulling Coronagraph Imager (DAViNCI) Imaging and Spectroscopy of Exo-planets from Nearby Stars*. Available at: http://davincimission.jpl.nasa.gov/IV_Mission_Description/ DAViNCI_NAS_RFI_Response-2009-04-01-rev03.pdf.

Trauger, J. and Traub, W. (2007). A laboratory demonstration of the capability to image an Earth-like extrasolar planet. *Nature*, 446: 771–73.

Tyson, R. K. (2000). *Introduction to Adaptive Optics*. Bellingham, WA: Society of Photo-Optical Instrumentation Engineers (SPIE) Press.

Wizinowich, P., Acton, D.S., Shelton, C., et al. (2000). First light adaptive optics images from the Keck II telescope: A new era of high angular resolution imagery. *Publications of the Astronomical Society of the Pacific*, 112: 315–19.

14 Scientific Opportunities for 30-Meter-Class Optical Telescopes

Richard S. Ellis

CONTENTS

INTRODUCTION: THE LEGACY OF LARGE ASTRONOMICAL TELESCOPES

The ground-based optical telescope is arguably one of the most important scientific inventions as it was the first to extend the range of our natural senses in exploring the Universe, thereby heralding a new era in scientific method aided by technological advances. A modern-day large telescope is also an excellent example of what our civilization does well. In the 400 years since its invention, the collecting area of the optical telescope has increased 100,000 times, and we are now poised to witness a further increase of a factor of 10 (Figure 14.1). Progressively larger and more powerful telescopes have led to numerous discoveries that have shaped our view of the Universe.

Although much of the early progress was driven by the natural curiosity and technical ingenuity of insightful and wealthy individuals who also served as the principal observers (e.g., William Herschel, 1738–1822), by the dawn of the 20th century, the concept of a large telescope servicing a community of astronomers had been established. Through national and international partnerships, the telescope has since played a vital role in the development and productivity of our astronomical communities. Two criteria in particular have been the focus of telescope development.

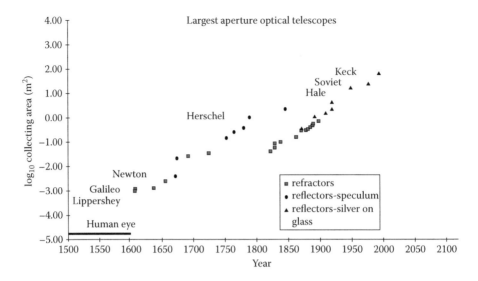

FIGURE 14.1   The inexorable growth of telescope aperture with time, commencing with Lipperhey and Galileo in the early 17th century and ending with the Keck 10 m (33 ft) telescopes commissioned in the 1990s. Large telescopes have led to discoveries that continue to shape our view of the Universe and, since the early 20th century, have led to the creation of national and international scientific communities. (Courtesy of Jerry Nelson, TMT.)

THE QUEST FOR LARGER APERTURE

The diameter D of a telescope's primary mirror (or lens) clearly governs its light-gathering power (which is $\propto D^2$) and thus has been key to progress in studying faint celestial sources. The drive to larger aperture telescopes began within only 20 years of the instrument's invention, with Johannes Hevelius (1611–1687) and Christiaan Huygens (1629–1695) grappling with unwieldy refractors with focal lengths up to 150 ft (45 m) in order to combat unwanted optical aberrations. William Herschel and William Parsons (3rd Earl of Rosse, 1800–1867) later championed ever-larger reflectors, both pushing the fabrication technology to achieve a deeper glimpse of space (King, 1955). It is hard to pinpoint when the drive toward ever-larger telescope apertures became a community issue, but the most well-documented advocate serving a community must surely be George Ellery Hale (1868–1938), the American solar astronomer who provided the leadership for realizing four successive generations of large optical telescopes (the Yerkes 40 in [1 m] refractor, the Mount Wilson 60 in [1.5 m] reflector, the 100 in [2.54 m] Hooker telescope on Mount Wilson, and the Palomar 200 in [5 m], now named the Hale telescope). A good summary of this remarkable legacy of large telescopes is the famous quote made in justification for construction of the 200 in reflector: "Starlight is falling on every square mile of the Earth's surface and the best we can do is to gather up and concentrate the rays that strike an area 100 inches in diameters" (Hale, 1928).

THE IMPORTANCE OF ANGULAR RESOLUTION

Although this chapter will concentrate on ground-based telescopes, the 2.4 m (7.9 ft)-diameter *Hubble Space Telescope* (*HST*) has shown the increasing importance of a 2nd criterion essential for the detailed study of celestial objects—*angular resolution*. Planned in the 1960s and 1970s, and modest in aperture by current ground-based standards, *HST* was launched in 1990 and has, until recently, demonstrated improved angular resolution over ground-based facilities by avoiding the blurring induced by turbulent layers in the atmosphere. On even the best ground-based sites, the native image quality (or "seeing") at optical wavelengths is typically 0.5 arcsec, corresponding to the diffraction-limited performance of a telescope only 0.25 m (9.8 in) in diameter.

HST has revealed the importance of angular resolution in many areas of astronomy: in locating faint compact sources such as distant supernovae (SNe) against the background light of their host galaxies, in resolving remote galaxies so their morphologies become apparent, and in resolving individual stars of various kinds in nearby bright galaxies.

The limited angular resolution of ground-based telescopes was recognized as a fundamental problem decades before *HST* was launched, and this led to the proposal that adaptive optics (AO) schemes be used to correct for atmospheric blurring (Babcock, 1953). Nearly a century ago, Hale (1908), pondering how large, ultimately, a telescope might someday be constructed, wrote: "It is impossible to predict the dimensions that reflectors will ultimately attain. Atmospheric disturbances, rather than mechanical/optical difficulties, are most likely to stand in the way. But perhaps even these, by some process now unknown, may at last be swept aside. If so, the astronomer will secure results far surpassing his present expectations."

TWO REVOLUTIONS IN REALIZING LARGE TELESCOPES

The main conclusion I wish to draw from the above introduction is that it is the combination of both aperture *and* resolution that is key to progress in astronomy. Fortunately, we finally stand on the threshold of realizing a new generation of large telescopes capable of delivering unprecedented photon-gathering power while delivering AO—corrected images close to the diffraction limit. Only in the last 10 years have two necessary technical components become available.

SEGMENTED MIRROR TELESCOPES

The history of large reflecting telescopes (Figure 14.1) has often been the struggle to manufacture primary mirrors. Hale and his colleagues suffered a number of failures in casting the famous Hooker 100 in (2.5 m) and Palomar 200 in (5 m) primaries. Supporting the increased mass of large primaries has led to both stiff honeycombed structures and thin "meniscus" designs supported with active control mechanisms. Materials with low thermal expansion coefficients were also developed.

A radical departure from the historic "monolithic" primaries was the segmented mirror concept pioneered by the Keck 10 m (33 ft) telescopes. Via the demonstration of good image quality using 36 actively controlled 1.8 m (5.9 ft) segments, this has opened the way for a new generation of extremely large telescopes (ELTs) whose primaries are heavily segmented, including the Thirty Meter Telescope project (TMT; http://www.tmt.org/) and the 42 m (138 ft) European Extremely Large Telescope (E-ELT; http://www.eso.org/public/teles-instr/e-elt.html).

ALL-SKY AO

The second component that enables the combined advance of larger apertures and improved angular resolution is laser (-assisted) guide star adaptive optics (LGSAO; see the chapter by Shao in this volume). AO is the technology used to improve the performance of a ground-based telescope by reducing the effects of rapidly changing atmospheric image degradation. Conventionally, this is done with a deformable mirror placed in the science light path (Figure 14.2). The surface of this mirror is adjusted in order to correct a distorted incoming wavefront. The signals that guide this real-time correction are deduced by analyzing, with a wavefront sensor, an incoming wavefront from a bright reference source (or guide star) within the science field. If a sufficiently bright star is available, it can be used as a reference (natural guide star AO [NGSAO]); otherwise, the signal reflected from atomic particles in the upper atmosphere induced by a laser beam launched from the telescope must be used.

Progress in making AO a routine tool for astronomers has been rapid over the last 10 years. Initial advances were impressive for Solar System and Galactic targets sufficiently bright to serve as their own reference sources. However, even with the most advanced AO systems, the sky coverage enabled by NGSs of adequate brightness (V ~ 12–15) became a limiting factor, especially for

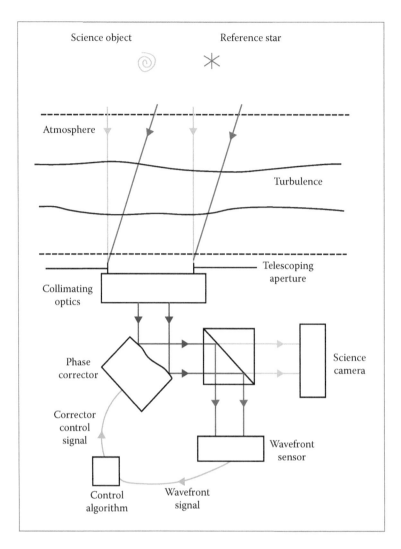

FIGURE 14.2 How AO works. Atmospheric turbulence induces wavefront and image quality degradations that can be compensated by a wavefront corrector. The relevant signal is supplied by observing a suitable reference star (or LGS) with a wavefront sensor. The correction is applied using a deformable mirror or "phase corrector." Only objects in a limited area of the sky can be corrected around the reference star that must therefore be close to the science target. The area of correction can be increased by using multiple lasers and more than one deformable mirror. (Courtesy of Gary Sanders.)

extra-Galactic targets that lie away from the Galactic plane. Only through the availability of LGSs has the technique become a general-purpose one, available to astronomers who are not specialists in using it. An LGSAO facility is, or will soon be, available on about a dozen 8–10 m-aperture (26–33 ft) telescopes worldwide, including the Keck, Subaru, European Southern Observatory's (ESO) Very Large Telescope (VLT), and Gemini telescopes.

A standard measure of the performance of an AO system is the strehl factor, which represents the ratio of the maximum intensity recovered by the system for a point source to that for a perfect system operating at the diffraction limit. As an example of recent progress, Wizinowich et al. (2006) have examined the performance of the Keck LGSAO system following a period of two years of community operations. Figure 14.3 shows that the strehl factor achieved in the 2 μ K band is typically 20%–40%, depending on the atmospheric seeing and the brightness of a tip-tilt reference

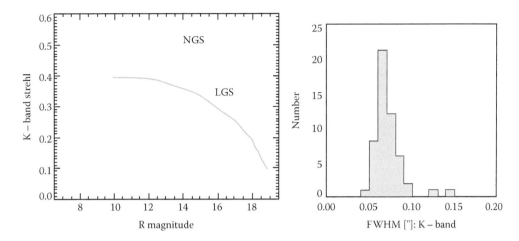

FIGURE 14.3   The performance of the Keck LGS system in the K band at 2.2 μm. The left panel shows the strehl factor, the peak brightness relative to that expected for diffraction-limited performance, as a function of the R-band magnitude of the tip-tilt reference star. The right panel shows a histogram of the achieved image quality (FWHM in arcsec). The system is routinely achieving better image quality than the *HST*. (Courtesy of Peter Wizinowich, Keck.)

star used to correct for low-order terms. Although the LGSAO system does not achieve the strehl factor for an NGS observation with a bright reference source, the sky coverage is vastly improved. The typical image quality within the K band is 0.06 arcsec FWHM (full width at half maximum), close to the diffraction limit. As a consequence of these developments, the publication output of the Keck AO facility has risen 10-fold in the past six to seven years. Highlights range from monitoring methane clouds on Neptune (NGS) to testing general relativity via the accelerated motions of stars in the vicinity of the black hole (BH) at the center of the Galaxy (LGS).

COMBINING APERTURE AND RESOLUTION

Finally, I examine the remarkable advances that are possible when we realize a 30 m (98 ft) aperture telescope that has a state-of-the-art AO facility. I will do this in terms of the sensitivity of a similarly equipped 8 m-aperture (26 ft) telescope, defining sensitivity as the inverse of the time required to reach a given signal-to-noise (S/N) ratio. If η is the system throughput and S is the recovered strehl factor, then for a telescope of aperture D, we can define two broad observing regimes:

1. *Seeing-limited observations and observations of bright, resolved sources.* Here there is no gain due to AO, and sensitivity gain arises solely from the larger light-gathering power, which scales as ηD^2. In this limiting and traditional case, a 30 m (98 ft) telescope achieves only a 14-fold gain in performance over an 8 m-class (26 ft) telescope.
2. *Background-limited AO observations of unresolved sources.* Here we achieve an additional gain of $S^2 D^2$ because AO permits us to reduce the size of the aperture within which the unwanted background signal is present. Together with the light-gathering factor, the resulting sensitivity is thus $\eta S^2 D^4$. Assuming similar strehl factors, the gain over an AO-equipped 8 m-class (26 ft) telescope is a factor of ~200. In the case of a high-contrast AO observation of an unresolved source, the sensitivity becomes $\eta[S^2/(1 - S)] D^4$.

The message from these simple examples is clear. The traditional "aperture-only" gain of a large telescope (example 1 above) is modest in comparison with dramatic advances that will come in exploiting AO in conjunction with a larger-aperture facility.

The Development and Readiness of AO

Given the pivotal role of AO in realizing the gains of future ELTs, it is natural to ask whether all of the key AO technologies are in place. The first generation of NGS systems were seriously restricted in their applicability because of limitations in sky coverage; for example, only a small fraction of interesting extra-Galactic targets have a sufficient bright star (V < 15) close by. LGSAO has overcome this restriction, but at the expense of a reduced strehl factor (Figure 14.3). This reduction in performance arises because of the "cone effect"—a single LGS illuminates the primary via a cone of radiation whose apex is not at infinity, but rather at a point ~90 km in the upper atmosphere. A consequence is that a column of residual turbulence over the primary is uncorrected in comparison with a NGS placed effectively at infinity. A second restriction is the limited field of view amenable to AO correction with a single deformable mirror.

Conventionally, a simple AO system can be considered one that is "conjugate" to a single turbulent layer in the atmosphere, and thus the AO-corrected field of view will represent the isoplanatic angle relevant for that height. These restrictions have led to the conceptual development of several enhanced AO modes whose goal is to remove or minimize their effect:

- *Laser tomography* AO (LTAO)—multiple lasers are used to defeat the "cone effect" and thereby recover the higher strehl factor currently possible only with a NGS system.
- *Multi-conjugate* AO (MCAO)—multiple deformable mirrors, optimized to correct for turbulence at various levels, are used to widen the corrected field beyond that appropriate for a single conjugate layer.
- *Ground-layer* AO (GLAO)—provides improved seeing over a very wide field of view via a low-order correction for turbulence close to the telescope itself.

Two further modes are particularly important in considering instrumentation for ELTs:

- *Extreme*-AO (ExAO)—maximizes the contrast over a very narrow field of view for detecting exoplanets. As the sensitivity gain for such a mode approaches D^6, this is clearly a highly advantageous application for an ELT.
- *Multi-object* AO (MOAO)—provides an independent AO correction for each of many spectroscopic units capable of roving within a large patrol field. MOAO differs from MCAO in that astronomical targets of interest are normally distributed sparsely over a given field, so providing an *in situ* AO correction is more practical for multi-object spectroscopic AO applications than attempting to correct the full field in a uniform manner.

We have seen evidence of great progress in LGSAO applications, but given the promise of MCAO, ExAO, and MOAO on ELTs, how close are we to realizing these more advanced modes? The necessary deformable mirrors to realize much of the hoped-for progress are now being prototyped and tested (e.g., by TMT). Likewise, the field coverage improvements made possible by MCAO have been recently demonstrated by ESO (Marchetti et al., 2007).

THE ERA OF ELTS

It is more than 15 years since the revolutionary Keck I 10 m (33 ft) telescope was commissioned on the summit of Mauna Kea, Hawaii. The success of both Keck telescopes, based on computer-controlled segmented primaries, paves the way for considering larger-aperture telescopes based on segmented primaries. And the Keck revolution came about 20 years after the commissioning of a previous generation of computer-controlled 4 m (13 ft) telescopes, including the Anglo-Australian Telescope (AAT). Noting that the development time for a new ELT is likely to be 15–20 years from the first concept study, three groups have embarked on detailed design studies for a next-generation facility. In chronological order of appearance, these include:

- The Thirty Meter Telescope (TMT; http://www.tmt.org/) has a 30 m f/1 primary composed of 492 segments each 1.44 m across. The collaboration involves Caltech, the University of California, and Canada as its founding members and Japan, China, and India as collaborating partners. The project has its history in three earlier concept studies beginning in 2000 (the California Extremely Large Telescope [CELT], the Canadian Very Large Optical Telescope [VLOT], and a study for a Giant Segmented Mirror Telescope [GSMT] undertaken by the US National Optical Astronomical Observatory [NOAO]). The TMT Project was formed in 2003 via the coalescence of these programs, and $80 million in partner funds has enabled a detailed design study, which began in 2004 and is now complete. Construction funds in the amount of $300 million have so far been pledged to the collaboration.
- The Giant Magellan Telescope (GMT; http://www.gmto.org/) has, as its founding partners, the Carnegie Institution of Science, the Harvard–Smithsonian Center for Astrophysics (HSCfA), The University of Arizona, The University of Texas at Austin, Texas A&M University, Astronomy Australia Ltd., The Australian National University, the (South) Korea Astronomy and Space Science Institute, and The University of Chicago. The GMT primary mirror comprises seven 8.2 m (27 ft) f/0.7 borosilicate mirrors providing a light-gathering power equivalent to a 21 m (69 ft) filled aperture telescope. Although only part way through a $50 million detailed design study, one of the 8.2 m (27 ft) "segments" has already been cast at the University of Arizona Mirror Laboratory and is now being polished as a demonstration piece.
- The European Extremely Large Telescope (E-ELT; http://www.eso.org/public/teles-instr/e-elt.html) is the largest currently planned telescope consisting of a 42 m (138 ft) f/1 primary comprising about 984 segments each 1.45 m in size. Although the primary mirror design is similar in many ways to that of the TMT, the E-ELT has a novel five-mirror optical design, including a 6 m (20 ft) secondary and two flat mirrors equipped to make AO correction. A €57 million design study is currently under way (2007–2012).

A montage of the three telescope designs is shown in Figure 14.4a through 14.4c. One is struck by the very different approaches. TMT is a logical extension of the twin Keck telescopes. The basic optical design is a Ritchey–Chretien hyperbolic f/1 primary with a convex hyperbolic 3.5 m (11 ft) secondary delivering an f/15 beam and a 20 arcmin field to instruments mounted on two Nasmyth platforms. The primary mirror segment size (1.4 m [4.5 ft]) is somewhat smaller than was adopted for Keck; this follows, in part, progress in improving the computational capability necessary for active segment control since the 1980s. Novel TMT features include a "calotte" dome, ensuring wind protection while maintaining adequate ventilation with an economic amount of steel, and an articulated tertiary mirror that can deflect the light with minimum overhead to a number of large instruments, ensuring maximum flexibility in science operations.

The GMT design follows from the successful fast f/1.25 parabolic primaries used in the twin 6.5 m (21 ft) Magellan telescopes and shares some of the features of the Large Binocular Telescope (LBT). These mirrors are lightweight borosilicate honeycombs manufactured in the Steward Observatory Mirror Laboratory, which now has the capability of casting and polishing equivalent mirrors up to a diameter of 8.4 m (27.6 ft). The GMT design requires the production of six off-axis circular mirrors of this size and a seventh in the center to form a segmented aperture equivalent to a 21.9 m (71.6 ft) monolith. Light passes through a hole in the center segment to an array of tertiary mirrors that can feed a number of compact instruments.

The GMT project has adopted an f/8 aplanatic Gregorian design that has a longer focal length than a conventional Cassegrain by virtue of a concave 3.2 m (10.5 ft) adaptive secondary mirror. The associated increase in the size of the telescope is mitigated by adopting a very fast primary

FIGURE 14.4   A new generation of ELTs is coming! This figure shows a montage of telescopes currently at the detailed design stage: (a) the TMT (http://www.tmt.org), (b) the GMT (http://www.gmto.org), and (c) the E-ELT (http://www.eso.org/public/teles-instr/e-elt.html). See text for further details of each. The standing figures in (b) and (c) give some indication of the scale.

focal ratio, f/0.7. This design has a number of practical advantages, for example, in the design of wide-field instruments. The main challenge will be to manufacture and figure the highly aspheric off-axis segments. It is for this reason that the consortium has already embarked on producing one.

The E-ELT is the most radical departure from its predecessors, the four 8 m (26 ft) telescopes comprising the ESO VLT. Although based on instruments placed on Nasmyth platforms as for TMT, the design incorporates adaptive optics into the main optical train, rather than within the science instruments (as for TMT) or via the secondary mirror (as for GMT). The optics involves a three-mirror anastigmat with a further two plane mirrors relaying the beam to a variety of focal stations while also serving to provide adaptive optics correction. The main focusing system involves a challenging 6 m (20 ft) diameter convex secondary and a concave 4.2 m tertiary. The overall focal ratio at the Nasmyth focus is f/17.7.

FUNDAMENTAL SCIENCE QUESTIONS FOR THE NEXT DECADE AND BEYOND

What science questions motivate the designs and fund-raising essential for realizing the above telescopes? And how practical is it to predict now what questions will be important 10 years later? In the latter respect, it is interesting to learn from history. We can go back to the mid-1980s when the science case for the Keck telescopes was planned and directly compare that vision with the first 10 years of the observatory's operation.

What we find are many surprises! The horizon for studying faint galaxies in 1985 was redshifts $z \sim 1$, whereas, in fact, Keck has pushed the frontiers to $z \sim 7$ and beyond. Synergy with *HST* was anticipated, but high-resolution imaging and spectroscopy were considered to be the sole province of *HST*; the gains now being achieved through AO were not in sight. Most of all, key discoveries such as the use of distant SNe to determine the cosmic acceleration, locating and understanding gamma-ray bursts, and detecting large numbers of extrasolar planets were not foreseen. Similar exercises can be done for earlier-generation telescopes, and the same conclusion always emerges: astronomers achieve more with new telescopes than their planners imagined! The key to success, therefore, is not so much to design rigidly around specific science projects that may no longer be relevant when the telescope is constructed, but to provide an appropriate balance of technical and science instrument *capability*, in terms of, for example, field of view, resolution, and wavelength coverage.

A future large telescope must also take account of powerful facilities at other wavelengths. These increasingly govern the way we do science on the ground, reshaping the traditional view that unusual sources are mostly found with nonoptical facilities and then "followed up" with optical/infrared (IR) large telescopes. AO on the new generation of ELTs will, for example, completely transform the synergy we can expect with space facilities such as the *James Webb Space Telescope* (*JWST*). TMT and E-ELT will outperform *JWST* in both angular resolution and light-gathering power for sources studied at near-IR wavelengths. Resolution is increasingly important in multiwavelength synergy and well catered for via future ground-based facilities such as the Atacama Millimeter Array (ALMA) and the Square Kilometre Array (SKA).

Science Directions

Here I list some of the key science questions that can help define the direction for the ELTs. The list is not exhaustive, and space does not permit a full discussion of each in terms of sample programs.

- What is the nature of dark matter and dark energy? What role do these ingredients play in structure formation? Although dedicated survey facilities are often more effective once a technique has been established for mapping these components, diagnostic studies are always needed to verify and improve the applicability of, for example, SNe as standard candles, or to measure faint redshifts to calibrate the power of weak gravitational lensing.
- What were the first luminous objects in the Universe, and when did they appear? When and how did the intergalactic medium become ionized? When and how did the most massive

compact objects form? Many future facilities can rightfully claim to wish to contribute to answering this question (*JWST*, ALMA, SKA), and this emphasizes a multiwavelength synergy referred to earlier.

- Can we chart a physically consistent picture of galaxy formation from high redshift to the present, including the formation and spatial distribution of heavy elements? At present, we have used our facilities to chart a variety of source categories (star forming, quiescent, dusty starbursts, etc.) without securing an overall picture of the evolutionary path. The key to progress is to move from transient properties of early galaxies (e.g., star formation rates and colors) to more physically based quantities, for example, those related to their internal dynamics and masses.
- How do stars and planetary systems form and evolve? Which planets are conducive to life, and can we verify the existence of life elsewhere in the Universe? This field is already opening up very rapidly, and there are tremendous opportunities in association with satellite projects, such as the *Convection, Rotation and planetary Transits* (*CoRoT*) and *Kepler Space Telescope* (*KST*) missions.

Before exploring a few specific applications from the above list, it is useful to return to the question of technical capability in the context of these areas. Figure 14.5 shows the requirements for some of the most promising science areas in terms of spectral resolution and angular resolution. The dotted line represents the boundary between seeing-limited and AO capability. What is striking is how many areas require angular resolutions of better than 0.2 arcsec, emphasizing the importance of AO in realizing major advances.

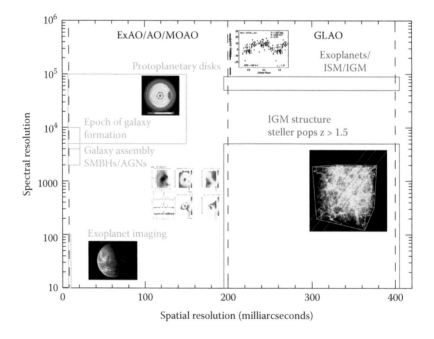

FIGURE 14.5 The scientific capabilities of a new-generation ELT in the parameter space defined by spectral resolution and angular resolution. Specific science projects (listed with illustrative subpanels; see the section "Fundamental Science Questions for the Next Decade and Beyond" in the text for further details of these) define instrument requirements and the need for AO (left side) and seeing-limited (right side) capabilities. (Courtesy of TMT Project.)

Galaxy Evolution

The highlights of progress during the last decade in multiwavelength surveys of the distant Universe can be encapsulated in the redshift dependence of the comoving density of star formation (Hopkins and Beacom, 2006) and assembled stellar mass (Wilkins et al., 2008). Multiwavelength studies of star-forming galaxies of various types (e.g., Lyman break and submillimeter galaxies) have revealed the important complementarity of observing both dust-obscured and ultraviolet (UV)-emitting sources. While there is clearly an approximate agreement between the integrated time-dependent star formation rate and the assembled stellar mass, the physical mechanisms that govern galaxy growth remain unclear. Key questions include: Do young galaxies accrete cold gas in a continuous process, or is assembly driven largely by mergers? What is the trajectory through which the various categories of high-redshift galaxies become present-day massive galaxies? What physical processes govern star formation and, in particular, lead to "downsizing"—the fact that low-mass galaxies appear to sustain activity much later than their massive counterparts?

To answer these questions, it is necessary to study selected high-redshift galaxies in much more detail than can be done with current facilities. As a typical redshift 3 galaxy is only 1–2 kpc across (corresponding to 0.5 arcsec), resolved studies demand sampling at 0.05 arcsec or better, corresponding to AO on a 20–30 m (66–98 ft) aperture. Integral field unit (IFU) spectrographs with such AO systems (Figure 14.6a) will deliver chemical and dynamical details on 100 pc scales, enabling us to address some key questions, such as:

- How does the age of the stellar population derived from spectral energy distributions and absorption-line features compare with the dynamical age of the system? At present, details of stellar populations are used to determine the luminosity-weighted age of a system without regard to its dynamical maturity. For example, old red galaxies are often assumed to have formed all of their stars in a single event in the past. Dynamical data allow us to decouple the history of star formation from that of mass assembly.
- How does the mode of star formation relate to the dynamical state? There is growing evidence that star formation is erratic and burst-like in many early galaxies. This could be because of intermittent accretion or regulation by feedback. As galaxies mature, it seems likely that their star formation rate settles to become roughly constant with time.

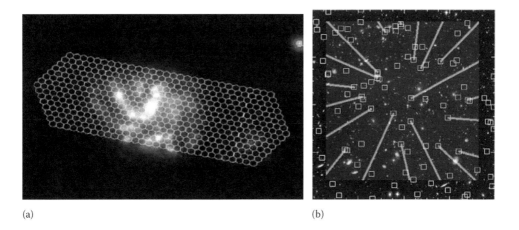

(a)   (b)

FIGURE 14.6   (a) The principle of using an IFU to spectroscopically dissect a distant galaxy. Each spaxial delivers an independent spectrum, allowing the internal properties of a source to be determined. Courtesy of Ian Parry, Cambridge University. (b) A multi-IFU spectrograph for TMT where each roving unit has its own *in situ* AO correction. Such an instrument would patrol an area of 2–3 arcmin and deliver resolved spectroscopic data for roughly 20 distant galaxies. (Courtesy of TMT Project.)

Dynamical data enable us to understand the transition from early systems to the more regular Hubble-sequence galaxies we see below a redshift of 1.

- Some progress has already been made in this direction using IFU systems on 8–10 m-class (26–33 ft) telescopes, but the sampling is nearly always suboptimal (typically 0.1–0.2 arcsec spatial resolution elements (spaxials) so that fine structure in the dynamical state is lost (Genzel et al., 2006; Law et al., 2007). Moreover, the paucity of photons even within such coarse spaxials means that only strong emission line targets can be selected with obvious biases to intensely active systems. The new generation of ELTs will enable improved sampling by a factor of 3–4, and the sensitivity will improve by a factor D^4 ($\sim \times 200$).

By reaching more efficiently to the "typical" distant galaxy, the exciting prospect of MOAO can be realized (Figure 14.6b); rather than studying distant galaxies individually, surveys of hundreds become possible. The appropriate technology comprises independently driven IFU units, each with *in situ* AO correction spanning a 2–3 arcmin field of regard, thus enabling simultaneous detailed internal studies of approximately 20 sources. Both the TMT and E-ELT teams have developed concepts for such a MOAO-fed IR spectrograph. However, such instruments are likely to be costly (~$60 million) and complex and serve to highlight the challenge of appropriately exploiting AO with future telescopes.

THE FIRST GALAXIES

The final frontier in piecing together a history of galaxy formation lies in locating the earliest sources seen at high redshift (see the chapter by Yoshida in this volume). The motivation is more profound than searching for record-breaking distant objects. After recombination, corresponding to a time when the Universe was ~300,000 years old, clouds of atomic and later molecular hydrogen clumped under their self-gravity and began to cool and collapse to ignite the first stellar systems and mark the end of the so-called Dark Ages. These early "baby galaxies" were devoid of heavy elements and produced prodigious amounts of UV radiation sufficient, perhaps, to reionize the intergalactic medium—an event called "cosmic reionization." Locating when this occurred and over what period the Universe was reionized is a major goal for future observers. Quite apart from being a landmark event in cosmic history, the demographics of early systems, for example, their distribution in luminosity and size, plays a key role in defining the subsequent evolution of galaxies.

Facilities such as *JWST*, ALMA, and SKA can rightfully claim that they will contribute to unraveling this period of cosmic history, but only *JWST* and the ELTs can identify the sources responsible for cosmic reionization. A key unknown is the expected angular sizes of the sources involved. The magnification afforded by gravitational lensing has enabled astronomers to demonstrate that at least some early star-forming sources at $z \sim 6$–7 have lensing-corrected sizes of only 150 pc, corresponding to <30 milliarcsec (Ellis et al., 2001; Figure 14.7a). In this case, ELT is equipped with AO will give a 10 to 100-fold gain in locating abundant, faint, small sources as compared with *JWST*. How would a primordial galaxy, free from heavy elements, be identified? Spectral synthesis models suggest that diagnostic lines of hydrogen and ionized helium can reveal the telltale signature of a stellar population skewed to high-mass stars (Schaerer, 2003; Figure 14.7b). AO spectroscopy with ELTs is capable of such work out to a redshift of $z \sim 15$, a limit set by the increasing thermal background beyond 2.5 μ. By contrast, the mid-IR capability of *JWST* will be unrivaled in searching for and categorizing systems beyond redshifts $z \sim 15$–20 where the Earth's atmosphere will obscure the relevant redshifted spectral features.

Finally, it is important to comment on how the oft-quoted synergy between, e.g., *HST* and a ground-based facility such as the W. M. Keck Observatory will change in the ELT era. Whereas *HST* produced dramatic images with angular resolutions of 0.1 arcsec, and ground-based telescopes were resigned to undertaking integrated light spectroscopy, the onset of AO has completely changed

(a) (b)

FIGURE 14.7 (a) A distant galaxy seen at a redshift $z = 5.7$ multiply imaged into a pair by a foreground cluster of galaxies acting as a gravitational lens (From Ellis, R.S., Santos, M.R., Kneib, J. P. et al., *Astrophysics Journal Letters*, 560, L119–22, 2001.). The magnification afforded by the gravitational lens enables us to resolve an early galaxy and measure its physical size. Such systems are smaller than 30 milliarcsec (corresponding to 150 pc) in diameter. If such dimensions are typical of early galaxies, AO on a 30 m (98 ft) aperture will be more effective in locating and studying such sources than *JWST*. (b) Predicted UV spectrum of a first-generation galaxy made up of only hydrogen and helium (From Schaerer, D., *Astronomy and Astrophysics*, 397, 527–38, 2003.). Vertical lines denote the origin of the emission lines: black for neutral hydrogen, blue for neutral helium, red for ionized helium. Diagnostic lines of ionized helium are visible to a redshift $z \sim 15$ with a ground-based ELT.

this partnership between ground and space. The 8–10 m (26–33 ft) telescopes with AO already have better angular resolution, admittedly over limited fields, than *HST*, and the situation will improve further with ELTs. The primary distinction between ELTs and *JWST* will be in wavelength coverage; ELTs will be unable to compete in sensitivity beyond 2.2 μ. ELTs will also be challenged in panoramic high-resolution imaging at optical wavelengths.

THE GALAXY–BLACK HOLE CONNECTION

The tight relationship observed between the mass of the central BH in a galaxy and the properties of the host galaxy itself indicates that there must be a strong connection between the assembly history of both (Magorrian et al., 1998). Presumably, energetic feedback processes originating near the BH somehow govern the rate at which incoming gas clouds cool and form new stars. Precision measurements of a BH mass can be determined only by resolving the region of influence around it and studying the velocity dispersion of stars within that region. An ELT equipped with AO offers the prospect of significantly extending the distance, and hence enclosed volume, for which such measures can be made. For example, the BH at the center of the Milky Way is $\sim 4 \times 10^6\, M_\odot$ *HST* can locate and measure such a system only to 3–4 Mpc, an 8 m (26 ft) telescope with AO to ~ 20 Mpc. By contrast, an ELT with AO could identify and measure such a BH to a distance of 70 Mpc, corresponding to a 40-fold increase in survey volume (Figure 14.8). Within such a huge local volume, other less direct measures for estimating BH masses can be calibrated and then applied to high-redshift galaxies. In this way, the growth rate of BHs and galaxy masses can be tracked independently for various populations (spirals, ellipticals, etc.). This is key to making progress in understanding how BH feedback works.

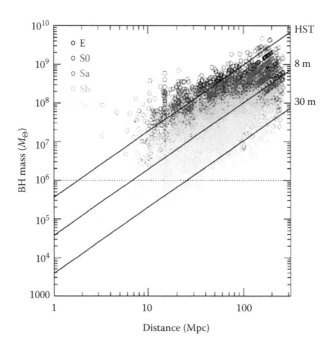

FIGURE 14.8 A simulation of the distribution of black hole (BH) mass in solar units (ordinate) and distance (abscissa) for galaxies in the local HSCfA redshift survey. Different colors refer to different morphological types. The top two solid lines represent the current distance limits for resolving the region of influence around the BH using stellar dynamics for the *HST* and a ground-based 8 m (26 ft) telescope equipped with AO. The lower solid line shows the corresponding limit for a 30 m (98 ft) telescope with AO. Clearly, the accessible volume increases dramatically, enabling calibration of other, less precise methods for measuring BH masses that can be extended to high redshift. In this way, the growth rate of BHs as a function of morphological type can be measured. (Courtesy of TMT Project.)

FORMATION OF EXTRASOLAR PLANETS

The past decade has seen an explosion in the discovery rate of extrasolar planets, the greatest number of which have been identified via Doppler surveys of solar-type stars. High-resolution IR imaging of dusty disks around young stars has likewise revealed tantalizing evidence for "gaps" in the dust distribution that may be associated with regions where planets are in the process of formation (see the chapters by Marcy, Seager, and Beichman in this volume). The key attributes of ELTs for making progress in this area are high angular resolution, increased sensitivity, and a mid-IR capability on a large aperture.

This science area is rapidly developing, and a major question of profound importance relates to the frequency of Earth-like planets and their atmospheric composition. Are the organic molecules thought to be essential precursors of life common in planet-forming disks?

A high-resolution spectrograph on a 30 m (98 ft) telescope increases the number of accessible host stars, for a given Doppler precision, by a factor of 30 or so. Moreover, the larger aperture enables the method to be extended to lower-mass stars so that detecting Earth-like planets in the so-called habitable zone (see the Beichman chapter in this volume) becomes feasible. Extreme AO offers the exciting prospect of direct imaging low-mass extrasolar planets. Reflected light at contrast levels of less than one part in 10^7 will probe inner Solar System scales. Finally, in the case where transiting planets have been located, e.g., from satellite surveys such as *CoRoT* or *KST*, the atmospheric composition of the transiting planet can be deduced from high-dispersion spectroscopy of good S/N ratio (Webb and Wormleaton, 2001; Figure 14.9).

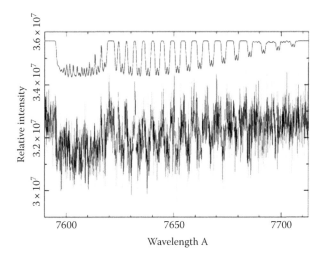

FIGURE 14.9 Detecting the absorption signatures of various gases in the atmosphere of a transiting planet (From Webb, J.K. and Wormleaton, I., *Publications of the Astrononomical Society of Australia*, 18, 252–8, 2001). Stars with transiting planets will be identified from precursor surveys such as those being conducted with the *CoRoT* or *KST* space missions. The simulation shows the difference between the spectrum of the star during and after a transit. With adequate signal-to-noise ratio (in this case 30,000 per 6 km/s resolution element), the signature of molecular oxygen in the planetary atmosphere can be identified.

SEEING-LIMITED APPLICATIONS—THE COSMIC WEB

Whereas the above programs have fully exploited AO, it is worth pointing out that the aperture gain of an ELT over the current generation of 8–10 m (26–33 ft) telescopes offers unique gains in the seeing-limited regime.

In the late 1970s, absorption-line spectroscopy of distant quasars revealed the presence of intergalactic clouds of neutral hydrogen along the line of sight, often termed the "Lyman alpha forest" from the ground-state transition most likely to be observed (Sargent et al., 1980). Subsequently, it was realized that many of these Lyman alpha absorbers had corresponding metal lines at the same redshift; thus, not only can the column density and sizes of the absorbing clouds be determined, but also their composition and ionization state can be tracked. This area of astronomy has become a valuable probe of the intergalactic medium and its association with nearby galaxies. Galaxies adjacent to the Lyman alpha clouds emit UV photons that maintain the ionization state, and outflows presumably enrich the medium chemically.

Yet distant quasars that are bright enough to provide the background signal for absorption-line spectroscopy are rare, so the valuable 3-dimensional cross-correlation signal between absorbing clouds and emitting galaxies cannot easily be created (Figure 14.10a). However, the gain in limiting magnitude afforded by a 30 m (98 ft) aperture brings a rapid increase in the number of distant galaxies suitable for high-dispersion spectroscopy (Figure 14.9b). By utilizing galaxies rather than quasi-stellar objects (QSOs) with a multi-object spectrograph, the cosmic web at high redshift can be properly explored.

SEEING-LIMITED APPLICATION—EXPANSION HISTORY OF THE UNIVERSE

The final application is very specific and relates to the mystery of the dark energy that causes the Universe to accelerate. (See the chapter by Sullivan in this volume.) Studies of distant SNe in the 1990s revealed this remarkable fact whose physical explanation remains unclear. The extant data (e.g., Astier et al., 2006) are most naturally explained by the presence of a constant vacuum

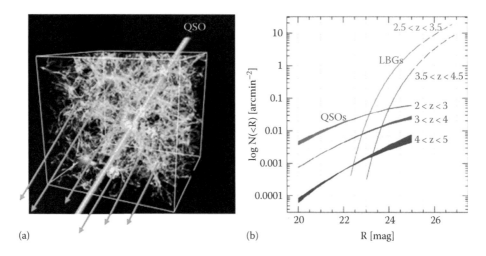

(a)   (b)   R [mag]

FIGURE 14.10   (a) Probing the cosmic web of high-redshift neutral hydrogen gas associated with dark matter via the absorption-line spectrum of background quasars with lines of sight probing various absorbing regions represented by blue lines. A high-dispersion spectrum of the bright background quasar reveals the detailed 1-dimensional distribution of hydrogen clouds along the line of sight. (b) Counts of quasars and Lyman break galaxies as a function of apparent magnitude. The fainter limiting magnitude accessible with an ELT brings many more suitable background sources into view, enabling full 3-dimensional tomography of the gas in the intergalactic medium. (Courtesy of TMT Project.)

energy—a negative pressure—that may herald a new property of spacetime. Alternatively, conceivably relativistic gravity requires some modification on large scales.

ESO astronomers have proposed an instrument—the COsmic Dynamics EXperiment (CODEX)—that would directly measure the expansion rate of the Universe in real time (Liske et al., 2008). Loeb (1998) proposed that the redshift drift in the Lyman alpha forest might be observable over a sufficient number of years and would yield a measure of the Hubble constant H(z) at early time. This is an extraordinarily weak signal corresponding to a shift of less than 1 cm/s/yr. The key requirements for such an observing parameter are a high-resolution spectrograph with excellent wavelength stability and continuous monitoring of a sufficiently large number of QSOs over a long period of time. Only the large collecting area of the 42 m (138 ft) E-ELT makes this possible. Across the sky, about 18 QSOs are bright enough with the E-ELT aperture for the required 20-year program. It is proposed that a high-frequency laser comb operating at 15 GHz can provide adequate calibration accuracy (Steinmetz et al., 2008). Why is such a measurement required given the likely investment in dedicated satellites for tracking dark energy (e.g., the *Joint Dark Energy Mission* [*JDEM*] and ESA *Euclid* program)? By studying the expansion rate at $z \sim 1.9$ and 3.5, dark energy is probed in a redshift range beyond the reach of the normal tracers such as Type Ia SNe and weak gravitational lensing (Figure 14.11).

CONCLUSIONS

In summary, we have seen how the success of segmented-mirror telescopes and LGSAO now makes it practical to move forward with the construction of a new generation of 30 m-class (98 ft) optical and IR telescopes. Many communities have presented compelling scientific arguments based on the unique combination of a large aperture and near-diffraction-limited image quality. In this regime, the gains in sensitivity are truly spectacular. Moreover, AO significantly changes the current synergy between space-based and ground-based facilities, emphasizing the importance of wide-field high-resolution survey imaging and mid-IR applications in space. Some technological hurdles remain, for example, in some of the more ambitious science instruments proposed, but detailed

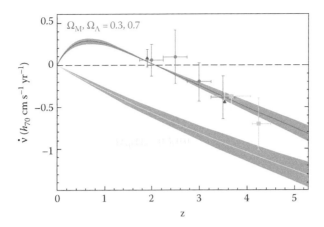

FIGURE 14.11   The CODEX project proposed for the E-ELT. The plot shows the velocity drift, i.e., the rate of change in the observed wavelength of an absorbing cloud (in cm/s/yr), for various redshifts in a universe without a cosmological constant Λ (lower curve) and in one with a dominant value ($\Omega_\Lambda = 0.7$). Different colored symbols relate to different sampling strategies in terms of the number of quasars monitored. The shaded areas represent the result of varying the present-day Hubble constant by $\pm10\%$. By monitoring roughly 20 quasars at a spectral resolution of R ($= \lambda/\Delta\lambda$) ~30,000 over 20 years, the experiment would provide the first direct measure of the rate of cosmic expansion beyond the redshift range accessible to traditional probes of dark energy, such as Type Ia supernovae and weak gravitational lensing (From Liske, J., Grazian, A., Vanzella, E., et al., *Monthly Notices of the Royal Astronomical Society*, 386, 1192–218, 2008.).

design work often in concert with industrial partners has shown the feasibility of moving forward with these new facilities. The financial requirements are daunting, as they have been throughout the history of realizing large telescopes, but the scientific discovery space is phenomenal.

REFERENCES

Astier, P., Guy, J., Regnault, N., et al. (2006). The Supernova Legacy Survey: measurement of Ω_M, Ω_Λ and w from the first year data set. *Astronomy and Astrophysics*, 447: 31–48.

Babcock, H. (1953). The possibility of compensating astronomical seeing. *Publications of the Astronomical Society of the Pacific*, 69: 229–36.

Ellis, R.S., Santos, M.R., Kneib, J. P. et al. (2001). A faint star-forming system viewed through the lensing cluster Abell 2218: First light at z~5.6? *Astrophysics Journal Letters*, 560: L119–22.

Genzel, R., Tacconi, L.J., Eisenhauer, F., et al. (2006). The rapid formation of a large rotating disk galaxy three billion years after the Big Bang. *Nature*, 442: 786–89.

Hale, G.E. (1908). *The Study of Stellar Evolution: An Account of Some Recent Methods of Astrophysics*, 242. Chicago: University of Chicago Press.

Hale, G.E. (1928). Science and the wealth of nations. *Harper's Monthly Magazine*. January, 243–51.

Hopkins, A.M. and Beacom, J. (2006). On the normalization of the cosmic star formation history. *Astrophysical Journal*, 651: 142–54.

King, H.C. (1955). *The History of the Telescope*. London: Charles Griffin & Co.

Law, D.R., Steidel, C.C., Erb, D.K., et al. (2007). Integral field spectroscopy of high-redshift star forming galaxies with laser-guided adaptive optics: Evidence for dispersion-dominated kinematics. *Astrophysical Journal*, 669: 929–46.

Liske, J., Grazian, A., Vanzella, E., et al. (2008). Cosmic dynamics in the era of extremely large telescopes. *Monthly Notices of the Royal Astronomical Society*, 386: 1192–218.

Loeb, A. (1998). Direct measurement of the cosmological parameters from the cosmic deceleration of extragalactic objects. *Astrophysical Journal*, 499: L111–14.

Magorrian, J., Tremaine, S., Richstone, D., et al. (1998). The demography of dark massive objects in galaxy centers. *Astronomy Journal*, 115: 2285–2305.

Marchetti, E., Brast, R., Delabre, B., et al. (2007). On-sky testing of the Multi-Conjugate Adaptive Optics Demonstrator. *European Southern Observatory (ESO) Messenger*, 129: 8–13.

Sargent, W.L.W., Young, P.J., Boksenberg, A., et al. (1980). The distribution of Lyman-alpha absorption lines in the spectra of six QSOs: Evidence for an intergalactic origin. *Astrophysical Journal Supplement Series,* 42: 41–81.

Schaerer, D. (2003). The transition from Population III to normal galaxies: Lyman alpha and Helium II emission and the ionizing properties of high redshift starburst galaxies. *Astronomy and Astrophysics,* 397: 527–38.

Steinmetz, T., Wilken, T., Araujo-Hauck, C., et al. (2008). Laser frequency combs for astronomical observations. *Science,* 321: 1335–338.

Webb, J.K. and Wormleaton, I. (2001). Could we detect O_2 in the atmosphere of a transiting extra-solar earth-like planet? *Publications of the Astrononomical Society of Australia,* 18: 252–58.

Wilkins, S.M., Trentham, N. and Hopkins, A.M. (2008). The evolution of stellar mass and the implied star formation history. *Monthly Notices of the Royal Astronomical Society,* 385: 687–94.

Wizinowich, P.L., Chin, J., Johansson, E., et al. (2006). Adaptive optics development at Keck Observatory. In: B.L. Ellerbroek and D.B. Calia (eds.), *Advances in Adaptive Optics II, Proceedings of the Society of Photo-Optical Instrumentation Engineers (SPIE)* Conference, Vol. 6272: 62709–2715.

Part V

Intellectual Impact of the Telescope on Society

15 The Impact of Astronomy on Chinese Society in the Days before Telescopes

Yi-Long Huang

CONTENTS

INTRODUCTION*

Throughout the course of Chinese history, astronomy has enjoyed a special position in society. Thanks to its political and social significance, astronomical records have been passed down through generations, providing a valuable research resource for modern-day historians. Artifacts and ancient monuments related to astronomy are also rich and abundant in China (Zhuang and Wang, 1988; Xu et al., 2000).

The Imperial Astronomical Bureau played an important role in almost every imperial house since at least the Han dynasty (206 BCE–220 CE). The obligations of ancient Chinese astronomers included sky observation, calendar calculation, and divination (through astrology, day-selection, feng shui, etc.). They observed the stars and constellations and then compared the results with the calculated movements of the Sun, the Moon, and the planets in hope of gaining a better understanding of their function.

Since the Han dynasty, the empire's support and control may have provided an important driving force for the institutionalization of astronomy in China. Moreover, the concept of astrology and the structure of the calendar we find in Han times shaped the subsequent Chinese astronomy up until at least the 17th century (Chen, 1987; Needham and Wang, 1959; Ho, 2003; Sivin, 2008). In the following sections, I will investigate the impact of astronomy on ancient Chinese society by looking at astrology and calendar compilation.

ASTROLOGY AND ANCIENT CHINESE SOCIETY

Ancient Chinese astronomers mapped all of the visible stars in a circle with the Polar Star as the center. They further divided the circle into 12 equally divided sectors, called *fen* 分 (kingdoms) or *ye* 野 (states), which corresponded to feudal regions in the Chinese empire (Pan, 1989). A traditional Chinese local gazette often has a section explaining the *fen* and *ye* corresponding to the county or prefecture. The celestial circle was also divided into 28 unequally divided sectors, called *xiu* 宿 (lunar mansions), which have different symbolic meanings (Schafer, 1977). For example, the lunar mansion *Wei* 胃 (stomach) is related to barn and food, and the lunar mansion *Zhen* 軫 (carriage) is related to army dispatch, etc.

---

* Also see the chapter by Xiao-Chun Sun in this volume for a detailed discussion of the impact of the telescope on astronomy and society in China.

The belief that sky phenomena were closely related to society or the ruling class began in China at least 2,000 years ago. The nomenclature of stars and asterisms in China reflects such character-istics because they are obviously a projection of the ruling class, government organizations, and the geographical areas of the Chinese empire. Traditional Chinese astrology was a highly elite and exclusive field, monopolized by the imperial astronomers and leaving no place for ordinary men who practiced other fortune-telling methods (Ho, 1966), which is quite different from horoscopic astrology in the West.

In ancient China, solar eclipses and certain other sky phenomena were seen as signs specifically related to the emperor, who was regarded as the Son of Heaven. Solar eclipses were treated espe-cially seriously during the Han dynasty given that they were believed to be a sign of bad admin-istration. When an eclipse occurred, the emperor usually had to send an edict to criticize himself and seek advice from wise and virtuous people. At this time, solar eclipses even provided notable leverage in power struggles, and prime ministers were occasionally removed from office because of them (Ban and Ban, 1975). However, in later periods, the value of this astronomical phenomenon decreased because eclipses had became more predictable.

Sky events were also sometimes used to justify the legitimacy of the emperor. In 419, a great comet appeared on the sky for more than 80 days. Because China was ruled by seven different regimes at that time, it was important to clarify which regime was responsible for this inauspicious phenomenon. The emperor of Northern Wei (386–534) was extremely concerned about this event, so he gathered scholars and astrologers and asked for their opinions. A famous astrologer, Cui Hao 崔浩, suggested that this meant that someone would soon usurp the throne of Eastern Jin. In less than two years, Cui Hao's predic-tion was realized when Liu Yu 刘裕 seized the power of Eastern Jin. As a result of this, the astrologer won the respect of his emperor (Wei, 1975). This incident shows that Cui Hao was successful not only because he was a good astrologer and was blessed with good luck, but also because he was very astute in politics.

Supported by the government, Chinese astronomers maintained an assiduous observation of the sky and became the most persistent and accurate reporters of celestial phenomena and natural anomalies in the world during ancient times. According to recent statistics, more than 10,000 astronomical records are preserved in ancient Chinese literature, including sunspots, eclipses, lunar occultations, novae and supernovae, comets, meteor showers, meteors, planetary and lunar movements, etc. These records in official literature were usually astrologically significant and were maintained by court astronomers and historians. Such a detailed collection is of immense value to astronomers and seismologists of later gen-erations for the study of stellar outbursts, comets, or the Earth's rotation (Stephenson and Clark, 1977). Among the ancient civilizations of the world, China is the only one with extensive and readily available astronomical records still preserved, especially from the 5th century BCE to the 10th century CE.

Given that most astronomers in ancient China were government employees, the majority of astronomical records were, therefore, found in bureaucratic archives. In most dynasties of imperial China, astronomers were allowed to work only for government institutes. This practice ensured the monopoly and continuity of astronomical activities. In order to guarantee the preservation of knowledge, the government enforced a hereditary system, mandating that only the descendant of an imperial astronomer could have a position at the Astronomical Bureau, although this may have lim-ited the development of astronomy overall because the group allowed to practice it was so exclusive.

Many previous studies on the history of Chinese astronomy were focused on its technical aspects. However, astronomy in ancient China was mainly used to prognosticate the fates of rulers and states, thus belonging to so-called portent astrology. Imperial astrologers often provided advice during military campaigns. For instance, when Emperor Zhenzong 真宗 (reigned 997–1022) of the Song 宋 dynasty initiated a war with the northern tribe Qidan (1004), an accompanying astronomer reported to him that there was yellow fog around the Sun, which showed a sign of reconciliation between the two countries. We do not know whether the report of the astronomer affected the deci-sion of the emperor, but a peace treaty was signed before long (*Song shi* 宋史; see Tuotuo, 1975).

More examples can be found during the early Ming 明 dynasty when Emperor Zhu Yuanzhang 朱元璋 (1328–1398) sent messages to his generals at the border to warn them of a possible invasion from enemies as a result of astrological interpretations (*Ming shilu* 明实录, 1962). In 1383, he even sent an army of several hundred thousand to the north to prevent the invasion of Mongolians because of a sky phenomenon related to the Moon (*Ming shilu*, 1962).

Astrology indeed played an important role in ancient Chinese society. Here, its significance will be illustrated by interpretations of two special celestial phenomena. The first is known as a five-planet conjunction, the most auspicious event in Chinese astrology, which implies that one will "amass the world ... the virtuous will be celebrated and the great men who change reigns will control the four quarters" (Huang, 1990). The second is the stationary point of Mars in the lunar mansion *Xin* 心 (heart; in Scorpio), a most inauspicious event (Chang and Huang, 1990; Huang, 1991b).

Through calculation of the positions of the five planets between 1000 BCE and 2000 CE, it was discovered that, during this period, a total of 27 conjunctions were visible to the naked eye within a span of 30 degrees. In short, observable five-planet conjunctions have occurred at a frequency of less than once every century, on average. Furthermore, only four of them ever gathered within an interval of 10 degrees.

Two of the most prominent and easily observed five-planet conjunctions over the last 3,000 years happened in 185 BCE and 710 CE. In 185 BCE, the minimum span of separation between the planets was no greater than 7 degrees (Figure 15.1). This auspicious phenomenon coincided with the time when the despotic Empress Lü Zhi 吕雉 (?–180 BCE) assumed the highest power of the Han empire.

Lü Zhi was the imperial consort of Liu Bang 刘邦 (247–195 BCE), founder of the Han empire. Her son, Liu Ying 刘盈 (211–188 BCE), ascended to the throne after the death of Liu Bang. In order to secure the succession of her son and to consolidate her political strength, Lü Zhi poisoned Liu Bang's most beloved son, Liu Ruyi 刘如意. Furthermore, she cut off the limbs of Ruyi's mother and also caused her to become blind, mute, and deaf. After Lu Zhi asked her son to take a look at her "work," Liu Ying was so frightened that he became mentally ill and was no longer able to manage government affairs.

When Liu Ying died in 188 BCE, Lü Zhi selected a son of his imperial concubine as the emperor because the empress did not give birth to a legitimate successor. Because the new emperor was too young, Lü Zhi acted as the actual ruler herself. The five-planet conjunction of 185 BCE occurred in the third year after Lü Zhi assumed power. Although no record of this remarkable astronomical event is extant, it surely attracted everybody's attention at that time. Lü Zhi, as the acting dictator of the country, was certainly interested in the interpretation of this event. Probably as an affirmation

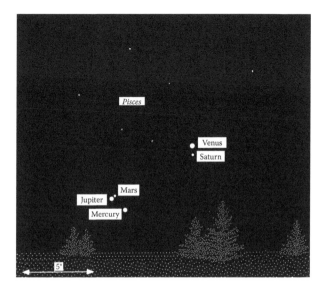

FIGURE 15.1   The five planets visible in the sky from Changan (108.9E, 34.3N) at 5 AM, March 25, 185 BCE.

of her political power, in the next year Lü Zhi decided to discredit the young emperor by claiming that he suffered from mental disease. No one dared to challenge her, and Lü Zhi's political power was upheld and assured (Ban and Ban, 1975).

The five-planet conjunction of 710 CE was even more interesting. It was visible to the naked eye starting around late May of that year. The five visible planets collected along a line on the northwestern sky and were quite easily observed just after sunset. At the time, Empress Wei 韦 was accused of plotting a rebellion against the emperor Li Xian 李显 (?–710) of the Tang dynasty. Wei became Li Xian's consort only after the death of his first wife, who had incurred the displeasure of Wu Zetian 武则天 (624–705), mother of Li Xian. In 684, Li Xian insisted on promoting Wei's father to the level of a high official, which was strongly against Wu Zetian's will. Wu Zetian therefore dethroned Li Xian and crowned another of her sons, Li Dan 李旦, as emperor. At the time, all affairs in the country were decided by Wu, who went even further to dethrone Li Dan eight months later. Wu Zetian founded the Zhou dynasty in 690 and became the first female emperor in China. In 705, the exiled ex-emperor, Li Xian, restored the Tang dynasty after a mutiny against his mother. The deposed empress, Wu, died quietly in her bed at the age of 80 later that year.

Wei and her secret lover, Wu Sansi 武三思, the nephew of Wu Zetian, played an important role in the restoration. Although China was ostensibly ruled again by a male, Empress Wei was, in fact, the mastermind behind her husband's throne. With Wu Zetian's example in mind, Empress Wei became more and more ambitious. First, she managed to execute informers of her rebellion plot. While the emperor started to suspect the misbehavior of his wife, Empress Wei had already set the wheels in motion. On July 8, 710, the emperor suddenly died after poison was put into his favorite cake by his wife, according to historians. The empress concealed her husband's death for two days until she had appointed her own relatives to key military posts. The only remaining son of Li Xian, Li Congmao 李重茂, a youth of 15, was enthroned on July 8. But Empress Wei announced herself as regent and held all power in her control.

It was quite a coincidence that a five-planet conjunction occurred during the plot in 710 CE. On June 26, the planets came even closer than in 185 BCE, the span of separation among the planets being within 6 degrees (Figure 15.2). This prominent sky phenomenon was visible for more than two months until Mercury moved too close to the Sun on July 3. Although we have no direct evidence, it is reasonable to assert that the astrological implication of these phenomena may have strengthened Empress Wei's intention to seize all power from her husband.

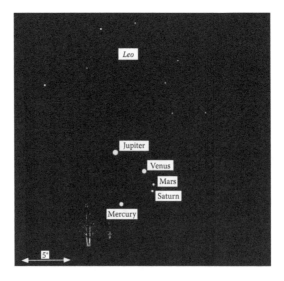

FIGURE 15.2   The five planets visible on the sky from Changan (108.9E, 34.3N) at 19:45, June 26, 710 CE.

On July 21 of the same year, Prince Li Longji attacked the palace at night with strong support from his aunt, Princess Taiping 太平公主, son of Wu Zetian, and beheaded Empress Wei and her henchmen. Li Longji's 李隆基 father, Li Dan, brother of Princess Taiping, succeeded to the throne, and Li Longji was chosen to be the crown prince. However, the real power was in the hands of Princess Taiping.

In midsummer 712 CE, regardless of the strong protest of Princess Taiping, Li Dan reacted to an observational report that recorded an irregular change in the star α Her representing the emperor, using it as an excuse to abdicate the throne, putting Li Longji in power. In the summer of 713, followers of Princess Taiping planned to stage a coup d'état to dethrone the new emperor, Li Longji. However, the plot was disclosed beforehand, and the emperor took the opportunity to kill the rebels. Princess Taiping was permitted a dignified suicide in her mansion. Her death marked the end of an era in which various women held the political power of the Chinese empire for more than two decades.

Empress Wei, Princess Taiping, and Li Longji might have all believed that they were the legitimate ruler because of the omen of a five-planet conjunction. This prominent sky phenomenon was believed to foretell the enthronement of a great ruler, but did not designate the name of the sovereign. Therefore, it left space for free interpretation and provided strong psychological support for various people to rationalize their political ambitions.

Although modern astronomical calculation shows that the five-planet conjunctions of 185 BCE and 710 CE were remarkable and lasted for a long time, so far no written records of these prominent phenomena have been found, while similar events with less satisfactory conditions were recorded several times in history and were all treated as extremely auspicious signs. In some cases, the event would be mentioned even if the five planets could not be seen simultaneously in the sky because of the interference of sunlight. The reason why the above-mentioned conjunctions were not recorded is perhaps because, in astrology, such a phenomenon was considered to be an omen of an enlightened ruler and of a change of dynasty, whereas Empresses Lü and Wei were marked as rulers of no virtue by subsequent historians. It is possible that, in order to avoid the embarrassment of such an explanation, the male-dominated historians who were responsible for creating the official history books simply omitted the two conjunctions.

While the five-planet conjunction was the most auspicious sky phenomenon, the stationary point of Mars in the lunar mansion *Xin* (in Scorpio) was regarded as most inauspicious in Chinese astrology. It implied that the prime minister would fall from power or that the emperor would die. In examples of the occurrence of this event, we see an even stronger interaction between astronomy and politics. According to the *Tianwen zhi* 天文志 (*Astronomical Treatise*) of *Han shu* 汉书 (*History of the Han Dynasty*), the rare event occurred in the spring of 7 BCE. On March 27, Prime Minister Zhai Fangjin 翟方进, who was charged with responsibility for the reported sky phenomena, committed suicide at the request of Emperor Liu Ao (52–7 BCE). The frightened emperor believed that this action could fulfill the omen and save his own life. Unfortunately, the healthy emperor did not escape from his destiny and suddenly died on April 17.

However, according to our calculations, there was no such sky event in 7 BCE. Mars was, in fact, stationary in Virgo in early February and passed Scorpio only in late August and early September. Obviously, this record was fabricated by the imperial astronomers. Although the stationary point of Mars usually remains visible for several nights, ancient people believed that it suddenly appeared and lasted for only a short time. Few people could question this belief because they thought that the movement of a stellar object could sometimes disagree with a prediction in accordance with Heaven's will.

The political struggle between the Wang and Zhai families may explain the fabrication. After the death of Emperor Liu Ao 刘骜, the Wang family was in power. Later on, Wang Mang 王莽 even overthrew the Han dynasty and founded the short-lived Xin dynasty (9–24 CE). There was no direct evidence as to who benefited from the fabricated sky report, using it to justify the removal of Zhai, or what caused the death of Emperor Liu Ao. However, Wang Mang is a possible suspect because he had previously poisoned a child emperor before he founded the new dynasty. Also, Zhai Fangjin's son was the first person to oppose the ruling of Wang Mang, perhaps to get his revenge both for his father and the late emperor.

Inspired by this case, modern astronomical analysis was made of 22 other records of "Mars staying in Xin" in official history books. It was found that 16 of them were probably false. On the other hand, about 40 such cases, which had occurred since the Western Han dynasty, failed to be recorded. Because such celestial phenomena were treated as the most inauspicious sky events in Chinese astrology, ancient astronomers might counterfeit some records to agree with current events, probably to argue for the validity of applying astrology in prophecy.

Furthermore, by studying the expressions of this phenomenon, we found that the astronomers in ancient China often made reference to what had happened after previous cases and added some more concrete descriptions into astrological books, thus expanding the flexibility of interpretation of astrology. By means of the counterfeit records, the credibility of astrology in the minds of its followers increased. This may be an important contributing factor to astrology penetrating ancient Chinese society, becoming a major part of traditional astronomy.

While most ancient Chinese astronomical records are genuine, several reports of the stationary point of Mars in Scorpio (including the one in 7 BCE) in ancient Chinese literature were fabricated, probably because of its strong astrological implications. On the other hand, the two excellent five-planet conjunctions discussed in this chapter were deliberately deleted by historians of later generations, who were mostly male, again probably because of the sharp contradictions between the astrological implications of the events and their historical accuracy.

THE CALENDAR AND CHINESE SOCIETY

Compiling a yearly calendar was one of the most important functions of the Imperial Astronomical Bureau in ancient China. Usually, several types of calendar would be issued each year: one was presented to the emperor, one to the crown prince, and one to the queen; another was for public use (the so-called civil calendar); and another specialized on the motions of the Sun, Moon, planets, etc.

A typical civil calendar was usually 32 pages (in the modern sense), which included two pages at the beginning listing the length of each month, then a two-page diagram showing the directions of some important spirits in the year (Figure 15.3), then two pages for each month listing the auspicious and inauspicious activities each day (Figure 15.4). The last four pages of the calendar assigned creatures to years (rat, ox, tiger, rabbit, dragon, snake, horse, ram, monkey, cock, dog, and pig) for people born in the last 60 years, rules for some spirits, and the names of imperial astronomers responsible for compiling the calendar.

Using the Qing dynasty (1644–1911 CE) as an example, every first day of the 10th month in the Chinese calendar, a grand ceremony presided over by the emperor was held to present the next year's calendars to important royal family members and all high-ranking officials. On the same day, millions of copies of civil calendars were distributed or sold by local governments to ordinary people all over the country.

Being an indispensable reference book for daily life activities, the civil calendar is perhaps the most universally circulated book in China, even today. Before and during the Tang dynasty (618–907 CE), the government usually needed to hire numerous people to copy calendars each year. But the production of these handwritten copies could hardly meet the needs of society. By the 9th century at the latest, some enterprising entrepreneurs started to adopt the newly developed printing technology to produce relatively cheap versions of the official civil calendar for profit. A certain official, Feng Su, stated in 835 that illicit printed copies had found their way to market areas even before the promulgation of the new civil calendar (Smith, 1992).

Because of the high demand of society, several different varieties similar to the state calendar existed in the market. One of the earliest almanacs is now held in the British Library and bears

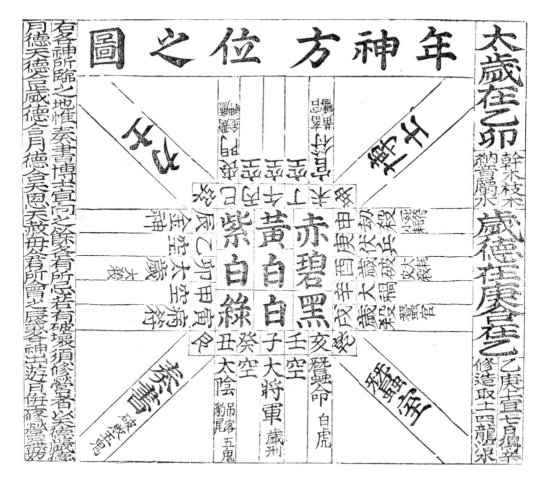

FIGURE 15.3   A diagram showing the directions of some important spirits for the year of 1855 in a Qing dynasty calendar printed in Korea.

the date 877 CE. This almanac is in a long poster form that shows day columns, zodiacal animals, and various fate-extrapolation systems. Its table-oriented design, coupled with numerous depictions, made it attractive, especially compared with the monotonous design of the official calendars.

Tongshu 通书 (the privately edited almanac) is another popular form that is usually several times thicker than a state calendar (Palmer, 1986). The contents of a *tongshu* comprise not only much detailed description on the auspicious and inauspicious activities for each day, but also rules for all kinds of fortune-telling methods, morality tales, and practical advice for daily life. Local eclipse data would also be provided in *tongshu*, but not in the civil calendar published by the government.

Although some people argued that once an error appeared on the wood block it would appear in all printed copies, the Song government decided, in 1024, to adopt the printing method to produce annual civil calendars. The cost of producing calendars dropped by 90% because of the introduction of printing technology (Li, 1986). After that, errors in the calendar became a serious problem. The director of the Imperial Astronomical Bureau was removed from office in 1184 because of the appearance of such typographic errors. The staff actually responsible for editing the calendar were also demoted, or hit 100 times with a stick (Xu, 1936).

Although a detailed sales report was usually not available for most years, we know that 3,123,185 copies were sold in 1328 from the fragmentary accounting records of that year, which are extant in *Yuan shi* 元史 (*History of the Yuan Dynasty*). From this, we can estimate that roughly one out of four

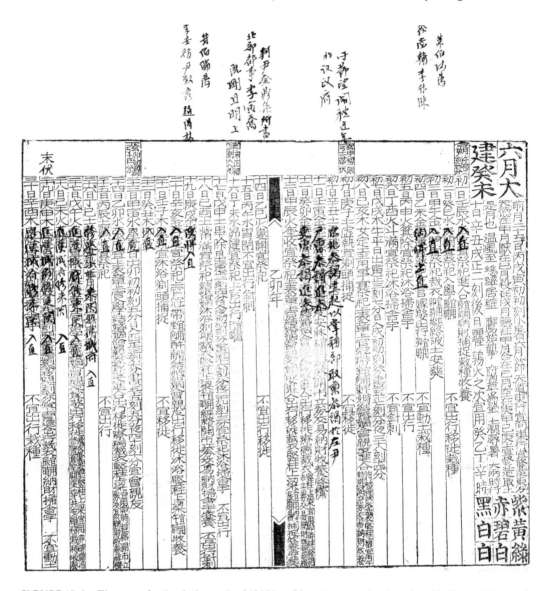

FIGURE 15.4 The pages for the sixth month of 1855 in a Qing-dynasty calendar printed in Korea. The auspicious and inauspicious activities are given for each day. For example, on the sixth day of the month, it states that the date is good for offering sacrifices, taking a bath, and cleaning the house, but no good for acupuncture. This calendar was owned by a high-ranking Korean official who wrote some of his personal activities on the calendar.

families in the country bought a legal copy at that time (Song, 1975). The popularity of calendars may even be wider because we have found that occasional prohibitions were issued by the Mongol Yuan (1279–1368) government against calendar piracy.

Selling the calendar could be quite profitable to some governments. It is astonishing to find that the total income accrued by selling the calendar amounted to roughly 0.5% of the state cash revenue in 1328! But this was not always the case for later periods. In 1382, Zhu Yuanzhang (1328–1398), the first emperor of the Ming dynasty, ordered the exemption of the cost of calendar production from calculations of the state budget (*Ming shilu*, 1962). Yongzheng, emperor of the Qing dynasty, made a clear statement in an edict of 1729 that his government treated the calendar distribution as a responsibility, but also as a profitable business.

During the early Song dynasty, a certain Hou family obtained the monopoly on the printing and selling of calendars according to the style and content of the official civil calendar. At the time, a pirated copy was worth one or two *qian* 钱 (5–10 *qian* was worth roughly 1 kg of rice at that time; 1,000 *qian* was roughly equivalent to one *tael* [*liang* 两] of silver). In 1071, the Prime Minister, Wang Anshi 王安石 (1021–1086), took back the distributing privilege and decided that the government would print the calendars itself. He even raised the price to several hundred *qian* a copy in order to increase state income (Li, 1986). Wang's economic reform failed in the end.

In spite of the large quantities printed each year, no more than 100 or 200 calendars before the end of Ming dynasty (in 1644) are extant out of billions of copies published. The main reason is because the content of a civil calendar is mostly related to the specific year and, therefore, is extremely time-sensitive. People either threw the used calendar book away or recycled it for other purposes (e.g., using its blank back or empty space to practice handwriting, using it as the paper lining for embroidery).

Apart from setting the beginning date, as well as the length of each month and year, traditional Chinese calendars were aimed at providing people with advice about what to do and what not to do on a given day. More than 200 spirits exist in the tradition of *zeri* 择日 (day-selection). Some of the spirits (those positive are called *shen* 神, those negative are called *sha* 煞) have a star-like or star-related name, but they usually have nothing to do with the actual location or movement of the stars. On the other hand, the locations of these spirits are usually based on the *ganzhi* 干支 (a sexagesimal dating system in China) of the day, the month, or the year.

In various examples from literature and classical novels, we often see how people made use of calendars for picking an auspicious date to cut out garments, to get married, to bury the dead, to have a haircut, to visit a patient, to take a bath, etc. They also would check a calendar to avoid an inauspicious date or direction to go on a long journey, to start building a house, to cultivate a garden, to receive acupuncture, to move house, etc. In 1817, Emperor Jiaqing 嘉庆 of the Qing dynasty even issued a decree to reprove his subordinates who submitted no memorial at all to the throne during certain days just because those days are indicated in the calendar as most suitable for submitting one (Huang, 1996a). In fact, some Chinese still use a calendar to select auspicious days for certain activities in their life.

The calendar was a very practical book for the Chinese (especially the Hans) in daily life. Therefore, it is no surprise that people in Dunhuang started to compile their own calendars after the occupation of Tibet in about 786 CE because they could no longer obtain calendars from the Chinese government. Because of their lack of acquaintance with calendar calculations, the length of each month, or the arrangement of an intercalary month, is frequently different from that in the Chinese calendar; however, calendars compiled in Dunhuang, in general, shared the same tradition of day-selection (Huang, 1992).

Starting from at least the Yuan dynasty, every civil calendar had to bear a government seal to prove its legal status. The law emphasized that anyone who compiled such a book illegally was subject to decapitation, and any informer would be awarded a large sum. The main reason for implementing such a heavy punishment was that the ancient Chinese treated the right of preparing a calendar as a symbol of having the highest political authority in the empire. In other words, all people needed to use the same calendar in a unified society. The economic interest of the proceeds of selling the calendars might have been another reason for some governments.

The strong symbolic meaning in politics meant that every party contending for hegemony regarded the calendar highly. For example, during the Ming–Qing transition, we find that almost every political power rushed out to issue a new calendar to claim its position of legitimacy. In fact, one of the first things a new dynasty in China usually did was to promulgate a new calendar.

The rulers of Korea and other neighboring countries that were treated as the vassal states of the Chinese empire had to send a special envoy to China each year in order to receive certain copies of the following year's calendar. In fact, the calendar became an important political symbol to prove the connection of a vassal state with China. These vassal states used the received calendar as a sample to print for their countrymen. In these calendars printed by vassal states, the dynasty title of the Chinese government had to be maintained instead of using their own reign title.

The tradition of selecting an auspicious date and direction in China was already quite mature in the 3rd century BCE. From the *rishu* 日书 (divination book) unearthed from a tomb sealed in 217 BCE at Shuihudi 睡虎地, Yunmeng 云梦, in the province of Hubei, we find abundant and complicated statements concerning the date selection for the procreation of babies, marriage, illness, burying the dead, hunting, planting, building, etc. (Loewe, 1994). Although the concept of *rishu* shares the same origin as the late-period calendar, it is probably a professional reference book for fortune-tellers. In contrast, the description in yearly calendars is much more direct and easy to understand by ordinary people.

From the content of old calendars, we can clearly see an evolutionary process and sometimes traces of cultural interaction. Those earliest calendars unearthed from Han tombs mostly gave the length of each month and the order in the sexagenary cycle of each day. Only a few of them indicated some spirits under certain days. However, those calendars discovered from Dunhuang 敦煌 (mostly compiled from the 8th to the 11th century) already provided abundant information on what to do and what not to do. After that, the main structure and content of calendars have no significant changes through the Song dynasty, although some details, emphases, or rules may vary from time to time.

Within the calendars discovered in Dunhuang, we find that many of them bear the character *mi* 灭 (translation of *mir* in Sogdiana) at the top of day columns every seven days. Those dates marked by *mi* are always Sundays according to the calculation. This system is, in fact, influenced by Manicheism, which was introduced to China as early as in 694 (Zhuang, 1960).

Through the distribution of Manicheism in China, the Western weekday system was absorbed implicitly into the Chinese calendar. In the 13th century, we find that each day in the calendar was assigned to one of the 28 lunar mansions whose order is fixed. The lunar mansion assigned to each day will start over after the end of the cycle is met. Moreover, each lunar mansion is also associated with one of the seven luminous celestial bodies (in the order of the Sun, the Moon, Mars, Mercury, Jupiter, Venus, and Saturn). Those days assigned to one of the four lunar mansions associated with the Sun are, in fact, Sunday, with the Moon, Monday, etc. Interestingly enough, the relationship between the day of the week and the associated celestial body is the same as that of the ruler of the day in Western astrology.

Probably starting from the Yuan dynasty, the content and style of the civil calendar experienced a new transformation. Those spirits usually appearing in the day columns and having too technical a meaning to ordinary people were omitted. An easy-to-read tabular form was employed. Such forms were followed through the Ming dynasty with almost no changes.

After the Manchus assumed the reins of government in 1644, the Jesuit (Johann) Adam Schall von Bell (1591–1666) took over the Imperial Astronomical Bureau and attempted to use his astronomical knowledge as an indirect means to preach Christianity. He had proposed making sweeping changes in the traditional calendars by the so-called new Western methods; the whole idea was not favored by the government. Nonetheless, certain changes, later on, were imposed on the civil calendars.

Probably affected by European advances in understanding, as well as the Church's view, the missionaries, in general, were hostile to the occult sciences involved in the Chinese calendars. Schall, as both a missionary and an imperial astronomer, however, had to make a compromise between meeting the requirements of Chinese society and not violating the attitude of the Church.

Within the calendars he compiled, Schall adopted similar forms of day columns as the traditional ones, but added some new materials (e.g., the sunrise, sunset, joint/midpoints for various locations in China, etc.) at the beginning of the calendar to convince his missionary fellows that the book was "scientific" and to convince Chinese people that it was a brand new calendar, designed specifically for the new dynasty.

The largest change Schall made was to use a new definition to subdivide each year (from the winter solstice to the next winter solstice) in 24 joint/midpoints. The traditional method employed an equal separation in time, but the new method employed an equal separation in solar longitude.

Schall's definition, which has no counterpart in Western astronomy, is still used today. Although Schall claimed that his definition was more "scientific," its rule of arranging an intercalary month is, in fact, much more complicated than the traditional one.

Schall also made some changes to dilute the level of superstitious content in his calendar. For example, he reversed the order of the two lunar mansions *Zi* 觜 and *Shen* 参 (Huang, 1991a). Their correct ascension order in the Chinese 28 lunar mansions was thought to be *Shen* following *Zi* (both in Orion). Yet this traditional order had changed since the late 13th century, because of the effect of precession, into *Zi* following *Shen* and became easier to be observed. In a Chinese calendar (even today), the 28 lunar mansions are arranged in correspondence with each day, although they have nothing to do with the astronomical positioning of the day. People born in dates of different corresponding lunar mansions are believed to have different fates.

Mainly because these reforms were contradictory to the traditional knowledge of fortune-tellers, Schall's new calendars incurred great skepticism from Chinese society. Although compiling a calendar was a serious crime, in the collections of Bibliothèque Nationale de Paris is an unofficial calendar of the 9th year of the Shunzhi 顺治 period (1652) that was printed in Fujian 福建. This calendar was quite similar to the official ones except that it still adopted the traditional order of *Zi* and *Shen* and provided the time of joint/midpoints according to both new and traditional definitions. The appearance of such a calendar reflects that the adoption of "new Western methods" may not have been so easy for Chinese society at the time. To some people, accepting "new Western methods" may have been more difficult than accepting the new reign of the Qing dynasty (Huang, 1996b).

Perhaps to advance his position in the Bureau, Schall decided to hold a more tolerant attitude toward the divination expressions applied in the traditional civil calendars. Practical as this attitude might have been, it had no way to get the overall approval of the missionaries and caused Schall to be severely reproached. Controversies about the divination expressions arose among the Jesuit missionaries. In 1664, the Church finally concluded that what was concerned with occult sciences in the Chinese calendars should be subsumed as superstitious.

Before giving any response to this conclusion, Schall was charged by Yang Guangxian 杨光先 (1597–1669) and the other Chinese conservatives for using an "incorrect" divination school to select a time to bury a prince who had died young; this accusation and others related to the confrontation between the traditional calendar and "new Western methods" resulted in the well-known Kangxi 康熙 Calendar Lawsuit. Consequently, Schall was put in jail, and traditional definitions were again adopted in the calendar. Not long afterward, Schall passed away without seeing the redress of the case by another Jesuit, Ferdinand Verbiest (1623–1688), who restored the "new Western methods" in 1669.*

While Jesuit astronomers were in charge of the Imperial Astronomical Bureau until 1826, the traditional order of *Shen* following *Zi* was adopted again in 1756 because of the continuing opposition from the conservative supporters. This case concretely indicated the close and undivided relationship of ancient astronomy and society in China.

Probably influenced by the heated arguments in events such as the Kangxi Calendar Lawsuit, Emperor Qianlong ordered his subordinates to publish an encyclopedia called *Qinding xieji bianfang shu* 钦定协纪辨方书 (*The Imperial Treatise of Time and Direction Selection*) to provide a "correct" way of working (Parker, 1888). Several such encyclopedic books with similar content published during the Yuan dynasty are still extant. In 1376, the Ming emperor, Zhu Yuanzhang, also ordered imperial astronomers to compile an encyclopedia called *Xuanze lishu* 选择历书 (*Time Selection for a Calendar Book*). Although *The Imperial Treatise of Time and Direction Selection* is not the first encyclopedia focused on time/direction selection, or the first one prepared by imperial authority, it is certainly the most influential divination book ever published in China. Most books published afterward would habitually claim, sometimes with no authenticity, that they were based on this book (Huang, 1996a).

---

* Also see the chapter by Peter Harrison in this volume.

In the early Qing dynasty, because official calendars were too simple to meet the needs of the people and were not distributed widely enough, some privately edited *tongshu* were introduced into the market. Because the printing of the *tongshu* challenged the right of government, in 1723 the Qing government reestablished the ban of the *tongshu* and pirated copies of the calendar and assigned certain businessmen to sell the official calendars in specific areas as a monopoly. However, this action bred popular discontent because the local officials and businessmen raised the price of the calendar significantly for profit. As a response, starting in 1730, the Qing government allowed ordinary people to use the official wood blocks of the calendar for printing.

However, pirated calendars were still circulated, especially in some out-of-the-way districts where legal ones were not easy to obtain in time. Those peddlers daubed the calendar cover with red color to mimic the government seal in order to hoodwink buyers. In 1751, the Qing government reexamined the laws relating to the calendar, which were usually too severe. According to the new regulations, ordinary people were allowed to reprint the calendar as long as it followed the official form, and a government seal was no longer needed on the cover of each calendar.

After 1751, many fortune-tellers and book printers began to scramble for this high-profit market. A very successful publisher in Quanzhou 泉州 named Jichengtang 继成堂 was said to sell several hundred thousand copies each year and circulate some overseas. This publisher even developed a voluntary chain-store system to sell its compiled *tongshu*, divination-related materials (e.g., incense for determining an exact time), and some popular medicines. This publisher owned 62 vendors spread over various counties in southern China in 1816 (Figure 15.5), thus antedating the first Western chain store (The Great American Tea Company), which appeared in the United States in 1859. The owners of Jichengtang were very business oriented. They utilized direct mailing to increase sales not long after post offices were set up widely in China in 1898.

To protect its popular annual *tongshu*, a red seal that indicated the genuine location of Jichengtang was hand stamped at the beginning of the section of day-columns of each *tongshu*. A serious warning to pirates can also be found at various places within the book.

The *tongshu* of Jichengtang started printing in roughly 1797. Although its influence is no longer so large in mainland China, it is still a standing firm in Taiwan, even after the dramatic changes of the last 200 years. According to a rough estimation, more than two-thirds of all the *tongshu* publishers in Taiwan now claim that their knowledge of compilation comes either directly or indirectly from Jichengtang.

Many editors of the *tongshu*, while pretending that their almanac corresponded with the official astronomy, claimed that their calendrical calculations were based on the "new Western methods" employed by the Astronomical Bureau and their auspicious explanations on *The Imperial Treatise of Time and Direction Selection*. However, because of the popularity the *tongshu* had gained, some calendars, in turn, partially assimilated the contents and the forms of the *tongshu*; this assimilation again reflects the interactive relationship between traditional astronomy and society in China.

Interestingly enough, almanacs in early modern Europe are in many respects like the Chinese calendars. Not long after the advent of printing technology in Europe, Gutenberg issued a printed almanac in 1448. By the 1470s, large numbers were being published. The European almanac in the early 16th century usually contained a calendar, weather forecasts, harvest predictions, lists of lucky and unlucky days, medical notes, the zodiac man, etc. Many people even used it in determining the proper times for surgery, letting blood, or taking medicine.

The English government, in the second half of the 16th century, made almanac publication a monopoly. Only a few stationers were granted the right to run this highly profitable business. Those who were making illegal and counterfeit almanacs were fined, and serious offenders were even prosecuted. In the 1660s, for which detailed figures survive, sales averaged roughly 400,000 copies annually in England, and roughly one out of three families bought one each year (Capp, 1979).

Both calendars and almanacs in China belong to the class of popular press that was usually considered of little academic importance. However, they do deserve more attention from the history community because they are a treasure from which we may learn a lot about the daily life or the

FIGURE 15.5   The vendors appeared in 1816 *tongshu* published by Jichengtang.

popular culture of the traditional Chinese society. From their content and distribution, we can also learn a great deal about the pattern and the scale of how divination interacted with a society.

CONCLUSION

Through the study of astrology and calendars in ancient Chinese culture, we may gain a more rounded picture of ancient Chinese astronomy and learn its richness in humanistic spirit, as well as, in some cases, its close connection to both politics and society in China.

REFERENCES

Ban, G. and Ban, Z. (ca. 85 CE) (1975). *Han shu* (*History of the Han Dynasty*). Beijing: Zhonghua shuju.
Capp, B. (1979). *Astrology and the Popular Press-English Almanacs 1500–1800*. London: Faber and Faber.
Chang, C.-F. and Huang, Y. (1990). Tianwen dui Zhongguo gudai zhengzhi de yingxiang: Yi Han xiang Zai Fangjin zisha wei li (The influence of astronomy toward ancient Chinese politics: Using the death of the Han prime minister Zhai Fangjin as an example). *Qinghua xuebao (Tsing Hua Journal for Chinese Studies)*, 20 (2): 361–78.
Chen, Z. (1987). *Zhongguo tianwenxue shi (History of Chinese Astronomy)*, 4 Vols. Shanghai: Renmin chubanshe.

Ho, P.Y. (1966). *The Astronomical Chapters of the Chin Shu.* Paris: The Hague, Mouton and Co.

Ho, P.Y. (2003). *Chinese Mathematical Astrology: Reaching Out to the Stars.* London and New York: RoutledgeCurzon.

Huang, Y.-L. (1990). A Study on five-planet conjunctions in Chinese history. *Early China,* 15: 97–112.

Huang, Y.-L. (1991a). Qing qianqi dui *Zi Shen* liang xiu xianhou cixu de zhengzhi (Confrontation on the order of two lunar mansions *Zi* and *Shen* in early Qing period). In T.-H. Yang and Y. Huang (eds.), *Jindai zhongguo kejishi lunji (Conference Proceedings on Science and Technology in Modern China).* Taipei: Zhongyang yanjiuyuan Jinshisuo & Hsinchu: Lishi yanjiusuo, National Tsing Hua University.

Huang, Y.-L. (1991b). Xingzhan shiying yu weizao tianxiang: Yi 'yinghuo shou Xin' weili (Astrology, implication, and fabricated sky events: Using "Mars Stationary in Scorpio" as an example). *Ziran kexue shi yanjiu,* 10 (2): 120–32.

Huang, Y.-L. (1992). Dunhuang ben juzhu liri xintan (A reexamination of Dunhuang's almanacs). *Xin shixue (New History),* 3 (4): 1–56.

Huang, Y.-L. (1996a). Tongshu: Zhongguo Chuantong tianwen yu shehui de jiaorong (Tongshu: Fusion between traditional Chinese astronomy and society). *Hanxue yanjiu (Chinese Studies),* 14 (2): 159–86.

Huang, Y.-L. (1996b). Cong Tang Ruowang shuobian mingli shixi Qing chu Zhong-Ou wenhua de chongtu yu tuoxie (Adam Schall's civil calendars: A case study on China-Europe cultural confrontation and compromise in the early Qing China). *Qinghua xuebao,* 26 (2): 189–220.

Li, T. (1986). *Xu zizhi tongjian changbian (Sequel to the Comprehensive Mirror to Aid in Government in Detailed Chapters).* Beijing: Zhonghua shuju.

Loewe, M. (1994). *Divination, Mythology and Monarchy in Han China.* Cambridge: Cambridge University Press.

Ming shilu (1962). *The Veritable Records of the Ming Dynasty.* Taipei: Zhongyang yanjiuyuan lishi yuyan yanjiushuo (The Institute of History and Philology of Academia Sinica).

Needham, J. and Wang, L. (1959). *Science and Civilization in China,* vol. 3: *Mathematics and the Sciences of the Heavens and the Earth.* Cambridge: Cambridge University Press.

Palmer, M. (1986). *T'ung Shu: The Ancient Chinese Almanac.* London: Rider and Co.

Pan, N. (1989). *Zhongguo hengxing guance shi (An Observational History of Stars in China).* Shanghai: Xuelin chubanshe.

Parker, A.P. (1888). Review of the *Imperial Guide to Astrology. The Chinese Recorder,* 19 (11/12): 493–99, 547–54.

Schafer, E.H. (1977). *Pacing the Void, T'ang Approaches to the Stars.* Berkeley: University of California Press.

Sivin, N. (2008). *Granting the Seasons: The Chinese Astronomical Reform of 1280, With a Study of Its Many Dimensions and a Translation of Its Records.* New York: Gardners Books.

Smith, R.J. (1992). *Chinese Almanacs.* Hong Kong: Oxford University Press.

Song, L. (1975). *Yuan shi (History of the Yuan Dynasty).* Beijing: Zhonghua shuju.

Stephenson, F.R. and Clark, D.H. (1977). *Applications of Early Astronomical Records.* New York: Oxford University Press.

Tuotuo (1975). *Song shi (History of the Song Dynasty).* Beijing: Zhonghua shuju.

Wei, S. (1975). *Wei shu (Book of the Wei Dynasty).* Beijing: Zhonghua shuju.

Xu, S. (1936). *Song huiyao jigao (Collective Drafts of Historical Material Compiled by Song Government).* Beijing: Zhonghua shuju.

Xu, Z., Pankenier, W., and Jiang, Y. (2000). *East-Asian Archaeoastronomy: Historical Records of Astronomical Observations of China, Japan and Korea.* Amsterdam: Gordon and Breach Science Publishers.

Zhuang, S. (1960). *Miri kao (An investigation on miri). Zhongyang yanjiuyuan lishi yuyan yanjiushuo jikan (Bulletin of the Institute of History and Philology Academia Sinica),* 31: 271–301.

Zhuang, W. and Wang, L. (1988). *Zhongguo gudai tianxiang jilu zongji (A Compilation of the Astronomical Records in Ancient China).* Nanjing: Jiangsu kexue jishu chubanshe (Jiangsu Scientific & Technological Publication).

16 The Impact of the Telescope in the West, 1608–1802

Owen Gingerich

Details of the invention of the telescope early in the 17th century are frustratingly hazy. As far as written records are concerned, we know that a Dutchman named Hans Lipperhey applied for a patent in October of 1608, but his petition was eventually denied on the grounds that the device was already common knowledge (van Helden, 1977). But that this happened in Middelburg should come as no surprise, as this town, along with Murano near Venice, was one of the few places where optically clear glass was made. Quite probably other spectacle makers had noticed that a combination of two lenses could magnify an image, but in the absence of quality lenses and because of the narrowness of the field of view, there was generally little interest in pursuing the matter.

The first telescopes were little more than carnival toys, magnifying perhaps three times, and any inventor seriously interested in turning such glasses toward the sky was obliged to improve the device. This was done, for example, by Thomas Harriot in England when he used a 6-power telescope to observe the Moon in the summer of 1609 (Harriot, 1609). His image of the Moon is the earliest surviving record of a telescopic astronomical observation, but his instrument had too low a magnification to reveal the mountains or craters. That astonishing discovery was made by Galileo Galilei late in the fall of 1609. Although he had a later start than Harriot, Galileo rapidly surpassed the Englishman, who never published anything about his own discoveries.

By August of 1609, Galileo amazed the Venetian senators with an 8-power spyglass (Drake, 1978, p. 140). (The name "telescope" would not be invented until April 1611; Rosen, 1943.) He kept improving his instrument, so that by November he had a 20-power device (Drake, 1976), and around this time he discovered that images would be significantly sharper if he added a diaphragm to the outer portions of the objective lens (Figure 16.1). Most of his astronomical discoveries were presumably made with that magnification. Thomas Harriot would not reach this level until the following summer, that is, in August 1610.

Among the Galileo papers in the Florence National Central Library are two sheets bearing seven ink-wash images of the Moon. Although Galileo did not record when he made these observations, it has nevertheless been possible to establish quite credible dates. This was first done by the selenographical cartographer Ewen Whitaker (1978), with minor adjustments made since his pioneering historical work (Gingerich, 2009, p. 159). The earliest pair of images, of the crescent Moon, were made on the evening of November 30, 1609. They show a partially cratered surface, but perhaps the most notable features are the small dots of light lying along the dark side of the terminator, that is, along the line dividing the illuminated portion of the Moon from the dark part where the sunlight had not yet fallen. Galileo, with his artistic training, must have almost instantly recognized that these points of light represented mountain peaks just beginning to catch the light as the Sun rose over these parts of the moonscape. In his excitement and enthusiasm, Galileo perhaps painted too many dots. It was nevertheless a dazzling discovery because it gave the first important hint that our Moon was Earth-like, with mountains and valleys, and not the pure ethereal sphere of Aristotle's cosmology.

Perhaps the most telling lunar image of Galileo's series is the waxing crescent Moon now dated December 2, 1609 (Figure 16.2). His depiction shows the terminator running through Mare

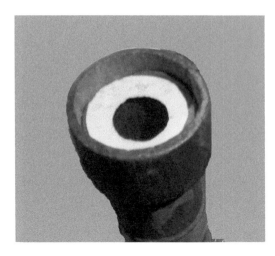

FIGURE 16.1 The objective glass in one of only two surviving telescopes believed to be from Galileo himself. The instrument belongs to the Galileo Museum in Florence. (Photo ©2009 by Owen Gingerich.)

Serenitatis with the horns of the surrounding mountain chains extending into the darkness. When the terminator reached this point a month later (29.5 days), this particular rapidly changing configuration of the shadows in Mare Serenitatis could be seen only from the opposite side of the Earth. However, in his *Sidereus nuncius*, the book in which Galileo described his earliest astronomical discoveries, this pattern with the shadow bisecting Serenitatis has been transformed to a first-quarter Moon so that Mare Serenitatis straddles into both the western and eastern hemispheres (Figure 16.2), an impossibility during the months when Galileo could have been making his lunar observations. Hence, this particular image tells us more about Galileo than about the precise mapping of the lunar surface. What was really important for Galileo was the fact that the Moon had topographic features, the heights and depths of mountains and plains. It did not matter to him precisely where the craters and maria were, but simply that the Moon exhibited these Earth-like features. This discovery directly contradicted the traditional Aristotelian view of eternal, ethereal heavens distinctly different from the terrestrial arena of change and corruptibility.

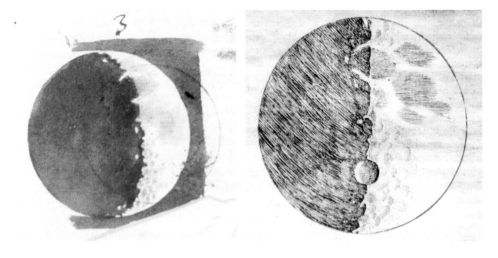

FIGURE 16.2 *Left:* Galileo's watercolor drawing of the Moon a day before first quarter, with the terminator bisecting Mare Serenitatis. *Right:* Galileo's transformed quarter-Moon image as it appears in his *Sidereus nuncius*.

The difference between Galileo's and Harriot's approaches to lunar mapping is quite striking. Apparently, Harriot finally had an instrument of sufficient power to distinguish the lunar craters sometime in 1610, after he had seen the pictures in Galileo's *Sidereus nuncius*. By August 5, he had a 20-power instrument. It is likely that in this month he recorded the features of the full Moon in remarkable detail. Because the set of Galileo's lunar drawings has a gap in the numbering, we can speculate that Galileo may also have recorded the full Moon, an image now lost. His final drawing, on the second Florentine sheet, corresponds to a lunar occultation of the star Theta Librae on the night of January 19, 1610, and it shows features in the dark area of the Moon that surely came from another source, such as a drawing of the full Moon. But the lack of shadows at full Moon might have deterred Galileo from attempting such a depiction, as his primary interest was not in mapping the Moon, but in using the shadows to demonstrate its topography. Harriot, who had had considerable surveying experience, including in coastal parts of what is now Virginia and North Carolina, was particularly attuned to mapmaking. Had he surveyed the full Moon a second time with comparable accuracy, he would surely have discovered its so-called libration or rocking motion.

The detailed examination of Galileo's lunar drawings clearly shows that he was primarily interested in lunar topography and its cosmological implications and far less in lunar cartography (Gingerich, 2009, p. 152).

Galileo admitted his Copernican sympathies in a letter to Kepler written in 1597, but when Kepler urged him to stand forth with his cosmological views, Galileo did not deign to reply. He remained a timid and less than fully convinced Copernican until his astonishing and unexpected celestial discoveries in the winter season of 1609–1610. Whether the light dawned on him during his lunar observations, or with the discovery that the heavens were far fuller of stars than anyone had imagined, or when he found Jupiter's moons that January, we cannot now ascertain with total certainty. But we can trace rather precisely one extraordinary "aha! moment" with respect to the little stars he observed next to Jupiter. On the night of January 7, 1610, the Moon hung very close to Jupiter in the southeastern sky, and perhaps this proximity led him from the Moon to the bright nearby planet. To his surprise, next to Jupiter were three small stars, all in a line. He returned to Jupiter the following night, "guided by what fates he knew not," and was puzzled to find the little stars rearranged in an unexpected way. Subsequently, he began keeping a more accurate record, and by the end of the week, on January 13, 1610, he discovered that there were four, not three, little stars. And when he saw that they lay not exactly in a straight line, he finally realized that they were in orbit around Jupiter (Gingerich and Van Helden, 2011). Almost immediately, the language of his notes abruptly changed from Italian to Latin. Why? Because Galileo realized that he had news for an international audience, and the international language of science was Latin (Galilei, 1610).

Galileo wrote up his discoveries and had them printed in an amazingly rapid two-month interval. His *Sidereus nuncius* ("*Starry Messenger*") was ready by March 13, 1610. For the first time, readers could see real images of the face of a distant astronomical body. In his book he allowed himself an interesting pro-Copernican statement. Some people found it incredible that the Earth could revolve about the Sun in an annual orbit and still keep the untethered Moon in tow. Neither Galileo nor anyone else had the physics to justify such a possibility, but Galileo pointed out that if Jupiter could do this, it might also be possible for the Earth to keep a satellite invisibly tied to itself.

In a profound way, the invention of the telescope opened the door to observational cosmology, and this was undoubtedly its greatest impact. Hanging in the balance were not just two, but three cosmological systems. There was the ancient geocentric system, laid out in most detail by Claudius Ptolemy around 150 CE. Then there was the heliocentric system with a fixed Sun and a moving Earth, proposed in detail by Nicolaus Copernicus in his *De revolutionibus* of 1543. But in 1588, the great observer Tycho Brahe advocated a geoheliocentric system in which the Sun with its retinue of planets moved around a fixed, central Earth.

Tycho (and Copernicus before him) knew that the heliocentric system predicted that there should be a small annual displacement of the stars (the so-called parallax) if the Earth was actually moving around the central point of the cosmos. Copernicus declared (correctly, as it turned out) that the failure to find the annual parallax was because the distances to the stars were so vast. Tycho, however, believed (falsely, as Galileo soon showed) that he could see the angular size of stars, and if they were so far away, they would have to be absurdly large. Tycho also knew that there was another fundamental difference between the Ptolemaic and the Copernican systems: Mars at its closest approach to the Earth in the Copernican system was half the distance of the Earth to the Sun, but in the Ptolemaic system it was slightly farther than the Sun. Tycho thought he could measure the distance to Mars, and this would have been just barely possible for his precise naked-eye astronomy if the planetary system were as small as commonly supposed. In a series of three major observational campaigns, with continually improved instrumentation, Tycho finally convinced himself that the measurements favored the Copernican arrangement (even though we know today that such a measurement with the naked eye was impossible) (Gingerich and Voelkel, 1998).

Tycho then faced a dilemma. An observer who had spent thousands of hours measuring the stars and planets, he was a thoroughgoing realist. He was not looking at just a hypothetical model (as most people considered the Copernican scheme to be), but at something real. However, in the Copernican system, Earth, "a lazy, sluggish body, unfit for motion" (in Tycho's words; Brahe, 1588), was thrown into dizzying motion. In his opinion, as a real description of the cosmos, the heliocentric motions offended both physics and Holy Scripture—after all, did not Psalm 104 say, "The Lord laid the foundations of the Earth that it not be moved forever"? This became the motivation for Tycho's own cosmology, a philosophy that saved common sense, physics, and Scripture all at the same time.

The Tychonic system was, however, a sterile one that would scarcely have opened the door to a new physics, and perhaps Galileo sensed this. By the time he published his *Starry Messenger*, Galileo knew that his telescope magnified the images of planets, but actually made stars look smaller, so that Tycho's argument (about the huge size of stars if they were as far away as the absence of observed parallax demanded) was false. Galileo undoubtedly saw the Tychonic system as an outlandishly jury-rigged compromise, but he did not dare attack it directly because it was a great favorite of the Jesuits, who had a firm lock on Catholic science education. Nevertheless, his observations of the small size of stars, as well as their plenitude, helped undermine the grounds for Tychonic cosmology.

Not only was Galileo's *Starry Messenger* a brilliant account of his astonishing discoveries, it was effectively a job application for a position in the Medici court in Florence. Within a few months of its publication, Galileo had secured the job and moved from Padua to Tuscany. And there, by the winter of 1610–1611, he made with his telescope yet another significant astronomical discovery. In Padua he had apparently never looked at the brightest planet, Venus, probably because it was shining in the morning sky and Galileo liked his sleep. Now, by the fall of 1610, Venus was easily accessible in the evening sky, and Galileo observed that the "mother of loves" (as he put it in an anagram of a Latin expression preserving his priority), like the Moon, exhibited a complete set of phases (Kepler, 1611). This was compatible with the Copernican system where Venus revolved around the Sun, but not the Ptolemaic system where Venus moved on an epicycle locked in the direction of the Sun. Because Venus never went beyond the Sun in the Ptolemaic system, it would never show a full phase. In a stroke, the Ptolemaic system was invalidated.

Had the telescope just proved that the Earth moved? No, because in the Tychonic system Venus also revolved around the Sun as the Sun carried the retinue of planets around the solidly fixed Earth. Although by Catholic standards the Lutheran Tycho was a damned author, the Jesuits fought to keep his work off the Roman *Index of Prohibited Books*, and thus both the Tychonic and Copernican systems hung in Urania's balance (Figure 16.3). Deciding between these two systems became the outstanding cosmological question of the 17th century, and while the telescope did not settle the

FIGURE 16.3   Detail from the frontispiece of Giovanni Battista Riccioli's *Almagestum novum* (Bologna, 1651). The slogan by the discarded Ptolemy proclaims, "I am raised up by being corrected." As a Jesuit, Riccioli gave precedence to his own Tychonic-type system, which is more weighty in Urania's balance than the Copernican system. Note the telescope in the hand of the observer, hundred-eyed Argus. (From the collection of Owen Gingerich.)

matter, its impact was to bring such a fresh perspective into traditional astronomy that cosmology became increasingly open for discussion.

That was where the matter stood when, in 1674, Robert Hooke, the inventive secretary of the Royal Society and Newton's nemesis, raised the issue with an unusual experiment, pioneering in the sense that perhaps for the first time a substantial experimental setup was designed to answer one specific question. Here is how Hooke introduced the state of play with respect to cosmological systems:

> 'Tis not only those [foolish people who "suppose the Sun as big as a sieve and the Moon as a cheddar cheese, and hardly a mile off"] but great Geometricians, Astronomers and Philosophers have also adhered to that side, yet generally the reason is the very same. For most of those, when young, have been imbued with principles as gross and rude as those of the Vulgar, especially as to the frame and fabrick of the World, which leave so deep an impression upon the fancy, that they are not without great pain and trouble obliterated. Others, as a further confirmation of their childish opinion, have been instructed in the Ptolemaick or Tichonick System, and by the Authority of their Tutors, overawed into a belief, if not a veneration thereof. Whence for the most part such persons will not indure to hear Arguments against it, and if they do, 'tis only to find Answers to confute them.

On the other side, some out of a contradicting nature to their Tutors, others by as great a prejudice of institution, and some few others upon better reasoned grounds, from the proportion and harmony of the World, cannot but embrace the Copernican Arguments. ...

I confess there is somewhat of reason on both sides, but there is also something of a prejudice even on that side that seems the most rational. For by way of objection, what way of demonstration have we that the frame and constitution of the World is so harmonious according to our notion of harmony, as we suppose? Is there not a possibility that things may be otherwise? nay, is there not something of probability? may not the Sun move as Ticho supposes, and the Planets make their Revolutions about it while the Earth stands still, and by its magnetism attracts the Sun and so keeps him moving about it, whilst at the same time Mercury and Venus move about it, after the same manner as Jupiter and Saturn move about the Sun whilst the Satellites move about them? especially since it is not demonstrated without much art and difficulty, and taking many things for granted which are hard to be proved, that there is any body in the Universe more considerable than the Earth we tread on. Is there not much reason for the Hypothesis of Ticho, when he with all the accurateness that he arrived to with his vast Instruments were not able to find any sensible Parallax of the Earth's Orb among the fixt Stars? (Hooke, 1674)

Hooke's instrument was a zenith telescope going up through his house toward the star Gamma Draconis, which passed directly overhead (Figure 16.4). Clever as it was, Hooke limited himself to only a handful of actual observations, and in the end his parallax measurements were not convincing.

Just over a decade later, in 1687, Isaac Newton produced his *Philosophiae naturalis principia mathematica,* a theoretical framework that resolved the physical questions that had paralyzed the

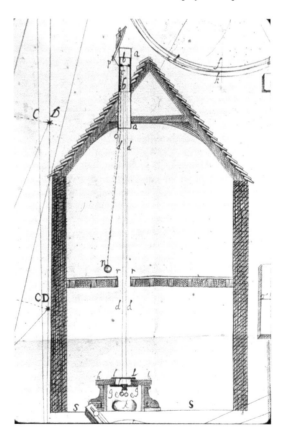

FIGURE 16.4 Robert Hooke's zenith telescope, with which he sought to find the annual stellar parallax. The light path goes through holes in his roof and his second floor. (From his *An Attempt to Prove the Motion of the Earth* (1674). [From the collection of Owen Gingerich.])

Copernican system. Universal gravitation showed why people were not flung off into space by a spinning Earth and how the Moon could remain in tow as the Earth revolved about the Sun. It accounted for the tides, and it demonstrated that comets should travel in conic-sectioned orbits. The *Principia* contains a long section on the observations and orbit of the great Sun-grazing comet of 1680, which, incidentally, was the first comet to be discovered with a telescope (by the German astronomer Gottfried Kirch). The book also contains the last known observations of this comet, made by Newton with a 7 ft (2 m) telescope after the comet had faded from naked-eye view. (Newton did not specify whether the telescope was a reflector of his own design.)

The *Principia* also predicted the oblate spheroid shape of a rotating Earth. In 1735, the French astronomer Pierre Louis de Maupertuis led an expedition to Lapland to measure the length of a degree of latitude, thereby proving Newton's theory and earning him the sobriquet "the man who flattened the earth" (Terrall, 2002). By that time, the telescope's use in surveying rivaled that in astronomy itself.

For practical purposes, the *Principia*, with the telescope as its handmaiden, settled the cosmological issue. Yet finding the annual parallax of stars remained the iconic *experimentum crucis* for demonstrating the motion of the Earth. So firmly was this locked in human imagination that when an analogous proof was obtained in 1728, made possible by telescopic observations, the discovery scarcely got the credit it deserved. This important discovery, by James Bradley, is called the aberration of starlight (Bradley, 1727–1728). Aberration was a subtle shift in the position of stars caused by the finite speed of light combined with the motion of the Earth. Its geometry was not as immediately comprehensible as the explanation of parallax—in fact, Bradley made a wooden model with moving parts to demonstrate how the aberration of starlight worked (Figure 16.5)—and perhaps this is why the annual parallax still remained the holy grail of observational cosmology.

Probably the telescopic discovery of Halley's comet in 1758 had even more impact than aberration of starlight in credentialing Newton's physics. In fact, Newtonian gravitation was not accepted

FIGURE 16.5   James Bradley's device to demonstrate the principle of aberration. The pair of diagonal lines on the movable block represents the telescope on the moving Earth. The beam of starlight, represented by the glass bead on the string, moves down (to the right) through the telescope as the telescope itself moves sideways. Thus, the starlight follows the diagonal line. In other words, the telescope has been tilted to compensate for the Earth's motion. (Courtesy of the Science Museum, Kensington, London.)

without controversy, and many of the French scientists in particular believed that it was downright superstitious to accept that invisible forces could keep the Moon coupled to the Earth. But Joseph-Nicolas Delisle, a French fan of Newton and Halley, was annoyed by his countrymen's lack of enthusiasm for Newtonian physics despite the work of Maupertuis, and he decided to teach them a lesson (Schaffer, 1990). Using Newtonian principles, he predicted the path of the returning comet and assigned his young assistant and telescopic observer, Charles Messier, to find it. When Messier succeeded, he was sworn to secrecy, so that the other French astronomers missed their chance to see Halley's comet until after it had passed to the other side of the Sun. They were of course furious at Delisle, but he had made his point! Indeed, as the 18th century marched on, the telescopic discovery of comets proved to be a particularly alluring playground for the celestial mechanicians who were following in Newton's footsteps.

Thus, by 1760, despite the fact that the annual stellar parallax had not yet been found, there were at least three observations, aided by the telescope, that helped credential the Copernican cosmology:

- Aberration of starlight (1728);
- Oblateness of the Earth (1739);
- Return of Halley's comet (1758).

With the ascendancy of Newtonian physics, the long-festering issues of whether the Earth moved withered away. When stellar parallax was finally found in 1837–1840, almost everyone already accepted a moving Earth.* Ironically, it was the observations by a brilliant autodidact who started by searching for the annual stellar parallax that turned the cosmological question in a totally new direction. In order to make these observations, he built the most remarkable telescopes the world had yet seen.

As a musician, William Herschel emigrated from his native Hannover to England, where he became increasingly fascinated by natural philosophy. Because he did not have the standard university education, he followed his own ideas about what might be interesting in astronomy. Refracting telescopes lay beyond his budget, so he decided to make his own reflector, which involved casting, grinding, and polishing a speculum-metal mirror. With his homemade telescope, he began in 1781 to use high magnification to find double stars. These he believed to be accidents of nature where a faint distant star just happened to be in nearly the same line of sight as a nearer and therefore brighter star. This circumstance offered a good opportunity to find the annual parallax because the nearer star would be more strongly affected by the Earth's motion, and over the course of a year the change in the position of the two components could possibly lead to the detection of stellar parallax (Herschel, 1782). At this young stage in his career, Herschel failed to appreciate that faintness did not necessarily mean farness. Although his parallax search never achieved his intended goal, he ultimately discovered that many of his double stars were actually physical binaries and not accidental pairings of two stars in the same line of sight. This in itself was a major finding. And, because he was using such high magnification in his search, he recognized something unusual when his systematic survey brought him to a previously unrecognized planet, a spectacular first for telescopic astronomy.

The fame and royal patronage brought by his discovery of the new planet, soon to be named Uranus, enabled Herschel to finance and build larger telescopes and to begin a far more ambitious project (Figure 16.6). Herschel not only wanted to determine the distances of individual stars through their annual parallax (a project that remained beyond the reach of even his largest

---

* Although the eventual discovery of annual stellar parallax falls beyond the chronological scope of this essay, it is useful to note here that this difficult observation was finally accomplished using two state-of-the-art achromatic instruments made by the distinguished Munich optician Joseph Fraunhofer. His 9.5 in (24 cm) refractor, made in 1824 and then the world's largest, was used by Wilhelm Struve in 1836 to measure positions of Vega, with results announced in 1837 for a parallax of 0.12 arcsec. Many false claims had been made previously, and the data were very noisy, so the results were not convincing until the following year, when Friedrich Bessel, using a 6 in (15 cm) Fraunhofer split-lens heliometer, announced a parallax of about 0.3 arcsec for the star 61 Cygni. Bessel is generally credited with discovering the first parallactic distance to a star because his results were far better defined than Struve's.

FIGURE 16.6   The Gold Medal of the Royal Astronomical Society shows Herschel's giant 40 ft (12 m) telescope on the verso. Awesome as the telescope was, it proved too clumsy for easy use, and Herschel's most productive instrument was his 20 ft (6 m) reflector. (Photo ©1975 by Owen Gingerich.)

reflectors), but wished to study the structure of the Milky Way itself. This he did by counting stars in selected zones as the Earth's rotation carried one sky area after another before his eyepiece. Where there were more stars, he deemed that the Milky Way extended farther in space, and in this way he constructed the first 3-dimensional model of our Galaxy. Wrong as it was—he had no concept of the dustiness of space that limited his vision—it nevertheless initiated and inspired a whole new field, the study of the sidereal Universe.

Meanwhile, William Herschel made a "comet finder" for his sister Caroline, who had been assisting him. With her telescope, Caroline eventually found eight comets—somewhat ironically, the only comet that William found turned out to be a planet! Caroline also began finding new nebulae, something that got William quite interested, and he began to find them too during his sweeps of the sky (Hoskin, 2006). In 1786, 1789, and 1802, Herschel published catalogs of nebulae and clusters totaling 2,500 deep-sky objects.

With Herschel's large telescopes, the closed Universe of the 15th century had truly given way to the vast structures of outer space. This was only the beginning of what telescopic astronomy would bring, but by the turn of the 19th century the impact was already enormous in changing our concepts of space and time. The stage was set for telescopes ultimately far greater than Herschel ever imagined and for a cosmological view even vaster than Galileo could have conceived. As Copernicus had declared, "So vast, without any question, is the divine handiwork of the Almighty Creator!"*

---

* These are the words with which Copernicus closed the cosmological Chapter 10 in Book 1 of his *De revolutionibus orbium coelestium* (1543).

REFERENCES

Bradley, J. (1727–1728). An account of a new-discovered motion of the fixed stars. *Philosophical Transactions*, 35: 637–61.

Brahe, T. (1588). In J.L.E. Dreyer (ed.), *Tychonis Brahe Dani Opera Omnia*, 4: 156. Copenhagen, 1922.

Drake, S. (1976). Galileo's first telescopic observations. *Journal for the History of Astronomy*, 7: 153–68.

Drake, S. (1978). *Galileo at Work: His Scientific Biography*. Chicago: University of Chicago Press.

Galilei, G. (1610). In A. Favaro (ed.), *Le Opere di Galileo Galilei*, 3: 427. Rome: G. Barbèra, 1890–1909. Reprint, 1968.

Gingerich, O. (2009). The curious case of the M-L *Sidereus Nuncius*. *Galilaeana*, 6: 141–65.

Gingerich, O. and Van Helden, A. (2011). How Galileo constructed the moons of Jupiter. *Journal for the History of Astronomy*, 42: 259–64.

Gingerich, O. and Voelkel, J. (1998). Tycho Brahe's Copernican campaign. *Journal for the History of Astronomy*, 29: 1–34.

Harriot, T. (1609). Manuscript in the possession of Lord Egremont, Petworth House, illustrated in Paolo Galluzzi (ed.), *Galileo: Images of the Universe from Antiquity to the Telescope*, 361. Florence: Giunti, 2009.

Herschel, W. (1782). On the parallax of the fixed stars. *Philosophical Transactions*, 72: 82–111.

Hooke, R. (1674). *An Attempt to Prove the Motion of the Earth from Observations*. London: T.R. for John Martyn.

Hoskin, M. (2006). Caroline Herschel's catalogue of nebulae. *Journal for the History of Astronomy*, 37: 251–5.

Kepler, J. (1611). *Dioptrice*, 22. Augsburg.

Rosen, E. (1943). *The Naming of the Telescope*. New York: Henry Schuman.

Schaffer, S. (1990). Halley, Delisle, and the making of the comet. In N.J.W. Thrower (ed.), *Standing on the Shoulders of Giants*, 254–98. Berkeley: University of California Press.

Terrall, M. (2002). *The Man Who Flattened the Earth: Maupertuis and the Sciences in the Enlightenment*. Chicago: University of Chicago Press.

Van Helden, A. (1977). The invention of the telescope. *Transactions of the American Philosophical Society*, 67: part 4. Reprint with a new introduction, 2009.

Whitaker, E. (1978). Galileo's lunar observations and the dating of the composition of *Sidereus nuncius*. *Journal for the History of Astronomy*, 9: 155–69.

17 The Impact of the Telescope on Astronomy and Society in China

Xiao-Chun Sun

CONTENTS

INTRODUCTION*

The telescope was introduced in China in the early 17th century, not long after Galileo made his telescopic observations in 1610. In Europe, Galileo's discoveries with the telescope provided substantial evidence supporting Copernicanism, shaking the very foundation on which the authority of the Roman Catholic Church was built, thus arousing fierce debates over cosmological issues. What impact, then, would the telescope have on Chinese astronomy and society at large?

Chinese astronomy consisted of two major parts, mathematical astronomy and astrology. A good astronomical system indicated the legitimacy of the imperial rule and symbolized good governance. The telescope was introduced in China when the Ming astronomers were engaged in a debate over methods to produce an accurate astronomical system and its associated calendar. This was a very practical and politically important issue. Using the telescope to observe the solar and lunar eclipses, Xu Guangqi 徐光启 (1562–1633; see Figure 17.1) and his Jesuit collaborators demonstrated that the "Western method" was superior to traditional Chinese methods in calendar making and led the Chongzhen Calendar Reform at the end of the Ming dynasty. Ironically, the "Western method" legitimatized by telescopic observations in China was not Copernican, but Tychonic (see Hashimoto, 1988, p. 19).

Galileo's major discoveries with the telescope were introduced to the Chinese as early as 1615 by the Jesuit Emmanuel Diaz in his *Catechism of the Heavens* (*Tian wen lue* 天问略). These discoveries, however, hardly came as a shock to the Chinese. The information the Chinese got from the Jesuits was partial (Sivin, 1973). The Jesuits did not even name Galileo, not to mention the cosmological and religious controversies related to his discoveries. For the Chinese, there was no categorical distinction between the celestial and the terrestrial domains. Substances in the Heavens were just as likely to be subject to changes and corruption as things on Earth. As a matter of fact, the Chinese had observed sunspots long before Galileo did. Some historians even claimed that the Chinese had observed the satellites of Jupiter before the invention of the telescope (Zezong, 1981). The Chinese considered these types of celestial phenomena as omens, indicating warnings from the

---

* Also see the chapter by Yi-Long Huang in this volume for a detailed discussion of astrology and calendar development in ancient China.

FIGURE 17.1   Xu Guangqi 徐光启 (1562–1633), who led the Chongzhen Calendar Reform at the end of the Ming dynasty by using the Western method. Also see Figure 17.2a,b.

Heavens to the rulers. Because the rulers could never be assured that they governed the country with complete perfection, they were on constant alert for strange occurrences. It was almost anticipated by the court officials and astronomers that some extraordinary phenomena would occur in the sky. So the Chinese eagerly accepted the findings of Galileo without interpreting them as a challenge to traditional Chinese philosophy.

The telescopic observations did stimulate some Chinese literati to make new speculations on cosmological issues, but this occurred only within the framework of traditional Chinese cosmology, which was essentially pluralistic and eclectic. For example, Fang Yizhi 方以智 (1611–1671) used the telescopic observation of the Milky Way to validate an ancient thought by Zhang Heng 张衡 (78–139 CE) that the stars were essentially water. Some also used Chinese precedence in observing some celestial phenomena to justify the claim of "the Chinese origin of the Western learning." The telescope was seen as just an aid to the naked eye. And it was on this account only that the Westerner had made some progress in learning that was originally Chinese. This claim may sound awkward to a modern observer, but it solved the psychological barrier to the acceptance of Western knowledge.

The telescope, however, played a more important role in Chinese astronomy than mentioned above. Astronomy was a secret science that was strictly guarded by the imperial government from private learning. More telescopes were brought into China, but they were either locked away in imperial storehouses or presented to dukes and princes as items of curiosity. The telescope also found some use in the military, as in Europe.

Japan and Korea exhibit histories similar to that of China in the 17th and 18th centuries. The telescope was brought for the first time to Japan in 1613 and to Korea in 1631. But the Japanese and Koreans in that period were much too preoccupied with Chinese astronomical systems to make substantial astronomical observations with the telescope.

ROLE OF THE TELESCOPE IN THE ASTRONOMICAL REFORM OF THE LATE MING DYNASTY

By the early 17th century, the Great Unity system (*Datung li* 大 统历) used by the Ming dynasty showed serious discrepancies in predicting celestial events, especially solar and lunar eclipses. The Great Unity system was based on the Season Granting system (*Shoushi li* 授 时历) by Guo Shoujing 郭守敬 in 1280 at the beginning of the Yuan dynasty. Although the Season Granting system was the most accurate system that had ever been produced in China, it went seriously out of step with the Heavens after being used for more than 300 years. In 1610, when the Imperial Astronomical Bureau again failed to predict a solar eclipse, the imperial authority realized that it was time to take steps toward reforming the astronomical system. The Ministry of Rites presented a memorandum (zouzhe **奏折**)* to the emperor suggesting that he summon those versed in astronomy to come to the capital to discuss matters related to calendar reform. Among those invited were Xing Yunlu 邢云路 and Fan Shouji 范守己, who had been working on improving the Great Unity system; Xu Guangqi and Li Zhizhao 李之藻, who proposed the use of the "Western method" to reform the calendar; and two Jesuit missionaries, Diego de Pantoja (1571–1618) and Sabbathinus (Sabatino) de Ursis (1575–1620), apparently recommended by Xu Guangqi and Li Zhizhao. Xu and Li had been in contact with Jesuits Mattio Ricci and others, studied mathematical astronomy with them, and worked together to translate some Western books on mathematics and astronomy. Both of them realized that the "Western method" was superior to the traditional Chinese ones in producing an accurate system, and they strongly advocated adoption of the former. This was the first time that Jesuit scholars were officially invited to participate in astronomical reform. But this suggestion was not immediately accepted by the emperor. He summoned only Xing Yunlu and Li Zhizhao to the capital to consult on calendrical matters.

Xing Yunlu was a Promoted Scholar of 1580 and was appointed to the local position of Inspector in Henan. In 1596, after the current Great Unity system had failed on several occasions to predict solar and lunar eclipses, he proposed changing the calendar in a memorandum to the emperor. The fact that an official outside of the Imperial Astronomical Bureau proposed changing the calendar was in itself arrogant because during the Ming dynasty studying astronomy in private was prohibited. Some criticized him on that account, but he managed to fend this off by arguing that it was justifiable for Confucian scholars to study the calendar. He could proceed unharmed also because the current astronomical system was indeed in bad shape.

Xing Yunlu's strategy for reform was to work in the traditional Chinese framework of calendrical science, i.e., to modify and improve the Season Granting system. For this purpose, he made a study of historical astronomical systems and wrote a book titled *Studies on Calendars and Harmonics from Past and Present* (*Gu jin lü li kao* 古今律 历考).

Li Zhizhao, on the other hand, proposed using Western astronomy to reform the calendar. In 1613, he submitted a memorandum pointing out mistakes in the predictions of solar and lunar eclipses by the Astronomical Bureau. He recommended employing Jesuit astronomers for the purpose of calendar reform. He mentioned in particular that the instruments used by Jesuit astronomers were extraordinarily precise.

The imperial authority took a pragmatic attitude toward astronomical reform. As long as a good calendar could be produced, it did not matter what methods, Western or Chinese, were used.

Calendar reform was initiated by four competing proposals. The first was based on the Great Unity system as proposed by Xing Yunlu and others. This was essentially the system made by Guo Shoujing at the beginning of the Yuan dynasty, now hopelessly outdated. The second was the Islamic *Huihui* 回回 system. Islamic astronomy had already been introduced in China under Mongul rule in the Yuan dynasty. A separate astronomical bureau, called the *Huihui* Department, had been established. During the Ming dynasty, this practice continued, and some astronomical

---

* The standard translation of "zouzhe **奏折**" is "memorial to the throne"; this was a kind of special report by very senior officials written on paper that was folded in accordion form and submitted to the emperor.

treatises on Islamic astronomy had been translated into Chinese. Now some astronomers proposed using the *Huihui* system, which in their eyes was not as alien as the Western system. The third method, which constituted a new Chinese system, was proposed by Wei Wenkui. The fourth was the Western astronomical system, which Xu Guangqi and Li Zhizhao strongly advocated.

To decide which system to adopt, the usual practice was to evaluate the proposed systems by checking predictions based on their methods against actual celestial events. From the long history of astronomy, the Chinese had already learned that the most accurate test of an astronomical system was the prediction of a solar eclipse. It was also politically important for the Imperial Astronomical Bureau to be able to predict solar eclipses because, if unpredicted, they were seen as warnings from Heaven that the virtue of the ruling house was failing. The imperial authority had been greatly alarmed when the current astronomical system failed to predict solar eclipses. For example, when the Astronomical Bureau again failed to predict the solar eclipse of May 1, 1629, the emperor admonished the astronomers: "The Astronomical Bureau has submitted wrong predictions of both the beginning time and the ending time of the solar eclipse. Astronomy is a serious matter. Tell them, I forgive them just this time. If they make mistakes again, surely I will have no mercy for them" (Wang, 1981, p. 325).

The Western method proved to be able to predict eclipses more accurately than the other systems. It gave more accurate values for the time and magnitude of an eclipse. On the occasions of several eclipses, Xu Guangqi would present to the throne beforehand the predictions made using the Western method, the Great Unity system, and the *Huihui* system. Then after the events, he would present another memorandum claiming that the Western method proved by observation to be the most accurate one (see Figure 17.2a,b). But this was still not convincing enough to the emperor. For example, in a reply to Xu Guangqi's memorandum of November 20, 1630, the emperor suggested that it was hard to verify Xu's claim that the Western method gave the most accurate predictions because "officials at the Astronomical Bureau used different instruments, thus obtaining different results concerning the time of the eclipse" (Wang, 1981, p. 355).

Now the problem that Xu Guangqi faced was to demonstrate the accuracy of the predictions made by the Western method. For this purpose, the telescope proved very useful. With naked-eye observations, the time and magnitude of the solar eclipse were difficult to determine. But by projecting the Sun's image through the telescope onto a piece of silk paper, this became very easy. This method had already been explained in 1626 by the Jesuit (Johann) Adam Schall von Bell (1591–1666) in his monograph *On the Telescope* (*Yuan jing shuo* 远镜说), which introduced the telescope to China (Schall von Bell, 1626). Using the telescope to observe solar eclipses, Xu Guangqi obtained more precise data, and the result could be demonstrated; thus, there was no doubt about the superiority of the Western method.

In 1631, in a memorandum concerning the upcoming solar eclipse of October 25, Xu Guangqi explained to the emperor the advantage of using the telescope to observe solar eclipses:

> Formerly [with the naked eye], solar eclipses only of magnitudes larger than one tenth were visible. … Because of the dazzling sunlight, only eclipses of four or five tenths of magnitude can be measured reliably to match what is predicted. The observed magnitude is always smaller than the calculated magnitude. If its magnitude is less than one tenth, the solar eclipse appears to be without eclipse at all. If its magnitude is a little above two tenths, it appears to be of magnitude less than two tenths. The observed magnitude being smaller than the actual, its observed time of first contact will also be later than the actual time. Now using the telescope in a dark room, we can see the sun's image clearly. In this way we can determine the eclipse's magnitude and time of first contact. These can be seen by all officials and students on site, nothing left in doubt. If we do not use the telescope but use the naked eye to observe, the dazzling sunlight will make the observation unreliable. Some may use a basin of water to observe. But because the surface of water is not stable, only eclipses of magnitude of one tenth and above can be observed. This is equivalent to three or four quarters in time. (Wang, 1981, pp. 392–3)

On that day, Xu and his assistants "observed the solar eclipse at the Astronomical Bureau. Two telescopes were installed, one on the terrace, the other indoors. Both measured the magnitude of the eclipse 1.5 *fen*, matching exactly what is computed [using the Western method]" (Xu et al., 1635, p. 26).

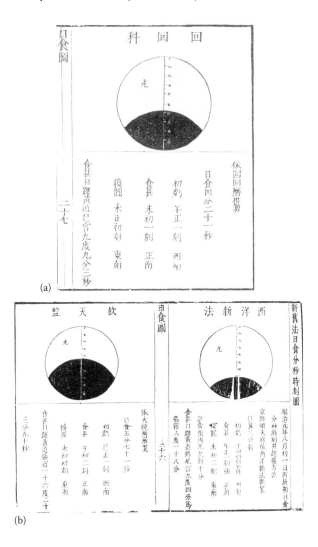

FIGURE 17.2a,b   A drawing from Xu Guangqi's *Xinfa suanshu* (*Treatise on Mathematical Astronomy According to the New Methods*) showing different predictions about the magnitude of a solar eclipse using the Islamic *Huihui* 回 回 system of calendrical astronomy (*a, top*), the Great Unity system (Datung li 大 统历) advocated by the Ming dynasty (*b, left bottom*), and the Western method promoted by Xu Guangqi (*b, right bottom*).

The use of the telescope improved the accuracy of measurements of solar eclipses considerably. The improvement in observing lunar eclipses was not as remarkable as for solar eclipses because lunar eclipses could be observed with the naked eye without any dazzling effect. Records of solar eclipses from the late Ming period show that the accuracy of observation increased remarkably after 1610. Before the use of the telescope, the accuracy of observations was unstable. The error in time averaged 15 min and sometimes deviated by as much as 47 min. After 1610, the error was reduced to 10 min (Shi, 2000). This was presumably the result of telescopic observation.

With all these convincing data obtained by telescopic observation of the eclipses, Xu Guangqi finally persuaded the emperor to use the Western method to reform the calendar.

To Xu Guangqi, the Western method meant more than just being useful for calendar making. He recognized that the Western method was built on rigid principles of deduction. With the Western method, one studied not only *what* it is, but also *why* it is what it is. Also, being a scholar-official who cherished ideas about "practical learning," Xu insisted that astronomy reform should achieve more than just producing a good calendar; it should produce knowledge that could be applied to practical and useful matters. He listed 10 applications of the Western method that could benefit from astronomical reform, including hydraulics, road and bridge construction, and military projects. The telescopic observations of eclipses played a critical role in persuading the emperor to agree to adopt the Western system.

IMPACT OF THE TELESCOPE ON COSMOLOGICAL THINKING

In his *Catechism of the Heavens*, Emmanuel Diaz (1615) reported these Galilean telescopic discoveries:

1. Venus changed phases like the Moon.
2. Saturn's shape was not completely round, but was oval like an egg, with two small stars attached to its sides.
3. Jupiter had four moons.
4. There were many more stars than could be observed with the naked eye. The Milky Way, which to the naked eye is like a silk ribbon, was a gigantic collection of stars.

Diaz did not mention Galileo's name, referring to him only as a famous astronomer who invented the telescope because his eyesight was weakened as a result of long observations. He kept the reader's attention by saying that he would talk about the telescope's "wonderful applications" when it was brought to China. The wonderful applications were actually introduced by Schall in *On the Telescope* 11 years later (Schall von Bell, 1626).

Schall explained how to use the telescope to observe the sky, particularly how to observe a solar eclipse. His account was fuller than that of Diaz, for he added sunspots and the Moon's rugged and cratered surface to the account of Galileo's discoveries.

Whereas in Europe telescopic observations provided forceful ammunition for advocates of the Copernican heliocentric system, Jesuits in China were not allowed by the Church to talk about the Copernican theory. Schall presented these discoveries as evidence supporting the Tychonic system. He did not even hint at the critical implications of Galileo's observations. Figure 17.3 shows how Schall explains the phases of Venus using Tychonic cosmology.

Presented this way, these discoveries did not make the Chinese aware of the cosmological and religious controversies related to them. However, the Chinese responded to these discoveries by elaborating on cosmological theories along the traditional line of thinking. It is particularly revealing to look at how the Chinese responded to Galileo's telescopic discoveries using a comparative

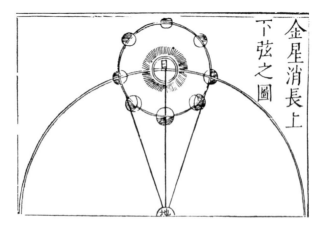

FIGURE 17.3 An illustration in Schall's *Yuan jing shuo* 远镜说 (*On the Telescope*, 1626) explaining the phases of Venus. Venus is shown as circling the Sun while the Sun moves around the Earth.

perspective. Their responses were very much shaped by the Chinese tradition of cosmologies and astronomical observations.

First of all, the telescope showed many more stars in the constellations than observed by the naked eye. For Westerners, this implied the infinitude of the Universe and led some to suggest a mystical vision of a pluralistic Universe (Kuhn, 1957, p. 220). The Chinese, however, had long ago recognized that the number of stars in the sky was not fixed. They had seen "guest stars" come and go. Zhang Heng in the 1st century CE even proclaimed that the number of faint stars counted was 11,520. The Milky Way was called the Celestial River and was not taken as a sublunary phenomenon. The telescopic observation that the Milky Way was a gigantic collection of stars was even seen by some Chinese scholars as evidence for some ancient Chinese cosmology.

In *The Spiritual Constitution of the Universe*, Zhang Heng says that the stars are "the quintessence of water" and that the Celestial River is of the same substance (Zhang, 120 CE). Fang Yizhi got his inspiration from telescopic observations and declared the following in his *Little Notes on the Principles of the Phenomena* (*Wu li xiao zhi* 物理小识):

> With the telescope, the Celestial River is seen as consisting of faint stars. It is also so with constellations Langwei and the Corpse star in the lunar lodge Gui. The commentary on the *Spiritual Constitution* says, the Celestial River is made of the quintessence of water. This is to say that stars are water's quintessence too. (Fang, 1664, p. 37)

Analogically, the telescopic observation of the Celestial River consisting of faint stars seemed to constitute a sort of proof for the old theory that stars are the quintessence of water. In his line of argument, there was no rupture or contradiction to traditional cosmology, only reinforcement.

Second, the telescope had a dramatic impact on the observation of the Moon and sunspots. The Moon's surface was observed to be covered with pits and craters, valleys and mountains. The observation of the Sun showed dark spots on its surface. In the West, this raised doubts about the distinction between the terrestrial and the celestial domains. The Heavens were not perfect and immutable at all, as assumed in Aristotelian cosmology. "The very existence of the spots conflicted with the perfection of the celestial region; their appearance and disappearance conflicted with the immutability of the heavens" (Kuhn, 1957, p. 221). Jesuit astronomers explained them as shadows caused by the occultation of stars close to the Sun. Again, this presented no problem at all to the Chinese, who had never made such a distinction between the heavenly and terrestrial regions. The theories used to explain phenomena on Earth were readily applied to phenomena in the sky. As a matter of fact, this followed naturally from the very logic of the theory of the correlation and interaction between Heaven and Man. The Chinese had already seen shadows on the Moon and had imagined them to be a toad, a tree, or a rabbit. They might have observed sunspots in the 12th century BCE, as recorded in the oracle bone inscriptions.* The earliest reliable record of a sunspot was from 28 BCE, as recorded in the *History of the Han* (*Han shu* 汉书; see Ban Gu 班固 and Ban Zhao 班昭, ca. 85 CE). The Chinese had even classified sunspots into categories according to their sizes and shapes. Hence, this telescopic discovery was not at all new to the Chinese and was readily accepted.

Jie Xuan 揭暄 (1610–1702) was one of the Chinese scholars who eagerly learned Western astronomical knowledge and made use of it to put forward his own theories. As one of the staunch Ming loyalists, he violently resisted the Manchu rule of the Qing dynasty and refused to pursue an official career under the new regime. He nevertheless had many contacts among scholars who knew about Western astronomy. Fang Yizhi was one of them. They met and discussed astronomical problems. Jie Xuan was probably the first Chinese who mapped the surface of the Moon with the aid of the telescope. From the observed movements of sunspots, he concluded that the Sun rotated on its own axis. This explanation of the movement of sunspots was quite different from what the Jesuits had offered. In the writings by the Jesuits, the sunspots were seen as occultations of small planets

---

* See T.D. Lee's chapter in this volume.

revolving very close to the Sun. By adapting the Tychonic cosmology to the cosmological ideas of Zhu Xi 朱熹 (1130–1200), a neo-Confucianist from the Southern Song, Jie Xuan proposed a cosmological theory of a spinning vortex. Not only was the Sun round and spinning, but so were all the other celestial bodies. The planets and the fixed stars were all round and spinning. The satellites of Jupiter, the "two tiny stars attached to Saturn," and the fact that the same side of the Moon always faces the Earth, all served as evidence for his cosmological theory.*

Finally, the discovery of the four moons of Jupiter had an immense impact on the Western cosmological imagination. If Jupiter could serve as the center of its own "planetary" system, why not the Sun? The observations of Jupiter provided a visible model of the Copernican Solar System. The argument against the Copernican system was much weakened by this discovery. Jesuit missionaries in China, however, presented these observations as evidence for the Tychonic system, which had the planets revolving around the Sun while the Sun moved around the Earth. With the other side of the cosmological debate completely concealed by the Jesuits, the Chinese could hardly imagine that a revolution in cosmology was implicated in these observations. Again, the Chinese did not consider these observations particularly surprising because they could be easily accommodated in traditional Chinese cosmology—not only imaginable, but probably actually observed by the Chinese long before the telescope. In a treatise on astrology from the 8th century, the *Astrological Treatises of the Kaiyuan Period of the Tang (Tang Kaiyuan zhanjing* 唐开元占 经), it is stated that "Jupiter has a tiny red star attached to it, forming an 'alliance'" (Xutan, ca. 724 CE, vol. (*juan* 卷) 23, p. 172). This was probably the earliest observation of one of Jupiter's four moons. The Chinese were also diligent observers of other planetary phenomena. Even more intriguing is that the Chinese might also have observed the rings of Saturn. In the same book, it says that "Saturn grows legs" and that "there are faint stars attached to Saturn" (Xutan, ca. 724, vol. (*juan*) 38, p. 203). Without further historical evidence, it is hard to conclude that the Chinese had already made these discoveries before Galileo's telescopic observations. But the point is that these discoveries did not go beyond the imagination of the Chinese. In Chinese cosmological thinking, everything was possible. The discoveries of the moons of Jupiter and similar phenomena came as no shock to the Chinese mind.

To summarize, while the telescopic discoveries caused a sensation in the West and posed a great challenge to the old cosmological ideas, they were placidly received in 17th-century China. This was partly because the information provided by the Jesuits was incomplete and distorted by their own ideology and partly because it presented no obvious contradiction to Chinese cosmological thinking, which emphasized the interaction between Heaven and Man and embodied the philosophy of change. The Chinese literati generally welcomed these discoveries made by the telescope and made use of them to rediscover and revive their own tradition. Some even put forward creative new theories by incorporating Western cosmologies with traditional Chinese ones.

IMPACT OF THE TELESCOPE IN CHINESE SOCIETY

In the West, the impact of the telescope on astronomical thought was profound. It changed humanity's outlook on the cosmos. Its later developments, the use of the micrometer for example, made the telescope a powerful astronomical measuring instrument, marking a watershed in positional astronomy. In China, as we have seen from the above account, the telescope also had a very important influence on calendar making and on cosmological thinking. But at the same time, we can see quite a different picture from what happened in the West. The telescopic discoveries did not bring about acceptance of the Copernican heliocentric theory, but helped strengthen the persuasiveness of the Tychonic system. Even after the Jesuit Michel Benoist (Jiang Youren 蒋友仁) (1715–1774) fully introduced the Copernican system in the *World Map with Illustrated Explanations* around 1760 (Benoist, ca. 1760), the Chinese refused to accept it because this was inconsistent with the established Tychonic system.

---

* Jie Xuan's most cosmological thoughts can be found in his *Xuan ji yi shu (Records of Ancient Arts of Astronomical Observations)*, published around 1675. See Shi (2000, pp. 136–47).

The telescope did not find further use in positional astronomy. In 1669, when the Jesuit astronomer Ferdinand Verbiest (1623–1688) proposed a large project of making instruments for the Imperial Observatory, he did not propose to mount telescopes on the astronomical measuring instruments. Instead, he constructed six classical instruments like Tycho Brahe's. Why was this so?

The question is related to the larger question of China's response to Western science. Some argue that China tended to reject Western scientific ideas because of ignorance and xenophobia. Others argue that the absence of a certain spirit in Chinese culture blocked the advancement of modern science in China.* Still others claim that there were defects in Chinese philosophical thinking that made modern scientific thinking in China impossible or doomed to failure. These arguments certainly touch on many important aspects of this question. But the picture was much more complicated than the one any reductionist approach would present. The case of the telescope shows that the transmission of scientific knowledge from the West to China was a process full of "misconception, ambiguity, reinterpretation, and selective adaptation," as pointed out by Nathan Sivin (1976, p. 1). It was much affected by the organization and motivations of the Jesuit missionaries. It was also conditioned by the type of Chinese imperial government and its motivations.

The issue is so complicated that it is impossible for me to elaborate on this in detail. But let me just outline a few points that are related to our discussion of the impact of the telescope in China.

First, on the part of Chinese intellectuals, I would not use "xenophobia" to describe the general mentality of late Ming and early Qing Chinese literati. As we have seen above, the Ming scholars were quite receptive to new astronomical and mathematical knowledge from the West. Not only did they see its immediate use in solving the problem of the calendar, but they also saw its applications in many practical matters, such as hydraulics, engineering, and the military. When the Jesuits introduced telescopic observations, the Chinese were not surprised at all, but took them as a new starting point to further their thinking about cosmology. Knowledge about Western astronomy actually spread very fast among Chinese literati. For example, we are told by Fang Yizhi that when he was nine years old he had already listened to Xiong Mingyu 熊明遇 (1579–1649) talking about Western astronomy on the occasion his father's visit to Xiong around 1620 (Hsu, 2008). Xiong Mingyu was a high official who was in contact with many scholars, as well as Jesuit missionaries. He wrote about telescopic observation as early as 1615 and published his work in 1634.

Second, on the part of the Jesuits, teaching Western astronomy was only part of their strategy of preaching Christianity in China. It was just by coincidence that they came at a time when the imperial government wanted to solve the calendar problem. But in their efforts at introducing Western astronomical knowledge to China, they were restrained by the Roman Catholic ideology on the one hand and by their own limited astronomical training on the other. The picture they presented was incomplete and distorted, and therefore inconsistent. Some clever scholars, such as Mei Wending 梅文鼎 (1633–1721), could discern that various theories were introduced by the Jesuits over time and could see that Western astronomy itself was developing, so it was normal that people would entertain different theories at the same time. But for some scholar-officials who, like Luan Yuan 阮元 (1764–1849), were used to the idea of unity and of knowledge, the inconsistency between theories, both introduced by the Jesuits from the same Western world, was simply unacceptable. This made them reject the Copernican theory when it was introduced in China.

Third, what the imperial government of China wanted was a reliable and accurate astronomical system that could produce useful almanacs and predict important celestial events. The Imperial Astronomical Bureau was put in charge of reforming the calendar. As long as a good mathematical astronomical system could be produced, the physical meaning of the adopted cosmological system did not really matter. In the early years of the Qing dynasty, some scholars refused to serve the Qing ruler, but nevertheless did some very creative work in astronomy. One of the most influential was Wang Xishan 王锡阐 (1628–1682). His criticism was that astronomers who were interested in only calendrical matters did not apprehend basic principles in establishing their

---

* See Peter Harrison's chapter in this volume.

theories. On his own, he made an effort to bring together mathematical techniques and cosmological principles.* But this was not what the imperial government wanted. Astronomers incumbent in the Astronomical Bureau, whether they were Jesuits or Chinese, did not feel motivated to adopt new theories if the problem of the calendar was solved.

Until the 1660s, the telescope still did not play an important role in positional astronomy. It did not have much to offer for increasing the accuracy of the calendar. Hence, astronomers in China did not find much use for the telescope in calendar making. More telescopes were brought to China, but usually they did not have much influence beyond the court and circle of high officials. Although the telescope aroused some curiosity and marvelous remarks among the Chinese—the emperors and some high officials wrote poems about telescopic observations—no new astronomical research problems were raised along with the introduction of the new instrument to China.

REFERENCES

Ban, G. 班固 and Ban, Z. 班昭 (ca. 85 CE). *Han shu* 汉书 (*History of the Han Dynasty*). Beijing: Zhonghua shuju. Reprint, 1962.

Benoist, M. (ca.1760). *World Map with Illustrated Explanations (Di qiu tu shuo* 地球图说). Zhengzhou: Daxiang Chubanshe. Reprint, 1998.

Diaz, E. (1615). *Tian wen lue* 天问略 (*Catechism of the Heavens*). Zhengzhou: Daxiang Chubanshe. Reprint, 1998.

Fang, Y. (1664). *Wu li xiao zhi* 物理小识 (*Little Notes on the Principles of the Phenomena*). Shanghai: Shangwu Yingshuguan, Reprint, 1937.

Hashimoto, K. (1988). *Hsü Kuang—Ch'I and Astronomical Reform*. Osaka: Kansai University Press.

Hsu, K.T. (2008). Evidential study on Fang Yizhi's first contact with Western scientific knowledge at the age of nine. Paper presented at the *12th International Conference of the History of Science in East Asia*. Baltimore, USA, July 14–18, 2008.

Jie, X. (ca. 1675). *Xuan ji yi shu* 璇玑遗述 (*Records of Ancient Arts of Astronomical Observations*). Zhengzhou: Daxiang Chubanshe. Reprint, 1998.

Kuhn, T.S. (1957). *The Copernican Revolution: Planetary Astronomy in the Development of Western Thought*. Cambridge, MA: Harvard University Press.

Schall von Bell, A. (1626). *Yuan jing shuo* 远镜说 (*On the Telescope*). Zhengzhou: Daxiang Chubanshe. Reprint, 1998.

Shi, Y. (2000). The eclipse observations made by Jesuit astronomers in Beijing: A reconsideration. *Journal of the History of Astronomy*, 31: 135–47.

Sivin, N. (1973). Copernicus in China. *Science in Ancient China: Researches and Reflections*, IV, 1–53. Aldershot, England: Variorum. Reprint, 1995.

Sivin, N. (1976). Wang Hsi-Shan. *Science in Ancient China: Researches and Reflections*, V, 1–28. Aldershot, England: Variorum. Reprint, 1995.

Wang, C. (ed.) (1981). *Xu Guanqi ji* 徐光启集 (*Collected Writings by Xu Guangqi [1562–1533]*). Shanghai: Guji Chubanshe.

Xu, G. (1635). *Xinfa suanshu* 新法算书 (*Treatise on Mathematical Astronomy According to the New Methods*). Zhengzhou: Daxiang Chubanshe. Reprint, 1998.

Xutan Xida 瞿昙悉达 (ca. 724 CE). *Tang Kaiyuan zhanjing* 唐开元占经 (*Astrological Treatises of the Kaiyuan Period of the Tang*). Beijing: Zhongguo Shudian. Reprint, 1989.

Zezong, X. (1981). The discovery of Jupiter's satellite by Ge De 2000 years before Galileo. *Chinese Astronomy and Astrophysics*, 2: 242–43.

Zhang Heng 张衡 (120 CE). *Ling xian* 灵宪 (*The Spiritual Constitution of the Universe*). In Fan Ye 范晔 (ca. 445 CE). *Hou han shu* 后汉书 (*History of the Later Han Dynasty*). Zhonghua shuju. Reprint, 1965.

---

* For a critical study of Wang Xishan's astronomical works, see Sivin (1976, p. 1).

Part VI

*"Big Questions" Raised
by New Knowledge*

18 Exoplanet Atmospheres and the Search for Biosignatures

Sara Seager

CONTENTS

INTRODUCTION

The search for our place in the cosmos has fascinated human beings for thousands of years. For the first time in human history, we have technological capabilities that put us on the verge of answering such questions as, Do other Earths exist? Are they common? and Do they harbor life?

Critical to inferring whether a planet is habitable or inhabited is an understanding of the exoplanetary atmosphere. Indeed, almost all of our information about temperatures and chemical abundances in stars and exoplanets comes from atmospheric observations. Ultimately, we would like an image of a twin of Earth as stunning as the Apollo images of our planet. For the foreseeable future, we are limited instead to observing exoplanets as spatially unresolved, i.e., as point sources.

Four hundred years after the invention of the telescope, the existence of exoplanets is firmly established. Figure 18.1 shows the known exoplanets as of January 2010, with symbols indicating the discovery technique.* The majority of the known exoplanets have been discovered by the Doppler technique, which measures the star's line-of-sight motion as the star orbits the planet-star common center of mass (see, e.g., Butler et al., 2006; Udry et al., 2007). While most planets discovered with the Doppler technique are giant planets, the new frontier is discovery of super-Earths (loosely defined as planets with masses between 1 and $10\,M_\oplus$). About a dozen radial velocity planets with $M < 10\,M_\oplus$ and another dozen with $10\,M_\oplus < M < 30\,M_\oplus$ have been reported. The transit technique finds planets by monitoring thousands of stars, looking for the small drop in brightness of the parent star that is caused by a planet crossing in front of the star. At the time of writing this article, around 50 transiting planets are known. As a result of selection effects, transiting planets found from ground-based searches are limited to small semimajor axes. Gravitational microlensing has recently emerged as a powerful planet-finding technique, discovering eight planets, two belonging to a scaled-down version of our own Solar System (Gaudi et al., 2008). Direct imaging is able to

---

* See the Extrasolar Planets Encyclopedia, http://exoplanet.eu/.

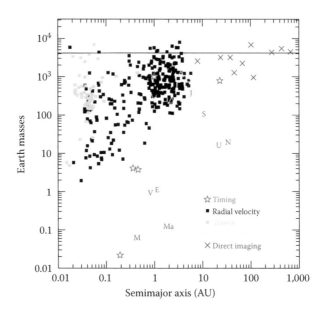

FIGURE 18.1 Known planets as of January 2010. The solid line is the conventional upper mass limit for the definition of a planet. (Data from http://exoplanet.eu/.)

find young or massive planets with very large semimajor axes. The mass of directly imaged planets (Kalas et al., 2008; Marois et al., 2008) is inferred from the measured flux based on evolution models and is hence uncertain. The timing discovery method includes both pulsar timing (Wolszczan and Frail, 1992) and time perturbations of stars with stable oscillation periods (Silvotti et al., 2007). See the chapter by Beichman in this volume for a more thorough description of exoplanet discovery techniques.

The next step beyond discovery is characterization of a planet's physical properties, such as density, atmospheric composition, and atmospheric temperature. In this chapter, I focus on the status and prospects for characterization of exoplanet atmospheres.

The most natural way to think of observing exoplanet atmospheres is by taking an image of the exoplanet. This would be akin to taking a photograph of the stars with a digital camera, although using a very expensive, high-quality detector. This so-called direct imaging of planets is currently limited to big, bright, young, or massive planets located far from their stars (see Figure 18.1; Kalas et al., 2008; Marois et al., 2008). Direct imaging of substellar objects is currently possible with large ground-based telescopes and adaptive optics (AO) to cancel the atmosphere's blurring effects. Solar-System-like small planets are not observable via direct imaging with current technology, even though an Earth at 10 pc is brighter than the faintest galaxies observed by the *Hubble Space Telescope* (*HST*). The major impediment to direct observations is instead the adjacent host star; the Sun is 10 million to 10 billion times brighter than Earth (for mid-infrared [IR] and visible wavelengths, respectively). No existing or planned telescope is capable of achieving this contrast ratio at 1 AU separations. The high planet-star contrasts are prohibitive. Fortunately, much research and technology development is ongoing for space-based direct imaging to enable in the future direct imaging of Earths and Jupiters in Solar Systems as old as the current Solar System.

At present, two fortuitous, related events have enabled observations of exoplanet atmospheres in a manner very different from direct imaging. The first event is the existence and discovery of a large population of planets orbiting very close to their host stars. These so-called hot Jupiters, hot Neptunes, and hot super-Earths have 1- to 4-day orbits and semimajor axes less than 0.05 AU (see Figure 18.1). The hot Jupiters are heated by their parent stars to temperatures of 1,000 K–2,000 K, making their IR brightness only one-thousandth that of their parent stars. While it is by no means an easy task to observe a 1:1,000 planet-star flux contrast, such an observation is possible—and is unequivocally more favorable than the 10^{-10} visible-wavelength planet-star contrast for an Earth twin orbiting a Sun-like star.

The second favorable occurrence is that of transiting exoplanets—planets that go in front of their star as seen from Earth. The closer the planet is to the parent star, the higher its probability to transit. Hence, the existence of short-period planets has enabled the discovery of many transiting exoplanets. It is the special transit configuration that allows us to observe the planet atmosphere without direct imaging (see Figure 18.2).

COMPARATIVE EXOPLANETOLOGY WITH HOT JUPITERS

TRANSITING PLANET OBSERVATIONS

Transiting planets are observed in the combined light of the planet and star (Figures 18.2 and 18.3). As the planet goes in front of the star (primary transit or primary eclipse), the starlight drops by the amount of the planet-to-star area ratio. If the size of the star is known, the planet size can be determined. During transit, some of the starlight passes through the optically thin part of the planet atmosphere (depicted by the transparent annulus in Figure 18.2), picking up some of the spectral features in the planet atmosphere. A planetary transmission spectrum can be obtained by dividing the spectrum of the star and planet during transit by the spectrum of the star alone (the latter taken before or after transit).

Planets on circular orbits that go in front of the star also go behind the star. Just before the planet goes behind the star (secondary eclipse), the planet and star can be observed together. When the planet disappears behind the star, the total flux from the planet-star system drops. The drop is related to both relative sizes of the planet and star and their relative brightness (at a given wavelength). The flux spectrum of the planet can be derived by subtracting the flux spectrum of the star alone (during secondary eclipse) from the flux spectrum of both the star and planet (just before and after secondary eclipse). The planet's flux gives information on the planet composition and temperature gradient (at IR wavelengths) or albedo (at visible wavelengths).

Observations of transiting planets exploit separation of photons in time rather than in space (see Figures 18.2 and 18.3). That is, observations are made in the combined light of the planet-star system. Primary and secondary eclipses enable high-contrast measurements because the precise on/off nature of the transit and secondary eclipse events provides an intrinsic calibration reference. This is one reason why *HST* and the *Spitzer Space Telescope* (*SST*, formerly the *Space Infrared Telescope Facility* [*SIRTF*]) have been so successful in measuring high-contrast transit signals that were not considered in their designs.

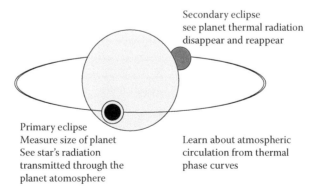

FIGURE 18.2 Schematic of a transiting exoplanet showing primary and secondary eclipse and what is learned from observations at those phases.

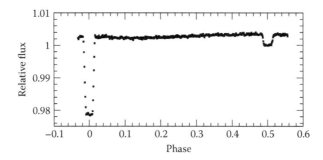

FIGURE 18.3   Infrared light curve of HD 189733A and HD 189733b. The flux in this light curve is from the star and planet combined. The first dip (from left to right) is the transit, and the second dip is the secondary eclipse. Error bars have been suppressed for clarity. (Data from Knutson, H.A., Charbonneau, D., Allen, L.E., et al., *Nature*, 447, 183–6, 2007.)

I describe one reason why *SST* has been so successful in detecting photons from exoplanets during secondary eclipse. Figure 18.4 shows the relative fluxes for the Sun, Jupiter, Earth, Venus, and Mars, each approximated as a blackbody. The planets also reflect light from the Sun at visible wavelengths, giving them two flux peaks in their schematic spectrum. We see from Figure 18.4 that at wavelengths <10 μm, the Solar System planets are more than seven orders of magnitude fainter than the Sun. The energy distribution of a generic hot Jupiter, with assumed geometric albedo of 0.05, equilibrium temperature of 1,600 K, and a radius of 1.2 R_J, is also shown in the same figure.

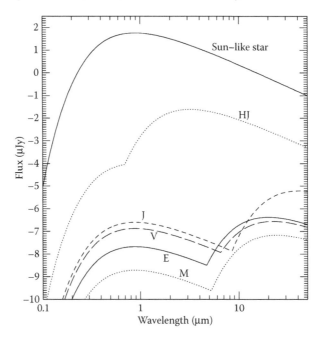

FIGURE 18.4   Blackbody flux (in units of 10^{-26} W m^{-2} Hz^{-1}) of some Solar System bodies as "seen" from 10 pc (or ~30 lt yr). The Sun is represented by a 5,750 K blackbody. The planets Jupiter, Venus, Earth, and Mars are shown and are labeled with their first initial. A putative hot Jupiter is labeled with "HJ." The planets have two peaks in their spectra: The short-wavelength peak is due to sunlight scattered from the planet atmosphere and is computed using the planet's geometric albedo. The long-wavelength peak is from the planet's thermal emission and is estimated by a blackbody of the planet's effective temperature. The HJ albedo was assumed to be 0.05 and the equilibrium temperature to be 1,600 K. (Temperature and albedo data were taken from Cox, A.N., *Allen's Astrophysical Quantities*, 4th edn. New York: Springer, 2000.)

This representative hot Jupiter is less than 3 orders of magnitude fainter than the Sun at some wavelengths. Equally important is that the planet-to-star flux ratio is favorable where the star and planet flux are high, i.e., plenty of photons are available to reach the telescope. The ~8 μm region is therefore a sweet spot for *SST* observations of hot Jupiter exoplanets.

At present, atmospheric studies are limited to hot Jupiters and hot Neptunes—big planets close to their host star that are far from "habitable." These stars are hot and therefore bright in the thermal IR, making their detection possible with existing telescopes. As a result of external heating from the host star, hot Jupiter atmospheres reach temperatures of 1,000 K–2,000 K. Hot Jupiters and Neptunes also have no solid surface to support life as we know it; studying their atmospheres not only is interesting in its own right, but can also be seen as a stepping stone to the future studies of planets that may support life.

Scientific Highlights of Hot Jupiter Atmosphere Discoveries

More than 24 exoplanet atmospheres have been detected. The exoplanet atmospheric data are dominated by *SST* observations (e.g., Charbonneau et al., 2005; Deming et al., 2005, 2006; Harrington et al., 2006, 2007; Knutson et al., 2007, 2008, 2009a,b). *HST* has successfully observed two exoplanet atmospheres (Charbonneau et al., 2002; Swain et al., 2008, 2009). Ground-based detections on favorable targets are still a major challenge, but are at last becoming successful (e.g., Redfield et al., 2008). To emphasize how many exoplanet atmospheres have actually been observed, we present a list of observations in Table 18.1.

Although many exoplanets have atmospheric measurements, not all of the observations or interpretations are robust. What have we learned about exoplanet atmospheres that we are certain of? The background to this question is that the observations are made difficult because planets are always much fainter than the host stars. The resulting data points may have large error bars, or in some cases error bars that are too small as a result of a lack of systematic effects. In other words, the field of exoplanet atmospheres is still very much a forefront science, pushing the technology being used for the observations.

The most significant and robust scientific highlight is identification of atoms and molecules in exoplanet atmospheres. Knowing some of the gases in a planet's atmosphere is a gateway to understanding the chemistry and physics of that planet. In a planet's atmosphere, however, the most abundant gases are not necessarily the ones that can be detected.

In hot Jupiters, sodium (Na), water vapor (H_2O), methane (CH_4), and carbon dioxide (CO_2) have been reported (Charbonneau et al., 2002; Tinetti et al., 2007; Grillmair et al., 2008; Swain et al., 2008, 2009). See Figure 18.5 for an example of H_2O and CH_4 in the transmission spectrum of HD 189733b (Swain et al., 2009). Escaping atomic hydrogen has also been reported (Vidal-Madjar et al., 2003), with more tentative reports of escaping atomic carbon and oxygen (Vidal-Madjar et al., 2004). At the high temperatures of hot Jupiters (~1,200 K–2,000 K), sodium is expected to be in atomic form. Water vapor is also expected, because a hot planet with plentiful hydrogen and oxygen will naturally have water (unless the unexpected situation of high elemental carbon compared with oxygen exists). Although carbon monoxide (CO) is the dominant form of carbon at high planetary temperatures, methane can be present if parts of the atmosphere are cool enough or result from nonequilibrium chemistry. Carbon dioxide was completely unexpected in high abundance in hot Jupiter atmospheres (Liang et al., 2003) because in an atmosphere dominated by hydrogen the conditions are not appropriate for abundant CO_2. A general consensus is that photochemistry initiated by the strong ultraviolet (UV) radiation from the host star splits up molecules that reform to CO_2. Studies are ongoing.

A second highlight of exoplanet atmospheric studies is the inference of atmospheric temperature inversions (Burrows et al., 2007; Knutson et al., 2008, 2009a). Near Earth's surface, the temperature decreases with increasing altitude. Mountain climbers are familiar with this because the air gets colder as one climbs higher. At an altitude of about 20 km (39,000 ft), however, Earth's atmosphere

TABLE 18.1

Tabulation of the Exoplanets That have been Observed in Secondary Eclipse at *SST* Infrared Array Camera (IRAC) Wavelengths, as of January 2010

Planet	3.6 μm	4.5 μm	5.8 μm	8 μm
HD 189733b	AAA	AAA	AAA	AAA
HD 209458b	AAA	AAA	AAA	AAA
HD 149026b	AA	AA	AA	AAA
HD 80606b		AA		AAA
GJ 436b	AA	AA	AA	AAA
CoRoT-1	AA	AA		
CoRoT-2	AA	AAA		AAA
HAT-1	AAA	AAA	AAA	AAA
HAT-2	A	A	AA	AA
HAT-3	A	A		
HAT-4	A	A		
HAT-5	AA	AA		
HAT-6	A	A		
HAT-7	AAA	AAA	AAA	AAA
HAT-8	A	AA		
HAT-10	A	AA		
HAT-11	A	A		
HAT-12	A	A		
TrES-1	AA	AAA	AA	AAA
TrES-2	AAA	AAA	AAA	AAA
TrES-3	AAA	AAA	AAA	AAA
TrES-4	AAA	AAA	AAA	AAA
WASP-1	AA	AA	AA	AA
WASP-2	AA	AA	AA	AA
WASP-3	AA	AA		AA
WASP-4	AA	AA		
WASP-5	AA	A		
WASP-6	AA	A		
WASP-7	A	A		
WASP-8		AA		AA
WASP-10	A	A		
WASP-12	AA	AA	AA	AA
WASP-14	A	AA		AA
WASP-17		AA		AA
WASP-18	AA	AA	AA	AA
WASP-19	AA	AA	AA	AA
XO-1	AAA	AAA	AAA	AAA
XO-2	AAA	AAA	AAA	AAA
XO-3	AAA	AAA	AAA	AAA

A = Observations officially planned by *SST*. AA = Data obtained by *SST* and analysis under way. AAA = Data are published. Ten exoplanet atmospheres with data under way are now published, and data has been obtained from a few planets with planned observations.

starts to become warmer with increasing altitude. This is called a temperature inversion and on Earth is known as the stratosphere. We can identify temperature inversions by the gross structure of absorption lines. Planets without temperature inversions should generally have absorption features in their spectrum (see Figure 18.6, *bottom panel*). Planets with temperature inversions would show

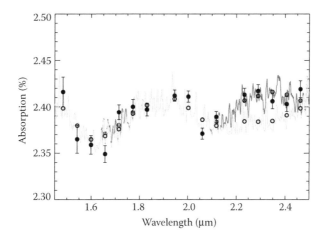

FIGURE 18.5 Transmission spectra of HD 189733. Data points with error bars are from *HST* (NICMOS). (From Swain, M.R., Vasisht, G., and Tinetti, G., *Nature*, 452, 329–31, 2008.) That the data strongly deviate from a straight line shows the presence of molecular absorption and is how individual molecules can be identified. The green curve is a model with just H_2O, and the red curve is a model with H_2O and CH_4. The open circles are theoretical points from the model convolved with the instrument point-spread function (PSF) and binned to the resolution of the data. The model with CH_4 fits the data much better than the H_2O-only model, showing the presence of CH_4. (The models are from Madhusudhan, N. and Seager, S., *Astrophysical Journal*, 707, 24–39, 2009.) The y-axis convention shows increasing absorption along the y-axis to help illustrate that the planet size increases at wavelengths with more molecular absorption. (Figure adapted from Swain, M.R., Vasisht, G., and Tinetti, G., *Nature*, 452, 329–31, 2008.)

the same features as emission lines. Planets with temperature inversions in one atmospheric layer, but not another, would show both absorption and emission in the same features (see Figure 18.6, *bottom panel*). We have been able to recognize temperature inversions in exoplanet atmospheres by fitting models to the data and finding that, for some planets, the only way to fit the data is by assuming a temperature inversion in the planet atmosphere (Burrows et al., 2007; Figure 18.7).

Aside from the identification of atmospheric gases and the inference of temperature inversions, many other interesting—but less conclusive—observations and observationally driven theoretical studies of hot Jupiter atmospheres are ongoing. These include 30+ hr "around-the-orbit" observations of an individual exoplanet showing a change in brightness as the planet goes through phases as seen from Earth (e.g., Knutson et al., 2007). Very detailed studies on the way that the absorbed stellar energy circulates in the atmosphere, including possibly supersonic winds, are ongoing.

Rather than describe the very detailed atmospheric circulation investigations, I present a summary based on an important question related to atmospheric circulation. Hot Jupiters are expected to be tidally locked to their parent stars—presenting the same face to the star at all times. This causes a permanent dayside and nightside. An intriguing question has been about the temperature difference from the dayside to the nightside. Are the hot Jupiters scorchingly hot on the dayside and exceedingly cold on the nightside? Or, does atmospheric circulation efficiently redistribute the absorbed stellar radiation from the dayside to the nightside?

Surprisingly, *SST* has found that both scenarios may be possible. It has measured the flux of the planet and star system as a function of orbital phase for several hot Jupiter systems (Harrington et al., 2006; Cowan et al., 2007; Knutson et al., 2007, 2009b). Assuming that the star has a constant brightness with time, the resulting brightness change is due to the planet alone. The HD 189733 star/planet system shows some variation at 8 µm during the 30 hr continuous observations of half an orbital phase (Knutson et al., 2007). This variation corresponds to about 20% variation in planet temperature (from a brightness temperature of 1,212 K to 973 K). In contrast, the non-transiting

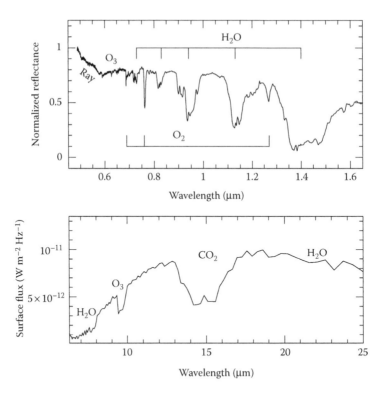

FIGURE 18.6 Earth's hemispherically averaged spectrum. *Top*: Earth's visible wavelength spectrum from earthshine measurements. (From Turnbull, M.C., Traub, W.A., Jucks, K.W., et al., *Astrophysical Journal*, 644, 551–9, 2006.) *Bottom*: Earth's mid-infrared spectrum as observed by *Mars Global Surveyor* en route to Mars. (From Pearl, J.C. and Christensen, P.R., *Journal of Geophysical Research*, 102, 10875–80, 1997.) Major molecular absorption features are noted; "Ray" indicates Rayleigh scattering.

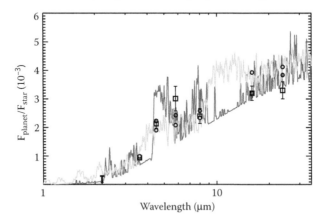

FIGURE 18.7 Evidence for thermal inversion in the atmosphere of HD 209458b. The squares with error bars are data points from *SST*. (From Burrows, A., Hubeny, I., Budaj, J., et al., *Astrophysical Journal*, 668, L171–4, 2007; Knutson, H.A., Charbonneau, D., Allen, L.E., et al., *Astrophysical Journal*, 673, 526–31, 2008.) The green curve is a model with absorption features only that corresponds to an atmosphere where the temperature decreases with increasing altitude. The red curve fits the data much better and is a model with emission features only, corresponding to an atmosphere with a vertical temperature inversion. The models are from Madhusudhan, N. and Seager, S., *Astrophysical Journal*, 707, 24–39, 2009, and the open circles are theoretical points from the model convolved with the instrument PSF and binned to the resolution of the data. (Adapted from Knutson, H.A., Charbonneau, D., Allen, L.E., et al., *Astrophysical Journal*, 673, 526–31, 2008.)

exoplanet Upsilon Andromedae shows a marked day-night contrast, suggesting that the dayside and nightside temperatures differ by more than 1,000 K (Harrington et al., 2006).

Once the stellar radiation is absorbed on the planet's dayside, there is a competition between reradiation and advection. If the radiation is absorbed high in the atmosphere, the reradiation timescale is short, and reradiation dominates over advection. In this case, the absorbed stellar radiation is reradiated before it has a chance to be advected around the planet, resulting in a very hot planet dayside and a correspondingly very cold nightside. If the radiation penetrates deep into the planet atmosphere where it is finally absorbed, the advective timescale dominates and the absorbed stellar radiation is efficiently circulated around the planet. This case would generate a planet with a small temperature variation around the planet.

Here I explain how hot Jupiters can exist both with and without large day-night temperature variations (see also Hubeny et al., 2003; Harrington et al., 2007; Burrows et al., 2007; Fortney et al., 2008). Hot planets such as Upsilon Andromedae are on one side of a temperature-driven chemical composition boundary, while cooler planets such as HD 209458b are on the cooler side. Specifically, if the hot Jupiter planet atmosphere is relatively cool, titanium oxide (TiO) is locked into solid particles that have little absorbing power in the atmosphere. In the hotter atmosphere, TiO is a gas that absorbs so strongly that it puts the planet in the reradiation regime, leading to a large day-night contrast. At the temperature of these hot dayside exoplanets, some elements will be in atomic (instead of molecular) form, and atomic line opacities may also play a significant absorbing role. The identification of the absorber (be it TiO or something else) is actually under debate (Spiegel et al., 2009). See Seager and Deming (2010) for a more detailed description of exoplanet atmospheric highlights.

EARTH VIEWED FROM AFAR

Earth's atmosphere is a natural starting point when considering planets that may have the right conditions for life or planets that have signs of life in their atmospheres. Earth from afar, with any reasonably sized telescope, would appear only as a point of light, without spatial resolution or surface features. We have real atmospheric spectra of the hemispherically integrated Earth (Figure 18.6) by way of earthshine measurements at visible and near-IR wavelengths (e.g., Turnbull et al., 2006) and from spacecraft that turn to look at Earth (e.g., Pearl and Christensen, 1997).

Earth has several strong spectral features that are uniquely related to the existence of life or habitability. The gas oxygen (O_2) makes up 21% of Earth's atmosphere by volume, yet O_2 is highly reactive and therefore will remain in significant quantities in the atmosphere only if it is continually produced. On Earth, plants and photosynthetic bacteria generate oxygen as a metabolic by-product; there are no abiotic continuous sources of large quantities of O_2. Ozone (O_3) is a photolytic product of O_2, generated after O_2 is split up by the Sun's UV radiation. Oxygen and ozone are Earth's two most robust biosignature gases. Nitrous oxide (N_2O) is a second gas produced by life—albeit in small quantities—during microbial oxidation-reduction reactions. See also the chapter by Beichman in this volume for a description of Earth's biosignatures.

Other spectral features, although not biosignatures (because they do not reveal direct information about life or habitability), can nonetheless provide significant information about the planet. These include CO_2 (which is indicative of a terrestrial atmosphere and has an extremely strong mid-IR spectral feature) and CH_4 (which has both biotic and abiotic origins). This implies that a detection of several of these features would provide credible evidence for the existence of life as we know it.

In addition to atmospheric biosignatures, the Earth has one very strong and very intriguing biosignature on its surface: vegetation. The reflection spectrum of photosynthetic vegetation has a dramatic sudden rise in albedo around 750 nm by almost an order of magnitude! Vegetation has evolved this strong reflection feature, known as the "red edge," as a cooling mechanism to prevent overheating, which would cause chlorophyll to degrade. On Earth, because of the presence of clouds, this signature is probably reduced to a few percent (see, e.g., Woolf et al., 2002; Seager et al., 2005a,b; Montanes-Rodriguez, et al. 2006; and references therein), but such a spectral surface feature could be much stronger on a planet with a lower cloud cover fraction. Recall that any observations of Earth-like extrasolar planets will not be able to spatially resolve the surface. A surface biosignature could be distinguished from an atmospheric signature by time variation; as the continents, or different concentrations of the surface biomarker, rotate in and out of view, the spectral signal will change correspondingly.

Earth viewed from afar would also vary in brightness with time, as a result of the brightness contrast of cloud, land, and oceans. As Earth rotates and continents come in and out of view, the total amount of reflected sunlight will change as a result of the high albedo contrast of different components of Earth's surface (<10% for ocean, >30%–40% for land, >60% for snow and some types of ice). In the absence of clouds, this variation could be an easily detectable factor of a few. With clouds, the variation is muted to 10%–20% (Ford et al., 2001). From continuous observations of Earth over a few-month period, Earth's rotation rate could be extracted, weather identified, and the presence of continents inferred. Palle et al. (2008) modeled Earth as an exoplanet using 3 months of cloud data taken from satellite observations. They showed that a hypothetical distant observer of Earth could measure Earth's rotation rate. This is surprising and means that, despite Earth's dynamic weather patterns, Earth has a relatively stable signature of cloud patterns. These cloud patterns arise in part because of Earth's continental arrangement and ocean currents. Beyond detecting Earth's rotation rate, Palle et al. (2008) found deviations from the periodic photometric signal, indicative to hypothetical distant observers that active weather is present on Earth.

Real data of the spatially unresolved Earth are available. Global, instantaneous spectra and photometry can be obtained from observations from Earth itself—by earthshine measurements. Earthshine, easily seen with the naked eye during crescent Moon phase (Figure 18.6, *top panel*), is sunlight scattered from Earth that scatters off of the Moon and travels back to Earth. Earthshine data are more relevant to studying Earth as an extrasolar planet than remote sensing satellite data. The latter are highly spatially resolved and limited to narrow spectral regions. Furthermore, by looking straight down at specific regions of Earth, hemispherical flux integration with lines of sight through different atmospheric path lengths is not available. Recently, the *EPOXI* spacecraft viewed Earth from afar—31 million miles distant. *EPOXI* is a NASA Discovery Mission of Opportunity, the result of the integration of the Extrasolar Planet Observation and Characterization (EPOCh) Investigation and Deep Impact eXtended Investigation (DIXI) that impacted and observed Comet Temple. *EPOXI* spent several months in 2008–2009 observing stars with known exoplanets, as well as observing Earth as an exoplanet. The *EPOXI* spacecraft obtained light curves of Earth at seven wavebands spanning 300–1,000 nm. Using two observations, each spanning a day, taken at gibbous phases, it was discovered that the rotation of Earth leads to diurnal albedo variations of 15%–30%, with the largest relative changes occurring at the reddest wavelengths. Using a principal component analysis of the multiband light curves, Cowan et al. (2009) found that 98% of the diurnal color changes of Earth are due to only two dominant eigencolors. The spectral and spatial distributions of the eigencolors correspond to cloud-free continents and oceans, enabling construction of a crude longitudinally averaged map of Earth.

PROSPECTS FOR STUDYING AN EARTH-LIKE EXOPLANET ATMOSPHERE

DISCOVERY AND CHARACTERIZATION PROSPECTS FOR A TRUE EARTH ANALOG

The holy grail of the research field of exoplanets is the discovery of a true Earth analog: an Earth-size, Earth-mass planet in an Earth-like orbit about a Sun-like star. Discovery of an Earth analog is a massive challenge for each of the different exoplanet discovery techniques (Figure 18.1) because Earth is so much smaller (1/100 in radius), so much less massive (1/10^6), and so much fainter (10^7 for mid-IR wavelengths to 10^{10} for visible wavelengths) than the Sun.

In order to observe a planet under such low planet-star flux contrasts, at visible wavelengths the diffracted light from the star must be suppressed by 10 billion times. A collection of related ideas are called "coronagraphs," a term that originated with instruments that artificially block out sunlight to allow observation of the faint corona of the Sun. Novel-shaped apertures, pupils, and pupil masks to suppress some or all of the diffracted starlight have been developed (see, e.g., Trauger and Traub, 2007 and references therein). A very promising idea is to use a novel-shaped external occulter that would fly tens of thousands of kilometers from the telescope in order to suppress the diffracted starlight (Cash, 2006).

Other exoplanet discovery techniques may have an easier time in discovering an Earth-mass planet. These other methods include space-based astrometry (Shao and Nemati, 2009 and references therein) or potentially ground-based radial velocity (Li et al., 2008). The transit technique will enable discovery of Earth-size, Earth-like orbit-transiting planets by NASA's *Kepler Space Telescope* (*KST*, which has discovered five transiting giant planets as of January 2010; Borucki et al., 2010). We emphasize that discovery of Earth-size or Earth-mass planets is not the same as identifying a habitable planet. Venus and Earth are both about the same size and same mass—and would appear the same to an astrometry, radial velocity, or transit discovery. Yet Venus is completely hostile to life as we know it because of the strong greenhouse effect and the resulting high surface temperatures (over 700 K), while Earth has the right surface temperature for liquid water oceans and is teeming with life. The absorption spectrums of Venus and Earth are quite different. This is why, in the search for habitable planets, we must hold on to the dream of a direct-imaging space-based telescope capable of blocking out starlight.

DISCOVERY AND CHARACTERIZATION PROSPECTS FOR A SUPER-EARTH ORBITING AN M STAR

The discovery and spectral characterization of a true Earth analog is immensely challenging, making it at present many years off in the future. Yet, a different kind of habitable planet is within reach—if only we are willing to extend our definition of Earth. The extension is a big Earth orbiting close to a small star.

All life on Earth requires liquid water, and so a natural requirement in the search for habitable exoplanets is a planet with the right surface temperature to have liquid water. Terrestrial-like planets are heated externally by the host star, so that a star's "habitable zone" (see Beichman's chapter in this volume for a definition) is based on the distance from the host star. Small stars have a habitable zone much closer to the star compared with Sun-like stars because of the small stars' much lower energy output than the Sun.

The chance of discovering a super-Earth transiting a low-mass star in the immediate future is huge. Observational selection effects favor their discovery in almost every way (see Figure 18.8). The magnitude of the planet transit signature is related to the planet-to-star area ratio. Low-mass stars can be 2–10 times smaller than a Sun-like star, improving the transit signal from about 1/100,000 for an Earth transiting a Sun to 1/25,000 or 1/1,000 for the same-sized planet. A planet's equilibrium temperature scales as $T_{eq} \sim T_* (R_*/a)^{1/2}$, where a is the semimajor axis of the orbit of the planet. The temperature of low-mass stars is about 3,500 K–2,800 K (for, respectively, a 2- and 10-times-smaller star than the Sun). The habitable zone of a low-mass star would therefore be 4.5–42 times closer to the star compared with the habitable zone of the Sun. To measure a planet mass, the radial velocity signal, K, scales as $K \sim (a M_*)^{-1/2}$, and the low-mass star masses are 0.4–0.06 times that of the Sun. Obtaining a planet mass is therefore about 3–30 times more favorable for an Earth-mass planet orbiting a low-mass star compared with an Earth–Sun analog. The transit probability scales as R_*/a. The probability for a planet to transit in a low-mass star's habitable zone is about 2.3%–5%, much higher compared with the low 0.5% probability of an Earth–Sun analog transit. Finally, from Kepler's third law a planet's period, P, scales as $P \sim a^{3/2}/M^{1/2}$, meaning that the period of a planet in the habitable zone of a low-mass star is 7–90 times shorter than the Earth's 1 yr period, and the planet transit can be observed often enough to build up a signal (compared with an Earth-analog once-a-year transit). A super-Earth larger than Earth (and up to about 10 $M_\oplus$ and 2 $R_\oplus$) is even easier to detect by its larger transit signal and mass signature than Earth. (See Nutzman and Charbonneau [2008] for more discussion on the benefit of targeted-star searches for transiting planets orbiting in

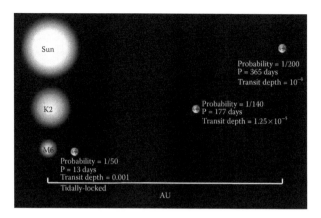

FIGURE 18.8 A schematic of transiting planets orbiting normal stars. The smaller the star, the closer the habitable zone is to the star, and the easier it is to detect.

habitable zones of M stars, and see Deming et al. [2009] for a description of the proposed *Transiting Exoplanet Survey Satellite* [*TESS*] space-based survey transiting planet yield.)

Recently, Charbonneau et al. (2009) discovered a 6.55 Earth-mass super-Earth orbiting the small M star GJ 1416. At 2.68 $R_\oplus$, this planet is not reminiscent of the rocky planets in our Solar System, but likely has a large gas envelope and possibly even a deep water-rich interior. Debate as to whether this planet is habitable is ongoing—the planet's equilibrium temperature is hotter than considered habitable, and the planet is likely too hot to have liquid water unless it has an unrealistically low internal energy (100 times less than Earth's; see Rogers and Seager, 2010).

Debate on how habitable a planet orbiting close to an M star can actually be is an active topic of research (see the reviews by Scalo et al., 2007; Tarter et al., 2007). Some previously accepted "show-stoppers" are no longer considered serious impediments. Atmospheric collapse due to cold nightside temperatures on a tidally locked planet will not happen as long as the atmosphere is thick enough (0.1 bar) for atmospheric circulation to redistribute absorbed stellar energy to heat the nightside (Joshi et al., 1997). Short-term stellar variability due to large-amplitude starspots could change the planet's surface temperature by up to 30 K in the most severe cases (Joshi et al., 1997; Scalo et al., 2007), but even some terrestrial life can adapt to freeze-thaw scenarios. Bursts of UV radiation due to stellar flares could be detrimental for life, but the planet's surface could be protected by a thick abiotic ozone layer (Segura et al., 2005), or alternatively life could survive by inhabiting only the subsurface.

Other concerns about habitability of planets orbiting close to M stars have not yet been resolved. Flares and UV radiation could erode planet atmospheres (Lammer et al., 2007), especially because the active phase of M stars can last for billions of years (West et al., 2008). Tidally locked planets in M star habitable zones will be slow rotators, and lack of a magnetic field due to a small dynamo effect will not protect the atmosphere from erosion. Planets accreting in regions that become habitable zones of M dwarf stars form rapidly (within several million years); the planet may not have time to accrete volatiles (e.g., water) that are present in the protoplanetary disk much farther away from the star (Lissauer, 2007).

Nevertheless, the transiting super-Earths in the habitable zones are highly valuable as the nearest-term potentially habitable planets. We anticipate the discovery of a handful of them in the next several years. With such rare, prize targets, we will strive to observe the transiting super-Earth atmospheres in the same way that we are currently observing transiting hot Jupiters orbiting Sun-like stars. The observations will be challenging, as a result of the thinner atmospheres on rocky planets compared with the puffy atmospheres of hot Jupiter (gas giant) planets. Also, the stars could be faint if Earths in a low-mass star habitable zone are rare. Fortunately, NASA's *James Webb Space*

Telescope (*JWST*), scheduled for launch in 2013, is capable of observing super-Earths transiting low-mass stars (see, e.g., Deming et al., 2009). We need to be patient with the tens to hundreds of hours of *JWST* time with the concomitant complex scheduling to cover many periodic transits in order to build up a decent signal of the atmospheric spectrum.

There is an exciting sense of anticipation in observing and studying super-Earth atmospheres, simply because we don't know quite what the diversity of super-Earth atmospheres will be. This is in contrast to Jupiter and the other Solar System gas giants, which have a "primitive" atmosphere. That is, Jupiter has retained the gases it formed with, and these gases represent the composition of the Sun. The super-Earth atmospheres, on the other hand, could have a wide range of possibilities for the atmosphere mass and composition, right down to the amount of hydrogen in the atmosphere (Elkins-Tanton and Seager, 2008). Hydrogen is a light gas and escapes from small, warm terrestrial planets. Earth and the other Solar System terrestrial planets have all lost whatever free hydrogen they may have started with. A hydrogen-rich atmosphere is very intriguing because it might have very different biosignatures than an oxidized atmosphere.

ATMOSPHERIC BIOSIGNATURES

The ideal atmospheric biosignature gas would have the following characteristics:

1. It would not exist naturally in the planetary atmosphere at ambient temperatures and pressures.
2. It would not be created by geophysical processes.
3. It would not be produced by photochemistry.
4. It would have a strong spectral signature.

Indeed, Earth's most robust biosignature, O_2, satisfies all four of the above criteria. Earth's biosignature gas N_2O satisfies the first three. Even though O_2 and N_2O are also generated by photochemistry, the amounts are miniscule.

It is interesting to consider the list of gases emitted by microbial life on Earth. A partial list includes O_2, H_2, CO_2, N_2, N_2O, NO, NO_2, H_2S, SO_2, and CH_4 (see, e.g., Madigan et al, 2001). Let us go through this list and consider the feasibility of each as a biosignature.

We have already briefly described Earth's biosignature gases O_2, N_2O, and CH_4. On Earth, the gas N_2 makes up 78% of our atmosphere and is therefore not a useful biosignature. Furthermore, as a homonuclear molecule, N_2 has no rotational-vibrational transitions and, hence, no spectral signature at visible and IR wavelengths. Like N_2, CO_2 is already present in Earth's atmosphere, making up about 0.035%. For Venus and Mars (the only other Solar System terrestrial planets with an atmosphere), CO_2 makes up more than 97% of their atmosphere. As such, CO_2 is considered a major planetary atmosphere gas that results from planet formation and evolution, and hence is not a useful biosignature. The gases H_2, H_2S, SO_2, NO, and NO_2 are produced on Earth either by volcanoes or by photochemistry, making them nonunique biosignatures. On Earth, these gases are produced in tiny quantities (both biotically and abiotically) and, hence, also lack any detectable spectral signature for a remote observer.

Many exoplanets may simply not have a unique biosignature like Earth's O_2 and N_2O. In our preparation for the search for exoplanet biosignatures, we must aim to consider the more common gases produced both as metabolic by-products and by either geophysical or photochemical processes. To understand their potential as biosignatures, we would have to model exoplanet environment scenarios to determine under which cases the common gases would be much more abundant than can reasonably be produced by geologic or photochemical processes. In other words, absent a unique biosignature such as oxygen, the main criterion for a biosignature is that the gas exists in such great quantities that its presence in a planetary atmosphere is well above the amounts that could be produced by any abiotic processes. In such a scenario, we may be relatively certain, but never 100% positive, that we have found signs of life on another planet. Alternatively, we may also be guided toward identifying biosignatures by a purely observational approach. One of the early ideas in biosignature research is to find a planet atmosphere that is out of chemical equilibrium, especially in redox disequilibrium (Lovelock, 1965).

CONCLUDING REMARKS

This is an exciting time in exoplanet research. The existence of exoplanets is firmly established. Hundreds of exoplanets are known to orbit nearby Sun-like stars. Six different exoplanet discovery techniques have reached success. The first 15 years of exoplanet discoveries have taught us to expect surprises because the random nature of planet formation and migration leads to many different planetary system architectures. We anticipate a similar diversity in super-Earth exoplanet atmospheres.

The Universe is filled with galaxies, likely more than 100 billion of them. Our own Milky Way Galaxy, in turn, has about 100 billion stars. With so many stars, there are bound to be a huge number of Earth-like planets in our Universe—regardless of whether Earth-like planets are common. Finding the Earths and searching for signs of life in their atmospheres will be a massive challenge—but one that humanity is ready for. In this way, we might finally answer the ancient question, Are we alone in the Universe?

REFERENCES

Borucki, W.J., Koch, D., Basri, G., et al. (2010). *Kepler* planet detection mission: Introduction and first results. *Science*, 327:977–80.

Burrows, A., Hubeny, I., Budaj, J., et al. (2007). Theoretical spectral models of the planet HD 209458b with a thermal inversion and water emission bands. *Astrophysical Journal*, 668:L171–74.

Butler, R.P., Wright, J.T., Marcy, G.W., et al. (2006). Catalog of nearby exoplanets. *Astrophysical Journal*, 646:505–22.

Cash, W. (2006). Detection of Earth-like planets around nearby stars using a petal-shaped occulter. *Nature*, 442:51–3.

Charbonneau, D., Brown, T.M., Noyes, R.W., et al. (2002). Detection of an extrasolar planet atmosphere. *Astrophysical Journal*, 568:377–84.

Charbonneau, D., Allen, L.E., Megeath, S.T., et al. (2005). Detection of thermal emission from an extrasolar planet. *Astrophysical Journal*, 626:523–29.

Charbonneau, D., Berta, Z.K., Irwin J., et al. (2009). A super-Earth transiting a nearby low-mass star. *Nature*, 492:891–94.

Cowan, N.B., Agol, E., and Charbonneau, D. (2007). Hot nights on extrasolar planets: Mid-infrared phase variations of hot Jupiters. *Monthly Notices of the Royal Astronomical Society*, 379:641–46.

Cowan, N.B., Agol, E., and Meadows, V.S. (2009). Alien maps of an ocean-bearing world. *Astrophysical Journal*, 700:915–23.

Cox, A.N. (2000). *Allen's Astrophysical Quantities*, 4th edn. New York: Springer.

Deming, D., Harrington, J., Seager, S., et al. (2006). Strong infrared emission from the extrasolar planet HD 189733b. *Astrophysical Journal*, 644:560–44.

Deming, D., Seager, S., Richardson, L.J., et al. (2005). Infrared radiation from an extrasolar planet. *Nature*, 434:740–43.

Deming, D., et al. (2009). Discovery and characterization of transiting super Earths using an all-sky transit survey and follow-up by the *James Webb Space Telescope*. *Publications of the Astronomical Society of the Pacific*, 121:952–67.

Elkins-Tanton, L. and Seager, S. (2008). Ranges of atmospheric mass and composition of super-Earth exoplanets. *Astrophysical Journal*, 685:1237–246.

Ford, E.B., Seager, S., and Turner, E.L. (2001). Characterization of extrasolar terrestrial planets from diurnal photometric variability. *Nature*, 412:885–87.

Fortney, J.J., Lodders, K., Marley, M.S., et al. (2008). A unified theory for the atmospheres of the hot and very hot Jupiters: Two classes of irradiated atmospheres. *Astrophysical Journal*, 678:1419–435.

Gaudi, B.S., Bennett, D.P., Udalski, A., et al. (2008). Discovery of a Jupiter/Saturn analog with gravitational microlensing. *Science*, 319:927–30.

Grillmair, C.J., Burrows, A., Charbonneau, D., et al. (2008). Strong water absorption in the dayside emission spectrum of the planet HD189733b. *Nature*, 456:767–69.

Harrington, J., Hansen, B.M., Luszcz, S.H., et al. (2006). The phase-dependent infrared brightness of the extrasolar planet Upsilon Andromedae b. *Science*, 314:623–26.

Harrington, J., Luszcz, S., Seager, S., et al. (2007). The hottest planet. *Nature*, 447:691–93.

Hubeny, I., Burrows, A., and Sudarsky, D. (2003). A possible bifurcation in atmospheres of strongly irradiated stars and planets. *Astrophysical Journal*, 594:1011–18.

Joshi, M.M., Haberle, R.M., and Reynolds, R.T. (1997). Simulations of the atmospheres of synchronously rotating terrestrial planets orbiting M dwarfs: Conditions for atmospheric collapse and the implications for habitability. *Icarus*, 129:450–65.

Kalas, P., Graham, J.R., Chiang, E., et al. (2008). Optical images of an exosolar planet 25 light-years from Earth. *Science*, 322:1345–348.

Knutson, H.A., Charbonneau, D., Allen, L.E., et al. (2007). A map of the day-night contrast of the extrasolar planet HD 189733b. *Nature*, 447:183–86.

Knutson, H.A., Charbonneau, D., Allen, L.E., et al. (2008). The 3.6–8.0 µm broadband emission spectrum of HD 209458b: Evidence for an atmospheric temperature inversion. *Astrophysical Journal*, 673:526–31.

Knutson, H.A., Charbonneau, D., Burrows, A., et al. (2009a). Detection of a temperature inversion in the broadband infrared emission spectrum of TrES-4. *Astrophysical Journal*, 691:866–74.

Knutson, H.A., Charbonneau, D., Cowan, N.B., et al. (2009b). Multiwavelength constraints on the day-night circulation patterns of HD 189733b. *Astrophysical Journal*, 690:822–36.

Lammer, H., Lichtenegger, H.I., Kulikov, Y.N., et al. (2007). Coronal mass ejection (CME) activity of low mass M stars as an important factor for the habitability of terrestrial exoplanets. II. CME-induced ion pick up of Earth-like exoplanets in close-in habitable zones. *Astrobiology*, 7:185–207.

Li, C.-H., Benedick, A.J., Fendel, P., et al. (2008). A laser frequency comb that enables radial velocity measurements with a precision of 1 cm s^{-1}. *Nature*, 452:610–12.

Liang, M.-C., Parkinson, C.D., Lee, A.Y.-T., et al. (2003). Source of atomic hydrogen in the atmosphere of HD 209458b. *Astrophysical Journal*, 596:247–50.

Lissauer, J.J. (2007). Planets formed in habitable zones of M dwarf stars probably are deficient in volatiles. *Astrophysical Journal*, 660:149–52.

Lovelock, J.E. (1965). A physical basis for life detection experiments. *Nature*, 207:568–70.

Madhusudhan, N. and Seager, S. (2009). A temperature and abundance retrieval method for exoplanet atmospheres. *Astrophysical Journal*, 707:24–39.

Madigan, M.M., Martinko, J., and Parker, J. (2001). *Brock Biology of Microorganisms*. Boston: Prentice Hall.

Marois, C., Macintosh, B., Barman, T., et al. (2008). Direct imaging of multiple planets orbiting the star HR 8799. *Science*, 322:1348–352.

Montanes-Rodriguez, P., Palle, E., Goode, P.R., et al. (2006). Vegetation signature in the observed globally integrated spectrum of Earth considering simultaneous cloud data: Applications for extrasolar planets. *Astrophysical Journal*, 651:544–52.

Nutzman, P. and Charbonneau, D. (2008). Design considerations for a ground-based transit search for habitable planets orbiting M dwarfs. *Publications of the Astronomical Society of the Pacific*, 120:317–27.

Palle, E., Ford, E.B., Seager, S., et al. (2008). Identifying the rotation rate and the presence of dynamic weather on extrasolar Earth-like planets from photometric observations. *Astrophysical Journal*, 676:1319.

Pearl, J.C. and Christensen, P.R. (1997). Initial data from the Mars Global Surveyor thermal emission spectrometer experiment: Observations of the Earth. *Journal of Geophysical Research*, 102:10875–880.

Redfield, S., Endl, M., Cochran, W.D., et al. (2008). Sodium absorption from the exoplanetary atmosphere of HD 189733b detected in the optical transmission spectrum. *Astrophysical Journal*, 673:L87–90.

Rogers, L.A. and Seager, S. (2010). Three possible origins for the gas layer on GJ 1214b. *Astrophysical Journal*, 716:1208–216.

Scalo, J., Kaltenegger, L., Segura, A., et al. (2007). M stars as targets for terrestrial exoplanet searches and biosignature detection. *Astrobiology*, 7:85–166.

Seager, S. and Deming, D. (2010). Exoplanet atmospheres. *Annual Review of Astronomy and Astrophysics*, 48:631–72.

Seager, S., Richardson, L.J., Hansen, B.M.S., et al. (2005a). On the dayside thermal emission of hot Jupiters. *Astrophysical Journal*, 632:1122.

Seager, S., Turner, E.L., Schafer, J., et al. (2005b). Vegetation's red edge: A possible spectroscopic biosignature of extraterrestrial plants. *Astrobiology*, 5:372–90.

Segura, A., Kasting, J.F., Meadows, V., et al. (2005). Biosignatures from Earth-like planets around M dwarfs. *Astrobiology*, 5:706–25.

Shao, M. and Nemati, B. (2009). Sub-microarcsecond astrometry with *SIM-Lite*: A testbed-based performance assessment. *Publications of the Astronomical Society of the Pacific*, 121:41–4.

Silvotti, R., Schuh, S., Janulis, R., et al. (2007). A giant planet orbiting the "extreme horizontal branch" star V 391 Pegasi. *Nature*, 449:189–91.

Spiegel, D.S., Silverio, K., and Burrows, A. (2009). Can TiO explain thermal inversions in the upper atmospheres of irradiated giant planets? *Astrophysical Journal*, 699:1487–500.

Swain, M.R., Vasisht, G., and Tinetti, G. (2008). The presence of methane in the atmosphere of an extrasolar planet. *Nature*, 452:329–31.

Swain, M.R., Vasisht, G., Tinetti, G., et al. (2009). Molecular signatures in the near-infrared dayside spectrum of HD 189733b. *Astrophysical Journal*, 690:114–17.

Tarter, J.C., Backus, P.R., Mancinelli, R.L., et al. (2007). A reappraisal of the habitability of planets around M dwarf stars. *Astrobiology*, 7:30–65.

Tinetti, G., Vidal-Madjar, A., Liang, M.C., et al. (2007). Water vapour in the atmosphere of a transiting extrasolar planet. *Nature*, 448:169–71.

Trauger, J.T. and Traub, W.A. (2007). A laboratory demonstration of the capability to image an Earth-like extrasolar planet. *Nature*, 446:771–73.

Turnbull, M.C., Traub, W.A., Jucks, K.W., et al. (2006). Spectrum of a habitable world: Earthshine in the near-infrared. *Astrophysical Journal*, 644:551–59.

Udry, S., Fischer, D., and Queloz, D. (2007). A decade of radial velocity discoveries in the exoplanet domain. In: B. Reipurth, D. Jewitt, and K. Keil (eds.), *Protostars and Planets V*, 685–99. Tucson: University of Arizona Press.

Vidal-Madjar, A., Lecavalier des Etangs, A., Desert, J.-M., et al. (2003). An extended upper atmosphere around the extrasolar planet HD 209458b. *Nature*, 422:143–46.

Vidal-Madjar, A., Desert, J.-M., Lecavalier des Etangs, A., et al. (2004). Detection of oxygen and carbon in the hydrodynamically escaping atmosphere of the extrasolar planet HD 209458b. *Astrophysical Journal*, 604:L69–72.

West, A.A., Hawley, S.L., Bochanski, J.J., et al. (2008). Constraining the age-activity relation for cool stars: The Sloan Digital Sky Survey Data Release 5 low-mass star spectroscopic sample. *Astronomical Journal*, 135:785–95.

Wolszczan, A. and Frail, D.A. (1992). A planetary system around the millisecond pulsar PSR 1257 + 12. *Nature*, 355:145–47.

Woolf, N.J., Smith, P.S., Traub, W.A., et al. (2002). The spectrum of earthshine: A pale blue dot observed from the ground. *Astrophysical Journal*, 574:430–33.

19 What New Telescopes Can Tell Us about "Other Worlds"

Charles A. Beichman

CONTENTS

INTRODUCTION

In the Western tradition, the "Many Worlds" concept goes back to pre-Socratic thought, with Epicurus asserting in his *Letter to Herodotus* (ca. 300 BCE) that "[T]here are infinite worlds both like and unlike this world of ours.... We must believe that in all worlds there are living creatures and plants and other things we see in this world...."

Such thoughts are to be found in Chinese history as well, with philosophers debating the nature of the Universe and the existence of other worlds. Around 100 BCE, the Hun Thien (浑天) School advocated a single celestial sphere, while Hsüan Yeh (玄烨) taught of infinite empty space. One modern-sounding explication of the "Many Worlds" hypothesis comes from the philosopher Deng Mu (邓牧) (1247–1306 CE):*

> Heaven and Earth are large, yet in the whole of empty space [hsü khungs] they are but as a small grain of rice.... It is as if the whole of empty space were a tree and heaven and earth were one of its fruits. Empty space is like a kingdom and heaven and earth no more than a single individual person in that kingdom. Upon one tree there are many fruits, and in one kingdom many people. How unreasonable it

---

* Part of this quotation also appears as an epigraph at the beginning of this volume. Also see the various chapters related to ancient Chinese scientific thought and culture in this book.

would be to suppose that besides the heaven and earth which we can see there are no other heavens and no other earths! (Needham and Wang, 1959, p. 221)

Renaissance and Enlightenment Europe offer numerous examples of natural philosophers proclaiming a belief in multiple worlds and even other worlds with life. Deng Mu sounds exactly like Italy's Giordano Bruno, who asserted in *De L'infinito universo e mondi* (*On the Infinite Universe and Worlds*) (1584) that "[T]here are countless suns and countless earths all rotating around their suns in exactly the same way as the seven planets of our system.... The countless worlds in the universe are no worse and no less inhabited than our Earth."

By the time of Newton, entire books were being written on the topic, with Christiaan Huygens asserting in his *Cosmotheoros: Or Conjectures Concerning the Planetary Worlds* (1698) that "[T]he Earth may justly liken'd to the planets ... [which have] gravity ... and animals not to be imagined too unlike ours ... and even Men ... [which] chiefly differ from Beasts in the study of Nature ... [and who] have Astronomy and its subservient Arts: Geometry, Arithmetick, Writing, Opticks."

These thoughts, however modern sounding, remained speculative. By taking advantage of the telescope, whose 400th anniversary we celebrate here, our modern era has advanced from speculation into the realm of science. By the time we reach the 500th anniversary of the telescope, we will have used some of the most advanced observatories imaginable (today) to address the existence and nature of other Earth-like worlds and to probe whether we are in fact alone in the Universe.

MODERN SCIENCE, BUT STILL SOME SPECULATION

The past decade has yielded spectacular increases in our knowledge concerning the "Many Worlds" hypothesis and the search for life beyond Earth. Yet many questions, some more speculative than scientific, remain. A useful way to encapsulate our knowledge and our ignorance is the famous Drake equation (Drake and Sobel, 1992), which relates the number of communicative civilizations to a series of factors:

$$N = R_* \times f_{\text{planet}} \times n_{\text{Earth}} \times f_{\text{Life}} \times f_{\text{Intelligence}} \times f_{\text{communicative}} \times \text{Lifetime}, \qquad (19.1)$$

where R_* is the rate of formation of suitable stars; f_{planet} is the fraction of those stars with planets; n_{Earth} is the number of Earth-like worlds per planetary system; f_{Life} is the fraction of those Earth-like planets where life actually develops; $f_{\text{Intelligence}}$ is the fraction of life sites where intelligence develops; $f_{\text{communicative}}$ is the fraction of communicative planets; and Lifetime is the "lifetime" of communicating civilizations. The first three terms are *astronomical* terms about which we are learning a great deal through telescopic observation, as discussed below. The fourth term, f_{Life}, is the *astrobiological* term about which the study of life in harsh terrestrial environments is expanding our conception of the minimum conditions for life. The remaining terms pertain to the evolution and survival of intelligent life (or at least of other electrical engineers capable of building radio telescopes) and will remain highly speculative until the success of one or more of the many SETI experiments to detect radio or other signals from other civilizations (Tarter, 2007). But the *astronomical* investigation of f_{planet} and n_{Earth} and the *astrobiological* investigation of f_{Life} are tasks of the highest scientific interest. The 2000 report of the US Decadal Survey of Astronomy and Astrophysics called this "so challenging and of such importance that it could occupy astronomers for the foreseeable future" (McKee and Taylor, 2001).

PLANET DETECTION METHODS

Table 19.1 and Figure 19.1 give a brief overview of planet detection methods and indicate the great progress since the initial detection of planets around a pulsar (Wolszczan and Frail, 1992) and 3 years later of a "hot Jupiter" (0.47 M_{Jup}) orbiting the main-sequence star 51 Pegasi with a 4.2-day

TABLE 19.1

Overview of Planet Detection Techniques

Technique	Date of First Discovery	No. Systems	No. Planets	Comments
Timing	1992	8	13	Pulsars and white dwarfs
Radial velocity	1995	424	506	Sin(i) ambiguity; favors small orbital radii
Transit	1999	126	134	Edge-on; favors small orbital radii
Microlensing	2004	11	12	Distant stars, no follow-up; probes outside habitable zone
Imaging	2004	21	24	Currently favors young planets on distant orbits

Source: Current as of June 15, 2011 from Jean Schneider's *Exoplanet Encyclopedia* (http://exoplanet.eu/).

period (Mayor and Queloz, 1995). Since then, the number and variety of planets have grown explosively, with planets found with orbits just grazing the surface of their stars to ones located well beyond the orbit of Jupiter and with sizes ranging from many times the mass of Jupiter to small rocky planets only a few times more massive than Earth itself. Researchers are racing to fill the critical *missing* area of this parameter space corresponding to the realm of "habitable terrestrial

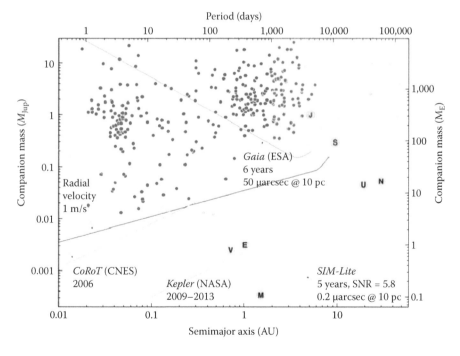

FIGURE 19.1 The incidence of planets detected using different techniques is shown for planets of different masses (*vertical axis*) and orbital semimajor axis (*horizontal axis*): radial velocity detections, blue dots; transit detections, red dots; microlensing, yellow dots; imaging, orange dots with large vertical error bars; timing, green dots. Also shown are sensitivity curves for various experimental techniques, including radial velocity, astrometry with *SIM* and *Gaia*, and transits with *CoRoT* and *Kepler*. The shaded areas in the upper right denote sensitivity for various imaging systems. Planets in our Solar System are indicated along with the habitable zone for a G2 star. (From Beichman, C.A., Krist, J., and Trauger, J.T. et al., *Publications of the Astronomical Society of the Pacific*, 122, 162–200, 2010.)

planets." *Terrestrial* planets are those between (roughly) 0.5 and 10 Earth masses ($M_{\oplus}$) on which water and other volatile molecules might be captured and retained, but not to such an extent as to allow a runaway accretion resulting in a gas giant or icy giant planet. Such a "rocky" planet would probably experience geophysical processes such as plate tectonics that would drive atmospheric evolution (Valencia et al., 2007). A *habitable terrestrial* planet is located in a stable, temperate orbit about a star* where liquid water is possible at or near the planet's surface. The location of the "habitable zone" (Kasting et al., 1993) scales with the stellar luminosity, L, roughly as $1\ AU \times (L/L_{\odot})^{1/2}$, e.g., 1 AU and a 1 yr orbit for a G2 star or 0.1 AU and 0.07 yr orbit for an M5 star. The exact location, width, and temporal evolution of the habitable zone depend on poorly understood processes such as the evolution of stellar luminosity and photospheric emission at all wavelengths (especially of high-energy ultraviolet [UV] and x-ray emissions that can threaten the integrity of complex biochemistry), geophysical processes (including volcanism and magnetic shielding from intense cosmic rays), and the evolution of planetary atmospheres and greenhouse effects.

The search for habitable, terrestrial planets is enormously challenging, but astronomers are developing techniques, on the ground and in space, that will bring us first a statistical understanding of the incidence of other Earths and then a census of habitable planets in our solar neighborhood. It is these nearest planets that will eventually be the targets of intense study looking for signs of life.

RADIAL VELOCITY

The radial velocity (RV) technique has given us our most complete census of planetary systems, covering a period range from a few days to more than 10 years and a mass range starting at <13 M_{Jup} of the largest gas giants, down through the icy giants and even into the super-Earth ($\leq 10\ M_{Earth}$) range. From the observations of 300 planets discovered by the time of the 400th anniversary of the telescope, we have learned much about the architecture of the gas and icy giants (Figure 19.2). Remarkably, in the period between the *NV400* anniversary celebration in 2008[†] and the time of this writing, another 100 planets were found by the RV method alone. See further information about exoplanet discoveries in the chapter by Marcy in this volume; also see Cumming et al. (2007) for a comprehensive summary and references to some of the properties described below:

- Approximately 10% of FGK stars have planets with masses above 0.3 M_{Jup} and periods <2,000 days. More than 10% of these planets are in multiple systems.
- The incidence of planets increases sharply with decreasing planetary mass, roughly as $dN/dM \propto M^{-1.3 \pm 0.2}$ until selection effects cause a drop-off below ~0.3 M_{Jup}.
- With the exception of objects in the closest orbits that are expected to be circularized by tidal effects, the distribution of orbital eccentricities is very broad, with circular orbits like those in our Solar System the exception rather than the rule. The incidence of gas giant planets increases dramatically with increasing metallicity, $\propto[Fe/H]^2$ (Fischer and Valenti, 2005).

As described elsewhere in this volume, the frontier of RV measurements is advancing dramatically, with observations breaking the 1 m/s barrier, observational time lines now exceeding a decade, fainter stars and/or later spectral types now coming under scrutiny, and surveys with multi-object spectroscopy just beginning (Ge et al., 2009). These advances have led to record holders across planet parameter space: the planet on the most distant, Jupiter-like orbit, HD 154345b, with a 0.95 M_{Jup} companion in a 9.2 yr, 4.2 AU circular orbit (Wright et al., 2007); super-Earths

---

* Or in an orbit about a giant planet that is itself in a temperate orbit. A moon such as Titan orbiting a giant planet located closer to its parent star than our Saturn might be an excellent abode for life.
† See http://nv400.uchicago.edu/.

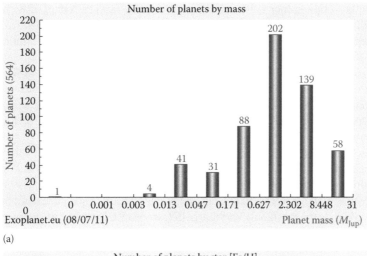

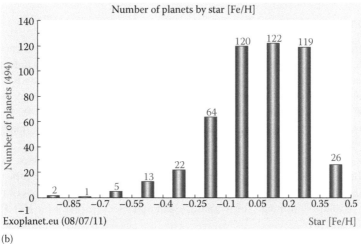

FIGURE 19.2 (a) Histogram showing number of planets of different masses; the census is incomplete below about 0.5 M_{Jup} as a result of selection effects. (b) Histogram showing number of planets of different host star metallicity. (From http://exoplanet.eu/ as of July 2011.)

such as GL 581, with a 5.2 M_{Earth} companion at 0.07 AU (Udry et al., 2007); and HD 156668, with a 4.1 M_{Earth} companion at 0.05 AU (Howard et al., 2010).* Searches for lower-mass planets are under way, and in the long run searches with a precision of 0.5 m/s might someday find an Earth in the 0.1 AU habitable zone of an M5 star (Udry et al., 2007). Of great promise are experiments to make high-precision RV observations in the near-infrared (IR) where late-type stars are brightest and signatures of even low-mass planets in the "habitable zone" are strong as a result of the low mass and luminosity of the host star.

There is considerable controversy as to the ultimate limit of the RV technique due to starspots, granulation, coherent oscillations, and other motions in stellar photospheres. If the typical RV "jitter" is ≥ 1 m/s (Moutou et al., 2009) with a correlation timescale of 1–2 weeks as a result of starspot

---

* RV masses are minimum masses, $M\sin(i)$, subject to unknown inclination, i. Thus, a brown dwarf companion (tens of Jupiter masses) of a bright star, if orbiting in a plane perpendicular to the line of sight, would show only a small RV perturbation. As seen from Earth, transiting systems must be within 1 or 2 degrees of exactly edge-on, thereby removing the $\sin(i)$ ambiguity.

lifetime and stellar rotation, then achieving a signal-to-noise ratio (S/N)* of ~5.8 on an Earth with an amplitude of 0.09 m/s would take more than $(1/0.09)^2 \times 5.8^2 \sim 4,150$ statistically independent observations separated by 1–2 weeks, or 80 years. Only in the case of the most quiescent K and M stars (perhaps 10%–20% of stars; Moutou et al., 2009) will it be possible to push into the habitable zone for Earth-sized planets. But for more typical stars, the practical limits to RV may be much higher, so that the detection of habitable Earths orbiting nearby solar-type stars may have to rely on other techniques, i.e., astrometry (see below). The scarcity of quiescent host stars means that the closest solar-type stars to Earth, i.e., those suitable for direct imaging searches, are unlikely to be identified as hosts for ~1 M_{Earth} planets in the habitable zone.

Despite these limitations, RV observations continue to be the workhorse of the planet detection enterprise, expanding the census, as well as confirming and providing masses and orbits of transiting planets. Europe in particular has expansive plans for new high-precision RV instrumentation, the Echelle SPectrograph for Rocky Exoplanet and Stable Spectroscopic Observations (ESPRESSO; Pasquini et al., 2009) on the 8 m (26 ft) Very Large Telescope (VLT) and for the proposed 42 m (138 ft) telescope.

TRANSITS AND PRECISION PHOTOMETRY

The first measurement of a transiting planet came as a follow-up observation to a known RV system, HD 209458 (Henry et al., 1999; Charbonneau et al., 2000). This first transit detection was based directly on a quantitative prediction from the RV ephemeris and was obtained with a completely different experimental apparatus: a small, backyard telescope. There are few more powerful examples of the power of the scientific method, which demands successful prediction for the validation of dramatic claims. The transit detection confirmed the planetary nature of RV detections at a time when some were arguing that these objects were mostly stellar companions seen in nearly face-on orbits and thus showing a low RV signature (Black and Stepinski, 2001). Once the transit result demonstrated that HD 209458b was edge-on, the ambiguity of the inclination of the orbit to the line of sight that affects the measured RV mass was removed, and the Jovian masses of the RV systems were no longer in doubt—although for a contrary example, see the result for HD 33636 discussed below.

But beyond simple confirmation and an unambiguous determination of planet mass, transits have proven to be an immensely powerful tool for learning about the physical properties of planets. Table 19.2 gives an incomplete listing of some of the key properties of planets that transits and precision photometry can reveal, in some cases in conjunction with dynamical measurements of mass. For a complete discussion on the power of transits, see Pont et al. (2009), in particular Charbonneau (2009) in that volume.

TABLE 19.2

Sample of Physical Characteristics Obtainable from Transits and Precision Photometry

Orbit inclination ~90 degrees gives unambiguous planet mass from RV	Planetary radius from transit depth	Rings and moons from transit profile	Vertical structure of atmosphere from mid-IR spectra
Star/planet spin-orbit alignment from stellar line profile variations during transit	Planet mass (from RV) + radius → bulk density and composition	Albedo from reflected light measurements	Global atmospheric circulation from thermal data over orbit
Stellar limb darkening from transit profile	Timing perturbations may signal other planets	Composition from primary and secondary transits in visible and IR	Stellar mass/density

---

* The minimum shown by simulation to be required for reliable detection (Catanzarite et al., 2008).

The transit effect is in principle simple. When the planet goes in front of a star in a primary transit, the total light detected from the system is reduced for a few hours by $(R_p/R_*)^2$, where R_p and R_* are the planetary and stellar radii, respectively. The diminution in brightness is 1.4% for a Jupiter–Sun analog and 0.01% for an Earth–Sun analog. In a secondary transit, the planet goes behind the star, and the diminution in brightness is given by the ratio of planet to stellar areas and the ratio of the brightness levels: $(R_p/R_*)^2 \times I_{v,p}/I_{v,*}$, where $I_{v,p}$ and $I_{v,*}$ represent the surface brightness of the planet and star, respectively. For "hot Jupiters," the secondary transit can be as strong as a few tenths of a percent.

When these simple photometric (or spectrophotometric) effects are measured with extremely high precision, an abundance of information about the planet and the host star can be derived (Seager and Mallén-Ornelas, 2003; Table 19.2). The transit geometry removes the uncertainties in planetary properties due to unknown inclination. The combination of mass from RV and radius from the transit yields density, a key indicator of internal structure and thus of planetary formation and evolution scenarios (Figure 19.3). Spectrophotometry at primary transit allows determination of atmospheric composition at visible wavelengths. Photometry at secondary transit allows IR energy from the planet to be observed directly (Charbonneau et al., 2005; Deming et al., 2005; Richardson et al., 2008), including emission in spectral lines such as CO_2 and CH_4 with the *Hubble Space Telescope* (*HST*; Swain et al., 2009), which allows determination of atmospheric temperature structure and chemical composition (Fortney et al., 2008; Burrows et al., 2008). The timing of transits could reveal the presence of other planets through orbital interactions. And long-term monitoring of transiting systems might reveal moons or rings around the known planet or additional coplanar planets. The intermediate orbital phases between primary and secondary transit are also observable at both visible and IR wavelengths. Excess emission over the purely stellar brightness can be interpreted in terms of albedo and phase function (in the visible) or surface temperature maps (in the thermal IR) and constraints on global energy circulation (Figure 19.4; Knutson et al., 2009).

The transit technique has entered a new realm with the successful launches of the Centre National d'Etudes Spatiales' (CNES) *Convection, Rotation and planetary Transits* (*CoRoT*) (Baglin et al., 2009) and NASA's *Kepler Space Telescope* (*KST*) satellites (Borucki et al., 2010), which will each survey more than 100,000 stars in areas of a few tens to ~100 square degrees. *CoRoT* has already

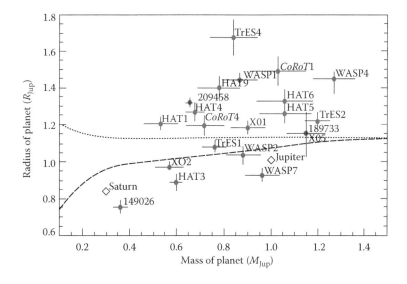

FIGURE 19.3 The radius and mass derived for transiting planets show a wide diversity, suggesting effects of *nature*, e.g., size of solid core (0 or 20 M_{Earth}; *dotted and dashed lines*), and/or *nurture*, e.g., tidal dissipation or insolation. Objects below the dashed line must have considerable contribution from elements other than just H and He (From Charbonneau, D., *The Rise of the Vulcans.* Cambridge: Cambridge University Press, 2009.).

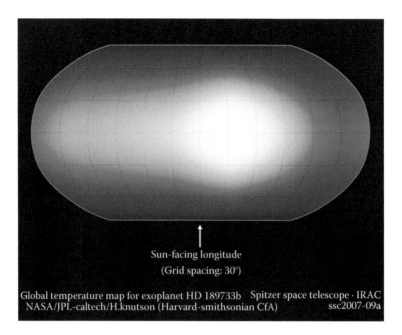

FIGURE 19.4 Continuous monitoring with the *Spitzer Space Telescope* of the star-planet sytem HD 189733 during a complete orbit of the planet around its host star led to this map of the IR surface emission from the planet HD 189733b. Violent winds shift the hottest spot from directly under the central point on the planet facing the host star, HD 189733. (From Knutson, H.A., Charbonneau, D., Cowan, N.B., et al., *Astrophysical Journal*, 690, 822–36, 2009.) Image from NASA/*Spitzer* Science Center.

detected a ~1.7 R_{Earth} planet with mass of roughly 5 M_{Earth} (Leger et al., 2009) in a scorching 20.5 hr orbit around a K0 star. *Kepler* has the sensitivity and mission duration consistent with the detection of more habitable systems, possibly even an Earth transiting a main-sequence G2 star in a 1 yr orbit. *Kepler*'s initial data release gives us great optimism that these ambitions will be achieved (Borucki et al., 2011). With just its first 4 months of data, the *Kepler* team has confirmed 16 planets, including some smaller than 2 R_{Earth}. *Kepler* also identified 1,235 candidate planets, not as yet confirmed via RV measurements or other techniques, including some planets smaller than 1 R_{Earth}. Since the vast majority of these are expected to be valid detections the *Kepler* data will ultimately lead to a dramatically improved understanding of the demographics of planets, including the critical parameter for the search for habitable planets: $\eta_\oplus$, the fraction of stars with Earth-sized planets (~1–2 R_{Earth}) orbiting in the habitable zones of their parent stars.

Beyond these two pioneering space transit missions and their associated follow-up using the *Spitzer Space Telescope* (*SST*, formerly the *Space Infrared Telescope Facility* [*SIRTF*]), *HST*, and eventually the *James Webb Space Telescope* (*JWST*), there are proposals to conduct space missions using small telescopes to look for transits over a much larger fraction of the sky. The transiting systems detected by these next-generation space observatories would be closer and brighter than those found by *CoRoT* and *KST*, thereby enabling much easier and much more powerful follow-up observations, including observations at high spectral resolution. The European Space Agency (ESA) is considering the *PLAnetary Transits and Oscillations of stars* (*PLATO*) mission to monitor more than 2,000 square degrees, while NASA is considering the *Transiting Exoplanet Survey Satellite* (*TESS*) mission (Ricker et al., 2010; Deming et al., 2009) to monitor the entire sky at visible wavelengths. Finally, a near-IR all-sky survey (*All Sky Transit Observer* [*ASTrO*]; or *Earth-Like Transit* mission [*ELEKTRA*]; see Beichman et al., 2011) would be particularly sensitive to transits of nearby, late-type stars by habitable super-Earths (~2 R_{Earth}), which might show H_2O and CO_2 and other signatures of habitability (Deming et al., 2009; Kaltenegger and Traub, 2009).

MICROLENSING

Microlensing takes advantage of a general relativistic property of light (Mao and Paczynski, 1991) whereby the light of a star at great distance ("the source," typically at 8 kpc) is amplified by the light of a star at intermediate distance ("the lens," typically at 1–4 kpc). In the case that the lensing star is accompanied by one or more planets, the resultant amplification curve can show deviations from simple theory that can be used to detect and characterize planets (Gould and Loeb, 1992). Intensive observations of individual microlensing events by networks of small telescopes have so far detected eight planets in seven systems with a wide range of characteristics: from 2.6 to 0.017 M_{Jup} (=5.6 M_{Earth}), spanning orbital distances around the *Einstein* ring of typical lensing stars, 2–5 AU (Gaudi et al., 2008). High-magnification events from low-impact parameter crossings increase the possibility of finding multiple planets and can be analyzed in sufficient detail to break many degeneracies in the theoretical analysis. One such system has been shown to consist of a 0.5 $M_\odot$ star (where $_\odot$ denotes the value of our Sun) at 1.5 kpc with two planets, 0.71 and 0.27 M_{Jup} in 2.3 and 4.6 AU orbits, respectively (Gaudi et al., 2008). The fact that microlensing systems are finding as many low-mass planets as high-mass planets, despite the much higher detection probability for the latter, suggests that Earth-mass objects in the 1–5 AU regime may be quite common (Gaudi et al., 2008). A space mission (Bennett et al., 2007) could find hundreds of planets via this mechanism, probing an important region just outside the typical habitable zones of main-sequence stars that would otherwise be hard to investigate. A disadvantage of the technique is that once the lensing event is over, there is no prospect for follow-up observations of the detected planets. The ASTRO 2010 decadal report, *New Worlds, New Horizons in Astronomy and Astrophysics* (Committee for a Decadal Survey of Astronomy and Astrophysics; National Research Council, 2010), included a microlensing survey as one of the goals of their highest priority new mission, *Wide-Field Infrared Survey Telescope* (*WFIRST*).

DIRECT IMAGING OF GIANT PLANETS

Observing from the Ground

During the time between the occurrence of the *NV400* conference held in Beijing in October 2008 and the writing of this chapter, the number of directly imaged planets has increased dramatically, with four objects being detected around two nearby, young A stars. Because the three planets around HR 8799 (Marois et al., 2008) and the single planet around Fomalhaut (Kalas et al., 2008) are young, their internal reservoir of gravitational energy generates enough luminosity to make the objects visible (Saumon et al., 1996). Planets older than about 100 Myr soon fade into obscurity and by 1 Gyr are invisible with existing coronagraphic capabilities. These young planets plus two earlier discoveries, 2M 1207 (Chauvin et al., 2005) and GQ Lup (Neuhäuser et al., 2005), are confirmed to be companions via their common proper motion with their host star and, in the case of Fomalhaut b, by orbital motion as well. What remains controversial, however, is the identification of these objects as planets (<13 M_{Jup}, the deuterium-burning limit), as opposed to brown dwarfs (13 $M_{Jup}< M < 70 M_{Jup}$) or even low-mass stars (>70 M_{Jup}). The relations between near-IR brightness, age, and mass are quite uncertain, and dynamical mass determinations are impractical for objects on long-period orbits. In fact, the models for young planets have been called into direct question. Marley et al. (2007) argued that core accretion models predict brightness levels 5–30 times lower at a given age than models that simply follow the luminosity evolution of a preexisting ball of gas. What is missing to resolve this controversy are objects of known age for which a combination of imaging (giving luminosity, effective temperature) and dynamical information (giving mass) is available to anchor the models. These combined data may become available with a combination of imaging using interferometers (Keck-I or VLT-I), coronagraphic imaging with ground-based telescopes or *JWST*, and astrometric mass measurements (see below). In the case of Fomalhaut-b, the interactions between the putative planet and the dust ring have been used to constrain the mass of the planet to be less than 3 M_{Jup} (Kalas et al., 2008).

Contrast ratio levels detectable with adaptive optics (AO) on 5–10 m (16–32 ft) telescopes are approaching ~10^{-4}–10^{-5} at 1 arcsec, which corresponds to tens of AU for nearby young stars. There are prospects for 1–2 orders of magnitude improvement in limiting contrast over the next few years as new instruments such as the Gemini Planet Imager (GPI; Macintosh et al., 2007) and P1640 at Palomar (Oppenheimer and Hinckley, 2009) come into operation. With coronagraphs on extremely large, *diffraction-limited* telescopes (30–42 m)/(98–138 ft), it should be possible to image young (10–100 Myr) gas giant planets orbiting within 2–3 AU of the closest young stars (25–50 pc) and possibly even detect mature planets orbiting the nearest low-mass stars (<5 pc) where the contrast ratio is favorable, e.g., GL 876 and GL 3522.

Observing Giant Planets with the *James Webb Space Telescope* (*JWST*)

While *JWST* has a diameter of "only" 6.5 m (21 ft) compared with existing 8–10 m (26–32 ft) telescopes and planned 30–42 m (98–138 ft) telescopes on the ground, and while *JWST's* wavefront error is relatively coarse, ~130 nm compared with the wavefront errors <50 nm possible with extreme AO systems on the ground, *JWST* is a *cooled* telescope operated in an extremely stable space environment. *JWST* will have enormous sensitivity at exactly the wavelengths where young planets are predicted to be very bright, i.e., at 4–5 μm, where the transparency of their atmospheres allows radiation from hot interior levels to emerge (Figure 19.5). The three imaging instruments on *JWST* each have a coronagraphic capability (Beichman et al., 2010): the Near Infrared Camera (NIRCam) has a traditional Lyot coronagraph (Krist et al., 2007) operating from 2 to 5 μm; the Canadian Tunable Filter Imager (TFI) (Beaulieu et al., 2008) has a traditional Lyot coronagraph plus an innovative non-redundant mask (NRM) imaging capability at 3–5 μm (Sivaramakrishnan et al., 2009); and the Mid-Infrared Instrument (MIRI) (Rouan et al., 2007) has four-quadrant phase masks operating around 10 μm. *JWST* should be able to observe planets more massive than ~0.2 M_{Jup} outside 1 arcsec

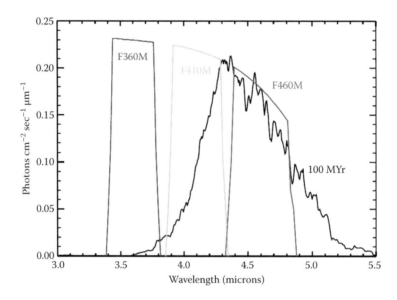

FIGURE 19.5 Models of young planets show that they are very bright in the 4–5 μm region of the spectrum, where *JWST* will have enormous sensitivity. The colored rectangular boxes show the extent of specific *JWST* filters superimposed on the predicted radiation from a hot, young (100-Myr-old) planet. (From Burrows, A., Sudarsky, D., and Lunine, J.I., *Astrophysical Journal*, 596, 587–96, 2003.) The brightness of a planet is the result of ongoing gravitational collapse. (From Krist, J.E., Beichman, C.A., Trauger, J.J., et al., 2007). Hunting planets and observing disks with the JWST NIRCam coronagraph. In D.R. Coulter (ed.), *Techniques and Instrumentation for Detection of Exoplanets III, Proceedings of the Society of Photo-Optical Instrumentation Engineers (SPIE)* Conference, Vol. 6693, 66930H–12.

with its Lyot coronagraphs and planets more massive than ~1 M_{Jup} inside 1 arcsec with its NRM interferometric mode. The different coronagraphic capabilities of *JWST* are summarized in Figure 19.6.

Figure 19.7 compares predictions for ground-based and space-based searches for giant planets based on a Monte Carlo simulation combining predictions of instrument performance with models of the brightness evolution of planets of various ages (1–100 Myr) and masses (0.1–10 M_{Jup}) orbiting more than 600 nearby young stars (Beichman et al., 2010). The brightness of the planet, given in terms of its contrast ratio to its host star, is taken from models appropriate to the planet mass and known age of the host star (Baraffe et al., 2003; Marley et al., 2007). Ground-based observations on the largest telescopes will find larger planets (>2 M_{Jup}) on smaller orbits (<10 AU), whereas *JWST* with its great sensitivity at 4.5 μm will find smaller planets across a range of orbital radii using its various coronagraphic modes. Taken together, these data sets will address formation and migration mechanisms for giant planets: core accretion in or around the region where volatile gases freeze out, i.e., the so-called snow line around 3–5 AU in our Solar System; disk fragmentation beyond 10 AU; or planetary migration at largest orbital separations. Spectroscopic data, particularly across the complete 1–10 μm band with *JWST*, will determine physical conditions on the planet. Adding astrometric masses to the imaging data will constrain evolutionary models of temperature and luminosity.

ASTROMETRY

The technique of astrometry uses the 2-dimensional measurement of positions on the sky to observe the dynamical effects of one or more planets of a host star. The smallness of the effect—amplitude ~500 μas $(a_{orbit}/5.2\ AU)\ (M_\odot/M_*)\ (M_{Planet}/M_{Jup})\ (10\ pc/Dist)$, where M_* is the mass of the parent star—compared with present-day measurement capabilities (milliarcsecond global astrometry and 50–100 μas for differential astrometry) accounts for the lack of results from this technique

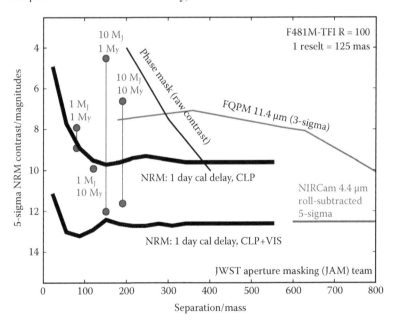

FIGURE 19.6 As discussed in the text, a comparison of different coronagraphs on *JWST*, showing magnitudes of contrast relative to the host star as a function of planet-star separation. The 12.5 mag of rejection correspond to a contrast ratio of 10^{-5}. The instruments include a classical coronagraph in the NIRCam, a NRM in the TFI, and a four-quadrant phase mask in the MIRI. (From Sivaramakrishnan, A., Tuthill, P., Ireland, M., et al., Non-redundant masking [NRM] on the *JWST* Fine Guidance Sensor-Tunable Filter Instrument [FGS-TFI]. *Proceedings of the Society of Photo-Optical Instrumentation Engineers* [*SPIE*] Conference, Vol. 7440, 74400Y, 2009. With permission.)

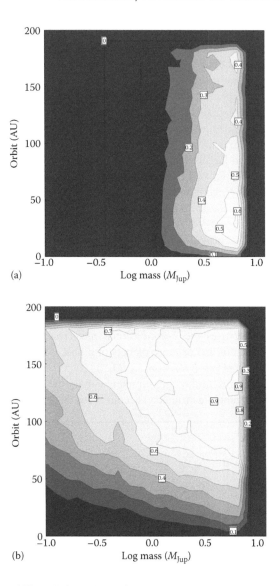

FIGURE 19.7 The detectability of giant planets in a survey of ~600 young stars in nearby star-forming groups as observed from a ground-based search at 1.6 μm (a) with an advanced coronagraph on the Palomar 5 m (16 ft) telescope and (b) with *JWST* at 4.4 μm using a standard Lyot coronagraph. The contours show the fraction of planets detectable of a given mass and orbital radius based on a Monte Carlo simulation (From Beichman, C.A., Krist, J., and Trauger, J.T. et al., *Publications of the Astronomical Society of the Pacific*, 122, 162–200, 2010.) using models of planet emission at a given mass and age from Baraffe et al. (Baraffe, I., Chabrier, G., Barman, T.S., et al., *Astronomy and Astrophysics*, 402, 701–12, 2003.)

to date. Massive Jupiters are barely in reach, while Earths are currently impossible to detect. Up until now, astrometry has been used to refine the parameters of the RV planet orbiting ε Eri using the Fine Guidance Sensor on *HST* (Benedict et al., 2006), while disproving the planetary nature of the RV-detected object orbiting HD 33636 by revealing the RV-detected companion to be an M star in a near–face-on orbit (Bean et al., 2007).

The lack of astrometric results will change dramatically with the advent of new space missions (Figure 19.8). ESA's *Gaia* mission will measure the positions of hundreds of millions of objects at the 25–100 μas level, adequate to detect Jovian-mass planets around hundreds of stars within 100 pc (Sozzetti et al., 2008) over the course of 5 years. More dramatically, NASA's *Space Interferometry*

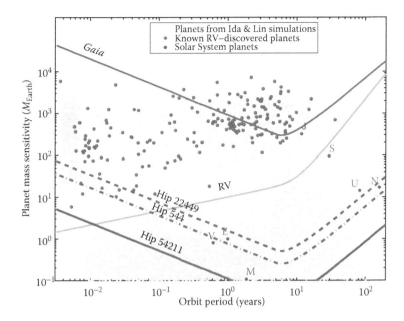

FIGURE 19.8 The discovery space for rocky Earth-like (1–10 M_{Earth}) planets in the habitable zone (0.7–1.5 AU) for a G star. The small yellow dots represent a theoretical planet distribution (Ida and Lin, 2005) for planets of 0.1–3,000 M_{Earth}. Known exoplanets (as of 2009) with semimajor axes >0.03 AU are shown as filled circles. Labeled curves represent the estimated sensitivity limits of indirect detection methods: radial velocity method and astrometry with *SIM* and *Gaia*. The *SIM* sensitivity in this space is a broad band, defined by the three lowest curves. The lowest curve shows the "best" star (as computed from star mass, luminosity, and distance), the middle curve represents the median star, and the upper curve shows the least favorable star in the sample. Also shown is the sensitivity of *Gaia* for stars at 50 pc, a typical distance for *Gaia* targets. (From Unwin, S.C., Shao, M., Tanner, A.M., et al., *Publications of the Astronomical Society of the Pacific*, 120, 38–88, 2008.)

Mission (SIM) will have the capability to measure differential, mission-averaged positions less than 0.2 µas (Catanzarite et al., 2008; Beichman et al., 2008; Traub et al. 2009), adequate to detect Earth-mass planets in the habitable zones of at least 50 nearby stars. With this capability, we will begin the census of potentially habitable worlds around the nearest solar analogs, which will someday be observed directly for signs of an atmosphere or life.

SIM's science goals related to planets are summarized here (see also the chapter by Shao in this volume):

- *A deep search for Earth analogs.* For an investment of ~40% of its mission time, *SIM* could search the closest, most favorable stars looking for terrestrial planets in the habitable zone (Figure 19.8). Depending on the results of transit surveys from *CoRoT* and *KST*, *SIM*'s survey will be tuned to probe a smaller number of stars very deeply, 50 stars to the 1 M_{Earth} level, or, if Earths are rare, then to look more broadly at ~150 stars to the 3 M_{Earth} level. At either extreme, *SIM* will locate specific planetary systems suitable for spectroscopic follow-up, as well as determine planetary masses and orbital parameters, which are critical for assessing habitability.

- *Architecture of planetary systems.* Our knowledge of planetary systems is limited by the 1-dimensional nature of RV measurements and by the fact that transiting systems typically reveal only one planet per star. Further, there are numerous classes of stars for which we have little or no information, e.g., binary stars, hot stars, and white dwarfs, which cannot be studied with high RV accuracy. *SIM* will undertake an astrometric survey of modest precision (4 µas) to investigate over 1,000 stars to complete our understanding of the architecture of planetary systems down to the level of Uranus-mass planets.

- *Planets around young stars.* Almost nothing is known about the presence of planets orbiting young stars because the active photospheres and rapid rotation of young stars make RV and transit detections difficult or impossible. Astrometric methods are superior in such cases. The few imaging results to date have been of planets on orbits of tens of AU, far from where planets are thought to form in the dense inner portions of the protostellar disk, i.e., near the snow line where gases such as water and methane freeze out. A *SIM* Key Project will perform astrometric measurements on more than 200 young stars (1–100 Myr old at distances of 25–140 pc from Earth) to look for planets in the critical 1–5 AU region (Beichman, 2001; Beichman et al., 2009; Figure 19.9). While *Gaia* or ground-based experiments may find a few of these planets, it will take the microarcsecond measurement capability of *SIM* to complete a census of young stars for planets in the range of Uranus to Jupiter.

THE SEARCH FOR LIFE ON OTHER WORLDS

Beyond the astronomical search for planets and a desire to better understand the processes governing their formation and evolution, there are strong scientific and societal desires to look for life beyond Earth. These urges raise the planet-finding program above the merely astronomical into realms addressing fundamental questions of biology, and even of human existence: *How did life originate? Are we alone?* Remarkably, the technology to address these questions is within our grasp—perhaps not with the current generation of telescopes (*SST, HST,* Keck), nor even with the next generation (*JWST, SIM,* ground-based 30 m (98 ft) telescopes), but with the generation after these. Investigation of potential biosignatures by astrobiologists (see the chapter by Seager in this volume) and the development of mission concepts by NASA and ESA have demonstrated that a

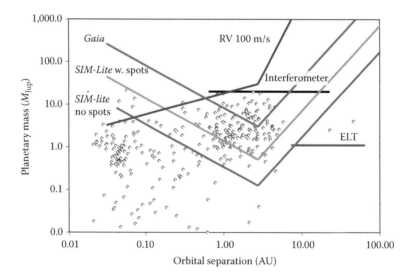

FIGURE 19.9 Sensitivity of different techniques to planets in the mass–semimajor axis plane for young 1 $M_\odot$ stars (<5 Myr) at a distance of 140 pc from Earth. A representative population of planets seen around nearby mature stars (*gray crosses*) indicates what we might find when looking at young stars. *SIM-Lite* sensitivity estimates are given for worst-case and best-case scenarios for starspot noise (From Beichman, C., Baratte, I., Crockett, C., et al., *Formation and Evolution of Planetary Systems: The Search for and Characterization of Young Planets.* White Paper to Astro 2010 Decadal Committee, 2009). Sensitivity estimates for *Gaia*, RV, a coronagraph on a 30 m (98 ft) Extremely Large Telescope (ELT) operating at 1.6 μm, and a multi-baseline interferometer operating in direct imaging mode (with closure phase) are shown for comparison.

search for habitable planets and for life itself via remote sensing is a credible scientific endeavor with a cost comparable to other flagship space missions.

WHAT TO LOOK FOR

Life manifests itself in many different ways here on Earth and will undoubtedly find new and surprising ways to survive in the widely varied environments possible on other worlds. But we can take advantage of certain attributes of life that are likely to be common wherever it is found. Metabolic processes from widespread life-forms should drive global environments out of thermodynamic equilibrium, e.g., the simultaneous existence of gases such as oxygen (O_2) and methane (CH_4) in an atmosphere, orders of magnitude out of simple chemical equilibrium (Lovelock, 1965). While it would be foolish to expect life to replicate itself only on Earth-analog planets, particularly given the variation of the Earth itself over geologic time (Kasting, 2004; Figure 19.10), we can appeal to basic physical principles to suggest what we might look for as signposts of habitability and even of life.

While deeply subterranean life—under a rocky crust, an icy surface, or water—may well be possible, it will be hard to detect via remote sensing. Our telescopic examination demands atmospheric or surface markers. Within our own Solar System, CO_2 is common to the three terrestrial worlds—Venus, Earth, Mars—with strong features at 15 μm (Figure 19.11) and in the near-IR. Detection of this feature alone would imply the presence of an atmosphere, a necessary if not sufficient criterion for habitability. Perhaps the next most important step in assessing habitability would be the detection of water vapor, again through strong features throughout the near- and mid-IR. Of the three rocky planets in our Solar System, only Earth shows water (Figure 19.11). Detection of an H_2O- and/or CO_2-rich atmosphere of a world of 1–10 M_{Earth} (as determined via RV or *SIM* astrometry) in the habitable zone of its host star would be cause for great excitement. Such a world containing water and other cosmically abundant gases in an atmosphere capable of evolving under geophysical or astrobiological forces would be a prime target for searches for life itself.

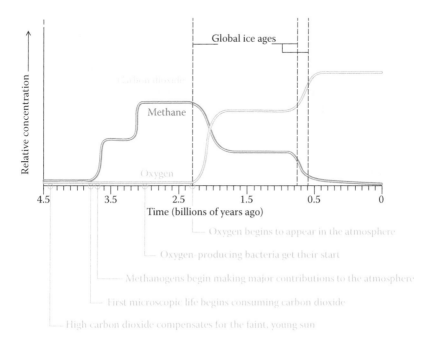

FIGURE 19.10 Evolution of Earth's atmosphere shows wide variation in the abundances of key molecular species (From Kasting, J., *Scientific American*, 291, 78–85, 2004. With permission.).

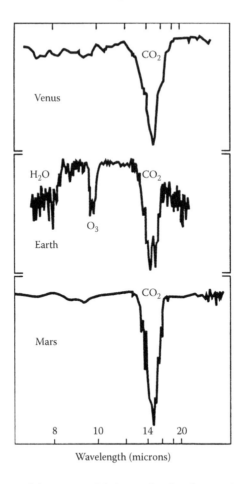

FIGURE 19.11 Mid-IR spectra of three terrestrial planets showing the complex atmosphere of Earth relative to Mars and Venus.

What would we look for next? Over the last decade, the astrobiology community has identified a handful of molecular species that might be signposts of primitive life, detectable at a low S/N ratio ($\sim$5–100) and at modest spectral resolution ($\lambda/\Delta\lambda \sim$ 5–100). These include O_2 and its proxy, ozone (O_3), as well as CH_4 and nitrous oxide (N_2O). All are potentially observable in spectra between 0.3 and 20 μm (Des Marais et al., 2002). O_2 bands in the visible and O_3 bands in the mid-IR or UV are perhaps the strongest biosignatures in a photosynthetic Earth analog, followed by CH_4, which might be much more abundant than in the present-day Earth (Kasting and Catling, 2004; Figure 19.10). Lines of CH_4 and N_2O in an exact Earth analog would be weak and masked by complex lines of water and other abundant species, but they might be more detectable in the presence of different metabolic processes. Surface chlorophyll might also be detected. Of course, it must be remembered that before claiming a biological origin for a molecular signature, it is critical (and difficult) to rule out geophysical processes such as volcanism, e.g., the ongoing debate about the detection of CH_4 on Mars (Mumma et al., 2009).

HOW TO LOOK FOR LIFE

Transit Experiments

The first spectrum of a habitable planet might come from a transit spectrum obtained for a super-Earth (2 R_{Earth}) orbiting a late M star. For such an object, the star-planet contrast ratio is relatively favorable, i.e., a super-Earth transiting an M8 star produces a 4% transit instead of the 0.01% transit of an Earth analog against a G2 star, making precision spectrophotometry feasible. A number of authors have examined these possibilities and come to a number of conclusions in common and a number on which there is still disagreement. First, since primary transit spectroscopy passes the light of the star through the atmosphere of the planet, maximizing the number of stellar photons is critical. Host stars must be bright and nearby. Second, while late-type stars are common but faint, transiting terrestrial planets could be identified by targeted searches (MEarth Project, looking for M dwarfs; Charbonneau, 2009) or all-sky surveys (*TESS, ELEKTRA*) and might be favorable abodes for life (Tarter et al., 2007). These objects could be followed up with *JWST* looking for spectral features of water, CH_4, or CO_2 during either the primary transit (planet going in front of the star) or the secondary transit (planet going behind the star).

There is disagreement on the observability of spectral features in terrestrial worlds. Transmission spectroscopy relies on the apparent size of a planet changing in different spectral channels as a result of optical depth effects in various atomic or molecular species. For cold, dense planets with strong surface gravity and for species of high molecular weight, the difference in apparent size will be small, making transit spectroscopy difficult (Kaltenegger and Traub, 2009). Alternatively, Deming et al. (2009) have argued that the presence of large amounts of hydrogen, as might be expected from photodissociation on an ocean planet, could puff up an atmosphere, making spectral features more observable. Habitable worlds with temperatures of ~300 K are much colder and would be harder to detect in secondary transit than the hot Jupiters observed to date (Deming et al., 2009; Kaltenegger and Traub, 2009).

Direct Imaging of Earths with the *Terrestrial Planet Finder* (*TPF*)

For more than a decade, there have been intensive studies of techniques to directly separate the light of a planet from its host star and to bring to bear the full power of spectroscopy on the characterization of potentially habitable planets. The debate has centered on what wavelength region to examine (visible or mid-IR); on what technique to use, internal coronagraphs or external occulters (*TPF*-C or *TPF*-O) or a nulling interferometer (*TPF*-I/*Darwin*); and on how big a telescope is needed to observe a reasonable number of planets. Constraints on the problem include the availability of technology to reject starlight at the appropriate level (Figure 19.12; 10^{-10} in the visible and 10^{-7} in the mid-IR) at

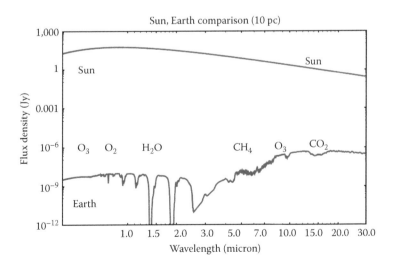

FIGURE 19.12 The relative brightness of the Earth compared to the Sun, seen from a distance of 10 pc, shows a contrast of 10^{-10} at visible wavelengths (<1 μm) and 10^{-7} at mid-IR wavelengths (~10 μm). Any direct imaging experiment must suppress this emission coming from a star sitting less than 0.1 arcsec away.

the appropriate angular scale (1 AU at 10 pc corresponds to 0.1 arcsec), and the permissible scope of the project in terms of budget and schedule.

It is impossible to summarize this debate here; refer to the ExoPlanet Community Report (Lawson et al., 2009) for a fuller discussion of these topics. But what is exciting is that remarkable progress has been made in the *physics* of starlight rejection at both visible and mid-IR wavelengths. Trauger and Traub (2007) have demonstrated visible rejection at the 10^{-10} level for 5%–10% bandwidths at an inner working angle of $4\lambda/D$. Other groups are showing promising test-bed results from alternative coronagraphic schemes. Ongoing studies are concentrating on such key variables as stellar rejection ratio, inner working angle (in units of λ/D, thus constraining the wavelength range of operation and telescope diameter), required stability for the entire optical system, numbers of targets, S/N ratio, and spectral resolution (Guyon, 2007). NASA currently has six variants of coronagraphs, including an external occulter ~50 m across flying tens of thousands of kilometers away from a ~4 m (13 ft) telescope (Cash et al., 2008; Vanderbei et al., 2008), under study for possible launch sometime after the year 2020.

The mid-IR is in some ways more favorable than the visible given the lower requirements on contrast (Figure 19.12). Excellent progress has been made in nulling interferometry with 10 μm rejection at a level of 10^{-5} over a 30% bandwidth (Peters et al., 2008), sufficient to detect terrestrial planets in the presence of local and exozodiacal emission. A variety of designs for nulling interferometers, known as *TPF*-I in the US and *Darwin* in Europe, have been developed, and technology studies are underway (Lawson et al., 2008).

TABLE 19.3

Synergistic Measurements are Needed to Characterize Planets and to Search for Life

	SIM	TPF-C/O	TPF-I
Orbital Parameters			
Stable orbit in habitable zone	*Measurement*	*Measurement*	*Measurement*
Characteristics for habitability			
Planet temperature	**Estimate**	**Estimate**	*Measurement*
Temperature variability due to eccentricity	*Measurement*	*Measurement*	*Measurement*
Planet radius	*Cooperative*	*Cooperative*	*Measurement*
Planet albedo	*Cooperative*	*Cooperative*	*Cooperative*
Planet mass	*Measurement*		
Surface gravity	*Cooperative*	*Cooperative*	*Cooperative*
Atmospheric and surface composition	*Cooperative*	*Measurement*	*Measurement*
Time variability of composition		*Measurement*	*Measurement*
Presence of water		*Measurement*	*Measurement*
Solar System characteristics			
Influence of other planets, orbit coplanarity	*Measurement*	**Estimate**	**Estimate**
Comets, asteroids, and zodiacal dust		*Measurement*	*Measurement*
Indicators of life			
Atmospheric biomarkers		*Measurement*	*Measurement*
Surface biosignatures (red edge of vegetation)		*Measurement*	

Note: **Measurement** indicates a directly measured quantity from a mission; **Estimate** indicates a quantity that can be estimated from a single mission; and *Cooperative* indicates a quantity that is best determined cooperatively using data from several missions.

CONCLUDING REMARKS

How will the search for exoplanets proceed? One can argue that we need information from astrometry, optical and IR photometry, and spectroscopy to fully characterize a planet as habitable and to look for signs of life. Masses, orbits, radius, density, and atmospheric composition are all important. Table 19.3 lists how each measurement contributes to some part of the whole (Beichman et al., 2007). While the euphoria associated with the prospect of an early launch of a *TPF* has faded with the realization of just how hard the search for life on other worlds is, the goal is achievable, if at greater cost and over longer times than we had originally hoped. Many different methods of looking for life, from transits to direct imaging at different wavelengths with coronagraphs, occulters, and interferometers, are being examined closely, and no limitations of physics or engineering have been found. Expensive? Yes. Impossible? No.

The path from Galileo's first telescope to *JWST*, some 400 years later, has been marked with many incremental steps, many of which have led to revolutionary insights about the cosmos. The path from *JWST* to the *TPF* telescopes of the future will be comparably exciting, but, we hope, somewhat quicker. By the time we celebrate the 500th anniversary of the telescope, we can confidently assert that we will have found many more planets, some habitable and perhaps, just perhaps, one with something we would all recognize as life itself. It is an exciting prospect.

ACKNOWLEDGMENTS

Some of the research described in this publication was carried out at the Jet Propulsion Laboratory (JPL), California Institute of Technology (Caltech), under a contract with the National Aeronautics and Space Administration (NASA).

REFERENCES

Baglin, A., Auvergne, M., Barge, P., et al. (The *CoRoT* Exoplanet Science Team) (2009). *CoRoT*: description of the mission and early results. In: F. Pont, D. Sasselov, and M. Holman (eds.), *Transiting Planets*. Proceedings of the International Astronomical Union Symposium No. 253, 71–81. Cambridge: Cambridge University Press.

Baraffe, I., Chabrier, G., Barman, T.S., et al. (2003). Evolutionary models for cool brown dwarfs and extrasolar giant planets: the case of HD 209458. *Astronomy and Astrophysics*, 402:701–12.

Bean, J.L., McArthur, B.E., Benedict, G.F., et al. (2007). The mass of the candidate exoplanet companion to HD 33636 from *Hubble Space Telescope* astrometry and high-precision radial velocities. *Astronomical Journal*, 134:749–58.

Beaulieu, M., Doyon, R., and Lafrenière, D. (2008). Performance results of the TFI coronagraphic occulting mask prototypes. In: J.M. Oschmann, Jr, M.W.M. de Graauw, and H.A. MacEwen (eds.), *Space Telescopes and Instrumentation 2008: Optical, Infrared, and Millimeter. Proceedings of the Society of Photo-Optical Instrumentation Engineers (SPIE)* Conference, Vol. 7010, 70103J–10.

Beichman, C.A. (2001). The search for young planetary systems. In: Ray Jayawardhana and Thomas Greene (eds.), *Young Stars Near Earth: Progress and Prospects*. Astronomical Society of the Pacific Conference Series Vol. 244, 376. San Francisco: Astronomical Society of the Pacific.

Beichman, C.A., Fridlund, M., Traub, W.A., et al. (2007). Comparative planetology and the search for life beyond the Solar System. In: B. Reipurth, D. Jewitt, and K. Keil (eds.), *Protostars and Planets V*, 915–28. Tucson: University of Arizona Press.

Beichman, C.A., Unwin, S.C., Shao, M., et al. (2008). Astrometric planet searches with SIM PlanetQuest. In: W.J. Jin, I. Platais, and M.A.C. Perryman (eds.), *A Giant Step: From Milli- to Micro-arcsecond Astrometry*. Proceedings of the IAU Symposium No. 248, 238–43.

Beichman, C. and the *ELEKTRA* Team (2011). The *ELEKTRA* Explorer Mission to Find Transiting Earth like planets. NASA Exoplanet Science Institute conference, *Exploring Strange New Worlds: From Giant Planets to Super Earths*, Flagstaff, AZ, May 1–6, 2011. Poster available at: http://nexsci.caltech.edu/conferences/Flagstaff/posters.shtml.

Beichman, C., Baratte, I., Crockett, C., et al. (2009). The *Formation and Evolution of Planetary Systems: The Search for and Characterization of Young Planets*. White Paper to Astro2010 Decadal Committee (8 pages) [arXiv:0902.2972]. Available at: http://mingus.as.arizona.edu/~bjw/astro2010/allpages_sort-rev.html.

Beichman, C.A., Krist, J., and Trauger, J.T. et al. (2010). Imaging young planets from ground and space. *Publications of the Astronomical Society of the Pacific*, 122:162–200.

Benedict, G.F., McArthur, B.E., Gatewood, G., et al. (2006). The extrasolar planet ε Eridani b: orbit and mass. *Astronomical Journal,* 132:2206–218.

Bennett, D.P., Anderson, J., and Gaudi, B.S. (2007). Characterization of gravitational microlensing planetary host stars, *Astrophysical Journal*, 660:781–90.

Black, D.C. and Stepinski, T.F. (2001). On the nature of extrasolar planetary candidates. *Bulletin of the American Astronomical Society*, 33:1198.

Borucki, W.J., Koch, D., Basri, G. et al. (2010). *Kepler* planet-detection mission: introduction and first results. *Science*, 327:977–80.

Borucki, W.J., Koch, D.G., Basri, G., et al. (2011). Characteristics of *Kepler* planetary candidates based on the first data set. *Astrophysical Journal*, 728:117 (20 pages).

Burrows, A., Budaj, J., and Hubeny, I. (2008). Theoretical spectra and light curves of close-in extrasolar giant planets and comparison with data. *Astrophysical Journal*, 678:1436–457.

Burrows, A., Sudarsky, D., and Lunine, J.I. (2003). Beyond the T dwarfs: theoretical spectra, colors, and detectability of the coolest brown dwarfs. *Astrophysical Journal*, 596:587–96.

Cash, W., Oakley, P., Turnbull, M., et al. (2008). The *New Worlds Observer*: scientific and technical advantages of external occulters. In: J.M. Oschmann, Jr, M.W.M. de Graauw, and H.A. MacEwen (eds.), *Space Telescopes and Instrumentation 2008: Optical, Infrared, and Millimeter, Proceedings of the Society of Photo-Optical Instrumentation Engineers (SPIE)* Conference, Vol. 7010, 70101Q–9.

Catanzarite, J., Law, N., and Shao, M. (2008). Astrometric detection of exo-Earths in the presence of stellar noise. In: M. Schöller, W.C. Danchi, and F. Delplancke (eds.), *Optical and Infrared Interferometry, Proceedings of the Society of Photo-Optical Instrumentation Engineers (SPIE)* Conference, Vol. 7013, 70132K–11.

Charbonneau, D. (2009). The rise of the Vulcans. In: F. Pont, D. Sasselov, and M. Holman (eds.), *Transiting Planets*. Proceedings of the International Astronomical Union Symposium No. 253, 1–8. Cambridge: Cambridge University Press.

Charbonneau, D., Brown, T.M., Latham, D., et al. (2000). Detection of planetary transits across a Sun-like star. *Astrophysical Journal*, 529:L45–8.

Charbonneau, D., Allen, L.E., Megeath, S.T., et al. (2005). Detection of thermal emission from an extrasolar planet. *Astrophysical Journal*, 626:523–29.

Chauvin, G., Lagrange, A.-M., Zuckerman, B., et al. (2005). A companion to AB Pic at the planet/brown dwarf boundary. *Astronomy and Astrophysics*, 438:L29–32.

Committee for a Decadal Survey of Astronomy and Astrophysics; National Research Council (2010). *New Worlds, New Horizons in Astronomy and Astrophysics*. Washington, DC: The National Academies Press (324 pages).

Cumming, A., Butler, R.P., Marcy, G.W., et al. (2007). The Keck planet search: detectability and the minimum mass and orbital period distribution of extrasolar planets. *Publications of the Astronomical Society of the Pacific*, 120:531–54.

Deming, D., Seager, S., Richardson, L.J., et al. (2005). Infrared radiation from an extrasolar planet. *Nature*, 434:740–43.

Deming, D., Seager, S., Winn, J., et al. (2009). Discovery and characterization of transiting super-Earths using an all-sky transit survey and follow-up by the James Webb Space Telescope. *Publications of the Astronomical Society of the Pacific*, 121:952–67.

Des Marais, D.J., Harwit, M.O., Jucks, K.W., et al. (2002). Remote sensing of planetary properties and biosignatures on extrasolar terrestrial planets. *Astrobiology*, 2:153–81. (Erratum 2003, Astrobiology, 3:216.)

Drake, F. and Sobel, D. (1992). *Is Anyone Out There? The Scientific Search for Extraterrestrial Intelligence*. New York: Delacorte Press.

Fischer, D. and Valenti, J. (2005). The planet-metallicity correlation. *Astrophysical Journal*, 622:1102–117.

Fortney, J.J., Lodders, K., Marley, M.S., et al. (2008). A unified theory for the atmospheres of the hot and very hot Jupiters: two classes of irradiated atmospheres. *Astrophysical Journal*, 678:1419–435.

Gaudi, B.S., Bennett, D.P., Udalski, A., et al. (2008). Discovery of a Jupiter/Saturn analog with gravitational microlensing. *Science*, 319:927–30.

Ge, J., Lee, B., Mahadevan, S. (2009). The SDSS-III multi-object APO radial-velocity exoplanet large-area survey (MARVELS) and its early results. American Astronomical Society, AAS Meeting 213, 336.02. *Bulletin of the American Astronomical Society*, 41:397.

Gould, A. and Loeb, A. (1992). Discovering planetary systems through gravitational microlenses. *Astrophysical Journal*, 396:110–14.

Guyon, O. (2007). A theoretical look at coronagraph design and performance for direct imaging of exoplanets. *Comptes Rendus Physique*, 8:323–32.

Henry, G.W., Marcy, G.W., Butler, R.P., et al. (1999). HD 209458. *International Astronomical Union Circular*, No. 7307.

Howard, A., Marcy, G., Fischer, D., et al. (2010). The Eta-Earth Survey for low-mass exoplanets. American Astronomical Society, AAS Meeting #215: 348.06. *Bulletin of the American Astronomical Society*, 42:530.

Ida, S. and Lin, D. (2005). Toward a deterministic model of planetary formation. III. Mass distribution of short-period planets around stars of various masses. *Astrophysical Journal*, 626:1045–60.

Kalas, P., Graham, J.R., Chiang, E., et al. (2008). Optical images of an exosolar planet 25 light-years from Earth. *Science*, 322:1345–348.

Kaltenegger, L. and Traub, W.A. (2009). Transits of Earth-like planets. *Astrophysical Journal*, 698:519–27.

Kasting, J. (2004). When methane made climate. *Scientific American*, 291:78–85.

Kasting, J.F. and Catling, D. (2004). Evolution of a habitable planet. *Annual Review of Astronomy and Astrophysics*, 41:429–63.

Kasting, J.F., Whitmire, D.P., and Reynolds, R.T. (1993). Habitable zones around main sequence stars. *Icarus*, 101:108–28.

Knutson, H.A., Charbonneau, D., Cowan, N.B., et al. (2009). Multiwavelength constraints on the day-night circulation patterns of HD 189733b. *Astrophysical Journal*, 690:822–36.

Krist, J.E., Beichman, C.A., Trauger, J.J., et al. (2007). Hunting planets and observing disks with the *JWST* NIRCam coronagraph. In: D.R. Coulter (ed.), *Techniques and Instrumentation for Detection of Exoplanets III, Proceedings of the Society of Photo-Optical Instrumentation Engineers (SPIE) Conference*, Vol. 6693, 66930H–12.

Lawson, P.R., Lay, O.P., Martin, S.R. et al. (2008). *Terrestrial Planet Finder Interferometer*: 2007–2008 progress and plans. In: M. Schöller, W.C. Danchi, and F. Delplancke (eds.), *Optical and Infrared Interferometry, Proceedings of the Society of Photo-Optical Instrumentation Engineers (SPIE) Conference*, Vol. 7013, 70132N–15.

Lawson, P.R., Traub, W.A., and Unwin, S.C. (eds.) (2009). *Exoplanet Community Report,* JPL Publication 09-3. Available at: http://ntrs.nasa.gov/archive/nasa/casi.ntrs.nasa.gov/20090012086_2009010616.pdf

Léger, A., Rouan, D., Schneider, J., et al. (2009). Transiting exoplanets from the *CoRoT* space mission VIII. *CoRoT*-7b: the first Super-Earth with measured radius, *Astronomy and Astrophysics*, 506:287–302.

Lovelock, J.E. (1965). A physical basis for life detection experiments. *Nature*, 207:568–70.

Macintosh, B.A., Graham, J.R., Palmer, D.W., et al. (2007). The Gemini Planet Imager. In: P. Kalas (ed.), *In the Spirit of Bernard Lyot: The Direct Detection of Planets and Circumstellar Disks in the 21st Century*. Proceedings of the conference *In the Spirit of Bernard Lyot: The Direct Detection of Planets and Circumstellar Disks in the 21st Century*. June 04–08, 2007. Berkeley: University of California.

Mao, S. and Paczynski, B. (1991). Gravitational microlensing by double stars and planetary systems. *Astrophysical Journal*, 374:L37–40.

Marley, M.S., Fortney, J.J., Hubickyj, O., et al. (2007). On the luminosity of young Jupiters. *Astrophysical Journal*, 655:541–49.

Marois, C., Macintosh, B., Barman, T., et al. (2008). Direct imaging of multiple planets orbiting the star HR 8799. *Science*, 322:1348–352.

Mayor, M. and Queloz, D. (1995). A Jupiter-mass companion to a solar-type star. *Nature*, 378:355–59.

McKee, C. and Taylor, J. (2001). *Astronomy and Astrophysics in the New Millennium*, 53. Washington, DC: National Academy Press.

Moutou, C., Mayor, M., Lo Curto, G., et al. (2009). The HARPS search for southern extra-solar planets. XV. Six long-period giant planets around BD -17 0063, HD 20868, HD 73267, HD 131664, HD 145377, and HD 153950. *Astronomy and Astrophysics*, 496:513–19.

Mumma, M.J., Villanueva, G.L., Novak, R.E., et al. (2009). Strong release of methane on Mars in northern summer 2003. *Science*, 323:1041–45,

Needham, J. and Wang, L. (1959). *Science and Civilization in China*, vol. 3: *Mathematics and the Sciences of the Heavens and the Earth*. Cambridge: Cambridge University Press.

Neuhäuser, R., Guenther, E.W., Wuchterl, G., et al. (2005). Evidence for a co-moving sub-stellar companion of GQ Lup. *Astronomy and Astrophysics*, 435:L13–16.

Oppenheimer, B. and Hinckley, S. (2009). High-contrast observations in optical and infrared astronomy. *Annual Review of Astronomy and Astrophysics*, 47:253–89.

Pasquini, L., Manescau, A., Avila, G., et al. (2009). ESPRESSO: A high resolution spectrograph for the combined coudé focus of the VLT. In *Science with the VLT in the ELT Era*, 395–99. Netherlands: Springer.

Peters, R.D., Lay, O.P., Hirai, A., et al. (2008). Progress in deep broadband interferometric nulling with the adaptive nuller. In: M. Schöller, W.C. Danchi, and F. Delplancke (eds.), *Optical and Infrared Interferometry, Proceedings of the Society of Photo-Optical Instrumentation Engineers (SPIE)* Conference, Vol. 7013, 70131V–8.

Pont, F., Sasselov, D., Holman, M. (eds.). (2009). *Transiting Planets*. Proceedings of the International Astronomical Union Symposium No. 253. Cambridge: Cambridge University Press.

Richardson, L.J., Harrington, J., Seager, S., et al. (2008). A *Spitzer* infrared radius for the transiting extrasolar planet HD 209458b. *Astrophysical Journal*, 649:1043–47.

Ricker, G.R., Latham, D.W., Vanderspek, R.K., et al. (2010). *Transiting Exoplanet Survey Satellite (TESS)*. American Astronomical Society, AAS Meeting #215: 450.06. *Bulletin of the American Astronomical Society*, 42:459.

Rouan, D., Boccaletti, A., Baudoz, P., et al. (2007). The coronagraphic mode of MIRI/*JWST*. In: P. Kalas (ed.), *In the Spirit of Bernard Lyot: The Direct Detection of Planets and Circumstellar Disks in the 21st Century*, Berkeley: University of California.

Saumon, D., Hubbard, W.B., Burrows, A., et al. (1996). A theory of extrasolar giant planets. *Astrophysical Journal*, 460:993–1018.

Seager, S. and Mallén-Ornelas, G. (2003). A unique solution of planet and star parameters from an extrasolar planet transit light curve. *Astrophysical Journal*, 585:1038–55.

Sivaramakrishnan, A., Tuthill, P., Ireland, M., et al. (2009). Non-redundant masking (NRM) on *JWST* FGS-TFI. *Proceedings of the Society of Photo-Optical Instrumentation Engineers (SPIE)* Conference, Vol. 7440, 74400Y.

Sozzetti, A., Casertano, S., Lattanzi, M.G., et al. (2008). Testing planet formation models with Gaia µas astrometry. In: W.J. Jin, I. Platais, and M.A.C. Perryman (eds.), *A Giant Step: From Milli- to Microarcsecond Astrometry*. Proceedings of the International Astronomical Union Symposium No. 248 256–9. Cambridge: Cambridge University Press.

Swain, M.R., Vasisht, G., Tinetti, G., et al. (2009). Molecular signatures in the near-infrared dayside spectrum of HD 189733b. *Astrophysical Journal*, 690:L114–17.

Tarter, J.C. (2007). The evolution of life in the universe: are we alone? *Highlights of Astronomy*, 14:14–29.

Tarter, J.C., Backus, P.R., Mancinelli, R.L., et al. (2007). A re-appraisal of the habitability of planets around M dwarf stars. *Astrobiology*, 7:30–65.

Traub, W.A., Ford, E., Laughlin, G., et al. (2009). Overview of the *SIM*-RV double-blind simulation to detect Earths in multi-planet systems. American Astronomical Society, AAS Meeting 213, 300.01. *Bulletin of the American Astronomical Society*, 41:267.

Trauger, J.T. and Traub, W.A. (2007). A laboratory demonstration of the capability to image an Earth-like extrasolar planet. *Nature*, 446:771–73.

Udry, S., Bonfils, X., Delfosse, X., et al. (2007). The HARPS search for southern extra-solar planets. XI. Super-Earths (5 and 8 M_{Earth}) in a 3-planet system. *Astronomy and Astrophysics*, 469:L43–7.

Unwin, S.C., Shao, M., Tanner, A.M., et al. (2008). Taking the measure of the universe: precision astrometry with *SIM* PlanetQuest. *Publications of the Astronomical Society of the Pacific*, 120:38–88.

Valencia, D., O'Connell, R.J., and Sasselov, D. (2007). Inevitability of plate tectonics on super-Earths. *Astrophysical Journal*, 670:L45–8.

Vanderbei, R.J., Cady, E., and Kasdin, N.J. (2008). Optimal occulter design for finding extrasolar planets. *Astrophysical Journal*, 665:794–98.

Wolszczan, A. and Frail, D. (1992). A planetary system around the millisecond pulsar PSR 1257 + 12. *Nature* 355:145–47.

Wright, J.T., Marcy, G.W., Fischer, D.A. et al. (2007). Four new exoplanets and hints of additional substellar companions to exoplanet host stars, *Astrophysical Journal*, 657:533–45.

20 Multiverse Cosmology*

Paul C. W. Davies

CONTENTS

INTRODUCTION

All cosmological models are constructed by augmenting the results of observations by a philosophical principle. Two examples from modern scientific cosmology are the principle of mediocrity and the so-called anthropic, or biophilic, principle. The principle of mediocrity, sometimes known as the Copernican principle, states that the portion of the Universe we observe is not special or privileged, but is representative of the whole. Ever since Copernicus demonstrated that the Earth does not lie at the center of the Universe, the principle of mediocrity has been the default assumption; indeed, it is normally referred to as simply "the cosmological principle." This principle underpins the standard Friedman–Robertson–Walker cosmological models.

In recent years, however, an increasing number of cosmologists have stressed the inherent limitations of the principle of mediocrity. Scientific observations necessarily involve observer selection effects, especially in astronomy. One unavoidable selection effect is that our location in the Universe must be consistent with the existence of observers. In the case of humans, at least, observers imply life. Clearly, we could not find ourselves observing the Universe from a location that did not permit life to emerge and evolve. This is the anthropic or—to use more accurate terminology—biophilic principle (Carter, 1974; Barrow and Tipler, 1986). Stated this way—that the Universe we observe

---

* Adapted from: Davies, P.C.W. (2004). Multiverse cosmological models. *Modern Physics Letters A*, 19(10), 727–43, ©World Scientific Publishing Company; reprinted with permission.

must be consistent with the existence of observers—the biophilic principle seems to be merely a tautology. However, it carries nontrivial meaning when we drop the tacit assumption that the Universe, and the laws of nature, *necessarily* assume the form that we observe. If the Universe and its laws could have been otherwise, then one explanation for why they are as they are might be that we, the observers, have selected them from a large ensemble.

This biophilic selection principle becomes more concrete when combined with the assumption that what we have hitherto regarded as absolute and universal laws of physics are, in fact, more like local bylaws (Rees, 2003a): they are valid in our particular cosmic patch, but they might be different in other regions of space and/or time. This general concept of "variable laws" has been given explicit expression through certain recent theories of cosmology and particle physics. To take a simple example, it is widely accepted that at least some of the parameters in the standard model of particle physics are not "God-given" fundamental constants of nature, but assume the values they do as a result of some form of symmetry-breaking mechanism. Their observed values may thus reflect the particular quantum state in our region of the Universe. If the Universe attained its present state by cooling from a super-hot initial phase, then these symmetries may have been broken differently in different cosmic regions. There is little observational evidence for a domain structure of the Universe within the scale of a Hubble volume, but on a much larger scale there could exist domains in which the coupling constants and particle masses in the standard model may be inconsistent with life. It would then be no surprise that we find ourselves located in a possibly atypical, life-encouraging domain as we could obviously not be located where life was impossible.

More generally, there may exist other spacetime regions, which we may informally call "other universes," that exhibit different physical laws and/or initial conditions. The ensemble of all such "universes" is often referred to as the Multiverse (Rees, 2003a).

VARIETIES OF MULTIVERSE

An early application of the biophilic principle was made by Boltzmann as a possible explanation for why the Universe is in a state far from thermodynamic equilibrium (Boltzmann, 1897; Albrecht, 2004). As there are vastly more states close to equilibrium than far from it, a randomly selected portion of the Universe inspected at a given time would be exceedingly unlikely to exhibit a significant departure from equilibrium conditions. (Boltzmann assumed a universe infinite in both space and time.) But, given that a significant departure from equilibrium conditions is an essential prerequisite for the existence of observers, our region of the Universe is *not* randomly selected; rather, it is selected by us. Boltzmann argued that statistical fluctuations will always create minor excursions from thermodynamic equilibrium and that major excursions, while exceedingly rare, are nevertheless possible in principle. In an infinite universe, there will always be astronomically large regions somewhere that, solely on the grounds of chance, exhibit sufficient departure from equilibrium to support the emergence of biological organisms. By hypothesis, ours is one such (exceedingly rare) region.

Boltzmann's original argument is unsatisfactory because the existence of human observers does not require a Hubble-sized nonequilibrium region of the Universe. Merely a Solar System region would suffice, and a fluctuation on this much smaller scale is overwhelmingly more probable than a cosmic-scale fluctuation. Today, however, Boltzmann's assumption of an infinitely old and uniform universe is discredited. Nevertheless, his basic reasoning may still be applied within the context of inflationary Big Bang cosmology, and the large fluctuation region objection possibility circumvented (Davies, 1983; Page, 1983; Albrecht, 2004).

Boltzmann's model universe provides an example of a restricted type of Multiverse. In this case, the laws of physics are uniform, but the thermodynamic circumstances are not because of random fluctuations. This ensures the existence of exceedingly rare atypical life-encouraging regions, which may then be selected by observers.

Although multiverse ideas have been discussed by philosophers and, to a lesser extent, by scientists for a long time, they have been propelled to prominence by two specific theoretical projects: string/M-theory and eternal inflation. String theory, and its development as M-theory, is an attempt to unify the forces and particles of physics at the Planck scale of energy, $(\hbar c^5/G)^{1/2}$. (Hereafter, unless stated to the contrary, we assume units $\hbar = c = 1$.) A seemingly inevitable feature of this class of theories is that there are no unique low-energy ("vacuum") sectors of the theory, but one estimate (Susskind, 2005) puts the number of distinct vacuum states at greater than 10^{500}. Each such sector would represent a possible world and possible low-energy physics. (The term "low-energy" is relative here; it means energies much less than the Planck energy, $\sim 10^{28}$ eV. That includes almost all of what is traditionally called high-energy physics.)

The problem arises because string theory is formulated most naturally in 10 or 11 spacetime dimensions, whereas the spacetime of our perceptions is 4-dimensional. The extra space dimensions are rendered unobservable by a process called "compactification": they are rolled up to a very small size. The situation may be compared with viewing a hosepipe. From a distance, it appears as a wiggly line, but on close inspection it is seen as a 2-dimensional tube, with one dimension rolled up to a small circumference. In the same way, what appears to be a point in 3-dimensional space may in fact be a circle going around a fourth space dimension. This basic notion may be extended to any number of extra dimensions, but then the process of compactification is no longer unique. In general, there are many ways of compactifying several extra dimensions. When additional degrees of freedom in string theory are taken into account, compactification may involve several hundred variables, all of which may vary from one region of the Universe to another. These variables serve to fix the low-energy physics by determining what sorts of particles exist, what their masses might be, and the nature and strengths of the forces that act between them. The theory also permits compactification to spaces with other than three dimensions. Thus, string theory predicts myriad possible low-energy worlds. Some might be quite like ours, but with slightly heavier electrons or a somewhat stronger weak force. Others might differ radically and possess, say, five large (i.e., uncompactified) space dimensions and two species of photons.

One can envisage an energy landscape in these several hundred variables (Susskind, 2005). Within this landscape there will be countless local minima, each corresponding to a possible quantum vacuum state and each a possible low-energy physical world. One of the parameters determined by each local minimum is the value of that minimum itself, which receives a well-known interpretation in terms of Einstein's cosmological constant, Λ. This parameter may be thought of as the energy density of the quantum vacuum.

Although string theory predicts a vast number of alternative low-energy physical worlds, the theory alone does not ensure that all such worlds are physically instantiated. The real existence of these other worlds, or "pocket universes," as Susskind has called them (Susskind, 2005), is rendered plausible when account is taken of inflationary Universe cosmology. According to this now-standard model, the Universe at or near its origin possessed a very large vacuum energy (or Λ). A Λ term in Einstein's gravitational field equations behaves like a repulsive force, causing the Universe to expand exponentially. This so-called inflationary episode may have lasted no longer than about 10^{-35}s in our pocket Universe before the enormous primordial vacuum energy decayed to the presently observed value, releasing the energy difference as heat. After this brief inflationary episode, the hot Big Bang model remains much as it was in the 1970s, involving the early synthesis of helium, eventual galaxy and star formation, etc.

In the fashionable variant known as eternal inflation, due to Vilenkin (1983) and Linde (1994), "our" Universe is just one particular vacuum bubble within a vast—probably infinite—assemblage of bubbles, or pocket universes. If one could take a "God's-eye view" of this Multiverse of universes, inflation would be continuing frenetically in the overall superstructure, driven by exceedingly large vacuum energies, while here and there "bubbles" of low- or, at least, lower-energy vacuums would nucleate quantum mechanically from the eternally inflating region and evolve into pocket universes. When eternal inflation is put together with the complex landscape of string theory, there is clearly a mechanism for generating universes with different local bylaws, i.e., different low-energy physics. Each bubble nucleation proceeding from a very large vacuum energy represents a symbolic "ball" rolling down the landscape from some dizzy height at random and ending up in one of the valleys, or vacuum states. So the ensemble of physical bylaws available from string theory becomes actualized as an ensemble of pocket universes, each with its own distinctive low-energy physics. The total number of such universes may be infinite, and the total variety of possible low-energy physics finite, but stupendously large.

According to Linde (1987), the size of a typical inflationary region is enormous, even by Hubble radius standards. The expansion rate is exponential with an e-folding time of $2\pi M_p^2/m^2$, where M_p is the Planck mass and m is the mass of the scalar inflaton field, the energy density of which drives the inflationary expansion. The resulting inflationary domains have a typical size $M_p^{-1}\exp(2\pi M_p^2/m^2)$. For a scalar field with GUT mass, $m \sim 10^{-5} M_p$, a typical inflation region has a staggering size M_p^{-1} $\exp(10^{11}) \sim 10^{10,000,000,000}$ cm, which should be compared with a Hubble radius of 10^{28} cm. Clearly, the observed Universe would almost certainly lie deep within such an inflation region, implying that the next pocket universe would be located exponentially far away and, therefore, will be decisively unobservable. In this model, the existence of other universes with differing physical laws has to be accepted on purely theoretical grounds.

The existence of a Multiverse does not rest on the validity of string theory or even inflationary cosmology as such. Rather, it is a generic property of any attempt to explain at least some low-energy physics as the product of particular quantum states combined with a model of the Universe originating in a Big Bang. Other multiverse theories exist in the literature. Perhaps the best known is the Everett interpretation of quantum mechanics (Everett, 1957). In its modern form, this so-called many-universes theory postulates that all branches of a wave function represent equally real universes existing in parallel (Deutsch, 1997). Thus, even quantum superpositions restricted to the atomic level are associated with ensembles of entire universes. Although the quantum Multiverse seems at first sight to be a completely different type of theory from the cosmological Multiverse, the two fuse when quantum mechanics is applied to cosmology. Thus, if one writes down a formal wave function for the Universe (Hartle and Hawking, 1983), then the different Everett branches of this cosmic wave function may be associated with different universes within an overall Multiverse. Tegmark has argued (Tegmark, 2004) that because theories such as eternal inflation stem from an application of quantum cosmology, Everett's interpretation of quantum mechanics does not actually increase the size or nature of the ensemble of universes that make up the already postulated Multiverse. The reason is the following: if the quantum fluctuations represented by the wave function of the Universe are ergodic, then the distribution of outcomes (i.e., distinct Everett branches within a given Hubble volume) is the *same* as a sample of different Hubble volumes described by a single branch of the wave function.

Another multiverse model has been discussed by Smolin (1997). He proposes that "baby universes" can sprout from existing ones via the mechanism of gravitational collapse. According to the classical picture, when a star implodes to form a black hole (BH), a spacetime singularity results in the interior of the hole. Smolin suggests that a quantum treatment would lead instead to the nucleation of a tiny new region of inflating space, connected to our space via a wormhole. Subsequent evaporation of the BH by the Hawking process severs the wormhole, thereby spatially disconnecting the baby universe from ours. Furthermore, following Wheeler (1980), Smolin (1997) proposes that the violence of gravitational collapse might "reprocess" the laws of physics randomly, producing small changes in values of parameters such as particle masses and coupling constants. Thus, the baby universe will inherit the physics of its parent, but with small random variations, similar to genetic drift in biological evolution. This process could continue *ad infinitum*, with baby universes going on to produce their own progeny. It would also imply that our Universe is the product of an earlier gravitational collapse episode in another universe. Those universes whose physical parameters favored BH production, for example, by encouraging the formation of large stars, would produce more progeny, implying that among all the universes, those with prolific BH production would present the largest volume of space.

This by no means exhausts the Multiverse possibilities; there are many logically and physically possible models in which an ensemble of universes can be described. Of recent interest are the brane theories, in which "our Universe" is regarded as a 3-dimensional sheet or brane embedded in a higher-dimensional space (Randall and Sundrum, 1999). Observers, along with most matter and radiation, are confined to a three-brane by a large potential gradient. In the ekpyrotic model of Steinhardt and Turok (2002), a brane collides with a confining 3-dimensional boundary to a 4-dimensional space to create what we interpret as the Big Bang. Conceptually, there is no impediment to foliating a 4-dimensional space with any number of branes, each of which (at least in the absence of collisions) constitutes a universe in its own right.

An extreme version of the Multiverse has been proposed by Tegmark (2003). Not content to imagine universes with all possible values of the fundamental "constants" of physics, Tegmark envisages universes with completely different laws of physics, including those describable by unconventional mathematics such as fractals. In fact, he suggests that *all* logically possible universes actually exist. Naturally, the vast majority of such universes would not support life and so go unobserved.

Tegmark justifies his extravagant hypothesis by appealing to the principle that the whole can be simpler than its parts. To take an elementary illustration, consider an infinite 1-dimensional array of equispaced points (a "crystal"). This structure, though infinite, is simply described by specifying the interval between adjacent points. Now extract from the array a random set of points. By definition, what remains is also a random set. We have replaced a simple set by two random subsets. But a random set requires a great deal of information to specify it, so it is more complex (in a manner that can be made mathematically precise using algorithmic information theory: Chaitin, 1987; Li and Vitanyi, 1997). Thus, the fusion of two complex sets can produce a simple set. In this manner, the set of all possible logically self-consistent universes might be regarded as simpler than most individual members of that set. In other words, Tegmark's extreme Multiverse might well be simpler than our single observed Universe, and, therefore, invoking Occam's razor, it should be favored as a description of reality.

It is clear that if physical reality is less than the set of all possible universes, then there has to be some rule or algorithm that separates the set of actually existing universes from the set of merely-possible, but in fact nonexisting universes. In the orthodox single-universe view of reality, there has always been the mystery of "why this Universe?" among all the apparently limitless possibilities. To use Hawking's evocative expression (Hawking, 1998, p. 174), what is it that "breathes fire" into one particular set of equations (i.e., the set describing *this* world) to single them out for the privilege of being attached to a really existing universe? This problem of what separates the actual from the merely possible persists in less extreme versions of the Multiverse, versions in which the members, even if infinite in number, do not exhaust the set of all possibilities. If there is a rule that divides the actual universe/s from the merely possible, one can ask, why that rule, rather than some other? Given the infinite number of possible rules, the application of one particular rule appears arbitrary and absurd. And one can always question where the rule comes from in the first place. But by embracing the Tegmark model of reality—all or nothing—these philosophical problems seem to be evaded. However, they are replaced by others (see "Arguments against the Multiverse Concept," below).

ANTHROPIC FINE-TUNING

The principal observational support for the multiverse hypothesis comes from a consideration of biology. As remarked in the Introduction, the Universe that we observe is bio-friendly, or we would not be observing it. This tautology develops some force when account is taken of the sensitivity of biology to the form of the laws of physics and the cosmological initial conditions—the so-called fine-tuning problem. It has been the subject of discussion for some decades that if the laws of physics differed, in some cases only slightly, from their observed form, then life as we know it, and possibly any form of life, would be impossible. Here are some well-known examples.

Carbon Production

An early example of anthropic fine-tuning was discussed by Hoyle et al. (1953). Life as we know it is based on carbon, which is synthesized in stellar cores via the triple-alpha reaction. This is a two-stage process that involves the formation of Be^8 from two alpha (He^4) particles followed by the capture of a third alpha particle via a resonant state. The energy of this resonant state happily coincides with the sum of the mass-energy of the three helium nuclei, ensuring an abundance of carbon production. This remarkable coincidence of energies arising from completely different branches of physics was in fact not known at the time Hoyle studied the nucleosynthesis problem; he deduced that such a felicitously placed resonance must exist, given that carbon-based observers exist. Subsequent experiments proved him right (Hoyle et al., 1953). The onward burning of carbon to oxygen is restrained by the absence of a similar resonant state at comparable energies. If the interplay of particle masses and the strong and electromagnetic forces were only slightly different, the position of this resonance would be displaced sufficiently for carbon production to decline dramatically. If the strong force were substantially weaker, or the electromagnetic force substantially stronger, the stability of carbon nuclei would be threatened.

A further fortuitous aspect to carbon production involves the weak nuclear force (Carr and Rees, 1979). Carbon is disseminated through the interstellar medium in part from supernova explosions, which are triggered by a pulse of neutrinos released from collapsing stellar cores of massive stars. The neutrinos couple to the surrounding stellar material via the weak nuclear force. If the weak force were weaker, this mechanism would be ineffective; if it were stronger, the neutrinos would be trapped in the dense imploding cores. In either case, the vital carbon would remain confined to the stars and not be available for life processes on planets. (The latter argument is weakened to some extent by the existence of other mechanisms for dispersing carbon, such as stellar winds.)

Hydrogen and Water

If the strong nuclear force were only about 4% stronger, the di-proton would be bound, but unstable against decay to deuterium via the weak force. This would have profound implications for primordial nucleosynthesis and the chemical makeup of the Universe. As pointed out by Dyson (1971), the primordial soup of protons would rapidly transform into deuterium, which would then synthesize 100% helium, leaving no hydrogen in the Universe. Without hydrogen, there would be no water, an essential ingredient for life. There would also be no stable hydrogen-burning stars such as the Sun to sustain a biosphere.

Space Dimensionality

It has been known for a long time that the stability of planetary orbits depends crucially on space having three dimensions (Whitrow, 1955). In a 4-dimensional space, for example, gravitation would obey an inverse-cube law, and planets would rapidly spiral into the Sun. It is hard to see how life as we know it could exist in more than three space dimensions. Life in two space dimensions is not obviously impossible, but problems with wave propagation might compromise information processing, precluding any advanced form of life (i.e., observers).

Strength of Gravity

Gravitation is famously much weaker than the other forces of nature:

$$\frac{e^2}{Gm_p^2} \approx 10^{40},$$

where e is the charge on the proton and m_p the proton mass. The strength of gravity sets the time and distance scale of the Universe: the Universe is big because gravity is so weak. But a big universe implies an old universe, and an old universe (billions of years) is a necessary prerequisite for the emergence of complex organisms. If gravitation were 100 times stronger, the Universe would collapse before observers had time to evolve.

PRIMORDIAL DENSITY PERTURBATIONS

The emergence of large-scale structure in the Universe depends on density perturbations in the early Universe to act as seeds of galaxies. Observations by the *Cosmic Background Explorer* (*COBE*) and the *Wilkinson Microwave Anisotropy Probe* (*WMAP*) show that the density contrast at the time of matter-radiation decoupling (380,000 years after the Big Bang) was about one part in 10^5. If the perturbations were significantly less than this, galaxies may never have formed, and the production of stars and planets would be unlikely. Conversely, stronger density perturbations would lead to the formation of supermassive BHs rather than galaxies. The observed density perturbations appear to be optimal as far as the eventual emergence of life is concerned (Rees, 2003a).

THE COSMOLOGICAL CONSTANT

A major unsolved problem of fundamental physics is the value of the cosmological constant Λ (i.e., the energy of the quantum vacuum). There is no satisfactory physical theory that explains why this parameter should be nonzero, yet so much smaller than the "natural" Planck value $M_p^4 : \Lambda_{obs} \sim 10^{-123} M_p^4$.

The possibility of an anthropic explanation for this staggering mismatch between theory and observation is now two decades old (Davies and Unwin, 1981; Linde, 1984; Weinberg, 1987), but has received more attention (Weinberg, 2000) since the observational confirmation that $\Lambda \neq 0$. The basic idea is that Λ is treated as a random variable that may change from one region of space to another. In the eternal inflationary model, it might, for example, take on different random values in each inflation region (i.e., each pocket universe). The range of values consistent with the emergence of biological organisms is fairly constrained. If Λ were an order or magnitude larger, the formation of galaxies would be seriously inhibited. There is no impediment to $|\Lambda|$ being smaller, but Λ should not be much larger than its observed numerical value and also negative, or it would bring about the collapse of the Universe (a "big crunch") before life and observers had time to emerge (Linde, 1984; Weinberg, 1987).

ARGUMENTS AGAINST THE MULTIVERSE CONCEPT

A variety of arguments have been deployed against both the multiverse concept and anthropic reasoning in general. These arguments are both physical and philosophical.

IT IS NOT SCIENCE

It is sometimes objected that because our observations are limited to a single universe (e.g., a Hubble volume), then the existence of "other universes" cannot be observed, and so their existence cannot be considered a proper scientific hypothesis. Even taking into account the fact that future observers will see a larger particle horizon, and so have access to a bigger volume of space, most regions of the Multiverse (at least in the eternal inflation model) can never be observed, even in principle. While this may indeed preclude direct confirmation of the multiverse hypothesis, it does not rule out the possibility that it may be tested indirectly. Almost all scientists and philosophers accept the general principle that the prediction of unobservable entities is an acceptable scientific hypothesis if those entities stem from a theory that has other testable consequences. At this stage, string theory does not have any clear-cut experimental predictions, but one may imagine that a future elaboration of the theory would produce testable consequences; similarly, future elaboration of the theory may have consequences for other multiverse models, such as brane theories or the production of baby universes. These theories are not idle speculations, but emerge from carefully considered theoretical models with some empirical justification.

A test of the multiverse hypothesis may be attained by combining it with biophilic selection (Rees, 2003a; Tegmark, 2003). This leads to statistical predictions about the observed values of physical parameters. If we inhabit a typical biophilic region of the Multiverse, we would expect the observed values of any biologically relevant adjustable parameters to assume typical values. In other words, if we consider a vast parameter space of possible universes, there will be one or more biophilic patches, or subsets, of the space, and a typical biophilic universe would not lie close to the center of such a patch. Thus, consider Boltzmann's "multiverse" hypothesis, where the parameter was entropy. Fluctuations in entropy are exponentially suppressed with departure from the maximum value, so that, if the biophilic explanation were correct, we would not expect to inhabit a region of the Multiverse in which the entropy was much less than the minimum necessary for the existence of observers. As discussed in "Varieties of Multiverse" above, the fact that we inhabit at least a Hubble volume of low entropy must be counted as strong evidence against Boltzmann's hypothesis.

Now suppose we apply the same reasoning to the value of the dark energy, Λ. In the absence of a good physical theory, it is not possible to know the probability distribution of values of Λ; but for the purpose of illustration, suppose it is uniform in the range $[-\Lambda_{max}, +\Lambda_{max}]$, where Λ_{max} is some maximum permitted value (e.g., the Planck value). Now consider a randomly selected observer. By hypothesis, the observer would have to inhabit a universe in which Λ lies within the biophilic range, say $[-\Lambda_b, +\Lambda_b]$, where $|\Lambda_b| \ll |\Lambda_{max}|$. If the observed value of the cosmological constant Λ_{obs} were determined to be $\Lambda_{obs} \ll \Lambda_b$, this would count as evidence against an anthropic explanation because there would exist many more habitable universes in which $\Lambda_{obs} \sim \Lambda_b$, and one could defend the multiverse hypothesis only on the unjustified assumption that humans occupied an atypical habitable universe. Calculations suggest that $\Lambda_{obs} \approx 0.1\Lambda_b$, which is consistent with a random selection from a uniform probability distribution (Rees, 2003a). If, contrary to observation, Λ were indistinguishable from 0, it would be reasonable to seek some deep principle of physics that fixes its value to be precisely 0.

In Smolin's theory (1997), there is a specific prediction that universes that maximize BH production dominate the total available spatial volume. Because star production (leading to BHs) is also a good criterion for carbon production and the establishment of habitable zones in planetary systems, these universes are also the ones likely to be inhabited. So randomly selected observers would be expected to find themselves in universes in which BH production is maximized. This is testable by determining whether or not star production depends sensitively on the observed values of physical parameters such as particle masses and force strengths. If, say, a small change either way in the proton mass were to markedly reduce star production, or the collapse of stars into BHs, it would provide support for Smolin's theory. Conversely, if it could be demonstrated that certain changes in the physical parameters might actually increase the rate of BH formation, it would falsify Smolin's theory.

IT IS BAD SCIENCE

Even if it is conceded that the Multiverse theory is testable in principle, one might still object to the theory on professional grounds. Some physicists have argued that the job of the scientist is to provide fundamental explanations for observed phenomena, without making reference to observers. Resorting to anthropic explanations may serve to undermine the search for a satisfactory physical theory, constituting a "lazy way out" of the need to account for features such as the apparent fine-tuning of parameters in relation to the existence of life. Thus, biophilic arguments have been criticized by some in very strong terms. Whatever one's predilection for anthropic reasoning, its supporters at least concede it should be an explanation of last resort (Weinberg, 1992). Set against this is the claim by some theorists (for example, Susskind, 2005) that *some* form of Multiverse is unavoidable, given our current knowledge of physics, and that observer selection effects are inevitable and must be taken into account in most sciences.

THERE IS ONLY ONE POSSIBLE UNIVERSE

It is occasionally argued that the observed Universe is the unique possible Universe, so that talk of "other" universes is *ipso facto* meaningless. Einstein raised this possibility when he said, in his typical poetic manner, that what really interested him was whether "God had any choice in the creation of the world" (Holton, 1998, p. 91). To express this sentiment more neutrally, Einstein was asking whether the Universe could have been otherwise (or nonexistent altogether). The hope is sometimes expressed that once a fully unified theory of physics is achieved, it will turn out to have a unique "solution" corresponding to the observed Universe. It is too soon to say whether string/M-theory will eventually yield a unique description (so far, the evidence is to the contrary), but the hypothesis of a unique reality would in any case seem to be easily dispatched. The job of the theoretical physicist is to construct mathematically consistent models of reality in the form of simplified, impoverished descriptions of the real world. For example, the so-called Thirring model (Thirring, 1958) describes a 2 spacetime-dimensional world inhabited by self-interacting fermions. It is studied because it offers an exactly soluble model in quantum field theory. Nobody suggests the Thirring model is a description of the real world, but it is clearly a *possible* world. So unless some criterion can be found to eliminate all the simplified models of physics, including such familiar constructs as Newtonian mechanics, there would seem to be a strong *prima facie* case that the Universe could indeed have been otherwise—that "God did have a choice."

MEASURES OF FINE-TUNING ARE MEANINGLESS

Intuitively, we may feel that some physical parameters are remarkably fine-tuned for life, but can this feeling ever be made mathematically precise? The fact that a variation in the strength of the strong nuclear force by only a few percent may disrupt the biological prospects for the Universe appears to offer a surprisingly narrow window of biophilic values; but what determines the measure on the space of parameters? If the strength of the nuclear force could in principle vary over an infinite range, then any finite window, however large, would be infinitesimally improbable if a uniform probability distribution is adopted. Even the simple expedient of switching from a uniform to a logarithmic distribution can have a dramatic change on the degree of improbability of the observed values, and hence the fineness of the fine-tuning. There will always be an element of judgment involved in assessing the significance of, or a degree of surprise that attaches to, any given example.

Many key parameters of physics do not seem to be very strongly constrained by biology. Take the much-cited example of carbon abundance. The existence of carbon as a long-lived element depends on the ratio of electromagnetic to strong nuclear forces, which determines the stability of the nucleus. But nuclei much heavier than carbon are stable, so the life-giving element lies comfortably within the stability range. The electromagnetic force could be substantially stronger without threatening the stability of carbon. Of course, if it were stronger, then the specific nuclear resonance responsible for abundant carbon would be inoperable; but it is not clear how serious this would be. Life could arise, albeit more sparsely, in a universe where carbon was merely a trace element, or abundant carbon could occur because of different nuclear resonances. Of course, if it could be shown that other, heavier elements are essential for life, this objection would disappear. (The prediction that much heavier elements are essential for life could be an interesting prediction of the Multiverse theory.)

These considerations of how to quantify the fine-tuning are worse to the point of intractability when it comes to assigning statistical weights to alternative laws, or alternative mathematical structures as proposed by Tegmark (2003).

MULTIVERSES MERELY SHIFT THE PROBLEM UP ONE LEVEL

Multiverse proponents are often vague about how the parameter values are selected across the defined ensemble. If there is a "law of laws," or meta-law, describing how parameter values are assigned from one universe to the next, then we have only shifted the problem of cosmic biophilicity up one level because we need to explain where the meta-law comes from. Moreover, the set of such meta-laws is infinite, so we have merely replaced the problem of "why this Universe?" with that of "why this meta-law?" This point was already made at the end of "Varieties of Multiverse" above. But now we encounter a further problem. Each meta-law specifies a different Multiverse, and not all multiverses are bound to contain at least one biophilic universe. In fact, on the face of it, most multiverses would not contain even a single component universe in which all the parameter values were suitable for life. To see this, note that each parameter will have a small range of values—envisage it as a highlighted segment on a line in a multidimensional parameter space—consistent with biology. Only in universes where all the relevant highlighted segments intersect in a single patch (i.e., all biophilic values are instantiated simultaneously) will biology be possible. If the several parameters vary independently between universes, each according to some rule, then for most sets of rules the highlighted segments will not concur. So, we must not only explain why there is any meta-law; we must also explain why the actual meta-law (i.e., the actual Multiverse) happens to be one that intersects the requisite patch of parameter space that permits life. And if the parameters do not vary independently, but are linked by an underlying unified physical theory, then each underlying theory will represent a different track in parameter space. Only in some unification theories would this track intersect the biophilic region. So one is now confronted with explaining why this particular underlying unified theory, with its felicitous biophilic confluence of parameter values, is the one that has "fire breathed into it," to paraphrase Hawking. In Tegmark's extreme Multiverse theory this problem is circumvented because in that case all possible meta-laws (or all possible unified theories) have "fire breathed into them" and describe really existing multiverses.

Sometimes it is claimed that there is no meta-law, only randomness. Wheeler, for example, has asserted (Wheeler, 1985, 1986) that "there is no law except the law that there is no law." In Smolin's version of the Multiverse (1997), gravitational collapse events "reprocess" the existing laws with small random variations. In this case, given a Multiverse with an infinity of component universes, randomness would ensure that at least one biophilic universe exists. (That is, there will always be a patch of parameter space somewhere with all highlighted segments intersecting.) However, the assumption of randomness is not without its own problems. Once again, without a measure over the parameter space, probabilities cannot be properly defined. There is also a danger in some multiverse models that the biophilic target universes may form only a set of measure zero in the parameter space and, thus, be only infinitesimally probable (Holder, 2004). Furthermore, in some models, various randomness measures may be inconsistent with the underlying physics. For example, in the model of a single spatially infinite universe in which different supra-Hubble regions possess different total matter densities, it is inconsistent to apply the rule that any value of the density may be chosen randomly in the interval $[0, p]$, where p is some arbitrarily large density (e.g., the Planck density). The reason is that for all densities above a critical value that is very low compared with the Planck density, the Universe is spatially finite, and so inconsistent with the assumption of an infinite number of finite spatial regions (Holder, 2004).

The need to rule out these "no-go" zones of the parameter space imposes restrictions on the properties of the Multiverse that are tantamount to the application of an additional overarching biophilic principle. There would seem to be little point in invoking an infinity of universes only then to impose biophilic restrictions at the multiverse level. It would be simpler to postulate a single Universe with a biophilic principle.

THE FAKE UNIVERSE PROBLEM

The Multiverse theory forces us to confront head-on the contentious issue of what is meant by physical reality. Is it meaningful to assign equal ontological status to our own, observed Universe and to universes that are *never* observed by any sentient being? This old philosophical conundrum is exacerbated when account is taken of the nature of observation. In most discussions of Multiverse theory, an observer is simply taken to mean a complex biological organism. But this is too restricted. Most scientists are prepared to entertain the possibility of conscious machines, and some artificial intelligence (AI) advocates even claim that we are not far from producing conscious computers. In most multiverse theories, although habitable universes may form only a sparse subset, there is still a stupendous number of them and, in many cases, an infinite number. (That is the case with Boltzmann's original model and eternal inflation, for example.) It is therefore all but inevitable that some finite fraction of habitable universes in this vast—possibly infinite—set will contain communities of organisms that evolve to the point of creating artificial intelligence or simulated consciousness. It is then but a small step to the point where the engineered conscious beings inhabit a simulated world. For such beings, their "fake" universe will appear indistinguishable from reality. So should we include these simulated universes in the ensemble that constitutes the Multiverse? At least two multiverse proponents have suggested that we might (Barrow, 2003; Rees, 2003b).

The problem that now arises is that any given "real" universe with world-simulating technology could simulate a limitless number of "fake" universes, so within the extended multiverse hypothesis, fake universes greatly outnumber real ones. (For strong AI proponents, who assert that consciousness may be simulated by universal discrete-state machines, this conclusion is reinforced by the Turing thesis, which implies that the simulations may themselves generate simulations, and so on.) This means that a randomly selected observer is overwhelmingly likely to inhabit a fake, rather than a real, universe. By implication, "our" Universe is very probably a simulation (Brooks, 2002; Bostrom, 2003). But if it is a simulation, then the application of physical theory to unobserved regions/universes is invalid because there is no reason to suppose that the simulating system will consistently apply the observed physics of our simulation to other, unobserved simulations. Thus, the multiverse hypothesis would seem to contain, Gödel-like, the elements of its own invalidity.

An additional philosophical problem that afflicts most multiverse models (e.g., Boltzmann's, Linde's) is the familiar one that, in an infinite universe, anything that can happen will happen and happen infinitely often, purely by chance. This is also discussed as the problem of duplicate beings (Ellis and Brundrit, 1979; Davies, 1995). Thus, eternal inflation predicts that 10 to the power 10^{29} cm away there will exist a planet indistinguishable from Earth with beings indistinguishable from us (Garriga and Vilenkin, 2001). By the same reasoning, there will be an identical Hubble volume to ours about 10 to the power 10^{115} cm away. Furthermore, there will be *infinitely many* such identical persons, identical Hubble volumes, or identical super-Hubble volumes in the Multiverse. Although there is no logical impediment to physical reality's being infinitely replicated in either space or time, any physical theory that predicts such a situation invites especially skeptical scrutiny.

WHY STOP THERE?

A final objection to the existing multiverse theories is a challenge to the criteria for defining universes. In most multiverse theories, universes are labeled by laws of physics and initial conditions. Even in Tegmark's extreme multiverse scheme, his chosen criterion is mathematical consistency. It might be objected that these terms are narrow and chauvinistic—indeed, just the sort of criteria to be expected from mathematical physicists. Other ways of categorizing universes are conceivable and could lead to even larger concepts of the Multiverse than Tegmark's. Examples might be the set of all possible artistic structures, morally good systems, or mental states. There may be criteria for categorization that lie completely beyond the scope of human comprehension. To suppose that the ultimate nature of reality is founded in 21st-century human physics seems remarkably hubristic.

CONCLUSION

Recent developments in particle physics, quantum mechanics, and cosmology lead naturally to the postulate of an ensemble of universes, or a Multiverse. Some extension to the restricted view that "what you see is what you get" would surely seem both inevitable and reasonable to all but the most out-and-out logical positivist because the limits imposed by the cosmological particle horizon are merely relative to our specific cosmic location. Although direct confirmation of other universes or regions of our Universe may be infeasible or even impossible in principle, nevertheless the Multiverse theory does make some observable predictions and can be tested.

For most people, somewhere on the slippery slope between being asked to accept the existence of regions of space that lie beyond our present particle horizon and Tegmark's "anything goes" Multiverse, credulity will dwindle. Some version of a Multiverse is reasonable given the current worldview of physics, but most physicists would stop well before Tegmark's Multiverse. They would also regard the prediction of a proliferation of artificially simulated universes ("fake" universes) as a *reductio ad absurdum* of the multiverse hypothesis.

Invoking the Multiverse together with the anthropic or biophilic principle in an attempt to explain fine-tuning is still regarded with great suspicion or even hostility among physicists, although it does have some notable apologists. There is consensus that such explanations should not impede searches for more satisfying explanations of the nature of the observed physical laws and parameters.

Multiverse theories raise serious philosophical problems about the nature of reality and the nature of consciousness and observation. Attempts to sharpen the discussion and provide a more rigorous treatment of concepts such as the number of universes, the probability measures in parameter space, and objective definitions of infinite sets of universes have not progressed far. Nevertheless, the multiverse idea has probably earned a permanent place in physical science, and as new physical theories are considered in the future, it is likely that their consequences for biophilicity and multiple cosmic regions will be eagerly assessed.

REFERENCES

Albrecht, A. (2004). Cosmic inflation and the arrow of time. In: J.D. Barrow, P.C.W. Davies, and C.L. Harper, Jr. (eds.), *Science and Ultimate Reality: Quantum Theory, Cosmology and Complexity*, 363–401. Cambridge: Cambridge University Press.
Barrow, J.D. (2003). Glitch! *New Scientist*, 7 June 2003: 44–5.
Barrow, J.D. and Tipler, F.J. (1986). *The Anthropic Cosmological Principle*. Oxford: Oxford University Press.
Boltzmann, L. (1897). Zu Hrn. Zermelos Abhandlung über die mechanische Erklärung irreversiler Vorgänge. *Annals of Physics*, 60: 392–98.
Bostrom, N. (2003). Are you living in a computer simulation? *Philosophical Quarterly*, 53: 243–55.
Brooks, M. (2002). Life's a sim and then you're deleted. *New Scientist*, 27 July 2002: 48–9.
Carr, B.W. and Rees, M.J. (1979). The anthropic principle and the structure of the physical world. *Nature*, 278: 605–12.
Carter, B. (1974). Large number coincidences and the anthropic principle in cosmology. In M.S. Longair (ed.), *Confrontation of Cosmological Theories with Observational Data*. Proceedings of the International Astronomical Union Symposium No. 63. Dordrecht: Reidel.
Chaitin, G.J. (1987). *Algorithmic Information Theory*. Cambridge: Cambridge University Press.
Davies, P.C.W. (1983). Inflation and time asymmetry in the Universe. *Nature*, 301: 398–400.
Davies, P.C.W. (1995). Appendix 2. *Are We Alone?* London: Penguin.
Davies, P.C.W. and Unwin, S.D. (1981). Why is the cosmological constant so small? *Proceedings of the Royal Society*, A377: 147–49.

Deutsch, D. (1997). *The Fabric of Reality*. New York: Penguin Group.

Dyson, F. (1971). Energy in the universe. *Scientific American,* 225: 25–32.

Ellis, G.F.R. and Brundrit, G.B. (1979). Life in the infinite universe. *Quarterly Journal of the Royal Astronomical Society,* 20: 37–41.

Everett III, H. (1957). "Relative state" formulation of quantum mechanics. *Review of Modern Physics,* 29: 454–62.

Garriga, J. and Vilenkin, A. (2001). Prescription for probabilities in eternal inflation. *Physics Review,* D64:043511 [DOI:10.1103/PhysRevD.64.023507].

Hartle, J.B. and Hawking, S.W. (1983). Wave function of the Universe. *Physics Review,* D28:2960 [DOI:10.1103/PhysRevD.28.2960].

Hawking, S.W. (1998). *A Brief History of Time*. New York: Bantam.

Holder, R. (2004). *God, The Multiverse and Everything*. London: Ashgate Publishing Ltd.

Holton, G. (1998). *The Advancement of Science, and Its Burdens*. Cambridge: Harvard University Press.

Hoyle, D.N.F., Dunbar, W.A., Wensel, et al. (1953). The 7.68-Mev State in C12 *Physics Review,* 92: 649 [DOI:10.1103/PhysRev.92.649].

Li, M. and Vitanyi, P. (1997). *An Introduction to Kolmogorov Complexity and Its Applications*. Berlin: Springer-Verlag.

Linde, A. (1983). Eternal chaotic inflation. *Modern Physics Letters,* A1: 81–5.

Linde, A. (1984). The inflationary Universe. *Reports on Progress in Physics,* 47:925 [DOI:10.1088/0034-4885/47/8/002].

Linde, A.D. (1987). Inflation and quantum cosmology. In: S.W. Hawking and W. Israel (eds.), *Three Hundred Years of Gravitation,* 604–30. Cambridge: Cambridge University Press.

Linde, A. (1994). The self-reproducing inflationary universe. *Scientific American,* 271: 48–55.

Page, D.N. (1983). Inflation does not explain time asymmetry. *Nature,* 304: 39–41.

Randall, L. and Sundrum, R. (1999). An alternative to compactification. *Physics Review Letters,* 83: 4690 [DOI:10.1103/PhysRevLett.83.4690].

Rees, M.J. (2003a). Numerical coincidences and "tuning" in cosmology. *Astrophysics and Space Science,* 285: 375–88.

Rees, M.J. (2003b). In the matrix. *Edge,* 116 [http://www.edge.org/documents/archive/edge116.html].

Smolin, L. (1997). *The Life of the Cosmos*. Oxford: Oxford University Press.

Steinhardt, P.J. and Turok, N. (2002). A cyclic model of the universe. *Science,* 296: 1436–1439.

Susskind, L. (2005). *The Cosmic Landscape: String Theory and the Illusion of Intelligent Design*. New York: Little, Brown.

Tegmark, M. (2003). Parallel universes. *Scientific American,* 288: 40–51.

Tegmark, M. (2004). Parallel universes. In: J.D. Barrow, P.C.W. Davies, and C.L. Harper, Jr. (eds.), *Science and Ultimate Reality: Quantum Theory, Cosmology and Complexity,* 459–91. Cambridge: Cambridge University Press.

Thirring, W. (1958). A soluble relativistic field theory? *Annals of Physics,* 3: 91–112.

Vilenkin, A. (1983). Birth of inflationary universes. *Physics Review,* D27:2848 [DOI:10.1103/PhysRevD.27.2848].

Weinberg, S. (1987). Anthropic bound on the cosmological constant. *Physics Review Letters,* 59: 2607 [DOI:10.1103/PhysRevLett.59.2607].

Weinberg, S. (1992). *Dreams of a Final Theory*. New York: Pantheon.

Weinberg, S. (2000). A priori probability distribution of the cosmological constant. *Physics Review,* D61: 103505 [DOI:10.1103/PhysRevD.61.103505].

Wheeler, J.A. (1980). From the black hole. In: H. Woolf (ed.), *Some Strangeness in the Proportion*. Reading, MA: Addison-Wesley.

Wheeler, J.A. (1985). Bohr's "Phenomenon" and "Law Without Law." In: G. Casati (ed.), *Chaotic Behavior in Quantum Systems: Theory and Applications*. Dordrecht: Reidel.

Wheeler, J.A. (1986). Quoted in: J.D. Barrow and F.J. Tipler (eds.), *The Anthropic Cosmological Principle,* 255. Oxford: Oxford University Press.

Whitrow, G. (1955). Why physical space has three dimensions. *British Journal of Philosophical Science,* 6: 13–31.

21 Universe or Multiverse?

Renata Kallosh and Andrei Linde

CONTENTS

INTRODUCTION

For most of the 20th century, scientific thought was dominated by the ideas of the uniformity of the Universe and the uniqueness of the laws of physics. Indeed, cosmological observations indicated that the Universe (on the largest possible scales) is approximately homogeneous and isotropic and that the same laws of physics operate everywhere. Uniformity of the Universe was somewhat of a mystery; but, instead of explaining it, scientists invoked the "cosmological principle," which said, ironically, that the Universe simply must be uniform because of the cosmological principle. Of course, we know that the Universe is not quite uniform; for example, it contains many galaxies that are crucially important for the existence of life. Thus, the cosmological principle could not be entirely correct; and if it is not entirely correct, it cannot be a fundamental principle of science.

A similar situation occurred with respect to the uniqueness of the laws of physics, although in a much more definite way. We knew, for example, that the electron mass is the same everywhere in the observable part of the Universe, so the obvious assumption was that it must take the same value everywhere and that it is a physical constant. Unlike the case of the uniformity of the Universe, no apparent violation of this law is known. Therefore, for a long time, one of the great goals of physics was to find a single fundamental theory that would unify all fundamental interactions and provide an unambiguous explanation for all known parameters of particle physics.

Physicists were looking for a fundamental explanation for why the electron mass could not be any different from its measured value. The strongest proponent of this idea was Albert Einstein, who said:

> I would like to state a theorem that at present cannot be based upon anything more than a faith in the simplicity, i.e. intelligibility, of nature: There are no arbitrary constants—that is to say, nature is so constituted that it is possible logically to lay down such strongly determined laws that within these laws only rationally completely determined constants occur (not constants, therefore, whose numerical value could be changed without destroying the theory). (Einstein, 1959, p. 63)

One of the best supports of Einstein's vision of a final theory is the computation of the anomalous magnetic moment of the electron and muon in quantum field theory. This is viewed as a triumph of theoretical physics and is considered the single best agreement between theory and experiment.

For example, the experimental value of the anomalous magnetic moment of the electron is $\alpha_e^{\text{exp}} = 0.0011596521884 \pm 0.43 \times 10^{-11}$, which is in excellent agreement with the theoretical prediction of QED: $\alpha_e^{\text{th}} = 0.0011596521535 \pm 2.40 \times 10^{-11}$. The theoretical value was obtained by computing a set of Feynman diagrams up to the fourth order in the fine structure constant α—that is to say, the diagrams up to the fourth loop order.

The agreement between this theory and experiment is at the fantastic level of 10^{-11}. The theoretical error is dominated by the error in the value of the fine structure α, taken from quantum Hall effect experiments. One could interpret this success as a suggestion that at least some of the constants of nature, such as the anomalous magnetic moment of the electron, can be computed exactly and do not require any additional assumptions.

However, the computations in the framework of quantum field theory require an experimental input for the fine structure constant α, as well as other coupling constants of the standard model. For example, the value of the fine structure constant $\alpha = 0.07297352570$, which is a part of the calculations, is not explained by the theory; it is taken from other experiments. In this sense, the success of quantum field theory is incomplete; its predictions are based partially on the values of experimentally measured parameters that the theory, by itself, cannot explain.

Values of some of these parameters are related to one another by certain symmetries, but most of these parameters look more like a collection of random numbers than a unique manifestation of a hidden harmony in nature. Meanwhile, it was pointed out long ago that a minor change (by a factor of 2 or 3) in the mass of the electron, the fine-structure constant α, the strong-interaction constant, or the gravitational constant G would lead to a universe in which life as we know it could never have existed. Adding or subtracting even a single spatial dimension of the same type as the usual three dimensions would make planetary systems impossible. Indeed, in spacetime with dimensionality $d > 4$, gravitational forces between distant bodies fall off faster than r^{-2}, which makes the planetary orbits unstable. Meanwhile, in spacetime with $d < 4$, the general theory of relativity tells us that such forces are absent altogether. This rules out the existence of stable planetary systems for $d \neq 4$. These facts, as well as a number of other observations, lie at the foundation of the so-called anthropic principle (see, e.g., Barrow and Tipler, 1986; Rees, 2000; Rozental, 1988). According to this principle, we observe the Universe to be as it is because only in such a universe could observers such as ourselves exist.

Many scientists are still embarrassed about the anthropic principle. This critical attitude is quite understandable. Historically, the anthropic principle was often associated with the idea that the Universe was created many times until it met with final success, although without explaining *who* created it and *why* it was necessary to make the Universe suitable for our existence. Moreover, it would be much simpler to create proper conditions for our existence in a small vicinity of the Solar System rather than throughout the entire Universe. Why is the extra effort necessary? Metaphysical questions aside, the anthropic principle did not make much sense in the context of the old scientific paradigm, which claims that the Universe is the same everywhere, and it is given to us as one "copy." In this context, comparing life in different universes is a meaningless exercise. There were several attempts to discuss "many universes" in the context of quantum cosmology and many-world interpretation of quantum mechanics (see, e.g., Barrow and Tipler, 1986; Davies, 2004 [also revised as a chapter in the current volume]). However, for a long time, the many-world interpretation of quantum mechanics remained very unpopular. Moreover, it did not seem plausible at all that quantum theory could possibly describe such a large object as our Universe.

The skeptical attitude toward "many universes" and anthropic considerations, which dominated scientific thought for many decades, began to crumble about 25 years ago. This change was triggered by the development of inflationary cosmology, by the discovery of an extremely large number of metastable vacua of string theory, and, finally, by the discovery of the extremely small vacuum energy (cosmological constant Λ) and its possible interpretation in the context of the Multiverse scenario.

INFLATIONARY COSMOLOGY: WHY UNIVERSE?

In the beginning of the 1980s, scientists suddenly realized that the observed uniformity of the Universe is not just a trivial fact following from observations, but a great theoretical problem. According to the standard Big Bang theory, the distant parts of the Universe (which we can now see through a powerful telescope) were not in causal contact at the time of the Big Bang and could not influence one another until the very late stages of the Universe's evolution. So one could only wonder what made these distant parts of the Universe so similar.

This problem was solved with the invention of inflationary theory, which claimed that the very early Universe expanded exponentially fast. This rapid expansion eliminated all previously existing inhomogeneities, rendering the *observable part* of the postinflationary Universe enormously large and completely uniform.

The history of inflationary cosmology is quite nontrivial. The first semirealistic model of the inflationary type was proposed in 1979–1980 by Starobinsky (1979, 1980). The Starobinsky model was rather complicated, and its motivation was not quite clear. The idea of inflation became really popular only after Guth proposed using exponential expansion of the Universe as a tool for solving many complicated cosmological problems (Guth, 1981). His scenario assumed that the hot Universe gradually cooled down and then was trapped in an unstable vacuum state with large potential energy, called false vacuum. In this state, the Universe expanded exponentially and became enormously large. Guth famously exclaimed: "It is said that there is no such thing as a free lunch. But the Universe is the ultimate free lunch." Indeed, in his scenario, all elementary particles in our Universe appeared as a result of the decay of an expanding false vacuum.

However, Guth immediately realized that his model, which is now called "old inflation," did not work. The Universe that was produced after old inflation was either very inhomogeneous or represented the interior of an empty, open universe. The problems of old inflation were addressed in the so-called new inflation scenario (Linde, 1982a; also see Albrecht and Steinhardt, 1982). This scenario suffered from several other problems and eventually was replaced by the chaotic inflation scenario (Linde, 1983c).

Inflationary theory is based on the investigation of various models of a scalar field (or many scalar fields). These fields, in a certain sense, are similar to the electrostatic potential: they may take different values in different points, but, unlike magnetic fields, they do not point in any direction. Particle physicists used these fields long ago for describing π-mesons. Another example of a scalar field is the Higgs field, which breaks symmetry between weak and electromagnetic interactions and makes some of the elementary particles massive. The search for the Higgs particles is one of the main goals of the famous Large Hadron Collider at CERN, Switzerland.

Scalar fields play a dual role in cosmology: their values determine the properties of elementary particles (i.e., the laws of low-energy physics) and the energy density of the Universe, which affects its rate of expansion.

To explain the main idea of chaotic inflation, let us consider the simplest model of a scalar field ϕ with a mass m and with the potential energy density $V(\phi) = m^2\phi^2/2$ (see Figure 21.1). (To simplify notation, in this chapter we use the system of units where the speed of light $c = 1$ and the Planck constant $\hbar = 1$. This system of units is often used in particle physics and cosmology.) Potentials of this type were used for a long time for a description of the π-meson field. Because this function has a minimum at $\phi = 0$, one may expect that the scalar field ϕ should oscillate near this minimum and gradually fall toward it. This is indeed the case if the Universe does not expand. However, one can show that, in a rapidly expanding Universe, the scalar field moves downward very slowly like a ball dropped in a viscous liquid, viscosity being proportional to the speed of expansion.

Two equations describe the evolution of a homogeneous scalar field in our model: the field equation

$$\ddot{\phi} + 3H\dot{\phi} = -m^2\phi, \tag{21.1}$$

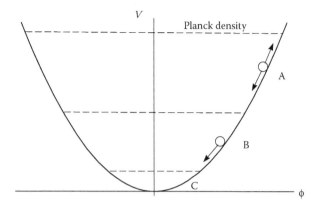

FIGURE 21.1 Behavior of the scalar field in the theory with $V(\phi) = m^2\phi^2/2$. If the potential energy density of the field is greater than the Planck density $\rho \sim M_p^4 \sim 10^{91}$ g/cm^3, quantum fluctuations of spacetime are so strong that one cannot describe spacetime in usual terms. Such a state is often called spacetime foam. At a somewhat smaller energy density (region A: $mM_p^3 < V(\phi) < M_p^4$), quantum fluctuations of spacetime are small, but quantum fluctuations of the scalar field ϕ may be large. Jumps of the scalar field due to quantum fluctuations lead to a process of eternal self-reproduction of the inflationary Universe. At even smaller values of $V(\phi)$ (region B: $m^2 M_p^2 < V(\phi) < mM_p^3$), fluctuations of the field ϕ are small; the scalar field moves downward very slowly like a ball dropped in a viscous liquid. Inflation occurs both in region A and region B. Finally, near the minimum of $V(\phi)$ (region C), the scalar field rapidly oscillates, creating pairs of elementary particles, and the Universe becomes hot.

and the Einstein equation

$$H^2 + \frac{k}{a^2} = \frac{1}{6M_p^2}\left(\dot{\phi}^2 + m^2\phi^2\right).$$

(21.2)

Here, $H = \dot{a}/a$ is the Hubble parameter in the Universe with a scale factor $a(t)$ (the size of the Universe); $k = -1,0,1$ for an open, flat, or closed universe, respectively; and M_p for the Planck mass, $M_p^{-2} = 8\pi G$. The first equation becomes similar to the equation of motion for a harmonic oscillator, where instead of $x(t)$ we have $\phi(t)$. The term $3H\dot{\phi}$ is similar to the term describing friction in the equation for a harmonic oscillator.

If the scalar field ϕ is initially large, the Hubble parameter H is large, too, according to the second equation. This means that the friction term is very large, and, therefore, the scalar field is moving very slowly. At this stage, the energy density of the scalar field remains almost constant, and expansion of the Universe continues with a much greater speed than in the old cosmological theory. Because of the rapid growth of the scale of the Universe and a slow motion of the field ϕ, soon after the beginning of this regime, one has $\ddot{\phi} \ll 3H\dot{\phi}$, $H^2 \gg k/a^2$, $\dot{\phi}^2 \ll m^2\phi^2$. So the system of equations can be simplified:

$$H = \frac{\dot{a}}{a} = \frac{m\phi}{\sqrt{6}M_p}, \quad \dot{\phi} = -mM_p\sqrt{\frac{2}{3}}.$$

(21.3)

The first equation shows that the size of the Universe $a(t)$ in this regime grows approximately as

$$a \sim e^{Ht},$$

(21.4)

where $H = m\phi/\sqrt{6}\,M_p$. This stage of exponentially rapid expansion of the Universe is called inflation.

When the field ϕ becomes sufficiently small, H becomes small, viscosity becomes small, inflation ends, and the scalar field ϕ begins to oscillate near the minimum of $V(\phi)$. As any rapidly oscillating classical field, it loses its energy by creating pairs of elementary particles. These particles interact with each other and reach a state of thermal equilibrium with some temperature T. From this point forward, the Universe can be described by the standard hot Universe theory.

The main difference between inflationary theory and the old cosmology becomes clear when one calculates the size of a typical inflationary domain at the end of inflation. One can show that, even if the initial size of the inflationary Universe was as small as the Planck size, $l_p \sim 10^{-23}$ cm, after 10^{-30} sec $l \sim 10^{10^{10}}$ cm. This makes our Universe almost exactly flat and homogeneous on the large scale because all inhomogeneities were stretched by a factor of $10^{10^{10}}$. This number is model-dependent, but in all realistic models the size of the Universe after inflation appears to be many orders of magnitude greater than the size of the part of the Universe that we can see now, $l_{\text{horizon}} \sim 10^{28}$ cm. This solves most of the problems of the old cosmological theory (Linde, 1990).

If the Universe initially consisted of many domains with chaotically distributed scalar fields ϕ, then the domains where the scalar field was too small never inflated, so they remain small. The main contribution to the total volume of the Universe is created by the domains that originally contained a large scalar field ϕ. Inflation of such domains creates huge homogeneous islands out of the initial chaos, each homogeneous domain being much greater than the size of the observable part of the Universe. That is why this scenario is called "chaotic inflation."

In addition to the scalar field ϕ driving inflation, which is called the inflation field, realistic models of elementary particles involve many other scalar fields φ_i. The final values acquired by these fields after cosmological evolution are determined by the position of the minima of their potential energy density $V(\varphi_i)$. In the simplest models, the potential $V(\varphi_i)$ has only one minimum. However, in general, the potential $V(\varphi_i)$ may have many different minima.

For example, in the simplest supersymmetric theory unifying weak, strong, and electromagnetic interactions, the potential energy V is a function of two scalar fields, Φ and φ. These fields are responsible for breaking symmetry between weak, strong, and electromagnetic interactions. The potential $V(\Phi, \varphi)$ in this theory has many different minima of equal depth. If the scalar fields Φ and φ fall to different minima in different parts of the Universe (the process called spontaneous symmetry breaking), the masses of elementary particles and the laws describing their interactions will be different in these parts. Each of these parts becomes exponentially large because of inflation. In some of these parts, there will be no difference between weak, strong, and electromagnetic interactions, and life as we know it will be impossible there. Some other parts will be similar to the part in which we live.

This means that, even if we find the final theory of everything, we will be unable to uniquely determine the properties of elementary particles in our Universe; the Universe may consist of many exponentially large domains where the properties of elementary particles and the properties of space may be quite different. Inhabitants of each of these exponentially large domains will not see other domains for an exponentially long time. Thus, for all practical purposes, a single inflationary universe becomes the "Multiverse," consisting of exponentially large independently existing "universes." Note, however, that in this context we are not talking about "parallel universes," a much more difficult and speculative concept, but about exponentially large parts of a single eternally inflating, self-reproducing universe, which we will call "the Multiverse."

This picture of the inflationary Universe divided into exponentially large parts with different laws of low-energy physics operating in each of them was proposed back in 1982 and 1983 (Linde, 1982b, 1983b), immediately after the development of the new inflationary scenario. The new cosmological paradigm was summarized in Linde (1983b): "As was claimed by Guth, the inflationary universe is the only example of a free lunch. Now we can add that the inflationary universe is the only lunch at which all possible dishes are available."

INFLATIONARY QUANTUM FLUCTUATIONS

A further step in the development of the theory of an inflationary Multiverse can be made if one takes into account quantum fluctuations produced during inflation.

According to quantum field theory, empty space is not entirely empty. It is filled with quantum fluctuations of all types of physical fields. The wavelengths of all quantum fluctuations of the scalar field ϕ grow exponentially during inflation. When the wavelength of any particular fluctuation becomes greater than H^{-1}, this fluctuation stops oscillating, and its amplitude freezes at some nonzero value $\delta\phi(x)$ because of the large friction term $3H\dot{\phi}$ in the equation of the motion of the field ϕ. The amplitude of this fluctuation then remains almost unchanged for a very long time, whereas its wavelength grows exponentially. Therefore, the appearance of such a frozen fluctuation is equivalent to the appearance of a classical field $\delta\phi(x)$ produced from quantum fluctuations.

Because the vacuum contains fluctuations of all wavelengths, inflation leads to the continuous creation of new perturbations of the classical field with wavelengths greater than H^{-1}. An average amplitude of perturbations generated during a time interval H^{-1} (in which the Universe expands by a factor of e) is given by $|\delta\phi(x)| \approx H/2\pi$ (Vilenkin and Ford, 1982; Linde, 1982c).

As a result of this process, the Universe becomes filled by an inhomogeneous distribution of classical scalar fields, which perturb the distribution of the density of matter in the Universe on an exponentially large scale. These perturbations grow and eventually lead to the formation of galaxies (Mukhanov and Chibisov, 1981; Hawking, 1982; Starobinsky,1982; Guth and Pi, 1982; Bardeen et al., 1983; Mukhanov, 1985, 2005). Simultaneously, they produce tiny perturbations of the cosmic microwave background (CMB) radiation. These perturbations were found by the *Cosmic Background Explorer* (*COBE*), *Wilkinson Microwave Anisotropy Probe* (*WMAP*), and other experiments. Most of the results of these experiments are in a very good agreement with predictions of inflationary theory (see, e.g., Komatsu et al., 2009; Sievers et al., 2009).

If the Hubble constant during inflation is sufficiently large, quantum fluctuations of the scalar fields may lead not only to the formation of galaxies, but also to the division of the Universe into exponentially large domains with different properties.

As an example, consider the simplest supersymmetric SU(5) theory unifying weak, strong, and electromagnetic interactions. As we already mentioned, in this theory there are two scalar fields Φ and φ breaking symmetry between weak, strong, and electromagnetic interactions. Different minima of the scalar potential $V(\Phi, \varphi)$ in this model are separated from one another by the distance $\sim 10^{-3} M_p$. The amplitude of quantum fluctuations of the inflation field ϕ and of the fields Φ and φ in the beginning of chaotic inflation can be as large as $10^{-1} M_p$. This means that at the early stages of inflation, the fields Φ and φ could easily jump from one minimum of the potential to another. Therefore, even if initially these fields occupied the same minimum all over the Universe, after the stage of chaotic inflation the Universe becomes divided into many exponentially large domains corresponding to all possible minima of the effective potential (Linde, 1983a). Each of these domains contains matter with dramatically different properties.

ETERNAL CHAOTIC INFLATION

The process of dividing the Universe into different parts becomes even easier if one takes into account the process of self-reproduction of inflationary domains. The basic mechanism can be understood as follows. If quantum fluctuations are sufficiently large, they may locally increase the value of the potential energy of the scalar field in some parts of the Universe. The probability of quantum jumps leading to a local increase of the energy density can be very small, but the regions where this happens start expanding much faster than their parent domains, and quantum fluctuations inside them lead to production of new inflationary domains that expand even faster. This surprising behavior leads to the process of self-reproduction of the Universe.

This process is possible in the new inflation scenario (Steinhardt, 1983; Linde, 1982b; Vilenkin, 1983). However, even though the possibility that this effect may divide the Universe into many different parts with different properties was realized already (Linde, 1982b), this observation initially did not attract much attention because the amplitude of the fluctuations in new inflation typically is smaller than $10^{-6}M_p$. This value seemed too small to probe most of the vacuum states available in the theory. As a result, the existence of the self-reproduction regime in the new inflation scenario was basically forgotten; for many years, this effect was not studied or used in any way even by those who found it.

The situation changed dramatically when it was found that the self-reproduction of the Universe occurs not only in new inflation, but also in the chaotic inflation scenario where the amplitude of quantum fluctuations can be as large as M_p (Linde, 1986). This process was called "eternal inflation" (Linde, 1986). In this scenario, the scalar field may wander for an indefinitely long time at a density approaching the Planck density. This induces quantum fluctuations of all other scalar fields, which may jump from one minimum of the potential energy to another for an unlimited time. Quantum fluctuations generated during eternal chaotic inflation can penetrate through any barriers, even if they have Planckian height, and the Universe after inflation becomes divided into indefinitely large numbers of exponentially large domains containing matter in all possible states corresponding to all possible mechanisms of spontaneous symmetry breaking—that is, to the different laws of low-energy physics (Linde, 1986).

In the theory supporting more than one scalar field, eternal inflation is even simpler to realize. For example, one may consider the situation where the scalar field was captured in the metastable false vacuum state, as in the old inflation scenario by Guth; but then the false vacuum decayed, and the stage of the slow-roll inflation began. Because the false vacuum never decays completely, the Universe always contains many parts in which the field is still trapped in a false vacuum. If there are many such false vacua (as is the case in string theory; see below), the field will eternally jump from one of them to another, producing the exponentially large parts of the Universe with different properties.

This means that inflation (and especially eternal inflation) may provide a perfect justification of the anthropic principle: we live in a part of the Universe with certain properties because the Universe has many exponentially large parts with different properties, and we can live only in those parts that have properties compatible with our existence. There is no need to assume that "somebody" was making one universe after another until final success: the inflationary Universe is perfectly capable of doing the job. Also, in this context, we can answer the most difficult objection to the anthropic principle: why would "anybody" need to work so hard at creating conditions congenial to life in the entire Universe if the only thing that is needed for life such as we find it on Earth is a small habitable part of the Universe in the vicinity of the Solar System? The answer is that inflation easily stretches the conditions in the vicinity of the Solar System to a size much greater than the observable part of our Universe.

STRING THEORY LANDSCAPE

Most of the advantages of this new picture of the world would not be so useful if the theory of fundamental interactions would allow only one vacuum state. In that case, even eternal inflation would not divide the Universe into parts with different laws of physics. This is where inflation finds a powerful ally: string theory.

String theory is the best known candidate for the theory of all interactions, including gravity. It is widely believed that this theory provides a consistent quantum theory of gravity coupled to matter in 10 spacetime dimensions. It is also believed that 6 out of 10 dimensions are compactified; that is to say, they are so small that we cannot move in them. Therefore, we view spacetime as 4-dimensional in agreement with all our observations.

Not all compactified spaces are possible; very often, string physicists consider so-called Calabi–Yau spaces, which have very complicated topological properties. These extra six dimensions have various kinds of magnetic fields, usually called fluxes. They are similar to the fluxes of the usual magnetic fields in our observable world. For each of the many possible topologies, this 6-dimensional space may contain a large number of fluxes. The total number of possible combinations of topologies and fluxes in extra dimensions can be exponentially large.

The laws of physics operating in our 4-dimensional spacetime depend crucially on what exactly happens in the compactified space. It is insufficient to say that our world is described by string theory. Depending on the properties of the compactified space, string theory can describe an exponentially large number of different 4-dimensional universes, which are also called string-theory vacua (Lerche et al., 1987; Bousso and Polchinski, 2000).

For a while, this abundance of possibilities was considered one of the most difficult problems of string theory. It contradicted Einstein's dream of a final, fully predictive theory. With invention of the eternal chaotic inflation scenario, the attitude to this issue gradually started to change. The first paper on eternal chaotic inflation contained the following manifesto:

> From this point of view, an enormously large number of possible types of compactification, which exist e.g. in the theories of superstrings, should be considered not as a difficulty but as a virtue of these theories, since it increases the probability of the existence of mini-universes in which life of our type may appear. (Linde, 1986)

However, for a long time we could not actually use this advantage because, until very recently, we did not really know whether string theory has *any* stable vacuum states. All vacua that string theorists were able to construct were unstable with respect to spontaneous generation of light scalar fields, called moduli, and also with respect to the growth of six compact dimensions. A 10-dimensional universe would not like to be imprisoned in 4-dimensional space; it would prefer to become large in all directions.

This problem was resolved in 2003 in the scenario proposed by Kallosh, Kachru, Linde, and Trivedi (KKLT) (Kachru et al., 2003), after many other efforts in this direction (Giddings et al., 2002; Silverstein, 2001; Maloney et al., 2002). Numerous developments and many constructions of the string-theory vacua have been proposed since then (see, e.g., a review of flux vacua in Douglas and Kachru, 2007), as well as interesting new ideas of de Sitter vacua constructions in Silverstein (2008). These constructions are very complicated; we cannot even describe them without using string theory terminology. Readers not interested in the mathematical details of this construction may skip it and go directly to the text following Equation 21.7.

In the KKLT scenario, one starts with a theory where fluxes were used to stabilize the so-called shape moduli. One still has a typical stringy runaway potential for the size moduli. One has to bend this potential using the nonperturbative quantum effects to fix the size moduli, including the volume of the compactified space. This allows the stabilization of all moduli in the state with negative vacuum energy. The effective 4-dimensional supergravity potential $V(\phi, \bar{\phi})$ is expressed in terms of the Kähler potential $K(\phi, \bar{\phi})$ and holomorphic superpotential $W(\phi)$, $\bar{W}(\bar{\phi})$.

$$V\left(\phi, \bar{\phi}\right) = e^{K\left(\phi, \bar{\phi}\right)}\left(|DW|^2 - 3|W|^2\right). \tag{21.5}$$

The first step, which is shown in the left panel of Figure 21.2, is to find vacua with negative vacuum energy in effective supergravity: one can prove a condition of unbroken supersymmetry, $DW = 0$, which can be solved by fixing the moduli and that leads to negative energy vacua

$$V\left(\phi, \bar{\phi}\right)\big|_{DW=0} = -3e^{K\left(\phi, \bar{\phi}\right)}|W|^2. \tag{21.6}$$

Therefore, one more crucial step is required to uplift the negative-energy minimum to the state with a positive vacuum energy. This is achieved by adding a positive energy of an anti-D3 brane in warped Calabi–Yau space. The total energy of the stabilized de Sitter vacua is a result of the almost

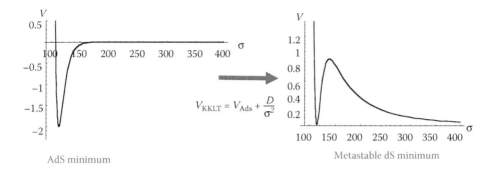

$$V_{KKLT} = V_{Ads} + \frac{D}{\sigma^2}$$

AdS minimum   Metastable dS minimum

FIGURE 21.2 In the left panel, one can see the first step in constructing stable vacua in string theory using fluxes as well as some quantum effects. Typically, this leads to a stabilization of the size moduli. The left plot shows the potential V depending on the volume σ of the six extra dimensions. The volume is stabilized in the so called anti-de Sitter (AdS) space with the negative cosmological constant (Λ) (negative vacuum energy). The transition from the left to the right panel shows the next step, called the "uplifting," which is required to get the de Sitter vacua with a positive Λ where the volume of extra dimensions is stabilized and we see our world as having only four large dimensions. The uplifting is achieved by adding the term D/σ^2, which is the energy of an anti-D3 brane. Depending on the details of this construction, one may obtain many different de Sitter vacua.

exact cancellation of the positive energy of the anti-D3 brane by negative vacuum energy (see the right panel in Figure 21.2).

Once the general mechanism of vacuum stabilization in string theory was established, it was found, in agreement with earlier expectations, that the total number of string theory vacua is incredibly large. For example, in type IIB string theory, the corresponding number was estimated (Douglas and Kachru, 2007) to be

$$N_{vac} \approx K^{L/2}. \tag{21.7}$$

Here, $K = 2b_3(X)$ is the number of flux quanta, where $b_3(X)$ is the third Betti number of the Calabi–Yau space X—that is, the number of homologically independent three-cycles. The second important number is related to the topological properties of the manifold $L = x/24$. It is believed that L ranges between 10 and 10^3 and K between 100 and 400. In specific well-defined models, one finds for N_{vac} the numbers 10^{307}, 10^{393}, and 10^{506}. The last case serves as a technical justification for the often-quoted number 10^{500} as the total number of string vacua. However, one should keep in mind that there are uncertainties in such counting, so the true number of vacua in string theory may be much smaller, or much larger, than 10^{500}.

One may wonder: In what sense does the choice of string theory vacua determine the laws of physics? Let us consider a simple example. In string theory, the effective gauge-field Lagrangian in four dimensions contains a factor depending on one of the moduli fields ϕ: $L = e^{\phi}F_{\mu\nu}^2$. The values of the moduli fields are determined by the positions of the minima of the potential energy of these fields $V(\phi)$, which in turn depend on numerous fluxes and other features of extra dimensions. Therefore, ϕ may take many different values, $\phi = \{\phi_1, \phi_2,..., \phi_n\}$. This leads to many different gauge couplings in different vacua (i.e., in different universes) because we get different gauge couplings $L_n = e^{2\phi_n}F_{\mu\nu}^2 \equiv (1/g_n^2)F_{\mu\nu}^2$. Thus, instead of a unique computable gauge coupling g, we find a large set of possible values $\{g_1, g_2,..., g_n\}$. In particular, we get different values for the fine-structure constant α in different parts of the Universe. Also, the value of the potential energy of these fields $V(\phi)$ determines the vacuum energy (Λ), which takes different values in different parts of the Universe.

Inflation makes each of such parts exponentially large and uniform, and the Universe becomes a Multiverse consisting of different exponentially large uniform parts with different effective laws of physics operating in each of them. Even if the whole Universe was initially in one of these vacuum states, inflationary quantum fluctuations and formation of expanding bubbles containing various string theory vacua divide the Universe into many different exponentially large parts. The Universe becomes the Multiverse. This is the essence of what is often called the string theory landscape (Susskind, 2003).

ANTHROPIC SOLUTION TO THE COSMOLOGICAL CONSTANT PROBLEM

The string theory landscape provided a unique framework for solving the Λ problem. Λ represents the vacuum energy density. Its natural value in the quantum theory of gravity would be

$\Lambda \sim M_p^4 \sim 10^{91} \text{g/cm}^3$.

For many years, cosmologists did not know whether vacuum energy Λ is 0, but they knew that $|\Lambda| \leq \rho_0 \sim 10^{-29}$ g/cm^3.

Here, ρ_0 is the total density of matter in the Universe. Given that 10^{-29} g/cm^3 is 120 orders of magnitude smaller than the naively expected value $\Lambda \sim 10^{91}$ g/cm^3, many physicists were trying to find some fundamental reason why Λ must be exactly 0. However, all of their attempts failed.

The situation became even more confounded about 10 years ago, when cosmological observations found that Λ is indeed extremely small, but not 0 (Perlmutter et al., 1999; Riess et al., 1998). According to the latest observational data, vacuum energy constitutes about 70% of all matter in the Universe (Kotmatsu et al., 2009). In other words, $\Lambda \sim 0.7\rho_0 \sim 0^{-29}$ g/cm^3, so Λ is not 0, but instead is positive and is 120 orders of magnitude below its expected value $\sim 10^{91}$ g/cm^3. If it is extremely difficult to explain why it is 0, it must be even more difficult to explain why it is almost 0, but not exactly. Indeed, 0 can be a fundamental number, because of symmetries, but not a 1 with 120 zeros between it and the decimal point. Moreover, even if one finds some approximate solution to this problem and explains why the vacuum energy must be so incredibly small, this property most probably will be destroyed by quantum corrections.

One should note that the acceleration of the Universe may also occur if the Universe is populated, for example, by some slowly changing invisible scalar fields that have properties very similar to the properties of Λ. This means that the dark energy responsible for the present stage of the acceleration of the Universe may be either time-independent (Λ) or it may very slowly change in time, just like the energy of the scalar field during inflation. However, the possibility of the time-dependence of dark energy does not help us to explain its incredible smallness.

The smallness of the vacuum energy has another puzzling aspect. In the early Universe, the vacuum energy was the same as it is now, but the density of usual matter was much, much higher. In the distant future, the vacuum energy will remain the same; but the density of normal matter will become much smaller. Thus, in addition to explaining why Λ is so small, we need to solve yet another problem: we must explain why we live in a particular epoch where the density of usual matter and the value of Λ are so close to each other. The only special feature of this epoch is that we happen to live in it. This makes it tempting to look for some kind of anthropic solution for both of the puzzles related to Λ.

The first attempt to solve the Λ problem using the anthropic principle in the context of inflationary cosmology was made back in 1984 (Linde, 1984b). In that paper, it was argued that if one considers antisymmetric tensor fields $F_{\mu\nu\alpha\beta}$, their fluxes contribute to the vacuum-energy density of the Universe $V(F)$, which in the classical approximation remains constant, just as energy density of a false vacuum. The total vacuum energy density is given by the sum of the energy of fluxes and the energy of scalar fields $V(\phi) + V(F)$. According to some popular versions of quantum cosmology, the probability of quantum creation of the Universe is extremely small unless the Universe is created at nearly Planckian density, $V(\phi)+V(F) \sim M_p^4$ (Linde,

1984a; Vilkenkin, 1994). Consider, for example, the simplest chaotic inflation scenario with $V(\phi) = m^2\phi^2 + V_0$. In this case, all classical universes emerge from spacetime foam with equal probability at the moment when $m^2\phi^2/2 + V_0 + V(F) \sim M_p^4$. After that, the scalar field rolls down to the minimum of its potential energy, and vacuum energy density in these universes relaxes to $\Lambda = V_0 + V(F)$. The sum $V_0 + V(F)$ by itself does not affect the probability of the quantum creation of the Universe because all universes are equally probable for $m^2\phi^2/2 + V_0 + V(F) \sim M_p^4$. Thus, one may conclude that the universes with all values of Λ are equally probable. It was argued that life of our type cannot exist in the universes with $|\Lambda| >> 10^{-29}$ g/cm^3 (Linde, 1984b). This, together with the flat probability distribution for creation of the universes with different Λ, may solve Λ problem.

Another way to solve Λ problem was proposed in 1984 by Andrei Sakharov (Sakharov, 1984). He argued that in a theory with many compactified dimensions, the number of different types of compactifications may be extremely large, forming a nearly continuous spectrum of values, with vacuum energy density levels separated by less than 10^{-29} g/cm^3. Then we would be able to live only in those universes (or parts of the inflationary Multiverse) where the absolute value of the vacuum energy is sufficiently small.

These considerations may work only if one can justify the anthropic constraints mentioned above. If Λ is large and negative, $\Lambda \leq -10^{-28}$ g cm^{-3}, the Universe collapses within a timescale much smaller than the present age of the Universe (Barrow and Tipler, 1986; Linde, 1987). In this case, the anthropic constraint is strong and unambiguous. The situation with a positive Λ is more complicated. In 1987, Weinberg pointed out that the process of galaxy formation occurs only up to the moment when the Λ begins to dominate the density of the Universe (Weinberg, 1987). The formation of large galaxies would be quite improbable if Λ were about 2 or 3 orders of magnitude greater than its present value. Subsequent investigation of this issue improved this bound (Efstathiou, 1995; Vilenkin, 1995; Martel et al., 1998; Garriga et al., 2000). In particular, according to Martel et al. (1998), the probability that an astronomer in any of the universes would find $\Lambda \leq 0.7\rho_0$ ranges from 0.05 to 0.12, depending on various assumptions.

In other words, if one does not use the anthropic principle, then the naive estimate of the probability to live in a universe with the present value of Λ would be about 10^{-120}. With an account taken of anthropic considerations, this probability grows by a huge factor of 10^{119} and becomes as large as O (10%).

The dependence of the vacuum energy on fluxes and topology (Linde, 1984b; Sakharov, 1984), and the anthropic constraints on Λ (Weinberg, 1987), form the basis of the possible solution of the Λ problem in the string landscape scenario. The main idea, as before, is that, in the inflationary Multiverse, during the process of eternal inflation, one can probe all possible vacua (Bousso and Polchinski, 2000; Suuskind, 2003). Among 10^{500} vacua, one can always find many that are anthropically allowed with the absolute value of the vacuum energy being smaller than $10^{-120} M_p^4$. We cannot live in the "typical" vacua where the absolute value of energy density is many orders of magnitude greater than $10^{-120} M_p^4$, which is why we live in one of those parts of the inflationary Multiverse where it is small enough. The existing anthropic constraints on Λ relate its value to the present value of the density of matter in the Universe, which provides a solution to both aspects of the Λ problem mentioned above. (For the latest discussion of this approach and the list of important references, see, e.g., De Simone et al., 2008; Bousso et al., 2009; Salem et al., 2009).

Note that, in order to solve the Λ problem, we need to know only some statistical properties of the distribution of various vacua rather than an exact value of energy density in each of them. Quantum corrections may easily change the value of the vacuum energy in each of these vacua. However, if the total number of vacua is large enough and the distribution of their number does not strongly depend on Λ at $\Lambda \ll M_p^4$, quantum corrections do not affect the possibility of solving the Λ problem by this mechanism. We illustrate this situation in Figure 21.3.

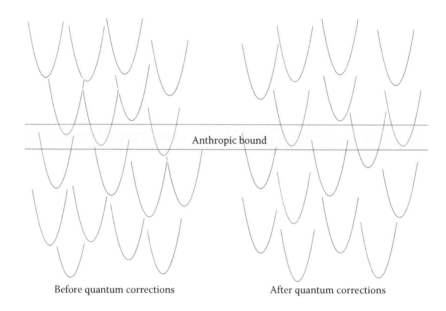

Anthropic bound

Before quantum corrections   After quantum corrections

FIGURE 21.3   The distribution of string theory vacua before and after quantum corrections. Each vacuum on this sketch corresponds to one of the many minima of the potential energy.

Thus, it looks plausible that a combination of inflationary theory, string theory, and anthropic reasoning can help us to solve the Λ problem. At the moment, we do not know of any alternative solutions.

DEBATING THE MULTIVERSE

The new cosmological paradigm changes the way we think about our place in the world, but it may take some time until we find an optimal way to use our new knowledge. Should we try to preserve at least some part of Einstein's dream by explaining why the Universe we live in, and the laws of physics operating there, are most natural and most probable, or at least not absolutely unnatural and improbable? Can we find a preferable way of calculating probabilities in the Multiverse consisting of an infinite number of different universes? Or should we simply say that we live in our part of the Universe by chance, learn the laws of physics operating around us, and try to predict our future on the basis of these laws?

The developments in Multiverse theory described in this chapter became a source of frequent and rather emotional debate among the leaders of modern physics, as well as the subject of several recent books on this subject (Susskind, 2006; Carr, 2007; Vilenkin, 2007). Weinberg, Rees, Vilenkin, Susskind, Wilczek, and other influential physicists participated in the development of the Multiverse theory for a long time. Gross is strongly against the anthropic principle associated with the Multiverse concept, arguing that we should not give up our hopes for the final theory too early. 't Hooft does not like the theory of the Multiverse, but says that he is afraid that it may be correct. Witten, one of the leaders of the string theory community, is seen as a symbol of the old Einsteinian ideal of a final theory. He seems to start accepting the Multiverse concept. He said in a 2008 *New York Times* interview: "I wasn't terribly enthusiastic the first, or even second, time I heard the proposal of a multiverse. But none of us were consulted when the universe was created. ... This interpretation of string theory might be close to the truth" (Overbye, 2008).

Several issues repeatedly arise during the debates of the Multiverse. Will we ever see any experimental evidence that we live in it if other parts of the Universe governed by different laws of low-energy physics are exponentially far away from us? And if we are not going to see them, is the Multiverse a scientific concept? Can we simply forget about it and return to the old idea of a single, globally uniform Universe?

In order to address these issues, let us go back in time to the moment before the invention of inflationary theory. We knew that the Universe is large, homogeneous, and isotropic, but it took us some time to realize that these are not merely trivial facts to be memorized, but important experimental data to be explained. For the last 25 years, inflationary theory remains the only well-developed framework in which these questions can be addressed. In addition, inflationary theory makes several predictions, many of which have been experimentally confirmed. This theory, at present, is the best theory explaining the uniformity of the observable part of the Universe. However, it shows that the Universe may consist of different exponentially large parts with different properties.

Similarly, before the invention in string theory of the mechanism of vacuum stabilization, we expected that this mechanism, when found, would unambiguously explain all of the properties of our world. However, now that we have found such a mechanism, we see that it may lead to an exponentially large variety of outcomes. A combination of the development of inflationary theory and string theory makes it easy to understand the formation of a Multiverse and simultaneously makes it very difficult to justify the old unwarranted assumption that all parts of the Universe should have similar properties.

An additional important consideration is that many coincidental relations between the properties of our world and the parameters of elementary particles so far have found their theoretical interpretation only in the framework of the theory of the Multiverse. This does not mean that we should stop trying to find a different explanation. Moreover, in the context of the Multiverse theory, one can make several important predictions, such as that of the possible nature of dark matter (Linde, 1988; Tegmark et al., 2006), which can be experimentally tested.

Paradoxically, instead of trying to justify the Multiverse scenario, one can formulate a similar, but opposite, question: Can we prove that the Universe does *not* consist of different parts with different properties? Why should we believe that it must look the same everywhere? Is it possible to return to the old idea of a single, uniform Universe and to avoid the concept of the Multiverse altogether? We believe that, in order to do so, one would need to take three very difficult steps simultaneously:

1. One would need to find an alternative to inflationary theory.
2. One would need to ignore recent developments in string theory, which is believed to be the best candidate for the theory of all fundamental interactions.
3. One would need to find an alternative solution to many coincidental problems, including the Λ.

These conditions are very hard to satisfy. Therefore, at the moment it does not seem easy to forget about all of the recent scientific developments discussed above and to return to the old picture of a single, globally uniform Universe.

In his Nobel lecture in 1979, Abdus Salam said: "Einstein knew that nature was not economical of structures: only of principles of fundamental applicability" (Salam, 1980). By following this line of thought, one may argue that nature is not economical with universes, only with the principles that allow their production. It seems that eternal inflation, in combination with string theory, provides a simple mechanism for the creation of the Multiverse consisting of infinitely many, locally uniform universes of many different types.

REFERENCES

Albrecht, A. and Steinhardt, P.J. (1982). Cosmology for grand unified theories with radiatively induced symmetry breaking, *Physical Review Letters*, 48:1220–223.

Bardeen, J.M., Steinhardt, P.J., and Turner, M.S. (1983). Spontaneous creation of almost scale-free density perturbations in an inflationary universe. *Physical Review D*, 28:679–93.

Barrow, J.D. and Tipler, F.J. (1986). *The Anthropic Cosmological Principle*. New York: Oxford University Press.

Bousso, R. and Polchinski, J. (2000). Quantization of four-form fluxes and dynamical neutralization of the cosmological constant. *Journal of High Energy Physics* 0006: 006 (30 pages) [arXiv:hep-th/0004134v3].

Bousso, R., Freivogel, B., and Yang, I.S. (2009). Properties of the scale factor measure. *Physical Review D*, 79:063513 (18 pages) [arXiv:0808.3770v2 [hep-th].

Carr, B. (ed.) (2007). *Universe or Multiverse?* Cambridge: Cambridge University Press.

Davies, P.C.W. (2004). Multiverse cosmological models. *Modern Physics Letters A*, 19 (10): 727–43. Also revised as a chapter in the current volume.

De Simone, A., Guth, A. H., Salem, M. P., et al. (2008). Predicting the cosmological constant with the scale-factor cutoff measure. *Physical Review D*, 78:063520 (16 pages) [arXiv:0805.2173v1 [hep-th]].

Douglas, M.R. and Kachru, S. (2007). Flux compactification. *Reviews of Modern Physics*, 79:733–96 [arXiv:hep-th/0610102v3].

Efstathiou, G. (1995). An anthropic argument for a cosmological constant. *Monthly Notices of the Royal Astronomical Society,* 274:L73–6.

Einstein, A. (1959). Autobiographical notes. In: P. A. Schilpp (ed.), *Albert Einstein, Philosopher-Scientist*. New York: Harper Torchbooks.

Garriga, J, Livio, M., and Vilenkin, A. (2000). The cosmological constant and the time of its dominance. *Physical Review D*, 61:023503 [arXiv:astro-ph/9906210v1].

Giddings, S.B., Kachru, S., and Polchinski, J. (2002). Hierarchies from fluxes in string compactifications. *Physical Review D*, 66:106006 (36 pages) [arXiv:hep-th/0105097v2].

Guth, A.H. (1981). The inflationary universe: A possible solution to the horizon and flatness problems. *Physical Review D*, 23:347–56.

Guth, A.H. and Pi, S.Y. (1982). Fluctuations in the new inflationary universe. *Physical Review Letters*, 49:1110–113.

Hawking, S.W. (1982). The development of irregularities in a single bubble inflationary universe. *Physics Letters B*, 115:295–97.

Kachru, S., Kallosh, R., Linde, A. et al. (2003). De Sitter vacua in string theory. *Physical Review D*, 68:046005 (10 pages) [arXiv:hep-th/0301240v2].

Komatsu, E., Dunkley, J., Nolta, M.R., et al. (2009). Five-year *Wilkinson Microwave Anisotropy Probe (WMAP)* observations: cosmological interpretation. *Astrophysical Journal Supplement Series,* 180:330–76 [arXiv:0803.0547v2 [astro-ph]].

Lerche, W., Lust, D., and Schellekens, A.N. (1987). Chiral four-dimensional heterotic strings from self dual lattices. *Nuclear Physics B*, 287:477–507.

Linde, A.D. (1982a). A new inflationary universe scenario: a possible solution of the horizon, flatness, homogeneity, isotropy and primordial monopole problems. *Physics Letters B*, 108:389–93.

Linde, A.D. (1982b). *Nonsingular Regenerating Inflationary Universe*. Cambridge: Cambridge University preprint Print-82-0554. Available at: http://www.stanford.edu/~alinde/1982.pdf.

Linde, A.D. (1982c). Scalar field fluctuations in expanding universe and the new inflationary universe scenario. *Physics Letters B*, 116:335–39.

Linde, A.D. (1983a). Inflation can break symmetry in SUSY. *Physics Letters B*, 131:330–34.

Linde, A.D. (1983b). The new inflationary universe scenario. In: G.W. Gibbons, S. W. Hawking, and S. Siklos (eds.), *The Very Early Universe*, 205–49. Cambridge: Cambridge University Press. Available at: http://www.stanford.edu/~alinde/1983.pdf.

Linde, A.D. (1983c). Chaotic inflation. *Physics Letters B*, 177–81.

Linde, A.D. (1984a). Quantum creation of the inflationary universe. *Lettere al Nuovo Cimento,* 39:401–5.

Linde, A.D. (1984b). The inflationary universe. *Reports on Progress in Physics*, 47:925–86. Available at: http://www.stanford.edu/~alinde/1984.pdf.

Linde, A.D. (1986). Eternally existing self-reproducing chaotic inflationary universe. *Physics Letters B*, 175:395–400.

Linde, A.D. (1987). Inflation and quantum cosmology. In: S. W. Hawking and W. Israel (eds.), *Three Hundred Years of Gravitation*, 604–30. Cambridge: Cambridge University Press.

Linde, A.D. (1988). Inflation and axion cosmology. *Physics Letters B*, 201:437–39.

Linde, A.D. (1990). *Particle Physics and Inflationary Cosmology*. Switzerland: Harwood, Chur [arXiv:hep-th/0503203].

Maloney, A., Silverstein, E. and Strominger, A. (2002). de Sitter space in noncritical string theory (23 pages) [arXiv:hep-th/0205316v2].

Martel, H., Shapiro, P.R., and Weinberg, S. (1998). Likely values of the cosmological constant. *Astrophysics Journal*, 492:29–40 [arXiv:astro-ph/9701099v1].

Mukhanov, V.F. (1985). Gravitational instability of the universe filled with a scalar field. *Journal of Experimental and Theoretical Physics Letters*, 41:493–6. [*Zhurnal Eksperimentalnoi i Teoreticheskoi Fiziki, Pisma*, 41:402.]

Mukhanov, V.F. (2005). *Physical Foundations of Cosmology*. Cambridge: Cambridge University Press.

Mukhanov, V.F. and G.V. Chibisov (1981). Quantum fluctuation and 'nonsingular' universe. *Journal of Experimental and Theoretical Physics Letters*, 33:532–5. [*Zhurnal Eksperimentalnoi i Teoreticheskoi Fiziki, Pisma*, 33:549.]

Overbye, D. (2008). Dark, perhaps forever, *The New York Times* (June 3). Available at http://www.nytimes.com/2008/06/03/science/03dark.html.

Perlmutter, S., Aldering, G., Goldhaber, G., et al. (1999). Measurements of Ω and Λ from 42 high-redshift supernovae. *Astrophysical Journal*, 517:565–86 [arXiv:astro-ph/9812133v1].

Rees, M. (2000). *Just Six Numbers: The Deep Forces that Shape the Universe*. New York: Basic Books, Perseus Group.

Riess, A.G., Filippenko, A.V., Challis, P., et al. (1998). Observational evidence from supernovae for an accelerating universe and a cosmological constant. *Astronomical Journal*, 116:1009–38 [arXiv:astro-ph/9805201v1].

Rozental, I.L. (1988). *Big Bang, Big Bounce*. Berlin: Springer Verlag.

Sakharov, A.D. (1984). Cosmological transitions with a change in metric signature. *Journal of Experimental and Theoretical Physics*, 60:214–18. [*Zhurnal Eksperimentalnoi i Teoreticheskoi Fiziki*, 87:375 (1984).] [See also the Special Issue of *Soviet Physics-Uspekhi*, 34:409–13 (1991).]

Salam, A. (1980). Gauge unification of fundamental forces. *Reviews of Modern Physics*, 52:525–38 [*Science*, 210:723–32].

Salem, M.P. (2009). Negative vacuum energy densities and the causal diamond measure arXiv:0902.4485v3 [hep-th].

Sievers, J.L., Mason, B.S., Weintraub, L., et al. (2009). Cosmological results from five years of 30 GHz CMB intensity measurements with the cosmic background imager (38 pages) [arXiv:0901.4540v2 [astro-ph.CO]].

Silverstein, E. (2001). (A)dS Backgrounds from asymmetric orientifolds (12 pages) [arXiv:hep-th/0106209v1].

Silverstein, E. (2008). Simple de Sitter solutions. *Physical Review D*, 77:106006 (37 pages) [arXiv:0712.1196v4 [hep-th]].

Starobinsky, A.A. (1979). Spectrum of relict gravitational radiation and the early state of the universe. *Journal of Experimental and Theoretical Physics Letters*, 30 (11): 682–85. [Spektr reliktovogo gravitatsionnogo izlucheniya i nachal'noe sostoyanie Vselennoi. *Zhurnal Eksperimentalnoi i Teoreticheskoi Fiziki, Pisma*, 30:719–23.]

Starobinsky, A.A. (1980). A new type of isotropic cosmological models without singularity. *Physics Letters B*, 91:99–102.

Starobinsky, A.A. (1982). Dynamics of phase transition in the new inflationary universe scenario and generation of perturbations. *Physics Letters B*, 117:175–78.

Steinhardt, P.J. (1983). Natural inflation. In: G.W. Gibbons, S. W. Hawking, and S. Siklos (eds.), *The Very Early Universe*. Cambridge: Cambridge University Press.

Susskind, L. (2003). The anthropic landscape of string theory (21 pages) [arXiv:hep-th/0302219v1].

Susskind, L. (2006). *The Cosmic Landscape: String Theory and the Illusion of Intelligent Design*. New York: Back Bay Books.

Tegmark, M., Aguirre, A., Rees, M., et al. (2006). Dimensionless constants, cosmology and other dark matters. *Physical Review D*, 73:023505 (29 pages) [arXiv:astro-ph/0511774v3].

Vilenkin, A. (1983). The birth of inflationary universes. *Physical Review D*, 27:2848–855.

Vilenkin, A. (1994). Quantum creation of universes. *Physical Review D*, 30:509–11.

Vilenkin, A. (1995). Quantum cosmology and the constants of nature (12 pages) [arXiv:gr-qc/9512031v1].

Vilenkin, A. (2007). *Many Worlds in One: The Search for Other Universes*. New York: Hill and Wang.

Vilenkin, A. and Ford, L.H. (1982). Gravitational effects upon cosmological phase transitions. *Physical Review D*, 26:1231–241.

Weinberg, S. (1987). Anthropic bound on the cosmological constant. *Physical Review Letters*, 59:2607–610.

22 Cosmos and Humanity in Traditional Chinese Thought

Yung Sik Kim

CONTENTS

INTRODUCTION: THE PROBLEMS OF GALILEO'S TELESCOPIC OBSERVATIONS*

Around the time of the Jesuits' arrival in China in the late 16th century, the Chinese saw and knew more or less as much about the world as the Europeans did (Peterson, 1973; Elman, 2005, pp. 24–149). To be sure, there were differences in their views about some key aspects of the natural world. Most notably, they had different views about "heaven" and "earth": for example, whether the earth was spherical or flat, whether heaven was filled with solid spheres or with *qi* (气), and so on. There also were differences in their views as to whether the basic elements constituting the world were material substances or "phases" (*xing* 行) and, related to this, whether the essential aspects of the human body were the anatomical organs or the physiological functions. Yet, the Europeans and the Chinese of the time were able to note, observe, and explain nearly all the important phenomena and objects. The difference, then, was in their attitudes toward those phenomena and objects. Europeans' responses to Galileo's telescopic observations highlight some such differences between the two traditional cultures' attitudes toward the natural world.

After Galileo had observed the heavens and announced the results, his discoveries caused many problems for Catholic astronomers and Aristotelian scholars in Europe at the time (Drake, 1978, pp. 134–76). At the core of the problems was the celestial-terrestrial (superlunar-sublunar) distinction that conditioned the Aristotelian worldview that had dominated European minds for centuries. The idea of the perfect, unchanging heavens opposed to the changing, imperfect sublunar world was challenged by Galileo's observations—in particular, the imperfect surface of the moon and the imperfect dark spots (sunspots) on the face of the principal heavenly body, which appeared to undergo an irregular movement (Kuhn, 1957, pp. 90–3). Use of the telescope—a mechanical tool devised for earthly purposes (warfare and navigation, for example)—for observing phenomena in the heavens created further problems: not only did it violate the celestial-terrestrial distinction, but it was in conflict with another Aristotelian distinction, namely, that between the natural and the artificial (Lloyd, 1970, pp. 105–6).

---

* Also see the chapters by Yi-Long Huang, Xiao-Chun Sun, and Peter Harrison in this volume.

Whether what people saw through the telescope represented the reality of the world, and not just images concocted by the tool composed of lenses in a tube, was also a problem for many scholars of the Church and astronomers in the universities. Reflecting the long European tradition of deep concern about the problems of reality versus mere image, real cause versus mere effect, etc., many of them refused even to accept the reality of what Galileo observed with the telescope, some of them even refusing to look through it (Drake, 1978, pp. 162, 165). Of course, there was also the problem of the conflict between the Copernican world system allegedly supported by Galileo's observations and the Aristotelian–Ptolemaic world system supported by the Church and biblical passages. Galileo's conviction that studying the world created by God and uncovering the laws hidden in natural phenomena was a service to God seems to have helped him, a devout Catholic, to overcome this problem. But the conflict between what is observed and what is believed must have continued to be a problem both for Galileo himself and for his opponents.

These problems were so serious to the Europeans of the time that to many of them the advantage of the telescope—the vast increase in mankind's ability to see things at a great distance—was not sufficient to enable them to overlook such problems. Yet none of these issues were a problem in China. When the report of Galileo's telescopic observations reached China, it did not create much reaction (d'Elia, 1960, pp. 17–19, 33–50).* This, however, was something that could be expected from the basic characteristics of traditional Chinese views of the natural world.

For example, there was no such sharp distinction in traditional Chinese thought between the celestial and the terrestrial worlds or between natural and artificial things. There was also no tradition, at least in the dominant Confucian school, of persistent questioning and doubting the reality of the actual world, the reality of what is perceived by the senses. In fact, Neo-Confucianism, the philosophical school that dominated the Chinese (and later the whole East Asian) intellectual world for the last several centuries of the premodern period, arose in reaction to Taoism and Buddhism, which rejected the reality of the actual world. In Neo-Confucianism what was observed was usually accepted in the way it was perceived by man.

In this chapter, I discuss the basic attitudes of the traditional Chinese toward the natural world and toward the objects and phenomena in it. I show the ways in which these attitudes influenced their views about the relation between the world and man, or between "the cosmos and humanity." Before setting out to do this, however, I have some words of caution. Although I use the expression "traditional Chinese," my discussions cannot do justice to the extreme diversity of their views, assumptions, and attitudes concerning the natural world. They were not uniform and often contradicted one another. This chapter concentrates mainly on the views and attitudes of Confucian scholars, especially those of the Neo-Confucian school. And given that even the Neo-Confucians themselves had quite diverse views and attitudes, it frequently focuses on those of Zhu Xi 朱熹 (1130–1200), who, as the author of the so-called Neo-Confucian synthesis, exerted an overwhelming influence on the East Asian intellectual world for many centuries.†

"COSMOS" AND HUMANITY: THE WORLD OF
HEAVEN-AND-EARTH, MAN, AND THE MYRIAD THINGS

There was no word in traditional China equivalent to the Western word, "cosmos." The word "yuzhou" 宇宙,‡ usually chosen as a modern translation of "cosmos," did not exactly have the meaning of "an orderly, harmonious universe." It is not that, for the Chinese, the world (or the universe) was not orderly or harmonious. But the order or harmony that could be seen in the world was a part

---

* There even were attempts by scholars to make observations using telescopes (Zhang, 1987, p. 193).
† Much of my discussion in this essay is based on Kim (2000).
‡ In the standard commentaries on the expression "yuzhou" in such ancient books as the Zhuangzi 庄子 and the Huainanzi 淮南子, the character "yu" 宇 was rendered as "heaven, earth and the four directions" (tiandi sifang 天地四方), and "zhou" 宙 as "the past and the present" (gujin wanglai 古今往来).

of an order or harmony of much greater scope, covering everything—man and a myriad of things, as well as the whole physical world. What was close to the notion of "cosmos" was the expression "heaven-and-earth" (*tiandi* 天地). This expression frequently designated the whole natural world (Peterson, 1982) and was often contrasted with the human world.

The relation between man and the world of heaven-and-earth had many sides. For one, there was the idea of the parallelism of microcosm-macrocosm, i.e., the notion that man as the small world is also the epitome of the great world, heaven-and-earth (Needham, 1956, pp. 294–303). A more pronounced aspect of the relationship between man and heaven-and-earth for the traditional Chinese was the idea of the triad of "heaven–earth–man," i.e., the notion of the world in which man lives harmoniously between heaven and earth.* The basic component of the triad was the idea that heaven-and-earth and man complement each other—heaven-and-earth does not do everything alone; there are some things that heaven-and-earth cannot do, which man does for heaven-and-earth (Kim, 2000, pp. 111–13).

It was not only man that inhabited the world of heaven and earth; there were the "myriad things" (*wanwu* 万物). Heaven-and-earth, man, and the myriad things thus represented the three basic constituents of the natural world for the traditional Chinese. And they stood in various relations to one another. For example, heaven-and-earth produces men and the myriad things. Produced by heaven-and-earth, men and things live—or exist—between heaven and earth. They receive the *qi* and the mind (*xin* 心) of heaven-and-earth and have them as their *qi* and minds.

The world made of these three constituents covered everything, not merely the things that could be characterized as "physical" or "material." Objects and phenomena involving life and mind, and even morality, were included in this world of heaven-and-earth, man, and the myriad things; there was no clear distinction between the "natural" and the "nonnatural" realms in that world. This lack of distinction was evident also in such basic concepts as *qi*, yin-yang 阴阳, and the five phases (*wuxing* 五行), which were endowed with characteristics of life, mind, and morality, as well as of matter (Kim, 2000, pp. 31–69). One consequence of this was that, for the traditional Chinese, the "natural" world existed in harmony with the "nonnatural"—human and social—world. The "natural" world was an integral part of the larger world that also contained human and social realms, with no boundary between them. Nothing was excluded from this world. Thus, there was no sharp boundary that separated man from the natural world of heaven-and-earth.

More pronounced was the distinction between man and the myriad (nonhuman) things (Kim, 2000, pp. 174–7). For the traditional Chinese, man was unique among all living beings. Man was not even considered as part of "the myriad things." With the character *wu* 物 (thing), they never referred to man. Man is endowed with *qi* that is correct, clear, complete, penetrating, balanced, numinous, and excellent, whereas the "things" are endowed with *qi* that is dark, turbid, muddled, one-sided, obstructed, and screened. In other words, because of these differences, man is bright, numinous, and complete, whereas the "things" are dark, ignorant, and one-sided. The key difference that distinguished man from nonhuman "things" lay in the human ethical virtues. For example, the five human relationships (*wulun* 五伦) were what distinguished man from animals. Animals could have some ethical virtues, but they could not have all of them. Similar one-sidedness could also be seen in the knowledge and the mental qualities of animals. Other features distinguishing man from animals were the ability to speak, the upright bodily stance, etc.

THE HEAVENLY *LI*: THE NATURAL WORLD AND MORALITY

The concept of *tian* 天 (heaven) went through a long and complicated historical development. Originally a term designating the ancestor-deity of the early Zhou 周 royal family, its development proceeded along two different lines (Lamont, 1973/1974; Eno, 1990).

---

* A source of this is the passage from the *Doctrine of the Mean* (*Zhongyong* 中庸, Chapter 22): "If [a man] can assist the transforming and nourishing of heaven and earth, he can form a triad with heaven and earth."

One line of development centered on the aspect of deity. It could be characterized by the gradual weakening of the anthropomorphic character of the concept, which in time came to represent a more general, abstract, and even conceptual deity. But *tian* continued to mean something that controls, rules, and presides over everything in the world. As development along this line proceeded further, *tian* was also thought to be something that produced men and things of the world. *Tian* even referred to "decree" (*ming* 命: fortune, life span, mandate) and "nature" (*xing* 性), which were endowed to men and things when they were produced. And in all this, the term retained the moral aspect present in its original meaning.

Another line of development resulted from the association of the word with the supposed site of the original ancestor-deity, the sky. It proceeded with the separation of the moral character from the concept. But this meaning was also expanded gradually, first to the whole natural world including the sky, then to events and processes of that world, and eventually to the *li* 理* underlying the workings of the natural world.

When they had proceeded this far, the outcomes of the two lines of development converged, and the physical, naturalistic aspect and the moralistic aspect of the concept of heaven were connected again. This connection of the two aspects came to provide a basis for the belief in the moral character of the natural world. Because of this connection, the concept of heaven could function as something that supplements the morally neutral character of the world composed of the *qi* processes of the natural world, covered by the term "heaven." "Heaven" could be taken even for a standard on which human ethical behavior should be based. There could be no possibility of tension between the morally neutral "natural" world and the human world that is governed by morality. On the contrary, the "natural" world was frequently invested with moral qualities (Metzger, 1977, pp. 82–5, 127–34). This, then, is what lay at the basis of the frequently found idea of a moral order underlying the natural world and providing a kind of "cosmic basis" for morality (Bol, 1989).

In the course of these developments, the term "heaven" came to have various meanings, which can be classified in the following four categories: (1) the physical sky; (2) the natural world, including things and events in it; (3) the *li* of both the natural and human realms; and (4) a concept or a being that rules or presides over the world (Fung, 1937, p. 31).

Of these categories, the most problematic—both for us and for the traditional Chinese—is the fourth, namely heaven as what "rules" or "presides over" the world, or "*zhuzai*" 主宰 in Chinese. We can find some hints about the meaning of the term "*zhuzai*" from Zhu Xi. After mentioning the regularities of the natural world, such as the succession of the four seasons, alternation of day and night, and the way plants flourish on the sunny side and wither on the shadowy side, he stated "This is clearly as if a person were in [heaven] and 'presided over' [these regularities]" (*Zhuzi yulei*, 4.4b; see Zhu Xi, 1962). But he was explicit that actually there is no such person up there: "If one now says that a person is in heaven and blames crimes [on him], it certainly is not acceptable" (*Zhuzi yulei*, 1.4b); "… it is as though a person were up there like that … But [the truth] is simply that the *li* is like this" (*Zhuzi yulei*, 4.6b). What Zhu Xi emphasized instead was that the "presiding" (*zhuzai*) aspect of heaven is somehow responsible for the "naturalness" and spontaneity of phenomena and processes in the natural world. This aspect of *zhuzai*, then, is not very far from the idea of heaven

---

* *Li* is a difficult term to define and even to translate. There is no single word in Western languages that would cover all facets of what it meant to the traditional Chinese mind. The existence of many translations for the term—"principle," "reason," "law," "pattern," etc.—often leaves transliteration as the only viable option, which bespeaks the difficulty. What is closest to a definition is the following remark by Zhu Xi: "When it comes to things under heaven, for each [of the things] there must be 'a reason (*gu* 故) by which [a thing is] as it is' (*so yiran zhi gu* 所以然之故) and 'a rule (*ze* 则) according to which [a thing] ought to be' (*so dangran zhi ze* 所当然之则). [These are] what are called '*li*'" (*Daxue huowen*, 1.11b–12a; see Zhu Xi, 1983). But he did not elaborate on these two different aspects of the concept. He spoke of them only in terms of particular examples. "In treating one's parents one ought to be filial, and in treating one's elder brother one ought to be brotherly. Such are 'the rules according to which [one] ought to be.' But in treating one's parents why should one be filial, and in treating one's elder brother why should one be brotherly? These are 'the reasons by which [a man is] as he is'" (*Zhuzi yulei*, 18.21b; see Zhu Xi, 1962).

as *li*. Eventually, the idea of the "*li* of heaven" (*tianli* 天理) emerged as the ultimate aim of the Confucian endeavor (Kim, 2000, pp. 109–10).

The *li* of heaven is a kind of "universal" *li* that covers, and is manifest in, all the individual *li* of things and events. In fact, an important aspect of the concept *li* is that there are many individual *li* for individual objects and phenomena and the one *li* for the whole of these objects and phenomena. Every object and phenomenon of the world—"a grass, a tree, or even an insect," to use Zhu Xi's expression (*Zhuzi yulei*, 15.12a)—has its individual *li,* and all individual *li* are manifestations of the one *li*.

The *li* of heaven is what the original human nature manifests in the form of such ethical virtues as humaneness (*ren* 仁) and righteousness (*yi* 义); it is because the heavenly *li* is obstructed and blocked by "human desires" that man loses the virtues and shows evil traits. This dichotomy of "the *li* of heaven" versus "human desires" was the basis of Confucian moral philosophy; thus, the state of mind free from human desires was, for Confucian scholars, the ultimate goal of man's self-cultivation. When that goal is attained, man's mind fully manifests the *li* of heaven.

It is not only the original human nature that manifests the *li* of heaven; every thing and event in the world has its *li*, which is but a manifestation of the one universal *li* of heaven. Therefore, man can also aim at the *li* of heaven through the method of "*gewu*" 格物 because the term, translated usually as "the investigation of things," to many Neo-Confucians meant reaching the *li* in "things." The *gewu* endeavor, however, did not primarily involve intellectual procedures. In fact, man's understanding of the *li* of things, achieved as the result of *gewu*, was considered a kind of "resonance" between the mind's *li* and things' *li*. This was so because the *li* of heaven resides both in man's mind (as the mind's *li*) and in things and events (as their *li*). Thus, when a man has reached the *li* of a thing or an event, it was described usually as "seeing" the *li*, rather than as "knowing" it. In other words, what has been gained is not so much knowledge of the *li* as insight into it. Also, for such "resonance" to take place, the mind needs to be "empty," "bright," and "tranquil." This is the original state of man's mind, manifesting the *li* of heaven fully, free from blockages created by human desires. In such a state, the mind spontaneously sees the *li* in things and events, which are nothing but manifestations of the *li* of heaven contained in the mind itself (Kim, 2000, pp. 21–2).

Gaining insight into the many *li* of individual things and events, however, was not the real aim of the *gewu* endeavor, the ultimate purpose of which was to reach the *li* of heaven via the many individual *li*. The key step in the *gewu*, then, lay in moving from those individual *li* to reach the one *li* of heaven. Yet, the connection between the many individual *li* and the one *li* of heaven was not quite traceable. It was never clear exactly how the grasp of many individual *li* could lead to the apprehension of the one *li* of heaven. However, all seemed to agree that the step must involve something more than a purely intellectual process. In Zhu Xi's words, one needs "laborious efforts" (*gongfu* 工夫) and "nourishing" (*yang* 养) in addition to "knowing" and "understanding." The mental state described by the term "reverence" (*jing* 敬) was important, for when a man is reverent, his mind is "bright," "transparent," and "alive"; all *li* are in the mind; and the heavenly *li* becomes "brilliant." Thus, the moral and intellectual endeavors of man converged in his search for *li*—the many individual *li* and the one *li* of heaven— through *gewu*. And in this convergence, the moral side was clearly the more important. To be sure, the intellectual aspect could not be ignored altogether, but, on the whole, the intellectual elements of the *gewu* endeavor were fused into its ultimately moral aims. It was to uphold morality and to avoid errors that one "investigated things" (Kim, 2000, pp. 22–5).

YIN-YANG, THE FIVE PHASES, ETC.: THE "CORRELATIVE" MODE OF THINKING

Whereas the combined expression "heaven-and-earth" designated the entire world as discussed above, its two parts, heaven and earth were contrasted by pairs of opposite qualities: clear–turbid, light–heavy, movement–rest, bright–dark, etc. Heaven and earth were characterized by various dichotomies: clear heaven vs. turbid earth; light heaven vs. heavy earth; moving heaven vs. still earth; bright heaven vs. dark earth; and so on.

These dichotomies, however, are a part of the yin-yang dichotomy that characterized everything of the world, and thus the contrast of heaven and earth reflected features of the yin-yang dichotomy. The distinction between heaven and earth was not absolute, reflecting the fact that opposition of the yin and yang characteristics was not absolute. For example, yin and yang undergo continuous cyclical repetition. Yin and yang alternate and follow each other continuously: at the extreme of yin emerges yang; as yang grows and reaches its extreme, yin reemerges. This process alternates continuously and is referred to as "the circulation" (*xunhuan* 循环) of yin and yang: because yin and yang keep following each other, there cannot be a distinction of what is earlier or later between them. It is frequently said that "movement and rest have no ends; yin and yang have no beginnings" (*Zhuzi yulei*, 1.1a).

Indeed, for the traditional Chinese, alternation of opposites was the most important characteristic of the phenomena in the natural world. They frequently spoke of cyclical alternations of movement and rest, going and coming, opening and closing, contracting and expanding, growing and vanishing, day and night, the sun and the moon, life and death, hot and cold weather, rainy and fine weather, exhalation and inhalation, and even of the bending and stretching movements of the worm. One consequence of such cyclical alternation was that one side of the pair was not devoid of the other side. There cannot be yin or yang completely devoid of each other: yin has to contain at least a trace of yang mixed with it, and vice versa.

Traditional Chinese also used other sets of categories in their discussions of natural phenomena, the five phases in particular. As categories, they are associated with various sets of characteristics. Different characteristics associated with a given category are connected to one another, thus giving rise to a network of mutual associations. This kind of association was a key mode of explanation in the traditional Chinese discussion of natural phenomena, and indeed, such categorical and associative character was a universal feature of the traditional Chinese discourse about the natural world, which many commentators have noted and referred to as "correlative thinking" (Porkert, 1974, 9–54; Graham, 1986).

We can see many examples showing this character of the traditional Chinese worldview. For one, the four seasons and the four cosmic qualities—*yuan* 元, *heng* 乙亨, *li* 利, and *zhen* 贞—formed a network of four sets of mutually associated characteristics through their respective five-phase associations with weather, compass directions, and the constant human virtues. Some sets of characteristics were associated directly with one another, without bringing in, as intermediaries, associations with the basic categories such as yin-yang and the five phases. Thus, the five musical notes (*wusheng* 五声) were directly associated with a group of five different things: the *gong* 宫 musical note with rulers, *shang* 商 with ministers, *jiao* 角 with people, *zhi* 征 with events (*shi* 事), and *yu* 羽 with things (*wu* 物). The 12 musical pitches (*shier lü* 十二律) were associated with the 12 months. There were also associations of man's perceptual organs with visceral organs, for example, eyes with liver and ears with kidneys (Kim, 2000, pp. 42–90).

These categorical and associative basic concepts are also cyclical: they come in cycles that repeat fixed sequences. Not only the yin-yang characteristics, but many sets of five-phase characteristics, the four seasons, the four cosmic qualities, and the life cycle of plants, for example, also repeat their fixed sequences endlessly. Such endless cyclical repetition was a universal feature of natural phenomena, which is not surprising because the cyclical nature of many natural phenomena—movement of the luminaries in the sky, the change of seasons, the ebbing and flowing of tides, and the plant life cycles—must have been obvious even to the most casual observer. In fact, the cyclical repetition became another key feature of traditional Chinese perceptions of natural phenomena.*

---

* It is in reference to such cyclical repetition that Needham has characterized "Chinese physics" by the concept of "wave" as opposed to "particle" (Needham, 1962, pp. 3–14).

STIMULUS AND RESPONSE: CAUSE AND EFFECT?

In this kind of correlative mode of thought, there is no cause-effect relationship of a mechanical sort. What we find instead is the concept of "stimulus and response" (*gan-ying* 感应). There were two kinds of stimuli and responses (Kim, 2000, pp. 122–8).

First, there was the stimulus-response interaction among things, events, and concepts that belonged to a single category and were associated with one another. Things of the same category, but in different cosmic realms, were supposed to affect one another by virtue of a mutual sympathy, to resonate like properly attuned pitch pipes: the same sounds respond to each other, the same *qi* seek each other. This kind of stimulus-response interaction was not restricted to things and events that can be called natural. More frequently mentioned cases of mutual interactions and influences were between men or between men and natural things and events. The most frequently mentioned example, however, is the idea that man's conduct can affect the course of natural events, which is referred to as "mutual stimulation of heaven and man" (*tianren xiang-gan* 天人相感). For example, Zhu Xi approved a disciple's remark that "if the ruler of people accumulates sins and mistakes, these stimulate and beckon inauspicious events, and bring about the calamities of eclipses of the sun and moon, collapse of mountains and drying of rivers, and drought and famine" (*Zhuzi yulei*, 62.19b). Of course, the traditional Chinese did not believe that all natural calamities were caused by men. Zhu Xi said, for example, that "there are times when human conduct brings them about, and also when they occur accidentally [i.e. by accident of natural events]" (*Zhuzi yulei*, 79.6b). Nor did they believe that all natural events could be affected by human conduct.

The second kind of stimulus-response relation occurs among things, events, and concepts that repeat continuously in fixed sequences, especially the dualistic ones that undergo continuous alternations. This kind of stimulus-response relation was far more frequently mentioned than the first kind. For example, the successive appearance and disappearance of the sun and moon, and of the cold and hot weather as well, was viewed as resulting from their mutual actions as stimuli and responses. Other alternating pairs were also viewed as stimulus and response upon each other: for example, expanding and contracting, movement and rest, vanishing and growing, going and coming, exhaling and inhaling, day and night, life and death, rainy and fine weather. Because stimulus and response follow each other continuously in this manner, there is no end to their cyclical alternation. What has acted as a response to a stimulus acts, in turn, as a new stimulus, which brings another response, and this cycle goes on without end. This second type of stimulus-response relationship was not restricted to what happens in the natural world, either. The frequent examples included alternations of sleeping at night and waking up in the morning, activity during the daytime and rest at night, speaking and keeping silent, goodness and badness in man's nature, flourishing and decaying, peaceful and troubled periods in history, and even the mutual reinforcement of the father's love and the son's filial piety.

It is possible to figure out the nature of the first kind of stimulus-response relation from its most typical example of the same musical notes responding to each other—or "resonating," to use a modern expression (Henderson, 1984, p. 20). On the other hand, it is difficult to see how pairs of continuously alternating things, events, and concepts act as stimulus and response to each other; in other words, it is difficult to see how what comes first in a stimulus-response sequence "causes" that which follows to occur. Again, there were some hints in Zhu Xi's sayings. For example, in explaining the bending and stretching movement of a worm, he said, "If the worm does not bend, then it cannot stretch" (*Zhuzi yulei*, 72.5a). He also said, "If the sun does not go, the moon does not come; if the moon does not go, the sun does not come. Cold and hot weather is also like this" (*Zhuzi yulei*, 72.4b). He even said that the going down of the sun can stimulate the moon to come up, and *vice versa*, and that "when the cold [weather] becomes extreme, it produces (*sheng* 生) the warm [weather]" (*Zhuzi yulei*, 24.24).

Sometimes Zhu Xi discussed more general aspects of stimulus-response relations. For example:

> Owing to this one event, there also comes about another event. They simply are stimulus and response. Owing to this second event, a third event also comes about. The second event is also a stimulus, and the third event is also a response" (*Zhuzi yulei*, 72.4a).

He also pointed out that a great or small stimulus brings about a great or small response, respectively (*Zhuzi yulei*, 53.6a). Expressed in this manner, the relationship between stimulus and response sounds very much like that between our modern ideas of cause and effect. In fact, of the pairs that Zhu Xi called "stimulus and response," some do appear to us as cause and effect. He said, "For example, coming [i.e. blowing] of the wind is a stimulus; moving of the tree is the response. Shaking the tree is also a stimulus; moving of the things below is also the response" (*Zhuzi yulei*, 72.3b). He viewed the sowing of seeds and the appearance of sprouts similarly as stimulus and response (*Zhuzi yulei*, 72.5b). His examples even included acquiring a concubine acting as a stimulus to which the response is purchasing silver utensils (*Zhuzi yulei*, 72.4a-4b). But he did not seem to be aware of the differences between these examples and the pairs that are simply sequences. For instance, after mentioning the stimulus-response relation of the wind and tree, he added the stimulus and response of day and night, as if the relation were the same kind: "When the day reaches the extreme, it must stimulate and get the night to come; when the night is at its extreme, it also stimulates and gets the day to come" (*Zhuzi yulei*, 72.3b).

It cannot be said, however, that the concept of cause and effect was absent in traditional Chinese thought. It is merely that their notions of how cause and effect act in the world were different from the modern concept of causality: cause and effect were never understood in mechanical terms, nor were they rigorously defined. Traditional Chinese ideas about stimulus and response, then, could be seen as their version of the cause-effect relation, different from the modern version in content, emphasis, and scope (Needham, 1956, 280ff; Kim, 2000, p. 305).

REGULARITIES IN NATURE: LAWS OF NATURE?

Many of the examples of "stimulus-response" interactions discussed in the last section were cyclical repetitions occurring regularly in the world. These could be called natural—or social and historical—"regularities." Traditional Chinese knew and spoke of many other regularities that are not cyclical, but which occur constantly (Kim, 2000, pp. 305–6). They frequently used the words "constant" (*chang* 常) or "constant *li*" (*changli* 常理) to refer to these regularities; for example, the "constant *li*" that winter should be cold and summer should be hot, the "constancy" (*chang*) in the frequency and magnitude of the tides, the constancy of heaven's movements, and even the constancy in the deviations of heavens' movements. The frequently quoted passage from the *Record of Rites*, "All living things must die. Having died they must return to earth" (*Liji*, Chapter 24), which was quoted repeatedly, can be seen as a similar example. Zhu Xi characterized the upward growth of a tree and the downward flow of water as "following the *li*" (*shun li* 順理) and the opposite cases as "against the *li*" (*ni li* 逆理), implying a constant *li* that the tree should grow upward and water should flow downward (*Zhuzi yulei*, 55.11b). The regularities mentioned by Zhu Xi also included certain social and historical instances. He even mentioned what sounds like Gresham's law, i.e., that "pure (good) coins are always rare; impure (bad) coins are always abundant" (*Zhuzi yulei*, 4.20b).

These regularities, however, were altogether different from the Western concept of laws of nature. The way the traditional Chinese treated these regularities was also different from the attitude of Western thinkers toward laws of nature. A regularity was considered by the traditional Chinese as a particular fact and not as a general law or principle covering many particular facts. They did not need to analyze the regularities or attempt to abstract simple, general laws from them. Even when they spoke of the regularities, what they were interested in was that such *li* exist, not in the detailed content of the *li*. Moreover, the regularities were frequently mentioned not for their own interest, but for analogies to similar regularities in human conduct and norms. For example, the fixed and unchangeable order of the four seasons was often mentioned to make the point that the three basic human relations (*san'gang* 三纲) and the five constant virtues (*wuchang* 五常) cannot be changed (*Zhuzi yulei*, 24.26a). Zhu Xi made the above-mentioned remarks about the *li* of the growth of trees and the flow of water while discussing the point that the teachings of Yang Zhu 杨朱 and Mo Di 墨翟 were "against the *li*." The Gresham-law-like remark was made by him to illustrate that great men are always rare while small men are always abundant. At times, this tendency led him even to offer a basically human-centered explanation for the regularities in the natural world:

> If there is day, there must be night. If it were day for a long time and there were no night, how could [one get] rest? And if there were no day, how could there be this brightness? ... If there were spring and summer but no autumn or winter, how could things be completed? If there were just autumn and winter with no spring and summer, then again, how could [things] come into being? (*Zhuzi yulei*, 72.4b).

We might consider in this connection a passage from the *Book of Poetry*, containing the word "*ze*" 则, meaning "a rule": "When heaven produces masses of people, if there is a thing there is a rule" (*Shijing*, no. 200). This passage appears to imply not only laws of nature, but also a supreme being responsible for them. But that impression does not survive a closer examination. For most of the examples considered for such rules were about norms of moral conduct, i.e., "the rules according to which one ought to be so" (*dangran zhi ze* 当然之则): for example, the rules according to which rulers and subjects ought to behave, the eyes ought to see, and the ears ought to hear. Moreover, it was never said explicitly that these rules were laid down by heaven. Indeed, it is far more likely that what the traditional Chinese had was the idea of a rule to be followed by heaven in producing men and things rather than that of a law provided by heaven to be followed by men and things.

We might also consider Joseph Needham's thesis that the concept of the laws of nature did not develop in Chinese scientific thought because of the absence of the idea of a Creator who also laid down "laws of nature" to be obeyed by His creatures (Needham, 1956, 518–83; Bodde, 1957, 1979). Needham reviewed occurrences of the term *ze* in various early Chinese texts and concluded, for reasons similar to those given above, that *ze* cannot mean laws of nature. My conclusion is also that the term was different from the laws of nature, but I cannot go along with Needham when he concludes, based on his survey of terms such as *ze* and *li*, that the idea of laws of nature was absent in China. For it cannot be denied that the traditional Chinese did recognize regularities in natural phenomena. Some regularities even led a man such as Zhu Xi to speak of the possible existence of a person in heaven presiding over them, although in the end he rejected the notion. To be sure, traditional Chinese ideas about these regularities were different from the Western concept of laws of nature (and they did not use such a term), but it is still possible to say that they represent their versions of laws of nature.

NATURE AS "NATURAL": ACCEPTING WHAT IS SEEN

The natural world—what exists in it, what happens in it—was "natural" for the traditional Chinese mind. Most of the objects and phenomena in the natural world were, to them, obvious. They took them for granted, in a "matter-of-fact" manner, and did not feel any need to explain them. In fact, the natural phenomena and objects were so "natural" and obvious to them that they frequently alluded to some common and familiar natural phenomena in the course of discussing moral and social problems, by adducing analogies between the obvious natural phenomena and the latter problems that were considered more problematic. Only rarely did they mention such common natural phenomena for themselves.

For example, Zhu Xi spoke of the fact that once a cart has started to move, no great exertion of force is needed to keep it moving; he argued that, in study also, a great exertion of effort is needed only at the beginning, after which it becomes easy (*Zhuzi yulei*, 31.8a). Similarly, to explain that when impurities enter the mind it loses "sincerity" (*cheng* 诚) and falls into self-deception, he used the analogy that when gold is mixed with a small amount of silver the whole bulk of gold loses its worth as gold (*Zhuzi yulei*, 16.19b). It is not impossible to learn from these examples something about Zhu Xi's views on natural phenomena—tendencies of moving objects, properties of a mixture of metals. But his real concern lay elsewhere—to argue for strong exertion of effort at the beginning stage of study, and for the importance of sincerity and purity of the mind, by showing that these points were analogous to natural phenomena. In neither of the examples were the phenomena themselves what Zhu Xi was really interested in. Many concrete natural phenomena and objects came up in this context in traditional Chinese discussions (Kim, 2000, pp. 3–4).

Thus, in their discussions of the natural world, the traditional Chinese did not go deeper than the surface phenomena of reality. Questions delving deeper into what one sees and accepts were not raised. Zhu Xi, for example, confessed that his influential friend, Zhang Shi 张拭 (1133–1180), did not approve his wish to write a commentary on Shao Yong's 邵雍 (1011–1077) theory about the outside of the world (*Zhuzi yulei*, 115.5a). The orthodox Neo-Confucians shunned discussions of questions such as these. Such an attitude can be seen in Zhu Xi's comment on the famous dialog between Cheng Yi 程颐 (1033–1107) and Shao Yong on where thunder comes from: on Cheng Yi's saying, "Thunder comes from where it comes from," which had been given in response to Shao Yong's question, "Where do you think [thunder] comes from?" Zhu Xi's comment was, "Why must one know where it comes from?" (*Zhuzi yulei*, 100.11a).*

To be sure, Zhu Xi did allow himself to discuss imaginary, theoretical situations. For example, he could imagine what a man, a "divine man" (*shen ren* 神人), placed outside heaven and earth, or between the sun and the moon, would see (*Zhuzi yulei*, 2.3b). While discussing the Buddhist doctrine of the other worlds, he speculated about a consequence of it, namely that the night would then have to be very long because the sun in its daily rotation would have to traverse around the other three worlds, all below the earth's horizon (*Zhuzi yulei*, 86.8b–9a). Nor was Zhu Xi alone in this. Many other Confucian scholars did engage themselves in speculations about imaginary situations. For example, a 17th-century Korean Confucian scholar, Chang Hyŏn-kwang 张显光 (1554–1637), in a treatise called the *Theory of the Universe* (*Ujusŏl* 宇宙说, 1631), speculated about such problems as the size of the heaven-and-earth, whether the world is only one, and whether the world could be destroyed.† But in these examples, Zhu Xi and Chang Hyŏn-kwang were not thinking of an actual possibility.‡ Moreover, even this much was an exception. On the whole, the traditional Chinese avoided abstract, theoretical—and useless—speculations. At any rate, even these exceptional cases could hardly match their discussions of moral and social problems in the level of elaboration, seriousness, and sophistication.

One can see some key aspects of the traditional Chinese basic ideas and assumptions that made them consider natural phenomena to be so "natural"—obvious, matter-of-fact.

For instance, qualities and activities of *qi* were considered innate, and thus once certain phenomena had been attributed to certain qualities and activities of *qi*, they were deemed sufficiently accounted for—without any need to look for external causes or hidden mechanisms (Kim, 2000, pp. 35–7). Zhu Xi's account of the formation of the earth, for example, will illustrate the point:

> In the beginning of heaven and earth there was only the *qi* of yin and yang. This *qi* moved and turned around continuously. When the turning became very rapid, a large quantity of the sediments of *qi* was compressed. And as there was no outlet [for the sediments] these consolidated to form the earth in the center (*Zhuzi yulei*, 1.4b).

---

* The dialog between Cheng Yi and Shao Yong is recorded in *Henan Cheng shi yishu*, Chapter 21a (see Cheng Hao 程颢 and Cheng Yi 程颐, 1981).

† Although he was persistent in pursuing these questions, his basic position was that such things cannot be known.

‡ I am not comfortable in calling them even a "thought experiment," although I do not have a convincing explanation for why they cannot be one.

He considered the rapid rotation of *qi* to be responsible for the formation of the earth, but he never paid attention to the cause of such rotation. It was almost as if rotation were the natural activity for *qi*.

The concept *li* also had an effect. *Li* of an object or phenomenon was merely something because of which the object exists or the phenomenon takes place as it actually does: when and only when there is *li* for it, does it exist or take place. Thus, *li* was not conceptually simpler or more fundamental than the object or phenomenon itself. *Li* referred to a given object or phenomenon as a whole in its totality; it was not what could be used in the explanation or analysis of the object or phenomenon in simpler terms. When *li* was mentioned, it was merely invoked to ensure the existence or occurrence of the object or phenomenon. Nor was the content of *li* analyzed; it is grasped as a whole. Thus, when the traditional Chinese noted regularities in nature, they were concerned only with their existence and not with concrete details of those regularities, which Zhu Xi sometimes referred to as *li* (Kim, 2000, pp. 25–7).

The dichotomy of what is "above physical form" (*xing er shang* 形而上) versus what is "below physical form" (*xing er xia* 形而下) also facilitated ready acceptance of natural phenomena.* Abstract and sublime concepts without manifest "physical forms" (*xing* 形)—the Way (*Dao* 道), *li*, mind, and human nature, for example—belong to the former, while concrete things with tangible physical forms are examples of the latter. Naturally, what was without physical form was difficult to understand and was thought to be important and worthy of further consideration, whereas what had physical form and was visible was easy to understand and was considered obvious and even trivial. Because most common natural phenomena are accompanied by tangible qualities and physical effects and are "below physical form," they were thought to be obvious and were simply accepted in the way they were perceived; no further investigation was attempted beyond the surface of the phenomenal realities of empirical data.

The perennial Confucian emphasis on the reality of the external world also seems to have reinforced the readiness in accepting commonly observed natural phenomena. Confucians considered their acceptance of the reality of the world to be what distinguished them from Taoists and Buddhists. They were not actively engaged in concepts such as "void" (*kong* 空, *xu* 虚) and "nothingness" (*wu* 无), which were too easily associated with Taoists and Buddhists who tended to lead men to concentrate on introspection without paying attention to the actual world. Thus many Confucians simply accepted natural phenomena rather than engaging themselves in abstract, theoretical discussions about them (Kim, 2000, pp. 307–9).

Nor did problems involving concepts such as elements, mixture, infinite, indivisibility, space, time, void, causality, law, and so on receive much attention from the traditional Chinese (Kim, 2000, pp. 302–7). To the Confucians these concepts could at best be imaginary. They did not appear to be of any help in reckoning with the reality of the actual world. They were not useful in dealing with the moral and social problems of the world, either. It must have appeared useless, for most Confucians, to be engaged in abstract, theoretical speculations involving them.

Yet, these were the problems that were pursued in great depth in the Western scientific tradition and generated intense debates and controversies. And it was such controversies and consequent close examinations that eventually led to the resolution of these problems during the European Scientific Revolution. Indeed, the contrast between the situation in the Western scientific tradition and that in China is both striking and interesting. Much of the medieval European discussion of the same concepts—"motion," "mixture," "space," "void," and "infinite"—was undertaken in theological-philosophical contexts (Murdoch, 1975). But while similar contexts did exist in Chinese Buddhism and Taoism, they were not picked up because both were opposed by the dominant Confucian literati (Kim, 1985). It was ironic that the basis of this rejection was an emphasis on the actual world,

---

* This dichotomy can be traced back to the famous passage of the "Commentary on the Appended Words" (*Xicizhuan* 系辞传, A12) of the *Book of Changes* (*Yijing* 易經): "What is 'above physical form' is called the 'Way' (*dao* 道); what is 'below physical form' is called the 'tools' (*qi* 器)."

and shunning of "useless" theoretical speculations, for abstract, theoretical speculations about such basic concepts could have contributed to a fundamental understanding of phenomena in the actual world. In the West at least, continuing debates over the interpretation of precisely such concepts helped bring into being modern science during the Scientific Revolution and, ultimately, the "useful" science that we have today (Grant, 1996, pp. 168–91). In contrast, excessive emphasis on the reality of the actual world, as well as on the necessity of the usefulness of discussions, made it difficult for men such as Zhu Xi to consider in detail those very concepts that could have proved useful in understanding, and living in, that actual world.

REFERENCES

PRIMARY SOURCES

Liji 礼记 *(Record of Rites)*.
Shijing 诗经 *(Book of Poetry)*.
Yijing 易 经 *(Book of Changes)*.
Zhongyong 中庸 *(Doctrine of the Mean)*.
Cheng Hao 程 颢 and Cheng Yi 程 颐. *Henan Cheng shi yishu* 河南程氏遗书 *(Surviving Works of the Chengs of Henan)*. In: *ErChengji* 二程集 *(Collected Works of the Two Chengs)*. Beijing: Zhonghua shuju 中 华 书 局, 1981.
Zhu Xi. *Daxue huowen* 大学 或 问 (Some questions on the *Great Learning*). *Wenyuange Siku quanshu* 文淵 阁 四库 全 书 edition. Taipei: Taiwan Shangwu yinshuguan 台 湾 商 务 印 书馆. Reprint, 1983.
Zhu Xi. *Zhuzi yulei* 朱子语类 *(Classified Conversations of Master Zhu)*. Taipei: Zhengzhong shuju 正中 书 局. Reprint, 1962.

SECONDARY SOURCES

Bodde, D. (1957). Evidence for "law of nature" in Chinese thought. *Harvard Journal of Asiatic Studies*, 20:709–27.
Bodde, D. (1979). Chinese "law of nature": A reconsideration. *Harvard Journal of Asiatic Studies*, 39:139–55.
Bol, P.-K. (1989). Chu Hsi's [Zhu Xi's] redefinition of literati learning. In: W.T. de Bary and J. W. Chaffee (eds.), *Neo-Confucian Education: The Formative Stage*, 151–85. Berkeley: University of California Press.
d'Elia, P.M. (1960). *Galileo in China*. Cambridge, MA: Harvard University Press.
Drake, S. (1978). *Galileo at Work: His Scientific Biography*. Chicago: University of Chicago Press.
Elman, B.A. (2005). *On Their Own Terms: Science in China*, 1550–1900. Cambridge, MA: Harvard University Press.
Eno, R. (1990). *The Confucian Creation of Heaven*. Albany: State University of New York Press.
Fung, Y.-L. (1937). *History of Chinese Philosophy*. Vol. 1, D. Bodde (trans.). Princeton: Princeton University Press.
Graham, A.C. (1986). *Yin-Yang and the Nature of Correlative Thinking*. Singapore: Institute of East Asian Philosophies.
Grant, E. (1996). *The Foundations of Modern Science in the Middle Ages: Their Religious, Institutional, and Intellectual Contexts*. Cambridge: Cambridge University Press.
Henderson, J.B. (1984). *The Development and Decline of Chinese Cosmology*. New York: Columbia University Press.
Kim, Y.S. (1985). Some reflections on science and religion in traditional China. *Han'guk Kwahak-sa Hakhoe-ji (Journal of the Korean History of Science Society)*, 7:40–9.
Kim, Y.S. (2000). *The Natural Philosophy of Chu Hsi [Zhu Xi] (1130–1200)*. Philadelphia: American Philosophical Society.
Kuhn, T.S. (1957). *The Copernican Revolution: Planetary Astronomy in the Development of Western Thought*. Cambridge, MA: Harvard University Press.
Lamont, H.G. (1973/1974). An early ninth century debate on Heaven: Liu Tsung-yüan's *T'ien Shuo* and Liu Yü-hsi's *T'ien Lun*. *Asia Major*, 18:181–208; 19:37–85.
Lloyd, G.E.R. (1970). *Early Greek Science: Thales to Aristotle*. New York: Norton.

Metzger, T.A. (1977). *Escape from Predicament: Neo-Confucianism and China's Evolving Political Culture.* New York: Columbia University Press.

Murdoch, J.E. (1975). From social into intellectual factors: An aspect of the unitary character of late medieval learning. In: J. E. Murdoch and E. D. Sylla (eds.), *The Cultural Context of Medieval Learning*, 271–339. Boston: Reidel.

Needham, J. (1956). *Science and Civilisation in China,* Vol. 2: *History of Scientific Thought.* Cambridge: Cambridge University Press.

Needham, J. (1962). *Science and Civilisation in China.* Vol. 4, pt. 1: *Physics.* Cambridge: Cambridge University Press.

Peterson, W.J. (1973). Western natural philosophy published in late Ming China. *Proceedings of the American Philosophical Society*, 117:295–322.

Peterson, W.J. (1982). Making connections: The "Commentary on the Attached Verbalizations" in the Book of Change. *Harvard Journal of Asiatic Studies*, 42:67–116.

Porkert, M. (1974). *The Theoretical Foundations of Chinese Medicine: Systems of Correspondence.* Cambridge, MA: MIT Press.

Zhang, Y.-T. (1987). *Mingmo Fangshi xuepai yanjiu chubian* 明末方氏学派研究初编 *(Preliminary Studies on the Fang School in the Late Ming Period).* Taipei: Wenjing.

23 Laws of Nature, Moral Order, and the Intelligibility of the Cosmos

Peter Harrison

CONTENTS

A number of questions about the rise of modern science have preoccupied historians of science for some time now. At the most general level is the issue of why modern science begins in the 17th century and in Western Europe. More specifically, historians have been interested in identifying the distinctive features of Western society that make possible the emergence and persistence of modern science. Equally challenging has been the comparative question of why modern science seems to take off in Europe and nowhere else. Although the latter question has been controversial—Why, critics might ask, should we define science in such a way that restricts its birth to the West?—interesting studies have been conducted that compare the scientific cultures and institutions of China, Islamic societies, and the West.

One partial answer to these questions has been the suggestion that the West had a unique conception of natural order that provided the necessary preconditions for the emergence of modern science. This chapter will explore this argument, drawing some brief comparisons between Western science and Chinese science and suggesting that an important difference lies in their respective conceptions of natural order. The idea of mathematical laws of nature, I will argue, is unique to the early modern West and is underpinned by theological considerations that arise out of Western monotheism.

JESUIT SCIENCE AND CHINESE ASTRONOMY*

In 1687, the Jesuit astronomer Ferdinand Verbiest (1623–1688) described for a European audience the inroads made by Western astronomy in 17th-century China:

> After Astronomy, marching like a venerable queen between the Mathematical Sciences and rising above all of them, had made her entry among the Chinese and had ever since been received by the Emperor with such an amiable face, all the Mathematical sciences also gradually entered the Imperial court as her most beautiful companions … However, the aim of their fervent desire to please was not to keep the Emperor's eyes only upon themselves, but to direct them fully towards the Christian Religion, whose beauty they all professed to worship, in the same way as smaller stars worship the sun and the moon. (Golvers, 1993, p. 101)

---

* Also see the chapters by Yi-Long Huang, Xiao-Chun Sun, and Yung Sik Kim in this volume.

Despite the rhetorical flourishes, Verbeist's celebratory account offers a reasonably accurate portrait of the status of Western astronomy in the imperial court during the late 17th century. In 1669, Verbiest had been appointed director of the Imperial Observatory in Beijing, a prestigious post that the Jesuits were to control for more than a century. Verbiest also became one of the emperor's favorites, tutoring him in the sciences, and during his tenure European astronomy gained wide acceptance among the Chinese literati.*

These spectacular successes were built on the foundations laid by previous generations of Jesuit scholars. Matteo Ricci (1552–1610) had initiated the Jesuits' contact with China in 1583, bringing with him Western maps of the world and various scientific instruments. On the basis of these, Ricci had achieved considerable renown as a mathematician and astronomer, and by the early 17th century he was installed in the Imperial Court in Peking. However, although Ricci had studied in Rome with the famous astronomer Christopher Clavius and indeed had translated parts of Clavius's works into Chinese, he was not primarily an astronomer and lacked the ability to compute the positions of the planets or determine the dates of eclipses. Realizing that Western science, and astronomy in particular, provided a means for extending Jesuit influence in China, Ricci wrote to Rome before his death in 1610 with a request that more astronomers be sent to China and that they bring with them the latest scientific works and instruments. His request was granted, and several able astronomers arrived in Peking to take over his work. Sabatino de Ursis (1575–1620) quickly made an impact by accurately predicting the date and time of the solar eclipse of December 15, 1610. Some years later, and less spectacularly, Johannes Terrentius (1576–1630), a sometime correspondent of both Galileo and Kepler, produced a new and comprehensive astronomical compendium that provided the foundations for calendar reform. The Jesuit (Johann) Adam Schall von Bell (1591–1666) introduced the telescope to China in his treatise *Yüan-ching shuo* 遠鏡說 (*On the Telescope*, 1626; Needham, 1986) and was appointed, in 1664, to the directorship of the Imperial Observatory—the position that Verbiest would subsequently assume.[†]

While the efforts of these individuals enhanced the status of Western astronomy in China, they also engendered considerable resentment. The elevation of Schall, for example, had resulted in the marginalization of the "Muslim" faction within the Observatory, and this exacerbated underlying tensions.[‡] The involvement of Islamic astronomers in the operations of the Observatory dated from the Mongol Yuan dynasty (1279–1368), when Persian and Arabian astronomers had helped establish the Observatory and make it a state-of-the-art institution. During the Ming dynasty (1368–1644) that followed, the initial vigor of the astronomical sciences had gradually waned, and by the time of the arrival of the Jesuits in China, astronomy was ripe for reformation. Not surprisingly, however, the successes of the Jesuits were resented by the incumbent astronomers. The leader of the Muslim group, Yang Kuang-hsien (Guangxian) (1597–1669), accused Schall and his Jesuit colleagues of spreading false religion, teaching erroneous science, and, most seriously, committing treason. In fact, the accusations of scientific incompetence were linked to a more serious political charge. An important element of Yang's case was the claim that Schall had chosen an inauspicious date for the burial of the empress's son—the selection of propitious dates for important events being the responsibility of Imperial astronomers—and that this had ultimately resulted in the deaths of the empress and of Emperor of Shunzhi. A protracted trial began in September 1665, involving tests of rival astronomical systems. Although Schall and Verbiest accurately predicted the solar eclipse of January 16, 1665 (their Chinese rivals were more than half an hour astray), Schall was sentenced to death, his Chinese assistants were executed, and other Jesuits were imprisoned or exiled. On the day scheduled for the execution, however, Peking was visited with a great earthquake, which persuaded the judges to reverse their ruling.

---

* For this and the following three paragraphs, I have drawn on Udias (2003, pp. 37–60); Po-chia Hsia (1998, pp. 186–93); Vogel (2006); and Chu (1999).

[†] The telescope was first mentioned in Chinese writings in 1615, and it was Terrentius's telescope that found its way into the possession of the emperor. See Needham (1986, p. 445).

[‡] On the Muslim and Chinese schools of astronomy, see Udias (2003, p. 38 and p. 44).

In the years that followed, Verbiest worked persistently to reestablish the reputation of Western astronomy, highlighting errors in the existing calendar and seeking to best his rivals in various astronomical competitions. In a specific test of competing astronomical systems, Verbiest challenged Yang Guangxian to predict the length of the shadow of a vertical rod on a particular day at a particular time. Verbiest did this successfully on several occasions before the emperor, while Yang was unable to do so. It was following his successes with these astronomical trials that Verbiest was appointed director of the Imperial Observatory. Eventually, the superiority of Western astronomy was secured, Yang was humiliated and exiled, and Verbeist was installed as director of the Imperial Astronomical Bureau (Udias, 2003, p. 46).* Following his death on March 11, 1688, Verbiest was buried near the graves of Ricci and Schall—on the present campus of the Beijing Administrative College.[†]

In addition to providing some local background for our discussions about the significance of the invention of the telescope and of cosmology, this brief history raises a number of related issues. One issue concerns what we might call the ideological use of astronomy as a means of propagating Western values in general and Christianity in particular. Another issue concerns the notion of *tian* 天 (heaven) and the extent to which it maps onto Western theological conceptions and ideas about natural order. But the two matters that I wish to pursue in more detail are these: (1) the possible relation of the physical sciences, such as astronomy, to human affairs and (2) the more general concern identified at the beginning of this chapter regarding the relative fortunes of Chinese science and Western science—and the eventual victory of the latter.

The first issue is the general question of the relation of celestial order in the heavens to terrestrial order in human societies. In Chinese thought, these connections have found expression in the notions of *tian* (heaven) and *tian ming* 天命 (the mandate of heaven). The apparent alliance between cosmic order and human affairs is evident in the account just provided: the injustice that had led to Schall's condemnation was manifested in the physical disturbances that led to the earthquake. The importance of using the calendar to select auspicious dates similarly suggests an important connection between the order and motions of the heavenly bodies and human affairs. Harmony in the social realm is reflected by harmony in the cosmos. In the West, these connections have been understood in terms of the relations between human laws and divine laws.

The second issue, Why the West and not China?, was the "grand question" of the eminent historian of science and specialist in Chinese science and technology Joseph Needham (1900–1995).[‡] In various ways, this same question has also been explored by scholars as diverse as Weber (2002), Brague (2002), Gaukroger (2006), and, most recently, Huff (2003, 2011).[§]

One approach to this important historical question has been to embark on a comparative exercise and examine other scientific cultures—in particular, Chinese and Islamic civilizations—in order to explore how different sets of cultural values might have had an impact on the development of science in different cultural contexts. It is significant, for example, that Chinese astronomy in the Han dynasty (206 BCE–8 CE) was considerably advanced: astronomers had developed sophisticated star maps and astronomical instruments; they utilized an equatorial-polar reference system; and they entertained the idea of celestial objects floating in an infinite void (Needham, 1986, p. 438). In these respects, Chinese astronomy was considerably superior to the Western astronomy of the same period. Further advances were made during the Yuan dynasty, which, as we have already noted, saw a fertile engagement with Islamic astronomy and the arrival of Persian and Arabian astronomers at the Imperial Court. These earlier periods also bear witness to Chinese ingenuity and technological accomplishment. The Chinese invented printing and paper, the wheelbarrow, the umbrella,

---

* There is some irony here in the fact that the astronomy that the Jesuits brought was essentially Ptolemaic, although they had the advantage of the possession of the Rudolphine tables—updated planetary tables based on the observations of Tycho Brahe and the theoretical heliocentric astronomy of Johannes Kepler.

† See "History of Zhalan" at http://www.bac.gov.cn/webnew/swdx/2en/about/listdetail.aspx?NodeID=25&ID=667.

‡ On the framing of this question, see Yung Sik Kim (2004); also see his chapter in this volume.

§ See also Harrison (1998).

gunpowder, the compass, stirrups, and suspension bridges.* As Voltaire observed in the 18th century, "4,000 years ago, when we could not read, the Chinese knew all the indispensably useful things of which we know today" (Voltaire, 1764, p. 115). With this background in mind, then, our more general question becomes: How did Western science in the 17th century come to overtake Chinese science and technology, given the rather more promising start of the latter?

Not surprisingly, there are a number of theories about how cultural differences of various kinds have played out in the development of science, and some have suggested that this is not even a question that we can sensibly ask. However, one influential line of thought refers us back to the first question that I raised, concerning different understandings of the relation between heavenly order in the physical realm and social order in the human realm. In the West, thinking about these relations gave rise to the idea of "laws of nature," and it has been argued, plausibly in my view, that *part* of what made modern science possible in the West was a new conception of the "laws of nature" that originally took as its point of departure the idea of divinely ordained moral laws. Thus, it is to exploring this issue that I turn next.

MORAL LAWS AND LAWS OF NATURE

The medieval West had a conception of natural laws, but these were typically understood to be universal *moral* laws, which human beings became cognizant of on account of their rationality. Thomas Aquinas (1225–1274) explained this as an analogy to a ruler who promulgates laws in his kingdom, whereas God enacts eternal law. Natural law (*Lex naturalis*) is simply the imprinting of that eternal law on human minds, which leads to participation of human agents in this eternal law (Aquinas, 1265–1274, 1a2ae. 91, 2). Specific human laws (*Lex humana)* and the law of nations (*Ius gentium*) are derived from this natural law. All of this was premised on the commonsense idea that only rational agents with a will can obey laws.

By way of contrast, the regularities observed in nature were understood not in terms of universal laws, but rather as arising out of the inherent natures of individual things (in keeping with Aristotle's understanding of motion and change and his conception of material, formal, efficient, and final causes). This Aristotelian position could be given a theological gloss as, for example, when Aquinas speaks of "the order *that God has implanted* in nature" (Aquinas, 1259–1264, 3, p. 100).† This order was not invariant or deterministic, for the natural powers of things could sometimes miscarry. Hence, the course of nature was accordingly understood as that which happened usually or "for the most part" (Aquinas, 1259–1264, 3, p. 99).

In the later Middle Ages, developments in both theology and natural philosophy led to modifications of this understanding of nature. Already in the 13th century, reactions against certain Aristotelian doctrines had led to speculations about how it was possible for God, on account of his omnipotence, to overrule the internal tendencies of things that, according to Aristotle, were essential to their natures (Piché, 1999). An increasing emphasis on the part of certain theologians on the omnipotence of God and on the primacy of His will promoted discussions of ways in which God might directly intervene in nature. In the early modern period, these tendencies culminated in the idea that God directly imposed His will on nature in the form of natural laws. This view was reinforced by a new matter theory—the corpuscular hypothesis, or what we would come to call atomic theory—according to which nature was made up of minute and inert (i.e., causally impotent) particles. At the same time, the idea that the world was like a machine was rapidly overtaking the older idea that the world was like an organism as the predominant model for understanding the operations of nature. In this new understanding of the natural world, "laws of nature" were for the first time conceptualized as regularities that God had imposed directly on the natural world. Henceforth, scientific explanations of natural events would be couched not

---

* For a complete list, see Simon Winchester's (2008) biography of Joseph Needham, pp. 267–78.
† See also Aquinas (1265–1266, q. 3, a. 8, ad 2).

in terms of the inherent causal properties of things, but in terms of universal laws that have been externally imposed on an essentially passive matter.*

One of the best-known examples of early modern formulations of laws of nature is Johannes Kepler's (1571–1630) laws of planetary motion. Kepler's third law, for example, states that the square of the orbital period of a planet (the time it takes to orbit the Sun) is proportional to the cube of its semimajor axis (its mean distance from the Sun). Kepler was quite explicit about the fact that his proposal that the planets are governed by mathematical laws was informed by, and indeed grounded in, specific theological convictions. He acknowledged also that his novel formulation of mathematical rules for describing the actual motions of the planets would attract the ire of traditional natural philosophers. "I shall have the physicists against me in these chapters," he confesses, "because I have deduced the natural properties of the planets from immaterial things and mathematical figures." The "natural properties" of things, as we have already noted, had been usually understood in terms of their intrinsic qualities, rather than in terms of mathematically imposed laws. Kepler's response to his imagined critics was to point out that "God the Creator, since he is a mind, and does what he wants, is not prohibited, in attributing powers and appointing circles, from having regard to things that are either immaterial or based on imagination" (Kepler, 1621, p. 123). Kepler also suggested that the reason Aristotelian science had overlooked the possibility of a divine imposition of mathematical order was because of Aristotle's conviction that the world was eternal and that the Aristotelian God did not create or impose order on the world. By way of contrast, Kepler pointed out, "our faith holds that the World, which had no previous existence, was created by God in weight, measure, and number, that is in accordance with ideas coeternal with Him" (Kepler, 1619, p. 115 and p. 146).[†]

It is possible to interpret Kepler's conception of a world based on mathematical ideas, which are "coeternal with God," as an endorsement of a kind of Christian Platonism. The Greek philosopher had famously held that the temporal and imperfect material world that we detect with the senses is based on an eternal and perfect world of ideas, accessible only by reason. But the notion of laws of nature was equally compatible with the decidedly non-Platonic idea that the universal mathematical truths were not eternally true, but true only by virtue of God's willing them. This was the view of the French philosopher René Descartes (1596–1650), who contended that all logical and mathematical truths derived from the will of God. In 1630, for example, Descartes wrote to his friend Marin Mersenne, arguing that "the mathematical truths, which you call eternal, were established by God and totally depend on him just like all the other creatures" (Cottingham et al., 1991, 3, p. 23).[‡] The early modern idea of laws of nature is therefore not a revival of an ancient Platonic conception of nature, but a radically new idea about how God directly exercises His will in the natural realm.

Descartes played a major role in inaugurating this new idea and in linking it to the divine will. In his *Principles of Philosophy* (1644), he identified three "laws of nature," which he described as constant and immutable. They derive these characteristics, he insisted, from their divine author, who likewise is said to be constant and immutable (Cottingham et al., 1985, 1, p. 240 and p. 286). English natural philosophers were to follow suit. Robert Boyle, the "father of chemistry" and discoverer of the eponymous gas law, wrote about "laws of motion" that "did not necessarily spring from the nature of matter, but depended on the will of the divine author of things" (Boyle, 1772, 5, p. 521). The Newtonian Samuel Clarke wrote in the same vein that "the *Course of Nature*, cannot possibly be any thing else, but the *Arbitrary Will and pleasure of God* exerting itself and acting upon Matter continually" (Clarke, 1738, 2, p. 698).

The idea that there are laws of nature—a fundamental assumption of much Western science—thus rested initially on the notion of divinely imposed ordinances, a notion that had been transposed

---

* On the emergence of the idea of laws of nature see Henry (2004), Harrison (2008a), Milton (1998), and Steinle (1995).

† The reference to "weight, measure, and number" comes from the Book of Wisdom 11:12, a favorite passage of Augustine. See *The Trinity* XI.iv; *The Literal Meaning of Genesis* IV, 3, 7–12; *Confessions* V.iv.7; *Answer to an Enemy of the Law and the Prophets* I, 6, 8; *Free Will* III, 12, 35.

‡ See also Descartes, *Objections and Replies*, in Cottingham et al. (1984, 2, p. 29).

from the moral realm to the physical. The necessity and universality of laws of nature were attributed to the immutability of their divine source, God. Over the next few centuries, some thinkers reversed the order of reasoning that had given rise to the idea of laws of nature, arguing from the existence of physical laws to the existence of moral laws. Already in the 16th century, the Protestant reformer Philipp Melanchthon (1497–1560) stressed the affinity between the order and lawfulness apparent in the heavens and the political and moral order of the human realm. "Like the order of the motions of the heavens," he wrote, "so too the whole of this political order, the bond of marriage, empires, the distinction of states, contracts, judgements, punishments, indeed all most true statutes originate from God" (Melanchthon, 1535, XI, p. 912).* Melanchthon's ideas provided a major theological incentive for the study of astronomy and had an important influence on the development of Johannes Kepler's conceptions of a divinely imposed celestial order (Barker and Goldstein, 2001, p. 25; Methuen, 1998, p. 209; Barker, 1997, p. 360).†

Indeed, in certain respects, the idea that the cosmic order portends moral order has its roots in antiquity. Plato had suggested that the lover of wisdom who becomes familiar with the divine order in the cosmos "will himself become orderly and divine" (Plato, *Republic* VI 500d). Hence, one reason for studying the regularities of nature was to effect personal transformation. This is in keeping with a classical conception of philosophy—including natural philosophy or what we would now call "science"—which held the basic philosophical task to be that of moral or spiritual formation (Hadot, 2002). It is not particularly surprising, then, to find the most influential astronomer of antiquity making a similar claim. Claudius Ptolemy (90–168) insisted that studying the mathematical regularities of the heavens "makes its followers lovers of this divine beauty, accustoming them and reforming their natures, as it were to a spiritual state" (Toomer, 1998, p. 37).

These connections between moral and cosmic order appear again in the moral philosophy of Immanuel Kant (1724–1804), albeit in a rather attenuated fashion. As is well known, Kant sought to find a rational basis for a universal *moral* law that was analogous to the universal *physical* law of gravitation that Newton had discovered in the heavens. This connection was expressed in the celebrated juxtaposition set out in the *Critique of Practical Reason* (1788): "Two things fill the mind with ever new and increasing admiration and awe, the more often and steadily we reflect upon them: the starry heavens above me and the moral law within me" (Guyer, 1992, p. 1). Neither should we forget that while Kant is best known for his philosophical achievements, he also made major contributions to speculative cosmological theory, and while his scientific ideas had their limitations, the nebular hypothesis that he outlined was in a sense the first statement of modern evolutionary cosmology.‡ However, by the time of Kant, the connection between moral and cosmic order was based on a loose analogy. Kant was attempting to effect a kind of Newtonian reformation in the realm of moral philosophy, but he did not suggest that the former is directly relevant to the latter.

In light of this discussion about the emergence of the idea of laws of nature, one answer to the question about the rise of science in the West—the question articulated by Joseph Needham and, more recently, by Remi Brague, Stephen Gaukroger, and Toby Huff—has to do with these theological ideas that gave rise to the distinctive notion of "laws of nature." As it turns out, this is not too far from the answer that Needham himself gave. He wrote:

> The available ideas of a supreme being, though certainly present from the earliest times, failed to retain enough personalized creativity to allow [for] the development of the conception of precisely formulated abstract laws ordained from the beginning by a celestial lawgiver for non-human Nature, and capable, because of his rationality, of being deciphered or re-formulated by other lesser rational beings. (Needham, 1951, p. 230)

---

* See also Barker and Goldstein (2001, p. 95).
† See also Kepler (1619, p. 146), (1600–1601, p. 144).
‡ See the essays in Butts (1986).

To summarize, the conception of a divine celestial law-giver never developed in China. As a consequence, they lacked a conception of laws of nature and did not develop an understanding of the natural world that was similar in structure to Western science.

Needham's account is plausible as far as it goes. However, I want to suggest that part of what was distinctive about the intellectual background of the West was not just the idea of "laws of nature" or even, more generally, the presupposition of the rationality of the cosmos and its transparency to the human intellect. While the notion of laws of nature is exclusive to the early modern West, these more general assumptions may be found, in various guises, in Pythagoreanism, Platonism, Stoic thought, and to some extent in the Chinese conception of *tian*. In addition to this assumption of the underlying rationality of the Universe, there are two crucial constraining principles, one to do with the role of divine choice, the other to do with the perceived limitations of the human mind.

HUMAN KNOWLEDGE AND NATURE'S LAWS

As we have seen, central to the idea of laws of nature is the notion of a divine legislator. Equally important, however, is the idea that the divine legislator is not constrained in the choices He might make when instantiating a particular cosmic order. This means that whatever we might intuit about the order of the cosmos, such rational intuitions will of themselves be insufficient and will invariably need to be supplemented by empirical investigation.* Descartes (who is often wrongly imagined to be an "armchair scientist") put it this way: "Since there are countless different configurations which God might have instituted here, experience alone must teach us which configurations He actually selected in preference to the rest" (Cottingham et al., 1985, 1, p. 256). In the preface to the second edition of Newton's *Principia* (written by Roger Cotes), we find a similar sentiment:

> … the business of true philosophy is to derive the natures of things from causes truly existent, and to inquire after those laws on which the Great Creator actually chose, to found this most beautiful Frame of the World, not those by which he might have done the same, had he so pleased. (Cotes, 1713, p. 393)[†]

Hence, while we can be assured that there *is* a rational pattern to nature, on account of the choices available to the divine legislator we still need to engage in careful empirical investigation in order to find out what that specific pattern is.

Consider how such a constraint might work in practice. A belief in the rationality and mathematical intelligibility of nature is completely consistent with, and indeed might well promote, the long-held assumption that planetary orbits will be perfect circles. It is the assumption of a Creator who exercises a degree of arbitrary choice in His designs within a range of options that enables speculation about the possibility of elliptical orbits. The idea of divine choice, as well as the conviction that it makes a difference in how we approach the study of nature, is explicit in the positions of Descartes, Boyle, Newton, and indeed many early modern natural philosophers.

The second constraining principle is a belief that the human mind and the human senses are somewhat limited in what they can know. According to a standard Western reading of human origins, the first human beings as created by God once enjoyed a perfect intuition of the structure of the world and its operations, but they lost this as a consequence of human sin. In this, the Christian idea of the fall, the first man was imagined to have had a complete knowledge of nature, on account of both his rational capacities and perfect sense organs (Harrison, 2007). As Martin Luther put it in his commentary on Psalm 127:

> Here it appeareth that at the beginning there was planted in man by God himself, a knowledge of husbandry, of physicke, and of other artes & sciences; Afterward men of excellent wit by experience & great diligence did increase those gifts which they had by nature. And this is but the strength of humane wisedome created in man at the beginning in Paradise. (Luther, 1577, p. 129)

---

* This is the so-called voluntarism and science thesis. See Klaaren (1977), Heimann (1978), and Osler (1994). For some reservations about this thesis see Harrison (2002).

† Paradoxically, this statement appears in the context of an attack on Descartes' philosophy.

The idea of a virtually omniscient Adam was commonplace during the 16th and 17th centuries, and the scientific enterprise was often understood as aiming to restore, at least in part, the scientific knowledge that Adam had possessed before his fall from grace. Francis Bacon (1561–1626) thus declared,

> For man by the fall fell at the same time from his state of innocency and from his dominion over creation. Both of these losses however can even in this life be in some part repaired; the former by religion and faith, the latter by arts and sciences. (Bacon, 1620, *Novum Organum* II, § 52)

A key motivation for the pursuit of science during this period, then, was the conviction that by engaging in scientific activity, the human race would gradually repossess a natural knowledge once lost, and by virtue of these gains, reestablish dominion and control over the natural world.

As part of this intriguing set of beliefs, a number of Western thinkers, up until the 18th century, thought that Adam had enjoyed the advantage of a kind of "telescopic vision" and that his knowledge of astronomy had been the equal of the best that modern astronomy could offer. Luther, again, contended that before his fall, Adam "could have seen objects a hundred miles off better than we can see them at half a mile, and so in proportion with all the other senses" (Luther, 1569, p. 57). Joseph Glanvill, an early member of the Royal Society, agreed that Adam had not needed "Galileo's tube"—the telescope—to have had knowledge of the heavenly bodies (Glanvill, 1661, p. 1 and p. 5). In this understanding of things, scientific instruments such as the telescope were prosthetic devices that enabled the human race to recapture knowledge of the world that they had once possessed through their natural sensory endowments alone.

More generally, new instruments and experiments, along with the idea that natural science should be a corporate and cumulative activity, were justified by appeals to the fallen, and hence weakened, condition of human bodies and minds. The conviction that knowledge could be improved was thus premised on a belief in the present mediocrity of human knowledge and of human knowledge-making capacities. As Robert Hooke, first curator of experiments at the Royal Society, expressed it in the preface of *Micrographia* (1665),

> … every man, both from a deriv'd corruption, innate and born with him, and from his breeding and converse with men, is very subject to slip into all sorts of errors. … These being the dangers in the process of humane Reason, the remedies of them all can only proceed from the real, the mechanical, the experimental Philosophy [i.e. the new science]. (Hooke, 1665, preface)

Thus was new experimental science commended as a palliative for human sin and the natural ignorance that was the consequence of that sin.

There are, then, two crucial constraining principles that modify the more general assumption about the intelligibility of the Universe: first, that God exercised his choice in the creation of a particular natural order; and second, that we need to be skeptical about the knowledge generated by fallen human minds. Together these assumptions promoted the idea that science is a difficult business, that we need to focus on small problems, and that science can aspire to success only if it is a disciplined, collective, and cumulative activity, assisted by instruments and experiments.

RELIGIOUS JUSTIFICATIONS OF WESTERN SCIENCE

The argument to this point has been that the idea of laws of nature is a distinctively Western idea and that it was one factor that led to the emergence of modern science. Moreover, the assumption of a natural correspondence between the lawfulness of the cosmos and the rational structures of the human mind was qualified by a commitment to the myth of a primeval fall and to the necessity of a critical approach to human knowledge. Once there had been a perfect reciprocal relation between the mind and the cosmos; however, this had been distorted by human sin. The role of experimental science was to attempt to bridge this gap and partially restore this original correspondence. Francis

Bacon, who in essence set out the justifications for the new approach of experimental science of the 17th century, thus wrote that the goal of the new science was to see

> … whether that commerce between the mind of man and the nature of things … might by any means be restored to its perfect and original condition, or if that may not be, yet reduced to a better condition than that in which it now is (Bacon, 1620, 4, n. 7).

At its birth, modern science was thus conceptualized as a set of remedial practices directed toward a reestablishment of the natural bond between the reason of the mind and the rationality of the cosmos. Science was a restorative exercise.

It should be acknowledged that some tension exists between these Baconian justifications for experimental science and the kind of justifications that we see in the writings of someone like Kepler. These tensions have at times been expressed in terms of a standard distinction between rationalism (which stresses the power and primacy of human reason) and empiricism (which emphasizes the importance of observation and experiment). These two traditions sit somewhat uneasily together in the work of someone like Newton, who, in spite of his spectacular successes, never really managed to resolve this essential tension between a commitment to the inherent rationality of the Universe on the one hand and human incapacity to grasp the mind of God on the other. Nevertheless, modern science seems to have been forged in the creative tension between a kind of optimistic rationalism and a critical empiricism.

There is perhaps one more thing to be said about the Baconian ideas, and that is that they provided the new sciences with a vital source of religious legitimation. Whereas we now tend to think that the practice of science is something that is self-evidently a good thing, it was by no means as clear-cut in the 17th century, when modern science was first emerging. The situation with regard to new science becomes apparent when we consider the controversy generated by the new experimental philosophy and by robust criticisms that were leveled against the fledgling Royal Society. A key criticism was this: What use is this new science? How is this knowledge socially useful? It is important to understand that during the course of the 17th century, science had yielded few practical or technological benefits, and even if it had, there was still a question of whether the provision of material comforts was an appropriate occupation for serious thinkers.

One vocal critic of the Royal Society, Henry Stubbe (1631–1676), thus complained that the new science was utterly incompetent to provide "that Moral discipline which instructs us in the nature of virtue and vice" (Stubbe, 1670a, p. 14). The derided philosophy of Aristotle, by way of contrast, was capable of promoting moral formation. For Stubbe, the new science consisted only in a crude "mechanical education" and such undignified pursuits as *planting of Orchards, making of Optick Glasses, magnetic and hortulane Curiosities*" (Stubbe, 1670b, p. 13). Others voiced similar concerns over what they regarded as the unsophisticated and unfulfilled utilitarianism of the mechanical and experimental sciences.*

What these criticisms suggest is that the new science needed not only discoveries and new applications, but also new social attitudes that valued those discoveries and their practical applications. The Baconian program provided just the kind of justification that the new natural philosophy needed by stressing that science provides the means by which human beings can regain the mastery of nature that they lost at the fall. The idea of science as a kind of redemptive process was vital for establishing its religious legitimacy in the 17th century and arguably gave modern science the religious and moral foundations that established its enduring importance as a central feature of Western society. The remarkable success of science in the West is often assumed simply to be the result of its being self-evidently the right way to pursue knowledge and of the practical benefits that it confers. What the history suggests is that modern science owes its success not merely to the brilliance and ingenuity of its first practitioners, but also to the emergence of a set of values, underpinned by religious considerations, that ensured it a permanent and central place in the societies of the modern West.

---

* On the Royal Society and its critics, see Harrison (2008b).

CONCLUSION

In the course of this chapter, I have considered two "big questions"—questions that I believe history can shed some light on. We have discussed the assumptions that make science *possible* along with the values that make science *desirable*. The tentative answers offered in this chapter have both involved reference to the unique religious landscape of the early modern West and hence shed light on a third question, implicit in the first two: Why the West?

On the first question, as we have seen, part of what makes science possible is the theologically informed assumption that there are laws of nature, promulgated by God and discoverable by human minds. The scientific program, however, requires more than just rational intuition because of the range of possible rational orders in the cosmos and, equally important, because of the limitations of the human mind. Science is thus made possible by the idea of an intelligible order of nature, along with a critical skepticism that demands that we constantly interrogate our methods and revise our findings. While science assumes that some ultimate rational truth about the cosmos lies out there, at the same time it also generally accepts that our versions of that truth at any one time will be partial, fragmentary, and provisional. These features of the modern scientific landscape are vestiges of its religious origins. Of course, in the years that have elapsed between the 17th century and now, we have largely lost sight of its original theological justifications. Nonetheless, intriguing questions remain about the extent to which we are warranted in still assuming that there are, for example, laws of nature, given that relatively few scientists still subscribe to the theological ideas that provided the foundation for these assumptions. Some have argued, more or less on these grounds, that we ought to abandon the classical notion of laws of nature (Cartwright, 2005).

The second question, regarding the value of science, is also often overlooked because the virtues of science seem so obvious. One way of getting some perspective on this question is to think about the old joke of the definition of the gentleman: a gentleman is someone who can play the bagpipes but chooses not to. *Mutatis mutandis*, in the case of the history of science, not becoming a scientifically advanced culture may not be a matter just of lacking the relevant capacities. It may also be a matter of choosing different cultural priorities—that is, of endorsing the position: yes, we can do science if we want to, but we believe other things to be equally or more important. Western historians have sometimes proceeded on the assumption that Western science serves as a measure of its superiority, but such a view rests on a number of uncritical assumptions about cultural values and priorities.

In thinking about this question, the definition of "the gentleman" actually turns out to be a historically relevant consideration. In 17th-century England, one important concern was whether experimental science was a suitable activity for a scholar and a gentleman. As we have seen, educational priority had traditionally been placed on the moral sciences rather than on the natural sciences. The former were regarded as the most useful both for individual edification and for society as a whole. The eventual success of science was achieved by appealing to the kinds of religious values identified by Francis Bacon, who emphasized the importance of world mastery over self-mastery and who pointed to the redemptive value of exercising that mastery over nature.*

These distinctive features of Baconian science, along with the factors that ensured its longevity, also point to key cultural and religious differences between China and early modern Western Europe and to why, comparatively speaking, scientific activity gained such an impetus in the West. Confucianism and Taoism, in different ways, acknowledge the order of nature. But, oversimplifying the matter somewhat, in the case of the former, priority is placed on social order and moral cultivation; in the case of the latter, the emphasis is on living in conformity to the natural state of affairs. Neither approach promotes a skeptical attitude to our commonsense intuitions about the world, and neither provides a justification for the pursuit of mastery over nature.

---

* This also makes for an interesting comparison with the Confucian concept of "the Gentleman" (*Chun-tzu*), for whom the cultivation of an inner natural virtue (*jen*) is the first priority.

To conclude, then, scientific success is not merely a matter of having the necessary presuppositions about the rationality of the cosmos, or knowledge of the requisite methods to uncover that rationality, or sufficiently brilliant minds capable of implementing those methods. The long-term success of science requires also a set of social values that promotes the goals of the sciences. Not only does a society need to be able to do science, it must wish to do so and be prepared to sacrifice other priorities in order to do so. The 17th century was the period that witnessed the first articulation of a set of values that gave priority to the natural sciences. Interestingly, it was also at this time that the quest for cosmic order first became disengaged from the quest for moral order.

REFERENCES

Aquinas, T. (1265–1266). *Quaestiones disputatae de potentia dei (Disputed Questions on the Power of God)*. English Dominican Fathers (trans.). London: Burns, Oates & Washbourne (1932).

Aquinas, T. (1259–1264). *Summa contra Gentiles*. English Dominican Fathers (trans.). 5 Vols. London: Burns, Oates & Washbourne (1934).

Aquinas, T. (1265–1274). *Summa theologiae*. English Dominican Fathers (trans.). London: Eyre and Spottiswoode (1964–1976).

Bacon, F. (1620). *The Great Instauration*. In: J. Spedding, R. Ellis, and D. Heath (eds.), *The Works of Francis Bacon*, Vol. 4. London: Longmans, Green and Co. (1857).

Barker, P. (1997). Kepler's epistemology. In E. Kessler, D. Di Liscia, and C. Methuen (eds.), *Method and Order in the Renaissance Philosophy of Nature*, 354–68. Aldershot: Ashgate.

Barker, P. and Goldstein, B. (2001). Theological foundations of Kepler's astronomy. In J.H. Brooke, M.J. Osler, and J. van der Meer (eds.), *Science in Theistic Contexts*, 88–113. Chicago: University of Chicago Press.

Besterman, T. (trans.). (1972). Voltaire, Christianity. *Philosophical Dictionary*. New York: Penguin Classics [orig. 1764].

Boyle, R. (1772). *The Christian Virtuoso*. In: T. Birch (ed.), *The Works of the Honourable Robert Boyle*, 6 vols. Hildesheim: Olms. Reprint, 1966.

Brague, R. (2002). *Eccentric Culture: A Theory of Western Civilization*. Chicago: St. Augustine's Press.

Butts, R.E. (ed.) (1986). *Kant's Philosophy of Physical Science*. Dordrecht: Reidel.

Cartwright, N. (2005). No God; no laws. *Dio, la Natura e la Legge. God and the Laws of Nature*. Angelicum-Mondo X:183–90.

Chu, P. (1999). Trust, instruments, and cross-cultural scientific exchanges: Chinese debate over the shape of the Earth, 1600–1800. *Science in Context*, 12:385–411.

Clarke, S. (1738). *The Works of Samuel Clarke*. D.D. London.

Cotes, R. (1713). Preface to the second edition of the *Principia*. In: *Isaac Newton: The Principia*. I. B. Cohen and A. Whitman (trans.). Berkeley: University of California Press (1999).

Cottingham, J., Stoothoff, R., Murdoch, D., et al. (trans.) (1984–1991). *The Philosophical Writings of Descartes*, 3 vols. Cambridge: Cambridge University Press [Vol. 1, 1985; Vol. 2, 1984; Vol. 3, 1991].

Gaukroger, S. (2006). *The Emergence of a Scientific Culture: Science and the Shaping of Modernity*. Oxford: Clarendon.

Glanvill, J. (1661). *The Vanity of Dogmatizing*. London.

Golvers, N. (1993). *The "Astronomia Europaea" of Ferdinand Verbeist, S.J.* [1687]. Nettetal: Steyler.

Guyer, P. (1992). Introduction: The starry heavens and the moral law. In: P. Guyer (ed.), *The Cambridge Companion to Kant*, 1–25. Cambridge: Cambridge University Press.

Hadot, P. (2002). *What Is Ancient Philosophy?* Cambridge, MA: Harvard University Press.

Harrison, P. (1998). *The Bible, Protestantism and the Rise of Natural Science*. Cambridge: Cambridge University Press.

Harrison, P. (2002). Voluntarism and early modern science. *History of Science*, 40:63–89.

Harrison, P. (2007). *The Fall of Man and the Foundations of Science*. Cambridge: Cambridge University Press.

Harrison, P. (2008a). The development of the concept of laws of nature. In: F. Watts (ed.), *Creation: Law and Probability*, 13–36. Aldershot: Ashgate.

Harrison, P. (2008b). Religion, the Royal Society, and the rise of science. *Theology and Science*, 6:255–71.

Heimann, P. (1978). Voluntarism and immanence: Conceptions of nature in eighteenth-century thought. *Journal of the History of Ideas*, 39:271–83.

Henry, J. (2004). Metaphysics and the origins of modern science: Descartes and the importance of laws of nature. *Early Science and Medicine*, 9:73–114.

Hooke, R. (1665). *Micrographia*. London.

Huff, T. (2003). *The Rise of Early Modern Science: China, Islam and the West*, 2nd edn. Cambridge: Cambridge University Press.

Huff, T. (2011). *Intellectual Curiosity and the Scientific Revolution: A Global Perspective*. Cambridge: Cambridge University Press.

Kant, I. (1788). *Critique of Practical Reason*. In: *Critique of Practical Reason and Other Works on the Theory of Ethics*. T.K. Abbott (trans.). London: Longmans, Green and Co. (1909).

Kepler, J. (1600–1601). *Apologia Pro Tychone contra Ursum*. In: N. Jardine (ed.), *The Birth of History and Philosophy of Science*. Cambridge: Cambridge University Press (1988).

Kepler, J. (1619). *Harmonice mundi (The Harmony of the World)*. E. J. Aiton, A. M. Duncan, and J. V. Field (trans. and intro.). Philadelphia: American Philosophical Society (1997).

Kepler, J. (1621). *Mysterium Cosmographicum*. A. M. Duncan (trans). Norwalk, CT: Abaris (1981).

Kim, Y.S. (2004). The "why not" question of Chinese science: The scientific revolution and traditional Chinese science. *East Asian Science, Technology, and Medicine*, 22:96–112.

Klaaren, E. (1977). *Religious Origins of Modern Science*. Grand Rapids: Eerdmans.

Luther, M. (1577). *A Commentarie vpon the Fiftene Psalmes*. London.

Luther, M. (1569). *Table Talk*. William Hazlitt (trans). Philadelphia: Lutheran Publication Society (1848).

Melanchthon, P. (1535). *De legibus. Corpus Reformatorum*. In: C. Methuen (2000). *Lex Naturae and Ordo Naturae. Reformation and Renaissance Review*, 3:110–25.

Methuen, C. (1998). *Kepler's Tübingen*. Aldershot: Ashgate.

Milton, J.R. (1998). Laws of nature. In: D. Garber and M. Ayers (eds.), *The Cambridge History of Seventeenth Century Philosophy*, vol. 1, 680–701. Cambridge: Cambridge University Press.

Needham, J. (1951). Human laws and the laws of nature in China and the West. *Journal of the History of Ideas*, 12:3–32, 194–230.

Needham, J. and Wang, L. (1959). *Science and Civilization in China*, Vol. 3: *Mathematics and the Sciences of the Heavens and the Earth*. Cambridge: Cambridge University Press.

Osler, M. (1994). *Divine Will and the Mechanical Philosophy: Gassendi and Descartes on Contingency and Necessity in the Created World*. Cambridge: Cambridge University Press.

Piché, D. (1999). *La Condamnation Parisienne de 1277*. Paris: Vrin.

Po-chia Hsia, R. (1998). *The World of Catholic Renewal: 1540–1770*. Cambridge: Cambridge University Press.

Steinle, F. (1995). The amalgamation of a concept—laws of nature in the new sciences. In: F. Weinert (ed.), *Laws of Nature: Essays on the Philosophical, Scientific and Historical Dimensions*, 316–68. Berlin: De Gruyter.

Stubbe, H. (1670a). *Campanella Revived*. London.

Stubbe, H. (1670b). *Plus Ultra Reduced to a Non Plus*. In: *Legends no histories, or, A specimen of some animadversions upon The history of the Royal Society…together with the Plus Ultra Reduced to a Non-Plus*. London.

Toomer, G.J. (trans). (1998). C. Ptolemy, *Almagest*. Princeton: Princeton University Press.

Udias, A. (2003). *Searching the Heavens and the Earth: The History of Jesuit Observatories*. Dordrecht: Springer.

Vogel, K.A. (2006). European expansion and self definition. In: K. Park and L. Daston (eds.), *The Cambridge History of Science, Vol. 3: Early Modern Science*, 818–39. Cambridge: Cambridge University Press.

Weber, M. (2002). *The Protestant Ethic and the Spirit of Capitalism*. New York: Penguin Classics [orig. 1905].

Winchester, S. (2008). *The Man Who Loved China*. New York: HarperCollins.

24 Why Are the Laws of Nature as They Are? What Underlies Their Existence?

George F. R. Ellis

CONTENTS

THE EXISTENCE OF THE LAWS OF NATURE*

A deep issue in both cosmology and human life is, *What underlies the existence of the laws of nature?* These laws define the possibility space within which the Universe and life come into being. We want to understand the existence and nature of causal laws that allow true complexity, such as ourselves—our bodies and minds, our actions and thoughts and emotions—to come into being. These laws are simply taken for granted in most biological and evolutionary discussions; but in the context of fundamental discussions concerning cosmology and life, one is entitled to query both their nature and their existence, for they shape the nature of physical existence and are the foundational reasons that any life whatever is possible. No life, and no biological evolutionary processes of any kind, would be possible if these laws were substantially different.

The fundamental laws of nature are based in mathematically describable variational principles for physical entities[†] invariant under symmetry groups and subject to quantum mechanical dynamics (Cottingham and Greenwood, 2007). The coupling constants that relate resulting forces and particles are such that complexity, including consciousness, can emerge out of these basic physical constituents; this is *a priori* a highly improbable state of affairs (Davies, 1982; Barrow and Tipler, 1986; Balashov, 1991; Rees, 2000, 2001).

A series of fundamental issues arise:

- Why do any such laws exist at all, and why do they have the nature they have, leading to our physical and mental existence?
- Is the ultimate reason pure happenstance, probability, necessity, or purpose?
- What is the nature of their existence—is it prescriptive or descriptive?

THE HIERARCHY OF COMPLEXITY

Our cosmological context (Harrison, 2000; Silk, 2005; Ellis, 2006) is an evolving Universe that is initially structureless, but eventually physical processes lead to the existence of galaxies, stars, and planets. The emergence of true complexity, including living beings such as ourselves, occurs on some suitable planets imbedded in this larger context. Complexity arises in modular hierarchical structures (Simon, 1982; Peacocke, 1989; Flood and Carson, 1990; Booch, 1994; Scott, 1995; Campbell and Reece, 2005; Ellis, 2008b), which underlie biological function, as indicated in Table 24.1. These structures emerge from the underlying physical basis in three very different, but interrelated ways:

TABLE 24.1

The Hierarchy of Structure and Complexity Underlying Human Existence Is Characterized by the Various Sciences Appropriate for Studying Each Level

Sociology/Economics/Politics

Psychology

Botany/Zoology/Physiology

Cell Biology

Biochemistry

Physical Chemistry

Atomic Physics

Nuclear Physics

Particle Physics

---

* Also see the chapter by Peter Harrison in this volume.
[†] One considers a quantity (the "action"), depending on the path from an initial to the final point, and determines on which path this quantity is either a minimum or maximum as one varies the path.

- *In evolutionary terms*, on very long timescales (>10^6 years). There was no life in the Universe when it was 300,000 years old or on Earth 4 billion years ago; living beings came into existence through an evolutionary process that lasted billions of years and generated complexity where none had existed before.
- *In developmental terms*, on medium timescales (hours to 10^2 years). Each higher-level animal, including human beings, starts off as a single living cell and then develops into a multicellular organism (~10^{13} interacting cells) through developmental processes that create functioning physiological structures (Gilbert, 2006).
- *In functional terms*, on short timescales (milliseconds to minutes). We are each made up of inanimate objects (protons, neutrons, and electrons) that are fashioned into physiological systems that together make a living being. Life emerges through the ongoing cooperation of these basic physical entities (Campbell and Reece, 2005).

Once life has been initiated, the developing organisms must continually keep functioning effectively at each stage of the developmental and evolutionary processes, despite the enormous changes in structure and complexity that occur as complexity develops. Thus, the functional processes evolve on both evolutionary and developmental timescales.

I suggest that at each level of the hierarchy of complexity, universal principles apply, describable as effective laws applicable at that level. Each level exists in its own right, even though it is based in lower levels. The way the lower-level laws work out is shaped by the higher-level contexts in which they act, deriving from action at a lower level, but with their own autonomy of operation, leading to effective laws at each level. These effective laws control what happens at each level in a way independent of time and place and independent of our understandings and descriptions, hence they may be thought of as having an existential reality describable in Platonic terms.* This is all possible because both bottom-up and top-down causation take place in the hierarchy of complexity.

BOTTOM-UP ACTION

A fundamental understanding we have attained through 300 years of scientific endeavor is that all complex objects are made of the same basic materials (atoms fashioned into molecules), with microforces determining what happens at the lower levels, and hence underlying properties at the higher levels. *Bottom-up action* is when what happens at the higher levels in the hierarchy is controlled by what happens at the lower levels (Figure 24.1, left). Examples are abundant:

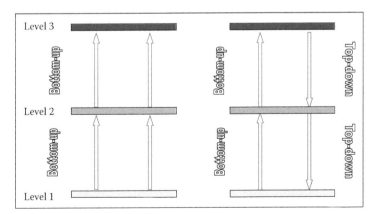

FIGURE 24.1 *Left:* Bottom-up causation. *Right:* Bottom-up and top-down causation. The fundamental importance of top-down causation is that it changes the causal relation between upper and lower levels in the hierarchy of structure and organization.

* A Platonic kind of existence is an existence in an abstract space, separate from physical reality.

- Microphysics underlies macrophysics, e.g., the kinetic theory of gases and the theory of solids, determining properties of gases (such as the temperature-pressure relation) and solids (such as electrical conduction and thermal capacity) from the underlying material properties (Goodstein, 1985; Durrant, 2000).
- Physics underlies chemistry, e.g., determining physical chemistry properties of matter and the nature of the chemical bond (Laidler et al., 2003).
- Molecular processes underlie organic life (Scott, 1989).
- Cells with their own internal function underlie all life (Harold, 2001).
- Physics and chemistry underlie the functioning of the brain (Scott, 1995).
- Individual human behavior underlies the functioning of society (Berger, 1963).

Bottom-up causation is the focus of reductionist accounts of life. However, although it is indeed a key factor in what occurs, it is not the whole story.

TOP-DOWN CAUSATION

Top-down causation takes place in the hierarchy of complexity (Campbell, 1974; Murphy and Ellis, 1995; Noble, 2007; Ellis, 2008b). The fundamental importance of top-down causation is that it changes the causal relation between upper and lower levels in the hierarchy of structure and organization: causes come from some upper levels to lower levels, as well as from lower levels to upper levels (Ellis, 2008a,b).

Through boundary effects (linking the system to the environment) and structural relations in the system itself, the higher levels of the hierarchy causally effect what happens at the lower levels (Figure 24.1, right) by determining the context of the lower-level processes. Through coordinating the action of the lower levels, the higher levels attain their causal effectiveness. Top-down causation is prevalent in the real physical world and in biology because no real physical or biological system is isolated. It is the occurrence of multiple top-down actions in conjunction with bottom-up action that enables the self-organization of complex systems. Here are some examples:

1. The synthesis of light elements in the early Universe (Silk, 2005): the amount of helium produced depends on the rate of change of temperature in the expanding Universe, which is controlled by the gravitational equations and the average amount of matter in the Universe. Thus, quantities defined at the cosmological level control the products of detailed nuclear reactions at the micro level.
2. The functioning of the molecular systems underlying cardiac behavior is determined in a top-down way by the state of the heart as a whole (Noble, 2007).
3. Training of artificial neural networks to perform a specific task (say, letter recognition) determines the interaction weights in the network (Bishop, 1999). This is a form of top-down causation from the pattern to be recognized (a high-level concept, as it is defined in terms of the relation between the elements) to the low-level property of network weights. Decision making is a property of the network rather than of any single cell.
4. The power of the human mind in the real world: for example, design and construction of a jumbo jet aircraft, thereby determining the disposition of myriads of atoms (Ellis, 2008b).

Numerous other examples are given in Ellis (2008b), which also looks at the key question: *How is effective top-down causation possible without violating the integrity of bottom-level causation?* The basic answer is that the lower-level physics does not uniquely determine higher-level outcomes, not only because of statistical effects, but because quantum theory does not determine a unique physical outcome, *even in principle* (Feynman, 1985; Penrose, 1989b, 2005; Polkinghorne, 2002; Al-Khalili, 2003). Physics gives the statistics of outcomes, but not the unique actual outcome. This opens up the space for selectional principles to operate as a form of top-down action.

THE CAUSAL EFFICACY OF NONMATERIAL ENTITIES

The overall importance of this discussion is that it is through top-down action that immaterial entities can have causal effects on the physical world (Ellis, 2008b).

The first important example is social constructions, such as the rules of chess and the value of money. These are indeed causally effective in the real world, but they are not the same as a state of any single physical object; in particular, they are not equivalent to any individual's brain state (although they are realized through such states).

The ontological status of rules, such as, say, for chess, is not dependent on any individual's brain or state of mind. Rather, such rules are an abstract social construction, shared by many people and arrived at through social interactions over the course of time, with many different embodiments: they are written down, can be talked about, are thought about, are embodied in computer programs, and so on. They are not physical entities, but they are causally effective. Similarly, physically, money is just coins or pieces of paper with patterned marks on them. This does not explain its causal significance. The effectiveness of money, which can cause physical change in the world such as the construction of buildings, roads, bridges, and so on, by top-down action of the mind to material objects, is based in social agreements that lead to the value of money (pricing systems) and exchange rates. These are abstract entities arising from social interaction over an extended period of time and are neither the same as individual brain states nor equivalent to an aggregate of current values of any lower-level variables (although they may be represented by, and are causally effective through, such states and variables).

Also, abstract entities of a Platonic nature can be causally effective in the real world through the action of the human mind. For example, mathematics comprehension and utilization is a case of causation from a Platonic world of mathematical abstractions to the human mind, being realized in details of neuronal connections and then into the real world, where it is causally effective in terms of creating patterns on paper and through underlying physics, engineering, commerce, and general planning.

The existence of a Platonic world of mathematical objects is strongly argued by Penrose (1997, 2005) and Changeux and Connes (1998), the point being that major parts of mathematics are discovered, rather than invented (rational numbers, zero, irrational numbers, and the Mandelbrot set being classic examples; see, e.g., Seife 2000 for the case of zero). They are not determined by physical experiment, but are rather arrived at by mathematical investigation. They have an abstract, rather than an embodied, character; the same abstract quantity can be represented and embodied in many symbolic and physical ways (Penrose, 2005), and these representations form an equivalence class. They are independent of the existence and culture of human beings; it is plausible that the same features will be discovered by intelligent beings in the Andromeda Galaxy as here, once their mathematical understanding is advanced enough (which is why they are advocated as the basis for interstellar communication). This Platonic world is, to some degree, discovered by humans and represented by our mathematical theories; that representation is a cultural construct, but the underlying mathematical features they represent are not—indeed, like physical laws, they are often unwillingly discovered, as, for example, the irrationality of the $\sqrt{2}$ and the number π. This world is causally efficacious in terms of the process of discovery and description: one can, for example, print graphic versions of the Mandelbrot set in a book, resulting in a physical embodiment in the ink printed on the page. The causal variables here are not coarse-grained lower-level variables and exist independent of the mind, even though they are discovered and comprehended through the mind.

THE NATURE OF EXISTENCE

It is the existence of these causally effective immaterial entities that enables us, indeed requires us, to contemplate kinds of existence other than that of merely physical entities. In discussing this, *I take as given the reality of the everyday world*—tables and chairs, and the people who perceive

---

TABLE 24.2

The Different Kinds of Reality Implied by Causal Relationships Can Be Characterized in Terms of 4 Worlds, Each Representing a Different Kind of Existence

- *World 1:* Matter and Forces
- *World 2:* Consciousness
- *World 3:* Physical and Biological Possibilities
- *World 4:* Abstract (Platonic) Reality

---

them—and then assign a reality additionally to each kind of entity that can have a demonstrable causal effect on that everyday reality. The problem then is to characterize the various kinds of independent reality that may exist in this sense. Taking into account the causal efficacy of all the entities discussed above, I suggest as a possible completion of the proposals by Popper and Eccles (1977) and Penrose (1997, 2005) that the 4 Worlds indicated in Table 24.2 are ontologically real (that is, they really do exist).* These are not different causal levels within the same kind of existence; rather, they are quite different kinds of existence, but related to one another through causal links. The challenge is to show, first, that each is indeed ontologically real and, second, that each is sufficiently and clearly different from the others that it should be considered as separate from them. I now discuss them in turn; see Ellis (2004) for a more detailed discussion.

MATTER AND FORCES

World 1 is the physical world of energy and matter, hierarchically structured to form lower and higher causal levels whose entities are all ontologically real.

This is the basic world of matter and interactions between matter, based at the micro-level on elementary particles and fundamental forces, and providing the ground of physical existence. It comprises *inanimate objects* (both naturally occurring and manufactured), *living things* (amoebae, plants, insects, animals, etc.), and *human beings,* who have the unique property of self-consciousness.

The hierarchical structure in matter is a real physical structuration and is additional to the physical constituents that make up the systems themselves. It provides the basis for higher levels of order and phenomenology, and hence of ontology. Ontological reality exists at each level of the hierarchy. Thus, we explicitly recognize as being real: quarks, electrons, neutrinos, rocks, tables, chairs, apples, humans, the world, stars, galaxies, and so on. The fact that each is made up of lower-level entities does not undermine its status as existing in its own right (Sellars, 1932). We can attain and confirm high representational accuracy and predictive ability for quantities and relations at higher levels, independent of our level of knowledge of interactions at lower levels, giving well-validated and reliable descriptions at higher levels accurately describing the various levels of emergent nonreducible properties and meanings.

CONSCIOUSNESS

World 2 is the world of individual and communal consciousness: ideas, emotions, and social constructions. This again is ontologically real (it is clear that these all exist) and causally effective.

This world of individual and communal consciousness is different from the world of material things and is realized through the human mind and society. Ideas, emotions, and social constructions are not brain states, although they can be represented as such, for they do not reside exclusively in any particular individual mind. They are the foundation for social interactions and intellectual activity. They are causally effective because, as discussed above, they do indeed have the capacity

---

* Ontology is the philosophical study of the nature of being or existence.

to change the state of the physical world. The existence of houses and automobiles, of books and computers, of cities and factories is evidence of this effectiveness.

PHYSICAL AND BIOLOGICAL POSSIBILITIES

World 3 is the world of possibilities. This world characterizes the set of all physical and biological possibilities world from which the specific instances of what actually happens in World 1 are drawn.

This world of possibilities is ontologically real because of its rigorous prescription of the boundaries of what is possible—it provides the framework within which World 1 exists and operates, and in that sense it is causally effective. This world is different from the world of material things, for it provides the background within which that world exists. In a sense, it is more real than that world because of the rigidity of the structure it imposes on World 1. There is no element of chance or contingency in it, and it is certainly not socially constructed (although our understanding of it is so constructed). It rigidly constrains what can happen in the physical world.

If one believes that physical laws are prescriptive, rather than descriptive (cf. the discussion below), one can view the world of all physical possibilities as being equivalent to a complete description of the set of physical laws (for these determine the set of all possible physical behaviors through the complete set of their solutions). The formulation given here is preferable in that it avoids making debatable assumptions about the nature of physical laws, but still incorporates their essential effect on the physical world. Whatever their ontology, what is possible is described by physical laws such as those identified in Table 24.3.

ABSTRACT (PLATONIC) REALITY

World 4 is the Platonic world of abstract realities that are discovered by human investigation, but that are independent of human existence. They are not embodied in physical form, but they can have causal effects in the physical world.

As discussed above, the existence of a world of mathematical objects is strongly argued by Penrose (1997, 2005) and Changeux and Connes (1998). It is a quite different kind of existence from that of physical entities: it is eternal and unchanging, and so is of a Platonic character.

EXISTENCE AND EPISTEMOLOGY

The major proposal (Ellis, 2004) is that *all these worlds exist—Worlds 1 through 4 are ontologically real and are distinct from one another*, as argued above. These claims are justified in terms of the effectiveness of each kind of reality in influencing the physical world.

What then of epistemology (the nature of our knowledge of how things are)? Given the existence of the 4 Worlds as described above, the proposal here is that *epistemology is the study of the relation*

TABLE 24.3
Three Fundamental Laws of Nature

The Second Law of Thermodynamics
Entropy always increases

Maxwell's Laws of Electromagnetism
Electric charges cause electricity and magnetism, which in turn determine the motion of the charges

Einstein's Law of Gravitation
Matter causes spacetime curvature

between World 2 and Worlds 1, 3, and 4. It attempts to obtain accurate correspondences to entities in all the worlds by means of entities in World 2.

This exercise implicitly or explicitly divides World 2 theories and statements into (1) true/accurate representations, (2) partially true/misleading representations, (3) false/poor/misleading representations, and (4) representations where we do not know the situation. These assessments range from statements such as "It is true her hair is red" or "There is no cow in the room" to "Electrons exist," "Newtonian theory is a very good description of medium-scale physical systems at low speeds and in weak gravitational fields," and "The evidence for UFOs is poor." This raises interesting issues about the relation between reality and appearance: for example, everyday life gives a quite different appearance to reality than microscopic physics—as Eddington (1928) pointed out, a table is actually mostly empty space between the atoms that make up its material substance; but in our experience, it is a real, hard object. As long as one is aware of this, it can be adequately handled.

THE NATURE OF THE LAWS OF NATURE

Given this understanding of the kinds of existence, what can we say about the laws of nature? These are impersonal dynamical principles operating on material entities in a spacetime that are needed for reliable emergence of complex orders of existence, such as galaxies, stars, planets, and life, as discussed above. This emergence can happen only if there are precisely ordered evolving relationships and processes governing the behavior of matter at the basic levels that are such as to allow the higher levels to emerge.

THE BASIC LAWS

The hierarchy of complexity is based at the lower levels in specific families of particles (Cottingham and Greenwood, 2007), interacting through four fundamental forces that are presumed unified at high energies. Their behavior is based in quantum mechanics, with interactions describable by variational principles subject to special relativity theory and to fundamental symmetries entailing conservation laws, but with the exact symmetries of the theory broken. These interactions involve specific masses and interaction strengths. All of this takes place in a 4-dimensional Riemannian spacetime, whose geometry is determined by the matter content of the spacetime.

EMERGENT LAWS

It is far from obvious that these fundamental laws can lead to the emergence of complexity. Regularities of behavior at the higher levels of the hierarchy are characterized by emergent laws at that level; for example, the perfect gas law, Ohm's law, Bernoulli's law, and so on are higher-level physical laws. Each has a domain of validity representing the conditions where it gives reliable results (for example, the temperature and density must be neither too high nor too low for each of these laws to be true). Within their domain of validity, these are equally as valid as the fundamental laws, for they can be thoroughly tested and shown to be accurate predictors of the relevant physical behavior. Similarly, Darwin's theory of evolution through natural selection is an emergent law at the biological population level, as are the laws of genetic inheritance, whereas the Hodgkin–Huxley equations for a nerve's action potential are an effective biological law at the physiological level. The issue of whether there are reliable psychological laws is more uncertain (for a discussion, see Silverberg, 2004); but there certainly are universal regularities applicable to human life, such as the need for food and the inevitability of death.

Actually, all the physical laws of which we are certain (such as those in Table 24.3) are emergent, rather than fundamental laws; for example, both Newton's and Einstein's laws of gravity are presumed to be approximations to some deeper underlying quantum theory of gravity. Indeed, we do not even know what the fundamental physical laws are (string theory/M theory is a popular contender, but it does not yet have a unique formulation and is not experimentally tested).

TABLE 24.4

Examples of the Geometrical Nature of the Laws of Physics

- Geometric conservation of particles or fields in a 3-dimensional space gives an inverse-square law.
- Parallel transport along curves underlies Yang–Mills theory, the basis of the standard model of particle physics.
- The Aharanov–Bohm effect and Feynman path integrals suggest that holonomy (based on parallel transport along closed curves) is a fundamental entity in quantum theory.
- Geodesic paths (corresponding to extreme distances) underlie variational principles, as a geometrical description of solutions of dynamical equations.

Their Mathematical Description

As mathematics describes ordered patterns of relationships, it is perhaps not surprising that these relationships and processes can be described mathematically. The very nature of mathematics is indeed to describe patterns (Devlin, 1996)—in space and in time, and indeed within patterns (leading to recursion and higher-order relations). A mathematically useful description is a probable outcome of reliable behavior, such as is described by physical laws. At a certain level, this is an answer to Wigner's (1960) famous question, *Why are physical laws so well describable by mathematics?*

What is surprising is that *fundamental physical relationships can often be described so accurately by very simple laws*, such as an inverse-square law. My suggestion is that *this is because the underlying nature of these fundamental laws is geometrical*, which results in their being accurately represented by simple analytic relations. Some examples are shown in Table 24.4.

It is a basic principle of mathematics that geometrical relations can be represented in analytic form, and vice versa; the proposal here is that the geometrical form of physical relationships is the more fundamental, and that is the reason for the relatively simple analytic forms for physical laws.

The Nature of the Laws of Physics

Two major issues arise in regard to the existence of the laws of physics. The first is the nature of these laws: are they *descriptive, just characterizing the way things are*, or are they *prescriptive, forcing them to be this way* (Carroll, 2004)?

For example, quantum field theory applied to the standard model of particle physics is immensely complex (Peskin and Schroeder, 1995; see Table 24.5). *What is the nature of existence of all this quantum apparatus? Are they just descriptions devised by our minds, or are they the way things really are?* Derived (effective) theories, including classical (nonquantum) theories of physics, equally have complex abstract structures underlying their use: force laws, interaction potentials, metrics, and so on. The same issue arises.

If laws are descriptive, this is just the way matter behaves; that is, the laws are phenomenological: they just describe what is. They are mathematical and physical constructs that happen to

---

TABLE 24.5

Entities Assumed to Exist When Quantum Field Theory Applies

- Hilbert spaces, operators, commutators, symmetry groups, higher-dimensional spaces;
- Particles/waves/wave packets, spinors, quantum states/wave functions;
- Parallel transport/connections/metrics;
- The Dirac equation and interaction potentials, Lagrangians, and Hamiltonians;
- Variational principles that seem to be logically and/or causally prior to all the rest.

---

characterize reasonably accurately the physical nature of physical quantities. That such descriptions should be accurate predictors of physical behavior is in itself remarkable. But then from where do the properties of matter derive, leading to these descriptions? How are its characteristic behaviors enforced? The specific issue arising is, *Why does all matter have the same properties wherever it exists in the Universe?* Why are electrons here identical to those at the other side of the Universe if the laws are only descriptive? We seem to have no handle with which to investigate such questions. But we can claim that even if this is their nature, the laws have an existence independent of human minds: the phenomenological descriptions are accurate and will be derived as such by any other intelligent beings in the Universe. In this sense, they are as ontologically real as the laws of mathematics, discussed above: they exist universally as descriptions of the nature of reality.

If they are prescriptive, the laws of physics somehow exist in a form that enables them to control the nature of existence. They then represent a fundamental underlying reality, as entities that have the power to control the behavior of physical quantities. Then matter will necessarily be the same everywhere because their behavior is determined by these laws (assuming that the laws themselves are invariable). The issue arising then, is, *In what way—where and how—do laws of physics exist and impose themselves on the matter in the Universe?* Do they, for example, have an existence in some kind of Platonic space that controls the nature of matter and existence?

The second issue, related to the first, is, *Do these laws in some sense precede the existence of the Universe, somehow governing its coming into being, or do they come into being with the Universe?* This is where this issue relates deeply to the nature of cosmology.

Many theories of the creation of the Universe assume that all these laws, or at least a basic subset of them, preexist the coming into being of the physical Universe because they are presumed to underlie the creation process; for example, the entire apparatus of quantum field theory mentioned above is often taken for granted as preexisting our Universe. This is, of course, an unprovable proposition, but it appears to be widely held. If it is true, it seems to me to support the idea that the laws are indeed ontologically real: indeed, more real than the Universe itself, which is then a transient manifestation of this underlying immutable reality. And it is difficult to see any other basis for the coming into being of matter effectively obeying physical laws.

THE ONTOLOGICAL NATURE OF EFFECTIVE (HIGHER-LEVEL) LAWS

As discussed in "Emergent Laws" above, we actually deal with effective laws only: we do not yet have access to the fundamental laws. Hence, my position is that the existential status of fundamental and effective laws has to be the same: they are both genuine representations of the way things behave at the different levels and also preexist the coming into being of any matter that materializes their nature or any mind that comprehends them. In this way, they are ontologically real (i.e., they actually exist), whether descriptive or prescriptive. And this applies at each level of the hierarchy: chemical laws and biological laws are just as real as physical laws, independent of the existence of any human mind.

This ontological reality can be shown by their causal effectiveness. Consider how Maxwell's *theory* of electromagnetism (an abstract entity, described by Maxwell's equations) led to the development of radio, and then to the existence of TV, cell phones, and so on. Maxwell's theory is not the same as any single person's brain state. It can be represented in many ways (on blackboards, in print, on computer screens, in spoken words, in neural connections) and many formalisms (via 3-dimensional vectors or 4-dimensional tensors, for example); these various representations together form an equivalence class, as they all lead to the same predicted outcomes. How do you demonstrate this kind of causation? Design an artifact such as a cell phone through use of Maxwell's theory, and then construct it and operate it. The abstract theory will have altered physical configurations in the real world and hence is causally effective. It is therefore an ontologically existing entity, successfully representing the nature of physical reality in a way that will be accessible to advanced civilizations anywhere in the Universe. The same is true, for example, of the laws of thermodynamics.

As noted in "Physical and Biological Possibilities" above, one can avoid talking about the laws of behavior *per se* by instead considering the *space of possibilities* underlying what exists physically, rigorously constraining the possible natures of what actually comes into existence. This space is more or less uniquely related to the underlying laws in the same way that the space of solutions of differential equations is related to the nature of the equations. This enables one to avoid the issue of the ontology of the laws, but does not solve it. It does confirm, however, that we can think of the space of possibilities, essentially representing the outcomes, and, hence, nature of the underlying laws, as genuinely existing: it is a transcendent eternal reality, governing the nature of what can actually come into existence.

ULTIMATE REASONS

Philosophers have debated for millennia whether the ultimate nature of existence is purely material or embodies some form of rationality *(logos)* and/or purpose *(telos)*. What in the end underlies it all? Is the ultimate nature of the Universe purely material, or does it in some way have an element of the mental? That profound debate is informed by physical cosmology, but cannot be resolved by the physical sciences alone. Given suitable lowest-level laws, with restricted structure and coupling constants (Davies, 1982; Barrow, 2003), a hierarchy of complexity with effective higher-level laws can emerge. But what essentially underlies the lowest-level laws on which the rest is based? Why do they exist, with the form they have that allows life to exist? Is the ultimate underlying reason pure chance, probability, necessity, or purpose? I consider each of these possibilities in turn.

Pure Chance, Signifying Nothing

The initial conditions in the Universe just happened, and they led to things being the way they are now, by pure happenstance. This is just the way it was: there is no suggestion that it was a probable outcome of some underlying dynamics. Probability does not apply. There is no further level of explanation that applies; searching for "ultimate causes" has no meaning.

This is a logically possible ultimate reason, but has no further explanatory power; indeed, it is denial that at a fundamental level there is any explanation and so is unsatisfactory to almost everyone (whether scientifically or religiously inclined), primarily because we know that explanations (both impersonal and personal) do indeed exist in the social and mental world. Furthermore, it is difficult to resist the argument that the outcome is so unlikely that pure chance simply is not credible as a reason. Not merely qualia and emotions, but also complex theories such as Einstein's theory of relativity and quantum field theory have come into existence as extraordinarily complex theoretical constructs. To suggest that these can all arise without existence of any underlying cause, or can come into existence out of pure chaos or nothingness without any further guiding structure, seems simply absurd; but if you have such a cause or guiding structure, you do not have pure chaos or nothingness.

This is certainly logically possible and, indeed, is philosophically unassailable; but it is not satisfying as an explanation, as we obtain no unification of ideas or predictive power from this approach. Nevertheless, some implicitly or explicitly hold this view.

High Probability

The idea is that although the structure of the Universe appears to be very improbable, in fact for physical reasons it is highly probable. This is often realized by some version of the idea of *universality*: "All that is possible, happens." The current embodiment of this idea is via the concept of a Multiverse: an ensemble of universes or of disjoint expanding universe domains is realized, in which all possibilities occur (Tegmark, 1998, 2004; Rees, 2000, 2001). It is then supposed that the anthropic principle is realized in both its strong form (if all that is possible happens, then life must happen: it is inevitable) and its weak form (life will occur only in some of the domains that are realized; these are picked out from the others by the fact that we can exist only where life is possible, so the anthropic principle is viewed as a selection principle in the Multiverse context). The favored cosmological setting for this idea is chaotic inflation (Linde, 1986, 1990; Vilenkin, 2006; Guth, 2007; Weinberg, 2007), sometimes integrated with the idea of the landscape of string theory (Susskind, 2005).

These arguments are only partially successful, even in their own terms. They run into problems if we consider the full set of possibilities: discussions proposing this kind of view actually implicitly or explicitly restrict the considered possibilities *a priori*, for otherwise it is not very likely that the Universe will be as we see it (inflation does not in fact solve the issue of probability; see Penrose, 1989a, 2005). Besides, we do not have a proper measure to apply to the set of initial conditions, enabling us to assess these probabilities. Furthermore, the multiverse hypothesis is not observationally or experimentally provable: the supposed other universe domains are not observable by any conceivable form of astronomical observation, as they lie beyond the visual horizon, and the assumed underlying physics is unproven and probably untestable (Carr and Ellis, 2008; Ellis, 2009).

Despite these problems, this approach has considerable support in the scientific community, particularly via the chaotic inflationary proposal. It is an attractive explanatory proposal, but is not testable physics; and it is, in fact, not an ultimate explanation. If it were to exist, the whole issue of probability arises again as regards the Multiverse (why this one, rather than another?), leading to the specter of infinite regress: we might explain the probability of a specific Multiverse by assuming an ensemble of multiverses. For the scientist, probability trumps a lack of any explanation. But probability by itself is always an incomplete explanation: for what underlies the laws that govern those probabilities? Why are the assumed laws underlying probabilistic calculations valid in the first place? In any case, probability is a good explanation for intermediate levels of explanation where the dynamical laws have already been established and some kind of variation takes place through either irreducible uncertainty (quantum mechanics) or statistical variation, but there is no evidence that it applies in the context of the Universe and ultimate causation; indeed, the very concept of probability is not applicable if there is only one object (the unique Universe) in existence. Application of probability arguments to the Universe itself is dubious because the Universe is unique (Ellis, 2006).

In brief: The unique Universe that actually exists may well not be probable, as was taken for granted in the past; indeed, it certainly is improbable because it is of such a nature as to allow life to exist. The whole Multiverse endeavor is an attempt to make the improbable appear probable. It only postpones the problem; it does not solve it.

An interesting variant is the idea of *cosmological natural selection* (Smolin, 1992). If a process of re-expansion after collapse to a black hole (BH) were properly established, it would open the way to the concept of evolution of the Universe not merely in the sense that its structure and contents develop in time, but in the sense that the Darwinian selection of expanding Universe regions could take place. The idea is that there could be collapse to BHs followed by re-expansion, but with an alteration of the constants of physics through each transition, so that each time that there is an expansion phase, the action of physics is a bit different. The crucial point then is that some values of the constants will lead to production of more BHs, while some will result in less. This allows for evolutionary selection favoring the expanding Universe regions that produce more BHs (because of the favorable values of physical constants operative in those regions), for they will have more "daughter" expanding Universe regions. Thus, one can envisage natural selection favoring those physical constants that produce the maximum number of BHs.

This is an intriguing effort to bring ideas of natural selection, as discussed in "The Causal Efficacy of Nonmaterial Entities" above, into cosmology; but again it is only partially successful (Rothman and Ellis, 1992) and does not solve the ultimate issues: if this indeed takes place, then why the laws of physics that allow it to take place?

NECESSITY

This is the idea that things have to be the way they are; there is no other option. The features we see and the laws underlying them are demanded by the unity of the Universe. Coherence and consistency require that things must be the way they are; the apparent alternatives are illusory. Only one kind of physics is self-consistent: all logically possible universes must obey the same physics.

To show that this is the case has been one of the most profound goals of theoretical physics, following up powerful unification principles with the aim of eventually devising a single theory of fundamental physics with no free parameter whatever. To really succeed would be a wonderful achievement, potentially leading to a self-consistent and complete scientific view of the foundations of physical existence. But we can imagine alternative universes!—why are they excluded? Indeed, this project seems to fail in any case because fundamental physics is presently going the other way: the hoped-for uniqueness of fundamental theories has evaporated and been replaced by the multiple billions of possibilities of string vacua (as explained by Susskind, 2005). Furthermore, here we run into the problem that we have not succeeded in devising a fully self-consistent view of physics: neither the foundations of quantum physics nor those of mathematics are on a really solid, consistent basis. Until these issues are resolved, this line cannot be pursued to a successful conclusion.

Additionally, if this approach ever succeeded, it would in fact worsen, rather than solve the fine-tuning problem. If there were just in the end one set of constants of nature that are consistent with one another, why should they take those precise improbable values that just happen to allow the existence of life? In the end, this is highly implausible. How could it be that the existence of love and pain and intellect is *necessarily* written into variational principles and mathematical symmetries such as SU(10) or E8 as the inevitable outcome of the only consistent possibility for their implementation?

In any event, just as in the previous case, this is not an ultimate answer: the attempt to implement necessity leaves unexplained the choice of those specific realized features of physics that then lead to the necessity. Why should physics have the specific restricted nature that leads to the necessity of particular high-level features?

This whole project is an attempt to implement the idea of necessity underlying existence and causation, but it necessarily has to be incomplete. What has to be explained includes the following: *Where do the very causal categories of chance, necessity, and purpose come from? How do these concepts arise and have meaning, and what underlying ontological entities or causation do they represent? Why are they themselves necessary? How can they even be relevant, as this whole discussion supposes, if there is no ontological referent that makes the dichotomy between them a meaningful issue?* That itself surely cannot be shown to be necessary: for it is the very category of necessity that has to be explained.

PURPOSE OR DESIGN

The symmetries and delicate balances we observe require an extraordinary coherence of conditions and cooperation of causes and effects, suggesting that in some sense they have been purposefully designed. That is, they may show evidence of intention, both in the setting of the laws of physics and in the choice of boundary conditions for the Universe, in such a way that life will inevitably come into being.

Unlike all the other options just discussed, this introduces an element of meaning, of signifying something, at the foundations. In all the other options, life exists by accident, as a chance by-product of processes blindly at work. Here, the view is that the totally different quality of existence that emerges in human life from the underlying physics, and the huge fine-tuning that is needed for this to occur, suggests an underlying intention or purpose that this should indeed be the case: *it was meant to be that way.* The modern version, consistent with all the scientific understandings of causation, would see some kind of purpose underlying the existence and specific nature of the laws of physics and the boundary conditions for the Universe in such a way that life (and eventually humanity) would then come into existence through the operation of those laws, then leading to the development of specific classes of animals through the process of evolution as evidenced in the historical record.

This is the only line of reasoning that does not just relegate the problem to a deeper level: assuming a higher intention is realized through physical reality by design enabling higher-level purpose and ethical principles to be embodied through the nature of the resulting possibility spaces. It is this higher-level set of purposive principles—the underlying *telos*—that is then the ultimate cause of both existence and its specific nature. It is unlikely that this kind of underlying intention could be effective, with emergence of a physical structure where purpose can be deployed so that ethical behavior is meaningful, unless on the one hand the lower-level laws had the kind of impersonal regular behavior that allows reliable higher-level behavior to emerge, thus allowing a mathematical description, and on the other hand something like quantum uncertainty were present so as to free the higher levels from total lower-level determinism. A layered structure emerges: purpose underlies impersonal laws that underlie the emergence of purpose. Two kinds of causation, intentional and impersonal, which undoubtedly both exist in the world around us, occur in an intertwined way, with chance events intervening and helping to lead to the richness of outcomes we see around us.

Thus, implementing this proposal necessarily invokes the other two: meaningful purpose entails both necessity and chance. Each of these kinds of causation (chance in the sense of probability, necessity, and purpose) does indeed occur in the world in various contexts, but the only one that seems to entail the possibility of being a deep foundation for the others is purpose. There will be some who will reject this possibility out of hand, as meaningless or as unworthy of consideration. However, it is certainly logically possible. Given an acceptance of evolutionary development, it is precisely in the choice and implementation of particular physical laws and initial conditions, allowing such development, that the profound creative activity occurs; and this is where one might conceive of design taking place.

This then relates to religious or spiritual views of the nature of reality, supported by a variety of evidence relative to those domains (that evidence certainly exists, with a wide variety of natures and in a wide variety of contexts; the argument is about its acceptability in each case).

THE NATURE OF EVIDENCE

What Kind of Evidence Is Relevant?

In examining these fundamental issues, one needs to take into account all the scientific evidence about the nature of physics, chemistry, and biology that comes from laboratory experiments, as well as about the nature of the Universe that comes from astronomical observations. In particular, I suggest that we are entitled to take the nature of the possibility space allowing existence of material entities supporting consciousness and purposive causation as evidence concerning the ultimate reasons the basic laws are as they are. But equally, as well as the discoveries attained by the scientific method, one should take into account data about the natures of our existence that come from our daily lives and the broad historical experience of humanity (our experiences of ethics and aesthetics, for example).

The claim I make is that the kinds of personal world experience we each have are certainly data on the nature of reality because we live in and indeed are part of reality. They do not have the quality of the strictly repeatable experiments that science engages in—they are much richer than that. As an example, many writings claim that there is no purpose in the Universe: it is all just a conglomerate of particles proceeding at a fundamental level in a purposeless and meaningless algorithmic way. But, I would reply, the very fact that those writers engage in such discourse undermines their own contention; they ignore the evidence provided by their own actions. There is certainly meaning in the Universe to this degree: *the fact that they take the trouble to write such contentions is proof that they consider it meaningful to argue about such issues*; this quality of existence has emerged out of the nature of the physical Universe. Indeed, the human mind is causally effective in the real physical world precisely through many activities motivated by meanings perceived by the human mind.

Any attempt to relate physics and cosmology to ultimate issues must take such real-world experience seriously; otherwise, it will simply be ignoring a large body of undeniable data. These data do not resolve the ultimate issues, but they do indicate dimensions of existence that indeed occur. I consider briefly here two types of such evidence that reinforce the arguments given above, namely, experiences related to ethical issues and experiences that many interpret as related to a spiritual or transcendent existence.

THE NATURE OF ETHICS

A key area where human experiences are important is ethics and morality: the issue of how we ought to conduct our lives. The origin of moral values has been the subject of debate for centuries. I will just state my own view here and give some references to literature where it is supported. I take the position of moral realism, which argues that we do not invent ethics, but discover it: there is a standard of morality that exists (in "reality," just waiting to be discovered) that is valid in all times and places, and human moral life is a search to understand and implement that true nature of morality. Thus, I believe that *moral choices relate to ethical values that are timeless and culture independent*, and this is based in the existence of a moral reality, as well as of a physical reality and a mathematical reality, that underlies the Universe.

For detailed arguments in support of this contention, based in the idea that morality is real and so not just a social construction or evolutionary invention, see Murphy and Ellis (1995), Gaita (2004), and Ellis (2008a). The only point I would make here is the following: there is in fact tacit agreement with this position in the writings of both Richard Dawkins (2006) and Viktor Stenger (2007), for both make strong claims about the evil caused by religion. In doing so, they are presuming to make a claim that is more than just their personal opinion: they are expressing this as irrefutable fact. Indeed, Stenger claims (p. 216) that it is a scientific fact that evil exists. So what is the experiment that establishes this as a *scientific* fact? There is none, as science does not comprehend the concepts of "right" and "wrong"; there are no scientific units ("milli-Hitlers") for degrees of evil. This is an ethical claim falsely dressed up as science. But the point is that this argument by both Dawkins and Stenger does show a belief in absolute standards of right and wrong, independent of culture and space and time, which is in agreement with my own position that underlying the Universe is a moral reality. We can then add to the 4 Worlds listed above in "The Nature of Existence" another one:

World 5 is the world of Platonic ethical forms, providing a foundation for our sense of ethics. This is perhaps related to a sense of beauty and aesthetics, and it is certainly linked to and based in concepts of purpose and meaning *(telos),* for these are what underlie ethical behavior.

Then the family of worlds we need to recognize becomes as shown in Table 24.6.

The core of being is then, in this view, the underlying ultimate purpose relating to meaning and morality. One can argue that this purpose, as identified by the spiritual traditions of all the major

---

TABLE 24.6

The Different Kinds of Reality Enhanced by a Layer Representing Ethical Values and Meaning

- *World 1:* Matter and Forces
- *World 2:* Consciousness
- *World 3:* Physical and Biological Possibilities
- *World 4:* Abstract (Platonic) Reality
- *World 5: Telos*—Ethical Reality Related to Purpose

world religions, is unselfish love, or *agape*. The kind of ethics that is compatible with this view is *kenotic* (self-emptying or self-sacrificial), invoking a loving attitude and respect for the freedom and integrity of others as a basic principle underlying the nature of existence, extending to sacrificing on their behalf, a free and willing gift which can have the effect of transforming the situation to a higher ethical level (Ellis, 1993, 2008a; Murphy and Ellis, 1995).

This quality of existence, whose transforming nature has a self-authenticating quality when experienced, is then seen, in this viewpoint, as the ultimate purpose of the whole, enabled by the nature of the laws of physics and the boundary conditions for the Universe that are set so as to make this possible, and indeed requiring that they be this way in order that these purposes can be realized (Ellis, 1993). The core of being, and its ultimate purpose, is then the possibility and the demonstration of caring love for others, as many religious standpoints have proclaimed; this is what leads to and underlies the nature of existence. The material is the vehicle by which this is made possible.

EXPERIENCES OF THE TRANSCENDENT

For many, a deeply religious worldview is crucial in understanding our lives and setting values, this worldview being based in our personal life experience, including our experience of a faith tradition and community, religious texts, and inspiring leaders. All of these are data that help us understand our situation and our lives, but their nature and significance are strongly contested by those with other beliefs. My focus here is on what I call "intimations of transcendence": significant experiences in the lives of many people that reinforce the worldview proposed above. They suggest that there is a copious abundance of being, a plenitude that is more than is necessary, underlying the reality of physical existence. This extra dimension of existence, which we sometimes can glimpse in fleeting ways, is in fact an experience of transcendence. I give a number of examples:

> I say to myself as I watch the niece, who is very beautiful: in her this bread is transmuted into melancholy grace. Into modesty, into a gentleness without words.... Sensing my gaze, she raised her eyes towards mine, and seemed to smile.... A mere breath on the delicate face of the waters, but an affecting vision.... I sense the mysterious presence of the soul that is unique to this place. It fills me with peace, and my mind with the words: "This is the peace of silent realms." I have seen the shining light that is born of the wheat. (From *Flight to Arras*, by Antoine de Saint-Exupéry, 1969)

> One day during my last term at school I walked out alone in the evening and heard the birds singing in the full chorus of song, which can only be heard at that time of year at dawn or sunset. I remember now the shock of surprise with which the sound broke upon my ears. It seemed to me that I had never heard the birds singing before and I wondered whether they sang like this all year round and I had never noticed it. As I walked I came upon some hawthorn trees in full bloom and again I thought that I had never seen such a sight or experienced such sweetness before. I came then to where the sun was setting over the playing fields. A lark rose suddenly from the ground beside the tree where I was standing and poured out its song above my head, and then sank still singing to rest. Everything then grew still as the sunset faded and the veil of dusk began to cover the earth. I remember now the feeling of awe which came over me. I felt inclined to kneel on the ground, as though I had been standing in the presence of an angel; and I hardly dared to look at the face of the sky, because it seemed as though it was but a veil before the face of God. (Bede Griffiths, quoted in Taylor, 2007, as quoted in Bellah, 2008)

> I call to mind that distant moment in [the prison at] Hermanice when on a hot, cloudless summer day, I sat on a pile of rusty iron and gazed into the crown of an enormous tree that stretched, with dignified repose, up and over all the fences, wires, bars, and watchtowers that separated me from it. As I watched the imperceptible trembling of its leaves against an endless sky, I was overcome by a sensation that is difficult to describe: all at once, I seemed to rise above all the coordinates of my momentary existence in the world into a kind of state outside of time in which all the beautiful things I had ever seen and experienced existed in a total "copresent"; I felt a sense of reconciliation, indeed of an almost gentle consent to the inevitable course of things as revealed to me now, and this combined with a carefree determination to face what had to be faced. A profound amazement at the sovereignty of Being became

a dizzying sensation of tumbling endlessly into the abyss of its mystery; an unbounded joy at being alive, at having been given the chance to live through all I have lived through, and at the fact that everything has a deep and obvious meaning—this joy formed a strange alliance in me with a vague horror at the inapprehensibility and unattainability of everything I was so close to in that moment, standing at the very edge of the "finite"; I was flooded with a sense of ultimate happiness and harmony with the world and with myself, with that moment, with all the moments I could call up, and with everything invisible that lies behind it and has meaning. I would even say that I was "struck by love," though I don't know precisely for whom or what. (Vaclav Havel, quoted in Bellah, 2008)

Perhaps more wonderful still is the way in which beauty breaks through. It breaks through not only at a few highly organised points, it breaks through almost everywhere. Even the minutest things reveal it as well as do the sublimest things, like the stars. Whatever one sees through the microscope, a bit of mould for example, is charged with beauty. Everything from a dewdrop to Mount Shasta is the bearer of beauty. And yet beauty has no function, no utility. Its value is intrinsic, not extrinsic. It is its own excuse for being. It greases no wheels, it bakes no puddings. It is a gift of sheer grace, a gratuitous largesse. It must imply behind things a Spirit that enjoys beauty for its own sake and that floods the world everywhere with it. Wherever it can break through, it does break through, and our joy in it shows that we are in some sense kindred to the giver and receiver. (Rufus Jones, 1920)

I believe with all my heart and mind that there is a spiritual dimension to all being that cannot be encapsulated in scripture or in creed; an essence that loses its creative force when its communication depends upon the use of words alone. It can I think be readily made manifest through metaphor in poem or story; yet I am deeply aware that even when presented in such a form the truth remains partial. For me, there is a reality that lies beyond our presently misdirected concern for the fruits of economic power. It is only when we acknowledge our deeper inner urge to discover meaning in existence that we begin to harvest the fruits of the spirit. Most of us are at least partially aware of epiphanies that come our way from time to time: the emergence when rounding a corner of a breathtaking panorama of mountain, forest and ocean; the sudden sensibility of a zephyr breeze rustling treetops; the scent of jasmine on a shower of rain. I believe there are illuminations far beyond these: intuitions, insights, divinations that are not shaped by the physical senses; the hand of a friend on one's shoulder in a time of trouble; the sudden recognition of a smile in a passing stranger; above all, the wondrous inspiration of the serendipity, synchronicity, and innate knowing in the fabric of our lives. More often than not gifts such as these, which indelibly inscribe themselves upon our memories, are regarded as gifts of God. (Lewis Watling, 2006)

Bernard d'Espagnat, winner of the 2009 Templeton Prize, said in an interview that science has helped him to "justify his impression of a link between beauty and the divine."* He explained: "When we hear great classical music or look at very great paintings, they are not just illusions but could be a revelation of something fundamental. I would accept calling it God or divine or godhead but with the restriction that it cannot be conceptualised for the very reason that this ultimate reality is beyond any concept that we can construct." That is a view I would concur with; it is elaborated in Ellis (2008a).

The Weight of Evidence

What does this all add up to? First, even in order to understand just the material world, it can be claimed that one needs to consider forms of existence other than the material only—for example, a Platonic world of mathematics, on the one hand, and a mental world, on the other, both of which can be claimed to exist and to be causally effective in terms of affecting the material. Our understanding of causation will be incomplete unless we take them into account. The fact that they exist shows that there is more to existence than just material things: as argued above, other kinds of things exist and have real causal powers.

---

* See: http://www.templeton.org/templeton_report/20090415/; also see http://www.timesonline.co.uk/tol/comment/faith/article5918050.ece.

Second, there are also extraordinary qualities of life, including purpose, ethics, aesthetics, and meaning. These certainly exist, but what is the ground of their existence? The possibility space arising out of the fundamental laws of existence must in some sense have these things built into it, for they certainly have come into existence. *And they can have the transcendent quality I have indicated through the quotations above. The fact that such a quality can exist is itself a statement about the nature of reality.* In brief, *ex nihilo nihil fit*: the possibility of meaning and ethics cannot arise out of genuine chaos or out of literally nothing; it has to have been built into the foundations that gave physical existence its structure.

In my view, they represent the ultimate nature of reality because it is extraordinarily unlikely that they could have this nature, this quality of existence, by pure happenstance. Invoking a Multiverse of necessity simply postpones facing the ultimate questions: How can such qualities come into being if they do not represent something about the nature of reality, something that even preexisted the existence of the Universe itself—just like the laws of physics, which are themselves abstract rather than physical entities?

THE NATURE OF THE ARGUMENT

So is the ultimate reason underlying the lowest-level laws on which everything else is based pure chance, probability, necessity, or purpose? They are all logically possible. Neither science nor philosophy can give a certain answer as to which is the deep underlying cause of things: metaphysical uncertainty remains (Ellis, 2006). However, if one wants to relate one's understanding to the deeper meaning of personal life, the last option has the most traction. The others in the end provide a more tentative relation to morality and meaning, although experience suggests that these exist.

This view cannot be *proved* to be true, but it is supported by much experience that has considerable persuasive power as a whole. We should take into account data from the whole of life, not just physics or astronomy: we are part of the Universe and live in it. When dealing with ultimate meaning, what is relevant is whatever seems to give ultimate meaning in human life. There is indeed purpose in the Universe (for example, we gathered in Beijing to understand its nature a bit better*). Either purpose emerges out of nothing, or it is there from the start as the foundation, then being reflected in life. The latter is indeed a possibility for the nature of ultimate causation and the underlying reason for why the laws of physics exist and are as they are (Ellis, 1993, 2008a; Murphy and Ellis 1995). This can provide a satisfying explanation for the full depth of life and human experience, in contrast to reductionist materialist explanations that explicitly or implicitly deny the full depth and reality of this profound experience.

Finally, it should be emphasized that this is not a scientific conclusion, nor is the argument presented one that can be sustained on scientific grounds alone; it is a philosophically based conclusion. The issues considered here (the nature of ultimate causation) are not amenable to scientific resolution, precisely because they go beyond the domain where scientific experiments or observations can give a reliable answer. The argument is thus a philosophical or metaphysical one, based securely on current science, but also taking into account wider philosophical and human issues than cannot be handled by science *per se*. Any attempt to adequately tackle the fundamental issues considered here will necessarily be of this nature. If one wishes to deal purely in terms of scientific argumentation, then the above will be beyond what one will consider as legitimate argument. But if one takes that stand, allowing scientifically rigorous explanation alone, one should also carefully refrain from making any statements about issues of ultimate causation, except if they are identified as purely the personal opinion of whoever is making the statement, and without any scientific standing.

I apologize here for not giving references to all the other books and articles that look at these issues and come to the same kinds of conclusions: space simply does not allow this. To pursue things further, one can read the works of Ian Barbour, Arthur Peacocke, John Polkinghorne, John Hick,

---

* At the *New Vision 400* conference: http://nv400.uchicago.edu/.

Keith Ward, Holmes Rolston, Nancey Murphy, Philip Clayton, John Haught, and many others. My views have benefited from them all. As an entrance point to this large literature, see Clayton and Simpson (2008).

REFERENCES

Al-Khalili, J. (2003). *Quantum: A Guide for the Perplexed.* London: Weidendfeld and Nicholson.

Balashov, Y.Y. (1991). Resource letter Ap-1: The anthropic principle. *American Journal of Physics*, 54: 1069.

Barrow, J.D. (2003). *The Constants of Nature.* New York: Pantheon Books.

Barrow, J.D. and Tipler, F.J. (1986). *The Anthropic Cosmological Principle.* Oxford: Oxford University Press.

Bellah, R.N. (2008). The rules of engagements: Communion in a scientific age. *Commonweal* (September 12): 15–21.

Berger, P.L. (1963). *Invitation to Sociology.* New York: Anchor Books.

Bishop, C.M. (1999). *Neural Networks for Pattern Recognition.* Oxford: Oxford University Press.

Booch, G. (1994). *Object Oriented Analysis and Design with Applications.* New York: Addison-Wesley.

Campbell, D.T. (1974). Downward causation. In F. Ayala and T. Dobhzansky (eds.), *Studies in the Philosophy of Biology: Reduction and Related Problems.* Berkeley: University of California Press.

Campbell, N.A. and Reece, J.B. (2005). *Biology.* San Francisco: Benjamin Cummings.

Carr, B. and Ellis, G.F.R. (2008). Universe or multiverse? *Astronomy and Geophysics*, 49: 2.29–37.

Carroll, J. (ed.) (2004). *Readings on Laws of Nature.* Pittsburgh: University of Pittsburgh Press.

Changeux, J.-P., and Connes, A. (1998). *Conversations on Mind, Matter, and Mathematics.* Princeton: Princeton University Press.

Clayton, P. and Simpson, Z. (2008). *The Oxford Handbook of Religion and Science.* Oxford: Oxford University Press.

Cottingham, W.N. and Greenwood, D.A. (2007). *An Introduction to the Standard Model of Particle Physics.* Cambridge: Cambridge University Press.

Davies, P.C.W. (1982). *The Accidental Universe.* Cambridge: Cambridge University Press.

Dawkins, R. (2006). *The God Delusion.* New York: Bantam Books.

Devlin, K. (1996). *Mathematics: The Science of Patterns.* New York: Holt.

Durrant, A. (ed.) (2000). *Quantum Physics of Matter.* Milton Keynes: Open University.

Eddington, A.S. (1928). *The Nature of the Physical World.* Cambridge: Cambridge University Press.

Ellis, G.F.R. (1993). The theology of the anthropic principle. In R.J. Russell, N. Murphy, and C.J. Isham (eds.), *Quantum Cosmology and the Laws of Nature*, 367–406. Vatican City: Vatican Observatory.

Ellis, G.F.R. (2004). True complexity and its associated ontology. In J.D. Barrow, P.C.W. Davies, and C.L. Harper, Jr. (eds.), *Science and Ultimate Reality: Quantum Theory, Cosmology and Complexity*, 607–36. Cambridge: Cambridge University Press.

Ellis, G.F.R. (2006). Issues in the philosophy of cosmology. In J. Butterfield and J. Earman (eds.), *Handbook of the Philosophy of Science: Philosophy of Physics, Part A*, 1183–285. Amsterdam: North Holland.

Ellis, G.F.R. (2008a). Faith, hope and doubt in times of uncertainty. *The James Backhouse Lecture: Australia Yearly Meeting of Quakers.* Friends Book Sales: sales@quakers.org.au.

Ellis, G.F.R. (2008b). On the nature of causation in complex systems. *Transactions of the Royal Society of South Africa,* 63: 69–84. Available at: http://www.mth.uct.ac.za/~ellis/Top-down%20Ellis.pdf.

Ellis, G.F.R. (2009). Dark matter and dark energy proposals: Maintaining cosmology as a true science? *Dark Energy and Dark Matter: Observations, Experiments and Theories.* E. Pécontal, T. Buchert, Ph. Di Stefano, et al. (eds.). European Astronomical Society Publications Series, 36: 325–36.

Feynman, R. (1985). *QED: The Strange Theory of Light and Matter.* Princeton: Princeton University Press.

Flood, R.L. and Carson, E.R. (1990). *Dealing with Complexity: An Introduction to the Theory and Application of Systems Science.* London: Plenum Press.

Gaita, R. (2004). *Good and Evil: An Absolute Conception.* London: Routledge.

Gilbert, S.F. (2006). *Developmental Biology.* Sunderland, MA: Sinauer Associates.

Goodstein, D.L. (1985). *States of Matter.* New York: Dover.

Guth, A. (2007). Eternal inflation and its implications. *Journal of Physics A*, 40: 6811–826.

Harold, F.H. (2001). *The Way of the Cell: Molecules, Organisms, and the Order of Life.* Oxford: Oxford University Press.

Harrison, E.R. (2000). *Cosmology.* Cambridge: Cambridge University Press.

Jones, R. (1920). Where the beyond breaks through. *The Friend*, 60 (new series): 26.

Laidler, K.J., Meiser, J.H., and Sanctuary, B.C. (2003). *Physical Chemistry.* Boston: Houghton Mifflin.

Linde, A.D. (1986). Eternally existing self-reproducing chaotic inflationary universe. *Physics Letters B*, 175: 395–400.

Linde, A.D. (1990). *Particle Physics and Inflationary Cosmology*. Newark: Harwood Academic.

Murphy, N. and Ellis, G.F.R. (1995). *On the Moral Nature of the Universe*. Minneapolis: Fortress Press.

Noble, D. (2007). *The Music of Life: Biology beyond Genes*. Oxford: Oxford University Press.

Peacocke, A.R. (1989). *An Introduction to the Physical Chemistry of Biological Organization*. Oxford: Oxford University Press.

Penrose, R. (1989a). Difficulties with inflationary cosmology. In E.J. Fergus (ed.). *14th Texas Symposium Relativistic Astrophysics*, Proceedings of the New York Academy of Science, New York.

Penrose, R. (1989b). *The Emperor's New Mind: Concerning Computers, Minds and the Laws of Physics*. New York: Oxford University Press.

Penrose, R. (1997). *The Large, the Small, and the Human Mind*. Cambridge: Cambridge University Press.

Penrose, R. (2005). *The Road to Reality: A Complete Guide to the Laws of the Universe*. New York: Knopf.

Peskin, M.E. and Schroeder, D.V. (1995). *An Introduction to Quantum Field Theory*. Reading, MA: Perseus Books.

Polkinghorne, J. (2002). *Quantum Theory: A Very Short Introduction*. Oxford: Oxford University Press.

Popper, K. and Eccles, J. (1977). *The Self and Its Brain: An Argument for Interactionism*. Berlin: Springer.

Rees, M.J. (2000). *Just Six Numbers: The Deep Forces That Shape the Universe*. New York: Basic Books.

Rees, M.J. (2001). *Our Cosmic Habitat*. Princeton: Princeton University Press.

Rothman, T. and Ellis, G.F.R. (1992). Smolin's natural selection hypothesis. *Quarterly Journal of the Royal Astronomical Society*, 34: 201–12.

Saint-Exupery, A. de (1969). *Flight to Arras*. San Diego: Mariner Books. (Originally published in French, 1942.)

Scott, A. (1989). *Molecular Machinery: The Principles and Power of Chemistry*. Oxford: Blackwell.

Scott, A. (1995). *Stairway to the Mind*. New York: Springer.

Seife, J. (2000). *Zero: The Biography of a Dangerous Idea*. London: Penguin.

Sellars, R.W. (1932). *The Philosophy of Physical Realism*. New York: Russell and Russell.

Silk, J. (2005). *On the Shores of the Unknown: A Short History of the Universe*. Cambridge: Cambridge University Press.

Silverberg, A. (2004). Psychological laws. *Erkenntnis*, 58: 275–302.

Simon, H.A. (1982). *The Sciences of the Artificial*. Cambridge, MA: MIT Press.

Smolin, L. (1992). Did the universe evolve? *Classical and Quantum Gravity* 9: 173–91.

Stenger, V.J. (2007). *God, the Failed Hypothesis: How Science Shows That God Does Not Exist*. Buffalo, NY: Prometheus Books.

Susskind, L. (2005). *The Cosmic Landscape: String Theory and the Illusion of Intelligent Design*. New York: Little, Brown.

Taylor, C. (2007). *A Secular Age*. Harvard: Harvard University Press.

Tegmark, M. (1998). Is "the theory of everything" merely the ultimate ensemble theory? *Annals of Physics*, 270: 1–51.

Tegmark, M. (2004). Parallel universes. In J.D. Barrow, P.C.W. Davies, and C.L. Harper, Jr. (eds.), *Science and Ultimate Reality: Quantum Theory, Cosmology and Complexity*, 459–91. Cambridge: Cambridge University Press.

Vilenkin, A. (2006). *Many Worlds in One. The Search for Other Universes*. New York: Hill and Wang.

Watling, L. (2006). What I believe. *Southern Africa Quaker News* 220: 47–50.

Weinberg, S. (2007). Living in the multiverse. In B. Carr (ed.), *Universe or Multiverse?* Cambridge: Cambridge University Press. Also see: arXiv:hep-th/0511037.

Wigner, E. (1960). The unreasonable effectiveness of mathematics in the natural sciences. *Communications in Pure and Applied Mathematics*, 13(1) (February 1960). New York: John Wiley & Sons, Inc. Available at: http://www.dartmouth.edu/~matc/MathDrama/reading/Wigner.html.

Appendix: The *New Vision 400* Conference

PURPOSE AND FOCUS OF THE *NV400* PROJECT

As noted in the Preface, this book is an important aspect of the *New Vision 400 (NV400)* project and grew from the international conference held in Beijing in October 2008 (see http://nv400.uchicago.edu/). This was the second in a series of international celebrations held worldwide to celebrate the discovery of the telescope, most of them held in 2009 in association with the great discoveries of Galileo that literally changed the direction of intellectual study in the Western world. The *NV400* conference (and hence the contributions in this book) differed from the other meetings in covering a broader range of scientific issues resulting from the use of telescopes by including discussions of the impact on both Eastern and Western cultures, as well as in addressing the "big questions" associated with the relationship between the conclusions of science (in particular, the laws of nature and their origin) and moral law in the two cultures. We know so much in the way we describe the Universe, yet the origin of the basic laws is based, in many cases, on empirical derivation, rather than on fundamental understanding of first principles. The understanding of science in the two cultures has historically depended on the way the questions were asked, and the impact of the telescope on worldviews and on knowledge has consequently been dramatically different. The conference poster following the Introduction celebrates the scientific and cultural contributions related to astronomy from both Eastern and Western traditions.

Professor Donald G. York (The University of Chicago) was the Principal Investigator for the meeting and recipient of the generous grant from the John Templeton Foundation (JTF) that supported much of the cost. Academician Jian-Sheng Chen (Peking University and the National Astronomical Observatories, Chinese Academy of Sciences [NAOC]) served as Co-Principal Investigator and was involved in all aspects of planning and executing the meeting and the public event. Sui-Jian Xue (NAOC), Xiao-Chun Sun (Institute for the History of Natural Sciences, Chinese Academy of Sciences), and Xiao-Wei Liu (Peking University) were especially helpful in advising on details of the program and logistics.

The Xiangshan Science Conference hosted the event as part of its overall discussion on matters of national interest and policy, partnering with NAOC and JTF, among others. This historic and successful meeting was attended by almost 400 registrants, including representatives of the primary government astronomy funding and support agencies of China (The National Natural Science Foundation of China [NSFC], The Chinese Academy of Sciences [CAS]); Europe (The European Space Agency [ESA], The European Southern Observatory [ESO]); the United Kingdom (The Science & Technology Facilities Council [STFC] of the Research Councils UK); and the United States (The National Science Foundation [NSF], The National Aeronautics and Space Administration [NASA], The National Optical Astronomy Observatory [NOAO]).

To summarize the motivation for the conference, quoting from the website (http://nv400.uchicago.edu/):

> In the last 400 years, the verifiable age of the Universe has increased by a factor of 100 million. The accepted size of the Universe has increased by that factor, cubed. Having known only a few planets for all that time, we now know of hundreds around other stars. We stand at the brink of addressing one of the most profound questions of civilization: are we alone in the Universe?

All of these changes of perception are attributable to the exploration of space with the telescope, the small instrument invented 400 years ago by Hans Lipperhey. The process that has led to these changes can be traced to accidental and purposeful technological advances; flashes of human insight; creative application of the technologies to extensive, targeted research programs; creative exploitation of accidental discoveries; and deep respect for the powerful description of nature by mathematics.

To capture the specifics of those processes that led to our current knowledge of the Universe, an international advisory committee was constituted to lay out a plan for the meeting, and a program committee was formed to find the best speakers for the areas decided upon. A local program committee handled the logistical arrangements, including meeting venues, hotels, food, transportation, advertising, and so forth. (See the committee listings at the end of this Appendix.)

CONTENT AND SCOPE OF THE *NV400* CONFERENCE

In keeping with the emphasis on the technology of telescopes that ran through the Beijing conference, a public event was held preceding the meeting that was based on the general theme of creativity and technology in scientific—particularly astronomical—discovery (see more information at http://nv400.uchicago.edu/pub_ev.html and http://www.nv400.org/cnindex.html). The purpose of the public event, which began with a welcome speech by Dr. John M. Templeton, Jr., was to share the key themes of the science meeting with the public, in this case 6,000 high school and college students who gathered in the Great Hall of the People on October 12, 2008, within a few days of the 400th anniversary of the accepted date of the patent application by Hans Lipperhey for the telescope. Professor Jian-Sheng Chen served as master of ceremonies.

Addresses at the public event were given by Nobel Laureates Tsung-Dao Lee and Riccardo Giacconi and by Shaw Prize Winner Geoffrey Marcy. The very personal stories told by the three famous speakers, which are presented as the first three chapters in Part I of this volume, made the public event exciting and memorable, especially for the high school students at which the event was aimed. (Videos of the presentations are available on the conference website: http://nv400.uchicago.edu/talks/index.html.)

We could not begin to exhaust the specific topics that were discussed over the three days of the meeting following the public event, corresponding to Parts II through VI of this book, in a single conference or in a single volume. Rather, the conference committees focused on well-recognized areas of exploration: the impact of telescopes on our knowledge of the Universe; some of the near-term challenges with which we are left after 400 years of using telescopes; the technologies needed and that are being developed to address those challenges in the future; the intellectual impact of the telescope on society; and specific "big questions" raised by the information we have obtained and how we will advance our knowledge into the future. Posters were displayed throughout the meeting relating to the same topics addressed by the plenary speakers: the impact of telescopes on our knowledge of the Universe, our near-term challenges in developing technologies needed to continue our explorations, and the intellectual impact of the telescope on society's "big questions," such as the origin and meaning of life, the Multiverse, the origin of physical laws, cultural issues, and so forth.

The event ended with a panel discussion on topics related to humanity's future vision and more on "life's big questions." Owen Gingerich moderated the panel, which consisted of Marco Bersanelli (University of Milano), Miao Li (Institute of Theoretical Physics, Chinese Academy of Sciences, Beijing), Timothy O'Connor (Indiana University, Bloomington), and Yun-Li Shi (University of Science and Technology of China, Hefei).

Two important awards ceremonies took place during the conference. The first honored the meritorious contributions of the many researchers who have advanced the state of astronomy into the 21st century (see Box A.1 [Lifetime Achievement Awards]). The second celebrated those who will continue to explore cosmic frontiers into the future (see Box A.2 [Young Scholars Competition]).

BOX A.1   LIFETIME ACHIEVEMENT AWARDS

The theme of creativity and technology at the *NV400* conference was further emphasized at the meeting by the presentation of plaques to 12 scientists whose work in astronomical technology, spanning 60 years, has had major impacts on the discoveries of the telescope as highlighted in this book:

Morley Blouke (Ball Aerospace & Technologies Corporation, now retired), **Richard Bredthauer** (Semiconductor Technology Associates), and **Craig Mackay** (Institute of Astronomy, University of Cambridge) for their design and fabrication of charge-coupled devices (CCDs), which, starting around 1978, qualitatively changed the effective sensitivity of telescopes and are instrumental to obtaining high-resolution images from ground-based telescopes;

Barry Clark (National Radio Astronomy Observatory, Socorro, New Mexico) for his work on radio interferometry over many decades, particularly in the area of computers and software;

Dan McCammon (University of Wisconsin, Madison) and **Samuel H. Moseley** (NASA Goddard Space Flight Center, Greenbelt, Maryland) for their development of cryogenic, microcalorimeters for x-ray spectroscopy, dramatically enhancing the sensitivity of astronomical x-ray instruments;

Thomas Phillips (California Institute of Technology, Pasadena) for his development of hot electron bolometers and semiconductor-insulator-semiconductor (SIS) detectors and their application to astronomical submillimeter astronomy;

V. Radhakrishnan (Raman Research Institute, Bangalore, India) for his pioneering work in radio spectroscopy;

Stephen Shectman (Carnegie Observatories, Pasadena, California) for his development of photon-counting, 2-dimensional detectors and their use on instruments for ground-based telescopes;

Oswald H. W. Siegmund (Space Sciences Laboratory, University of California, Berkeley) for his contribution to the development of photon-counting detectors for space ultraviolet and x-ray instruments;

Govind Swarup (Tata Institute of Fundamental Research, now retired) for his major contributions to developing two of the largest meter-wavelength telescopes in the world, the Ooty Telescope and the Giant Metrewave Radio Telescope; and

Ernest Joseph Wampler (Santa Cruz, California, retired, formerly of Lick Observatory, the Anglo-Australian Observatory, and the European Southern Observatory) for his contributions to constructing the first multiplexing electronic scanner in regular astronomical use and the use of the device (the Image Dissector Scanner) and its successors on various astronomical research programs.

BOX A.2   YOUNG SCHOLARS COMPETITION

Another important aspect of the *NV400* conference was the Young Scholars Competition (see http://www.nv400.org/young_scholars.html), which was held for Chinese students from Chinese institutions to encourage young researchers as they pursue the scientific discoveries of the future. An international panel of judges reviewed the 16 oral and written presentations, the latter distributed throughout the science meeting. Five cash prizes were awarded as follows:

Two first prizes:
- "A Holistic View of the Dark Universe" (Topical Area: Astronomy and Cosmology) by Xue-Lei Chen, Yan Gong, Feng Huang, Xin Wang, and Feng-Quan Wu (National Astronomical Observatories, Chinese Academy of Sciences)
- "Research on LAMOST Wavefront Sensing" (Topical Area: Technological Innovations) by Yong Zhang (Nanjing Institute of Astronomical Optics and Technology, National Astronomical Observatories, Chinese Academy of Sciences)

Three second prizes:
- "The Rosette Eye: The Key Transition Phase in the Birth of a Massive Star" (Topical Area: Astronomy and Cosmology) by Jin-Zeng Li, Michael D. Smith, Roland Gredel, Christopher J. Davis, and Travis A. Rector (Li is an astronomer with the National Astronomical Observatories, Chinese Academy of Sciences)
- "Eternal Inflation: Prohibited by Quantum Gravity?" (Topical Area: Astronomy and Cosmology) by Yi Wang (Institute of Theoretical Physics, Chinese Academy of Sciences)
- "The Debate over the Theory of Precession in the Late Ming and Early Qing China" (Topical Area: History of Astronomy) by Guang-Chao Wang (Institute for the History of Natural Science, Chinese Academy of Sciences)

In addition, the National Astronomical Observatories, Chinese Academy of Sciences (NAOC), in partnership with the John Templeton Foundation (JTF), which supports innovative projects that focus on big and foundational questions in science that seek a deeper understanding of the world and that are unlikely to be supported by conventional funding sources, will make available $2 million for science awards that address the most difficult questions at the frontiers of astronomical knowledge. The request for proposals was distributed in China in April, 2010. Professor Donald G. York and Professor Gang Zhao are the principal investigators for the program.

NV400 CONFERENCE SUPPORT AND SPONSORS

As already mentioned above and in the Acknowledgments at the front of the book, we particularly wish to acknowledge the following people: Jian-Sheng Chen (Peking University and NAOC) for his collaboration with Donald York in organizing this conference. They were ably supported by Sui-Jian Xue (NAOC), who served on all of the committees, interfaced with all of the local Chinese Institutions, and handled all of the financial arrangements in China, and by Xiao-Chun Sun (Institute for the History of Natural Sciences, CAS) and Xiao-Wei Liu (Peking University), who, provided continuous assistance and advice to the conference organizers and to the editors of this book on matters of Chinese culture, language, and science.

A number of officers and staff members of JTF also gave great support to Donald York in setting up the conference with the many hours they devoted to thinking through and discussing ideas for the conference, in addition to traveling to Beijing to assist with planning. We particularly wish to acknowledge the following principals and staff:

- Dr. John M. Templeton, Jr., President and Chairman, who provided generous interest and support throughout the project.
- Charles L. Harper, Jr., who served as one of the original project developers in his former role as Senior Vice President and Chief Strategist.
- Hyung S. Choi, Director of Mathematical and Physical Sciences, who played an integral role in developing the academic program for the symposium.
- Judith B. Marchand, Vice President of Special Projects, and Pamela P. Thompson, Vice President of Communications, who through their tireless efforts explored and organized the venues for the meeting and assured the general well-being of the attendees.

Finally, we reiterate our deep gratitude to the sponsors who supported the development of this project and the international committees who strove to ensure the quality and merit of both the conference and this book; we list them here.

CONFERENCE COSPONSORS

United States—The John Templeton Foundation (JTF), Philadelphia, Pennsylvania

China—Various Organizations, Beijing
- China Xiangshan Science Conference
- Chinese Ministry of Science and Technology
- Chinese Academy of Sciences
- National Natural Science Foundation of China
- Chinese Astronomical Society
- China Center of Advanced Science and Technology
- National Astronomical Observatories, Chinese Academy of Sciences
- Peking University

ADVISORY COMMITTEE

- Catherine Cesarsky, Service d'Astrophysique, CEA Saclay, Gif-sur-Yvette, France
- Freeman J. Dyson, Institute for Advanced Study, Princeton, New Jersey, United States
- George F.R. Ellis, University of Cape Town, South Africa
- Cheng Fang, Nanjing University, Nanjing, P. R. China
- Owen Gingerich, Harvard-Smithsonian Center for Astrophysics, Cambridge, Massachusetts, United States
- Wen-Han Jiang, Institute of Optics and Electronics, Chinese Academy of Sciences, Chengdu, P. R. China
- K.Y. Lo, National Radio Astronomy Observatory, Charlottesville, Virginia, United States
- Tan Lu, Purple Mountain Observatory, Nanjing, P. R. China
- Frank Shu, University of California, San Diego, United States
- Ding-Qiang Su, Nanjing University, Nanjing, P. R. China
- Robert Williams, Space Telescope Science Institute, Baltimore, Maryland, United States
- Ze-Zong Xi, Institute for the History of Natural Sciences, Chinese Academy of Sciences, Beijing, P. R. China*

---

* In memoriam; Professor Xi passed away following the conference.

PROGRAM COMMITTEE

- Donald G. York, The University of Chicago, Illinois, United States (Principal Investigator)
- Jian-Sheng Chen, National Astronomical Observatories, Chinese Academy of Sciences, and Peking University, Beijing, P. R. China (Co-Principal Investigator)
- John D. Barrow, University of Cambridge, United Kingdom
- Jacqueline Bergeron, Institut d'Astrophysique de Paris, France
- Chris Carilli, National Radio Astronomy Observatory, Charlottesville, Virginia, United States
- Dun Liu, Institute for History of Natural Science, Chinese Academy of Sciences, Beijing, P. R. China
- Xiao-Wei Liu, Peking University, Beijing, P. R. China
- Shu-De Mao, University of Manchester, United Kingdom
- Saul Perlmutter, University of California, Berkeley, United States
- Joseph Silk, University of Oxford, United Kingdom
- Michael Shao, Jet Propulsion Laboratory, Pasadena, California, United States
- Shuang-Nan Zhang, Institute of High Energy Physics, Chinese Academy of Sciences, and University of Alabama, Huntsville, United States
- Gang Zhao, National Astronomical Observatories, Chinese Academy of Sciences, Beijing, P. R. China

LOCAL ORGANIZING COMMITTEE (BEIJING, P. R. CHINA)

- Jun Yan (Chair), National Astronomical Observatories, Chinese Academy of Sciences
- Jian-Hua Lin (CoChair), Peking University
- Sui-Jian Xue, National Astronomical Observatories, Chinese Academy of Sciences
- Yan-Song Li, Peking University
- Zhen-Ya Chen, Peking University
- Jin Zhu, Beijing Planetarium
- Bing-Xin Yang, Xiangshan Science Conference
- Zhen-Yu Wang, Chinese Academy of Sciences
- Jin-Xin Hao, Chinese Academy of Sciences
- Guo-Xuan Dong, National Science Foundation China

Index

T - #0475 - 071024 - C452 - 254/178/20 - PB - 9780367382094 - Gloss Lamination